GEOLOGICAL HISTORY OF BRITAIN AND IRELAND

EDITED BY

Nigel Woodcock
Earth Sciences, University of Cambridge, UK

Rob Strachan
Geology, Oxford Brookes University, UK

b
**Blackwell
Science**

© 2000 by
Blackwell Science Ltd
Editorial Offices:
Osney Mead, Oxford OX2 0EL
25 John Street, London WC1N 2BS
23 Ainslie Place, Edinburgh EH3
    6AJ
350 Main Street, Malden
    MA 02148 5018, USA
54 University Street, Carlton
    Victoria 3053, Australia
10, rue Casimir Delavigne
    75006 Paris, France

Other Editorial Offices:
Blackwell Wissenschafts-Verlag
    GmbH
Kurfürstendamm 57
10707 Berlin, Germany

Blackwell Science KK
MG Kodenmacho Building
7–10 Kodenmacho Nihombashi
Chuo-ku, Tokyo 104, Japan

First published 2000

Set by Best-set Typesetter Ltd., Hong
Kong
Printed and bound in Great Britain
at the University Press, Cambridge

A catalogue record for this title
is available from the British Library

ISBN 0 632 03656 7

Library of Congress
Cataloging-in-publication Data

Geological history of Britain and
Ireland/edited by Nigel H.
Woodcock, Rob A. Strachan.
        p. cm.
    Includes index.
    ISBN 0-632-03656-7
      1. Geology—British Isles.
I. Woodcock, N. H.   II. Strachan,
R. A. (Robin A.)

    QE261 .G45 2000
    554.1—dc21

                00-034313

DISTRIBUTORS
Marston Book Services Ltd
PO Box 269
Abingdon, Oxon OX14 4YN
(*Orders*: Tel: 01235 465500
        Fax: 01235 465555)

USA
    Blackwell Science, Inc.
    Commerce Place
    350 Main Street
    Malden, MA 02148 5018
    (*Orders*: Tel: 800 759 6102
                781 388 8250
            Fax: 781 388 8255)

Canada
    Login Brothers Book Company
    324 Saulteaux Crescent
    Winnipeg, Manitoba R3J 3T2
    (*Orders*: Tel: 204 837 2987)

Australia
    Blackwell Science Pty Ltd
    54 University Street
    Carlton, Victoria 3053
    (*Orders*: Tel: 3 9347 0300
            Fax: 3 9347 5001)

For further information on
Blackwell Science, visit our website:
www.blackwell-science.com

Contents

Contributors

Roger Anderton
BP/Amoco Exploration, Farburn Industrial Estate, Dyce, Aberdeen AB2 0PD, UK

Sarah J. Davies
Department of Geology, University of Leicester, University Road, Leicester LE1 7RH, UK

Stephen P. Hesselbo
Department of Earth Sciences, University of Oxford, Parks Road, Oxford OX1 3PR, UK

Robert E. Holdsworth
Department of Geological Sciences, University of Durham, Science Laboratories, South Road, Durham DH1 3LE, UK

Andrew S. Gale
Natural History Museum, Cromwell Road, London SW7 5BD, UK and School of Earth Sciences, University of Greenwich, Medway Campus, Chatham Maritime, Kent ME4 4AN, UK

Paul D. Guion
Geology, Oxford Brookes University, Headington, Oxford OX3 0BP, UK

Peter Gutteridge
Cambridge Carbonates Ltd., 11 Newcastle Drive, The Park, Nottingham NG7 1AA, UK

Alastair H. Ruffell
School of Geography, Queen's University of Belfast, University Road, Belfast BT7 1NN, UK

Rob G. Shelton
Shell U.K. Exploration & Production, Seafield House, Sleat of Rubislaw, Aberdeen, AB15 6BL

Rob. A. Strachan
Geology, Oxford Brookes University, Headington, Oxford OX3 0BP, UK

Laurence N. Warr
Geologisch-Paläontologisches Institut, Ruprecht-Karls-Universität, INF 234, 69120 Heidelberg, Germany

Nigel H. Woodcock
Department of Earth Sciences, University of Cambridge, Downing Street, Cambridge CB2 3EQ, UK

Preface

It is now twenty years since publication of an undergraduate-level textbook on the complete geological history of Britain and Ireland (R. Anderton *et al.* 1979, *A Dynamic Stratigraphy of the British Isles*, George Allen & Unwin). The challenge of producing an updated concise synthesis of this history has become more difficult year by year, as the regional geology literature has expanded rapidly. Offshore exploration for hydrocarbons, deep seismic reflection profiling and advances in isotopic dating are but three of the many new sources of data and ideas over recent years. Gone are the days when Leonard Wills, Kingsley Wells and Dorothy Rayner were individually able to produce substantial summaries of British and Irish geology. To distil the knowledge of the 1990s into a readable yet authoritative form, a multi-author approach has proved necessary, despite the editorial complexity of contrasting styles and conflicting deadlines.

This book was conceived in 1992 but the project was only launched with its present roll of contributors in 1997. The past two years have been productive and stimulating, not least because of the need to merge individual contributions into a story that is, if not seamless, at least not marred by too many inconsistencies. We have tried to strike a balance between the need to maintain a smooth narrative and the obligation to interrupt this story with the many uncertainties that bear on it. Although not attempting a reference book on 'stratigraphy', we have tried to provide enough maps and lithological logs for the reader to appreciate the nature and quality of the data on which the geological history is based. That reader is likely to be an undergraduate student with at least a year of introductory geology training, a postgraduate or professional geologist needing a regional overview, or an informed amateur wanting more than an introductory guide to the regional geology. To make the text more accessible, it contains no citations. Rather, a list of suggested reading is provided at the end of each chapter. The sources of figures, or of the data from which they were constructed, are given in the captions and listed at the end of the chapters. We wish to acknowledge the societies and publishing houses who have permitted illustrations to be used.

We want first to thank John Underhill and Rob Gawthorpe for the time and thought that they gave the initial phase of this project. Our editors from Blackwell Science have been supportive and patient throughout. Simon Rallison steered the early phase of the project, Ian Sherman saw us through our mid-project reorganization, and Ian Francis has seen the book to its completion. All the book's authors have contributed to reviewing other chapters, and their comments have made the editors' task more stimulating, if not always more easy! Our warm thanks also go to those colleagues who took the time to review individual chapters: Ian Alsop, Brian Bell, Alan Clark, Dan Evans, Clark Friend, Peter Friend, Rod Gayer, Phil Gibbard, Ken Hitchen, Helen Morgans Bell, John Morris, Graham Park, Tim Pharaoh, John Rippon, Denis Smith, Jack Soper, Finn Surlyk, Dan Tietsch-Tyler, Geoff Warrington and John Winchester. Angela Coe is thanked particularly for her contribution to Chapter 17. Finally, Nigel Woodcock wishes to thank Tjeerd Van Andel for his wise reassurance that geology is enriched by being seen from a historical as well as a scientific perspective.

Nigel Woodcock   Rob Strachan
Cambridge   Oxford

Part 1
Introduction

Illustration overleaf: 'Hutton's unconformity' at Siccar Point, Cockburnspath, Berwickshire (reproduced by kind permission of Landform Slides). Vertical Silurian (Llandovery) greywackes are overlain unconformably by gently dipping sandstones of Upper Old Red Sandstone (Devonian) age. This site has been described as one of the most important geological sites in the world—it is arguably the place where the science of modern geology was born. Its importance arises from the work of James Hutton, who discovered the unconformity in 1788. Hutton argued that the rocks beneath the unconformity represented 'the ruins of an earlier world', a vanished world that after passing though stages of mountainous landscapes had been worn down to a surface of upturned rocks as can now be seen under the Old Red Sandstone. Hutton realized that 'one land mass is worn down while the waste products provide the materials for a new one'. He argued for the existence within the Earth of powerful forces which were capable of tilting and uplifting the older rocks to result in their subsequent erosion.

1 Regional geological history: why and how?

N. H. WOODCOCK AND R. A. STRACHAN

1.1 Development of historical geology in Britain and Ireland

An understanding of the Earth requires a blend of two distinct scientific methods. *Causal* scientific enquiry attempts to understand the fundamental processes of the Earth, irrespective of their age, typically using the analytical methods of physics or chemistry. The geometrical rules of plate tectonics, the fluid mechanics of sediment transport or the chemical evolution of a magma chamber are all examples of such causal principles. Characteristic of geology, however, is its use of *historical* analysis, which recognizes that complex Earth processes depend on their place in geological time. In particular, geological processes may be strongly influenced by preceding events. So, a new plate boundary may preferentially follow an old weakness in the lithosphere, or the chemistry of a magma will be influenced by the compositional history of the source mantle or crust. Thus *historical geology* can be carried out either on a global scale or, more typically, on a regional scale.

The regional focus of this book, comprising Britain, Ireland and its surrounding crust, has a remarkably varied geology for so small a fragment of continent. This region contains a fine rock record from all the geological periods from Quaternary back to Cambrian, and a less continuous but still impressive catalogue of events back through nearly 2500 million years of Precambrian time. This protracted geological history would have been interesting enough to reconstruct if it had been played out on relatively stable continental crust. However, Britain and Ireland have developed instead at a tectonic crossroads, on crust traversed intermittently by subduc-

tion zones and volcanic arcs, continental rifts and mountain belts. The resulting complexity makes the geological history of this region at once fascinating and perplexing.

The coincidence of this complex geology with a scientifically inquisitive human culture was the catalyst for a period of prodigious geological discovery and understanding in the late 18th and early 19th centuries. James Hutton was persuaded, by the unconformities that mark Britain's major tectonic phases, of both the longevity and cyclicity of geological events. His *Theory of the Earth* (1795), and its development by John Playfair and Charles Lyell, mark the beginning of modern historical geology. The foundation in London of the world's first Geological Society (1807) provided a forum in which Britain's geological history was rapidly pieced together. William Smith soon published his geological map of England and Wales (1815), based on the principles of correlating strata and their fossils, principles that have become modern stratigraphy. The Geological Survey of Great Britain was formed (1835) to produce a more detailed geological map of the country, and a few years later came Richard Griffiths' geological map of Ireland. At about the same time began the collaboration, and later confrontation, between Roderick Murchison and Adam Sedgwick, which resulted in the establishment of the Cambrian (1835), Silurian (1835) and Devonian (1839) Systems and, through Charles Lapworth's eventual mediation (1879), the Ordovician System.

For a century and a half, therefore, Britain, Ireland and the European countries that host the type areas for Carboniferous and higher systems have provided a reference region for global geological history. This is justification enough to maintain an up-to-date understanding of the regional geological history; but there are other reasons.

1.2 Why study historical geology?

The rapid 19th-century growth of historical geology was partly stimulated by the search for and exploitation of geological resources, mainly metallic minerals and coal. This economic stimulus was revitalized in the 1960s with the exploration for offshore oil and gas resources. Finding petroleum requires an accurate knowledge of the stratigraphy and present geometrical structure of a prospective area. However, any programme of exploration and production also includes detailed reconstructions of the geological history of the area, particularly of successive sedimentary environments and of the subsequent diagenetic, thermal and deformation history of the resulting rock sequence. The geological histories that are available for all onshore and most offshore areas of Britain and Ireland provide the essential basis for planning local investigations for geological resources. The results from exploration reveal, in turn, new details or even major modifications to the regional geological history. This modern symbiosis between academic and commercial interests has advanced the geological history of Britain and Ireland at a rate unprecedented since the 19th century.

Older even than the study of historical geology has been the search for the fundamental causal laws of how the Earth works. Hutton and Lyell tried to use specific examples of local or regional geological history to diagnose and illustrate global geological principles. However, despite Lyell's uniformitarian principle that 'the past is the key to the present', our understanding of Earth processes relies heavily on knowledge of their historical context. The atmosphere, oceans, climate, biota, crust and mantle have all evolved through Earth history, so that many of their more complex processes depend on what came before them. A regional geological history, such as that of Britain and Ireland, provides the temporal framework for understanding the context of process-orientated research. Moreover, the feedback from this research stimulates the refinement of geological histories in the same way as the exploration for economic resources.

There is a third, less tangible, reason for studying regional geological history: the distinctive intellectual challenge and discipline of the study itself. Philosophers of science have struggled to characterize the way that a geologist works and thinks. Having identified physics as the quintessential science, they have typically measured other sciences against its supposed objectivity, predictability and precision. Geology has therefore been viewed merely as a derivative and imprecise form of physics. In practice, geologists rarely work solely by the process of deductive and inductive logic that we call the scientific method. Geology has an essential historical dimension, which distinguishes it from pure physics, chemistry or biology. The geological record is inevitably complex and incomplete, and deciphering it requires an interpretative reasoning similar to that applied to human history. A geological history is charted that best fits the available observations; then it is iteratively improved, revised or rejected as new data become available. Interpretative reasoning is criticized, particularly by pure scientists, for its circularity. However, it has the over-riding strength of being the way in which our human understanding of most everyday problems is built up. It should be that trained geologists are therefore particularly well equipped to solve complex problems that occur at the interface between science and society.

1.3 Arranging and dating events: stratigraphy

How is a regional geological history reconstructed? Sections 1.3–1.6 outline the methodology used to reconstruct the history of any one part of the Earth's crust. This introduction is deliberately brief, because a comprehensive introduction would become a textbook of geology in itself. Any history is potentially a synthesis of all available geological information, and no type of evidence is irrelevant. This chapter stresses only general principles, but provides pointers to more detailed topics together with suggestions for further reading.

The rock record is made of igneous, sedimentary and metamorphic rocks, with a geometrical arrangement due partly to the original formation of these rocks and partly to their later deformation. The first step in charting any geological history is to translate this rock geometry into a sequence of geological events through time. This interpretation uses a set of assumptions (Fig. 1.1) sometimes described, perhaps

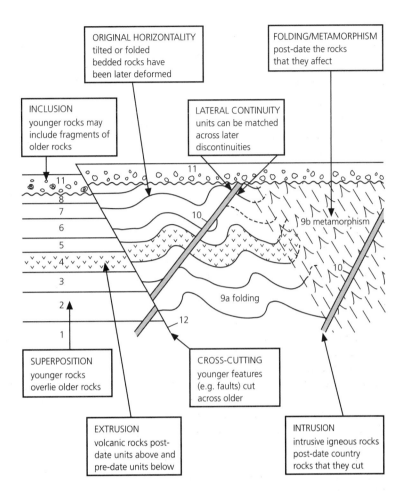

Fig. 1.1 Cross-section illustrating the rules of stratigraphy, which allow rock geometry to be translated into a sequence of events (numbered 1–12).

too formally, as the laws of stratigraphy. So, the principle of *superposition* is that younger rocks in a bedded sequence overlie older rocks, and that of *inclusion* is that younger rocks may include fragments of older rocks. *Cross-cutting* relationships are generally such that younger features—typically rock bodies or faults—cut across older features. The assumption of *lateral continuity* allows units to be matched across intervening faults, intrusions or other interruptions. Igneous rocks are dated by the rule of *extrusion*—that the age of an extrusive unit in a bedded sequence lies between that of underlying and overlying units—and the rule of *intrusion*—that intrusive bodies are younger than any country rocks that they cut. The assumption of *original horizontality*, at least for many sedimentary units, determines

that most tilted or folded rocks have been later deformed. The principles of *deformation and metamorphism* are that these events are younger than the rocks that they affect.

A catalogue of successive geological events can be reconstructed for most regions of the Earth, however complex. More sophisticated stratigraphic subdivision can be achieved in areas dominated by bedded sedimentary or volcanic rocks, and their deformed equivalents. In the example shown in Fig. 1.2, the arranged strata are shown as a graphic log, subdivided into named distinctive units. This *lithostratigraphic* division is typically based on rock characteristics observable in the field, for instance composition, grain size, colour, primary textures and structures. The basic subdivision is into *formations*,

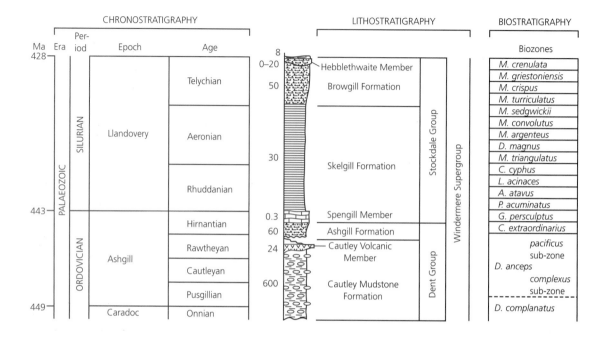

Fig. 1.2 The different components of formal stratigraphic procedure in bedded rocks, illustrated with part of the Windermere Supergroup, north-west England.

which are units distinctive enough to be mapped over at least tens of kilometres. Formations are just one level in an optional hierarchy of lithostratigraphic units, usually named after typical geographical localities. So, two of the formations in Fig. 1.2 contain locally mappable, though regionally discontinuous, *members*. Several formations are gathered into a *group*, and several groups into a *supergroup*. Intrusive and metamorphic rocks may be too complicated to subdivide on this scheme, and may be organized into mappable *complexes*.

Rock sequences can be subdivided on criteria other than their lithological character, for instance by their magnetic or chemical properties. An example is the oxygen isotope *chemostratigraphy* used to calibrate and subdivide Quaternary sequences. However, fossil content is the prime discriminator in most sedimentary rocks. In the example in Fig. 1.2, graptolites have been used to define a *biostratigraphic* division into *biozones*. Each zone is named after a distinctive species of index fossil, typically in a larger assemblage of other species. Other available fossil groups might yield their own independent zonation.

In Cambrian and later rocks, fossils usually provide the most important tool for correlating the local rock sequence with a global or regional stratotype sequence, and assigning a *chronostratigraphic* age. Chronostratigraphic names are arranged in two hierarchies (Fig. 1.2), one used for the standard intervals of time (*era*, *period*, *epoch*, *age*) and the other for the volumes of rock formed during those intervals (*erathem*, *system*, *series*, *stage*). In the example in Fig. 1.2, the top of the Spengill Member of the Skelgill Formation is equated with the base of the Llandovery Series, because the global stratotype for this boundary, in the Southern Uplands of Scotland, defines it at the base of the *persculptus* biozone. International agreement is being attempted on the names and type sequences of all stages. However, this task is not yet complete and, even when it is, difficulties of correlation with the type sequences will ensure that many local chronostratigraphic names will still be used. A chronostratigraphic chart for Britain and Ireland (Fig. 1.3) reflects the mix of international and local names used in this book, and is supplemented by more detailed stratigraphic diagrams in individual chapters.

The chronostratigraphic scale can be calibrated in millions of years, mainly based on radiometric dating of the type sequences or other sequences that can be

Eon	Era	Period	Epoch/Age	Ma
				0
PHANEROZOIC	CENOZOIC	Quaternary	Holocene	0.01
			Pleistocene	1.8
		Tertiary – Neogene	Pliocene	5.3
			Miocene	23.8
		Tertiary – Paleogene	Oligocene	33.7
			Eocene	54.8
			Paleocene	65.0
	MESOZOIC	Cretaceous (Late)	Maastrichtian	71.3
			Campanian	83.5
			Santonian	85.8
			Coniacian	89.0
			Turonian	93.5
			Cenomanian	98.9
		Cretaceous (Early)	Albian	112
			Aptian	121
			Barremian	127
			Hauterivian	132
			Valanginian	137
			Berriasian	144
		Jurassic (Late)	Tithonian	151
			Kimmeridgian	154
			Oxfordian	159
			Callovian	164
		Jurassic (Mid)	Bathonian	169
			Bajocian	177
			Aalenian	180
		Jurassic (Early)	Toarcian	189
			Pliensbachian	195
			Sinemurian	202
			Hettangian	206
		Triassic (Late)	Rhaetian	210
			Norian	221
			Carnian	227
		Triassic (Mid)	Ladinian	234
			Anisian	242
		Triassic (E)	Scythian	248

Eon	Era	Period	Epoch/Age	Ma
				248
PHANEROZOIC	PALAEOZOIC	Permian (L)	Zechstein	256
		Permian (E)	Rotliegendes	290
		Carboniferous (Silesian)	Stephanian	305
			Westphalian	315
			Namurian	327
		Carboniferous (Dinantian)	Visean	342
			Tournaisian	362
		Devonian (Late)	Famennian	377
			Frasnian	383
		Devonian (Mid)	Givetian	388
			Eifelian	394
		Devonian (Early)	Emsian	410
			Pragian	414
			Lochkovian	418
		Silurian (Late)	Pridoli	419
			Ludlow	423
		Silurian (Early)	Wenlock	428
			Llandovery	443
		Ordovician (Late)	Ashgill	449
			Caradoc	462
		Ordovician (M)	Llanvirn	470
		Ordovician (Early)	Arenig	485
			Tremadoc	495
		Cambrian	Late	505
			Mid	518
			Early	545
	PROTEROZOIC	Neoproterozoic		1000
		Mesoproterozoic		1600
		Palaeoproterozoic		2500
	ARCHAEAN	Late Archaean		3000
		Middle Archaean		3500
		Early Archaean		4000
		Hadean		4500

Fig. 1.3 A chart of chronostratigraphic names commonly used in Britain and Ireland, together with their chronometric calibration in millions of years before present (Ma) (epochs from Gradstein & Ogg 1996; updated for the Devonian from Tucker *et al.* 1998).

correlated with them. A number of methods are now available (Table 1.1) for the dating of geological events as diverse as the age of high-grade metamorphism in Precambrian gneisses (using uranium and lead isotopes in zircon) and the formation of peat and charcoal in Quaternary sedimentary rocks (using carbon-14). Recent improvements in analytical tech-niques mean that it is now possible to date the crystallization of minerals such as zircon to a precision of 1 or 2 million years, even in the oldest Precambrian rocks.

The resulting *geochronometric* age can be derived from a local rock sequence only if it contains radiometrically datable rocks. The radiometric ages of such rocks as lava flows and volcanic ashes, for example, will correspond closely to the ages of the sediments with which they are interbedded. The dating of bentonites, clays derived from volcanic ash, has allowed accurate calibration of that part of the Early Palaeozoic timescale in Fig. 1.2. Radiometric

Table 1.1 The principal methods of radiometric age determination referred to in this book.

Isotopic system	Useful age range	Commonly dated materials
Carbon-14	<40 000 years	Wood, peat, charcoal, carbonate shells
Rubidium–strontium (Rb–Sr)	>5 million years	Microcline, muscovite, biotite, whole igneous or metamorphic rocks
Potassium–argon (K–Ar)	>100 000 years	Muscovite, biotite, hornblende, glauconite, whole volcanic rocks
Uranium–lead (U–Pb)	>5 million years	Monazite, titanite, zircon
Samarium–neodymium (Sm–Nd)	>250 million years	Pyroxene, garnet, whole igneous or metamorphic rocks

dating is the only usable stratigraphic method in many unfossiliferous successions, particularly in the Precambrian where the boundaries of chronostratigraphic units are defined as precise but arbitrary ages in millions of years. Even in the Phanerozoic, the radiometric calibration of chronostratigraphic boundaries is necessarily subject to errors, both from the analytical procedures and from interpolation to the boundary from dated rocks below and above it. These errors are reflected in the use, in other parts of this book, of some ages that differ from those on the time chart (Fig. 1.3), and in the certainty that future data will amend this calibration yet further.

The stratigraphic methods described so far have primarily been developed onshore, where limits of exposure typically provide a one- or at best two-dimensional transect through rock successions. A method particularly applied to the offshore successions around Britain and Ireland is *seismic stratigraphy*, based on the successive geometry of surfaces visible on reflection profiles. To be interpreted correctly, these reflectors must be tied to recorded lithostratigraphy in a borehole. Then, arrays of seismic profiles provide at least a two-dimensional view, and often a three-dimensional glimpse, of stratigraphic relationships. Key surfaces are those that appear to truncate adjacent sets of reflectors. If reflectors are

taken to be time-surfaces, these discordant surfaces must be unconformities, marking times of non-deposition or erosion.

Although best visualized in seismic sections, the method of subdividing rock successions using their contained unconformities can be applied to an even wider range of evidence from bedded rocks, and forms the basis of *sequence stratigraphy*. A sequence, in this strict sense, is a packet of strata, bounded above and below by unconformities or their laterally continuous surfaces in conformable successions (Fig. 1.4). The unconformities can be due to nondeposition or to a component of erosion. In either case the time-gap (hiatus) is generally taken to represent a relative lowering of sea level. The hypothesis that important changes of sea level may be eustatic and of global extent has provided a basis for interregional correlation, and for the construction of global reference curves of sea-level change. Alternatively, the sequence boundaries offer important evidence for more regional tectonic events.

Sequence stratigraphy is applied most widely in the weakly deformed Upper Palaeozoic, Mesozoic and Cenozoic successions that can readily be seismically imaged around Britain and Ireland. However, the subdivision of successions along their contained unconformities can be applied to earlier successions too and, on a large scale, forms the basis for the subdivision of this book (see Section 1.7).

1.4 Locating events in space: palaeogeography

One of the main challenges of reconstructing geological histories is that rock successions are rarely preserved at the site of their original formation. The latitude, longitude and orientation of the crustal fragment on which the rocks lie will have changed because of global plate motions, and its relationship to adjoining fragments may have altered because of regional tectonic displacements. Proper portrayal of geological history requires palaeogeographic maps, showing the shifting arrangement of the continental crustal pieces that mainly host the preserved geological record.

The most conspicuous indicators of changed palaeolatitude in a sedimentary sequence are those lithologies that form preferentially in a restricted climatic belt. Most obviously, glacial tillites are deposited in

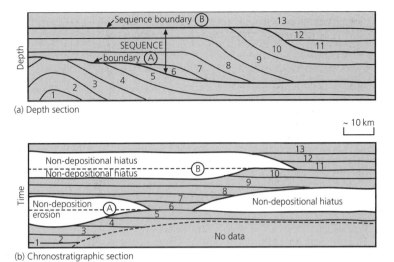

(a) Depth section

~ 10 km

Fig. 1.4 The essential features of a depositional sequence in sedimentary rocks, shown in a depth section (a) and a corresponding chronostratigraphic (time) section (b).

(b) Chronostratigraphic section

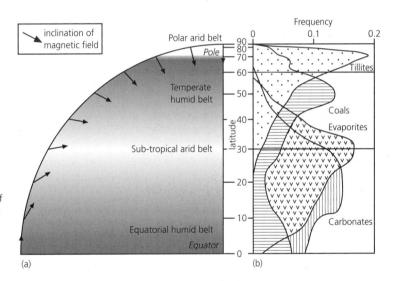

Fig. 1.5 (a) The main climatic belts of the Earth, with the varying inclination of the magnetic field with latitude. (b) The frequency of climatically sensitive lithofacies formed at different latitudes (Mesozoic and Cenozoic data from Scotese & Barrett 1990).

polar and high temperate latitudes (Fig. 1.5). Coals form from the abundant vegetation of both the equatorial and the temperate rainy belts, whereas evaporites need the warmth and low rainfall of the intervening arid subtropical belt. Limestones are most common in the equatorial, subtropical and warm temperate belts where sea-water temperatures are high and sunlight can penetrate effectively. Phosphorites and continental red-beds also tend to occur in low latitudes. The northward drift of Britain and Ireland through much of Phanerozoic time is evident from such facies changes. Particularly clear is the

transit from the southern arid subtropics, producing Devonian red-beds, across the Carboniferous equator, with its abundant coals, to the northern arid belt recorded by Permo–Triassic red-beds and evaporites. However, many more subtle latitudinal influences will be noted in this book.

A seemingly more precise measure of palaeolatitude can be obtained from rocks containing mineral grains that align with the Earth's magnetic field. Because this field is a dipole, on average centred at the geographical pole, the direction of the measured palaeomagnetic field records the rotation of the site with respect to longitude lines. Because the inclination of the present field varies systematically with latitude (see also Fig. 1.7), the measured inclination of an old field with respect to a palaeohorizontal, such as sedimentary bedding, yields the palaeolatitude. In deformed rocks, the identification of the appropriate palaeohorizontal datum may be problematical. Errors may also arise by overprinting of one magnetic record by another, typically during metamorphism. However, the main uncertainty in using palaeomagnetic data to make maps is the unconstrained longitude of each crustal fragment. The global palaeogeographic maps in this book should be viewed critically in the knowledge that each fragment could, on palaeomagnetic evidence, lie anywhere along the same latitude band. In practice, geological matches or mismatches between continents can be used to further constrain their positions, but these qualitative criteria often prove to be controversial.

A third way of diagnosing palaeolatitude is from the global distribution of faunal and floral assemblages. Climatic control of present biogeography is particularly evident in land plants, shallow marine benthic organisms, and in the plankton that live high in the marine water column. These groups are more responsive to latitudinal climate change than is the deep-water benthos that occurs on the outboard margins of continents. Distributions of appropriate fossil organisms may therefore reveal biofacies restricted to particular palaeolatitudes. The cold-climate *Glossopteris* flora was one of the first clues that the southern continents once formed a unified Gondwana in high southern latitudes. Latitude-controlled assemblages are now most used to check and refine palaeomagnetically derived continental maps (e.g. Fig. 1.6), particularly in the Palaeozoic, where reconstructions are more loosely constrained than for later eras.

More useful even than the control on biofacies by latitude is that by continental separation. In the Early Ordovician example (Fig. 1.6) distinct benthic trilobite assemblages occur on different continents in the same equatorial latitudes. Hence Laurentian faunas are distinct from those at the low-latitude end of Gondwana. Based on the biofacies evidence, Siberia and North China might be arranged closer to Laurentia, with an oceanic barrier between them and the South China component of Gondwana. Such arguments depend crucially on knowledge of the

Fig. 1.6 Early Ordovician (Arenig–Llanvirn) platform trilobite assemblages, showing control both by palaeolatitude and continental separation (modified from Cocks & Fortey 1990, with permission from the Geological Society, London 1999).

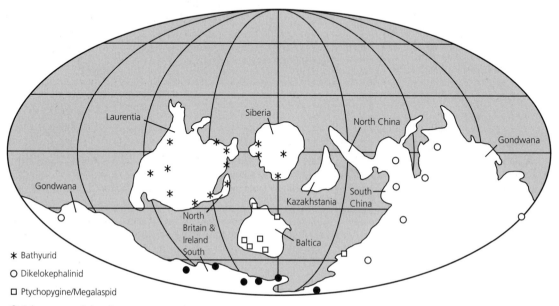

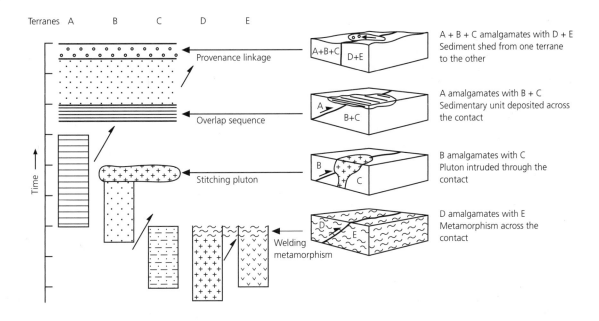

Fig. 1.7 Hypothetical assembly diagram for five terranes, showing four major ways of recognizing terrane linkage (modified from Howell 1989).

palaeoecology of the organisms concerned. For example, Early Ordovician pelagic trilobites were not confined by oceanic barriers and show a simple latitudinal control.

Both palaeomagnetic and faunal criteria can identify not only displacement of continents with respect to one other but also mismatches between smaller crustal fragments. Strong contrasts in stratigraphic or tectonic sequences between two areas, now juxtaposed across a major fault, may also suggest that they formed some distance apart. A fault-bounded volume of crust with an internally coherent geological history is called a *tectonostratigraphic terrane*. Britain and Ireland are commonly divided into about 10–15 Neoproterozoic to Early Palaeozoic terranes (see Chapter 2, Section 2.3.1), although large separations cannot be proven between all adjacent terranes. The date of amalgamation of two terranes can be recognized from a distinctive event common to both areas (Fig. 1.7). Examples are a shared metamorphic event, a pluton intruded or a sedimentary sequence deposited across the dormant terrane boundary, or a sedimentary unit on one terrane containing distinctive clasts derived from the other. The sense of displacement on the terrane-bounding faults is a guide to the former relative locations of the terranes, but the magnitude of the original separation may be highly uncertain.

The palaeogeographic maps of Britain and Ireland in this book try to show the major oceanic separations between continents, although not usually to scale. The maps also indicate known terrane boundaries, although typically not the restored positions of the terranes. All maps up to the time of the Caledonian Orogeny, and maps of southern Britain and Ireland up to the Variscan Orogeny, must be viewed with the possibility of subsequent terrane displacements in mind. Most maps do not attempt to correct for more local crustal deformation, such as the extension as sedimentary basins form or the shortening or shear across them as they are destroyed. These displacements are on a comparable scale to the sedimentary systems and igneous or metamorphic provinces that form the next level of detail on a palaeogeographic map. By ignoring such local tectonic deformation, the size and shape of geological elements may be distorted on a map, but not their internal topology or their relationship to other elements.

1.5 Specifying geological environments

Once geological events are adequately located in

space and time, the next stage in reconstructing a history is to identify the proper geological environment for each event. Were the components of a sedimentary rock deposited on a delta or a turbidite fan? Was an intrusive rock formed beneath a continental volcanic arc or within ocean crust? It is the spatial pattern of these environments that provides the detail on a palaeogeographic reconstruction at any one time, and it is the changing patterns through successive time intervals that bring geological history to life. This section briefly reviews the sorts of environmental classifications that are useful on the scale of regional geological reconstructions.

1.5.1 Sedimentary environments

The first level of interpretation of sedimentary rocks is into *facies*, comprising units with similar specified characteristics; for instance a graded sandstone facies or a laminated mudstone facies. However, individual facies are generally too localized or change too frequently to be useful in describing regional geological history. Facies typically occur with others in *facies associations*, which can better be assigned to a particular depositional environment. Hence, a facies association of massive sandstone, graded sandstone and laminated mudstone might be regionally extensive and diagnosed as the product of a deep-sea turbidite system. This is the scale of environmental system that usually appears in regional reconstructions. There is no one definitive classification of depositional environments, but a typical subdivision is shown in Table 1.2a.

Outcrop-scale observations, or their equivalent in cores or downhole log records, are the primary data for diagnosing the environment of deposition of a sedimentary rock. By contrast, petrographical or geochemical observations can help to identify the source of the detritus supplied to a sedimentary basin, and can act as a guide to the broader tectonic setting of that basin. The relative proportions of quartz, feldspar and lithic grains in sandstones have proved to be particularly useful (Table 1.2b). Sand derived from large continents tends to be mineralogically mature, and therefore rich in quartz and poor in rock fragments. By contrast, arc-derived sandstones are immature, and rich in feldspar and volcanic rock fragments. Sandstones derived from

Table 1.2 Classification schemes for: (a) depositional environments (based on Reading, 1996); (b) provenance of clastic source material (scheme of Dickinson & Suczek 1979).

(a) Depositional environments

Environment	Sub-environment examples
Glaciers	Sub- and supra-glacial Glaciofluvial Glaciolacustrine Glaciomarine
Alluvial systems	Rivers Levees and floodplains Alluvial fans
Lakes	Clastic systems (Bio)chemical systems
Desert aeolian systems	Dune fields Sand sheets
Clastic coasts	Deltas Non-deltaic coasts
Arid shorelines and basins	Sabkhas Evaporite basins
Shallow seas	Clastic systems Carbonate systems
Deep seas	Clastic systems (Hemi)pelagic systems
Volcanoes	Pyroclastic systems

(b) Environment of sediment provenance

Tectonic setting	Provenance type
Continental (quartz + feldspar, low lithics)	Craton interior Transitional Uplifted basement
Magmatic arc (feldspar + lithics, low/moderate quartz)	Dissected Transitional Undissected
Recycled orogen (quartz + lithics, low/moderate feldspar)	Subduction complex Collisional orogen Foreland uplift

active orogenic belts—subduction complexes, collision zones and foreland uplifts—have an intermediate maturity with a mixture of quartz and sedimentary rock fragments.

1.5.2 Igneous environments

A first level of interpretation of igneous rocks is into plutonic and volcanic types, each of which may be further subdivided on the basis of their mineralogy. However, as for sedimentary rocks, identification of rock types alone is of limited use in reconstructing ancient geological environments because some igneous rocks can occur in several different tectonic settings. Typically, several igneous rock types may occur together to form a petrogenetically related *magma series*, which may be diagnostic of a particular tectonic environment (Table 1.3). Magmas of the calc-alkaline series are at present restricted in their occurrence to magmatic arcs located above subduction zones. The recognition of calc-alkaline characteristics in the geochemistry of ancient igneous rocks may therefore be an important indication of their likely tectonic setting. Such sequences tend to be characterized by a wide variety of basic, intermediate and acidic rock types, both volcanic and plutonic. By contrast, oceanic ridge systems have a more restricted range of magma types dominated by the low-potassium, subalkalic basalts known as tholeiites. Tholeiitic basalts are also found in intraplate tectonic settings such as oceanic islands and intracontinental rifts. In these cases, they commonly form part of a bimodal igneous suite together with alkaline types such as trachytes. Detailed study of the nature, relative proportions and geochemistry of ancient magma types may therefore shed light on their likely tectonic setting.

1.5.3 Metamorphic environments

Metamorphic mineralogies and textures are controlled by temperature, pressure and deviatoric stress, which in turn are related intimately to tectonic setting. Some specific metamorphic minerals are diagnostic of certain tectonic settings and/or processes. Glaucophane, for example, is invariably found within the low-temperature, high-pressure metamorphic sequences characteristic of subduction zones. Coesite is diagnostic of the ultra-high pressures typical of either the deep roots of mountain belts or impact craters. In general, however, most metamorphic minerals can be found in a number of different tectonic settings. One way of classifying metamorphic rocks is by use of mineral assemblages to define *metamorphic facies*. An alternative and potentially more useful approach from a regional tectonic viewpoint is to use a combination of field mapping and petrological studies to differentiate between the different settings which have led to metamorphism. Broad categories of local and regional metamorphism can be distinguished according to whether the metamorphic rocks are clearly limited in area and related to a localized event, or are of large areal extent. Each of these categories can be subdivided according to setting (Table 1.4).

1.5.4 Structural regimes

Structural observations, when integrated over a large area, typically allow a phase of crustal deformation to

Table 1.3 Characteristic magma series associated with specific tectonic settings (based on Wilson 1989).

Tectonic setting	Plate margin		Within plate	
	Convergent	Divergent	Intra-oceanic	Intracontinental
Volcanic feature	Island arc, active continental margin	Mid-ocean ridge, back-arc spreading centre	Oceanic island	Continental rift zone, continental flood-basalt province
Characteristic magma series	Tholeiitic Calc-alkaline Alkaline	Tholeiitic	Tholeiitic Alkaline	Tholeiitic Alkaline
SiO_2 range	Basalts and differentiates	Basalts	Basalts and differentiates	Basalts and differentiates

Table 1.4 Classification scheme for metamorphic environments (based on Duff 1993).

Type of metamorphism	Diagnostic features
Local metamorphism	
Contact metamorphism	Metamorphic rocks are found adjacent to and clearly related to an igneous body
Dynamic metamorphism	Metamorphism is associated with a zone of strong deformation (fault or shear zone)
Impact metamorphism	Metamorphism is associated with the impact on Earth of an extraterrestrial body
Regional metamorphism	
Orogenic metamorphism	Metamorphic rocks are associated with zones of mountain building
Oceanic metamorphism	Metamorphism of ocean crust by injection of basic magmas and circulation of hydrothermal fluids at spreading ridges
Burial metamorphism	Metamorphism due to sufficient burial in a thick sedimentary basin succession

Table 1.5 Classification scheme for: (a) regional deformation regimes; (b) plate tectonic settings.

(a) Tectonic regimes

Tectonic regime	Examples of structures
Contractional	Reverse faults and thrusts Flexural or fault-related folds Steep low-grade fabrics Reverse ductile shear zones High-grade fabrics
Extensional	Planar/listric normal faults Fault-related folds or tilt blocks Low-dip ductile shear zones Low-dip high-grade fabrics
Transcurrent	Strike-slip or oblique-slip faults Folds ± oblique fabrics Ductile strike-slip shear zones Steep fabrics and folds

(b) Plate tectonic settings

Kinematic setting	Examples of components
Divergent plate boundary	Ocean ridge/transform Continental rift Aulacogen or failed rift
Convergent plate boundary	Trench/subduction complex Fore-arc basin Inter-arc/back-arc Foreland basin
Continental collision zone	Foreland basin Hinterland basin
Conservative plate boundary	Oceanic transform Continental margin transform Continental transform
Intraplate regions	Ocean floor Continental craton

be assigned to a particular, regionally consistent, structural regime (Table 1.5a). The three end-member regimes are contractional, extensional and transcurrent, although intermediate transpressional and transtensional regimes are common. Certain types of locally observed structures are common in each regime, although heterogeneity of the deformation can give, for instance, locally extensional zones in a regionally contractional regime. The most complex structural settings are orogenic belts, which may display an evolutionary history lasting several tens of millions of years. Studies in young mountain belts, for example the Himalayas, show that the fold and thrust nappes related to contraction during the early stages of collision may be overprinted by transcurrent structures developed during the later 'locking-up' of colliding blocks. Furthermore, it has become apparent that the central parts of many mountain belts have undergone a late phase of extensional deformation that results from gravitational instability following the main contraction and crustal thickening. Older mountain belts have typically undergone uplift and erosion, and the orogenic roots so exposed give a valuable insight into the history of the mid-crust.

1.5.5 Plate tectonic setting

One large-scale goal of historical geology is to diagnose a particular plate tectonic environment or subenvironment for each chapter of a regional story (Table 1.5b). This diagnosis is only possible, if at all, from a synthesis of all the available sedimentary, igneous, metamorphic and structural evidence. There are a number of problems inherent in such a synthesis. A major difficulty is that plate tectonics operates at a much larger scale than that at which we are trying to solve geological problems. This theme can be developed by superposing an outline of Britain and Ireland onto a present-day tectonic map of Asia (Fig. 1.8) where different local tectonic settings (varying from convergence, through strike-slip to extension) are all fundamentally related to one overall process, which is the collision of India and Asia. By only studying one part of the Asian region we would diagnose only one component of the overall plate tectonic

setting. Any plate tectonic analysis of Britain and Ireland must therefore consider the regional context by examining adjacent areas.

A further problem is that some plate tectonic settings are inevitably destined to be destroyed or overprinted. For instance, ancient ocean floor or ridge/transform settings are usually only likely to be preserved as minor ophiolitic fragments in later orogenic belts. Indeed, the consensus is that most ophiolites are likely to represent back-arc or fore-arc basin crust than oceanic crust *sensu stricto*. Finally, it is apparent that in many mountain belts the original relations between adjacent crustal blocks have been obscured by major strike-slip displacements, of the order of hundreds of kilometres, complicating any resulting plate tectonic reassembly.

1.6 Telling the story: analogues and models

This book will deploy a number of techniques for illustrating interpretations of the geological history of

Fig. 1.8 Structural map of present-day Asia, with Britain and Ireland superimposed in analogous tectonic positions (suggested by Dewey 1982, with permission from the Geological Society, London 1999) for particular periods of geological time. The juxtaposition highlights the scale problem in diagnosing the plate tectonic setting of small areas.

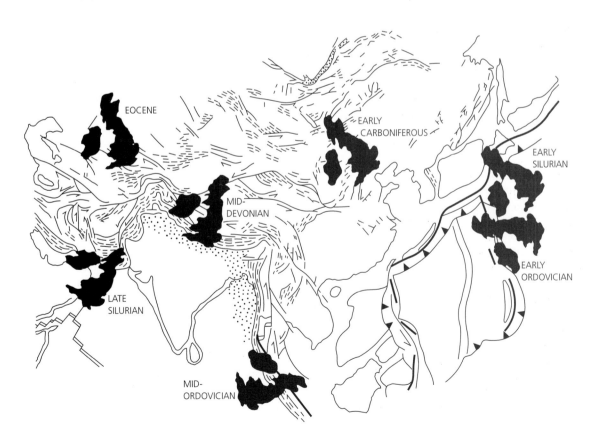

Britain and Ireland, particularly regional palaeogeographic maps and crustal-scale cross-sections. At this scale, no interpretation is more than partly constrained by the surviving or accessible evidence. However, any acceptable interpretation must also be geologically realistic in comparison with recent or well-understood ancient analogues and, less crucially, with results of theoretical models.

Examples of appropriate recent analogues are the subduction systems of south-east Asia, which have commonly been used to guide reconstruction of the systems that bordered the Early Palaeozoic Iapetus Ocean. The hypothesis is that the ancient systems would have had an outboard to inboard zonation from subducting ocean floor, through accreted subduction complex and fore-arc basin to active volcanic arc, and perhaps a back-arc basin and extinct arcs. Mismatch of the ancient with the recent systems can promote a search for missing components, perhaps removed further along the system by strike-slip faulting, or focus attention on the need for particular further data.

Theoretical models of geological processes provide a second type of test for regional reconstructions. An influential example is the rift-and-sag model of sedimentary basin formation, in which rapid subsidence due to lithospheric thinning is followed by decaying thermal subsidence as the temperature structure of the lithosphere was re-established. This model has been widely applied to the upper Palaeozoic, Mesozoic and Cenozoic basins of onshore and offshore Britain, so that their basin fill is commonly divided into syn-rift and post-rift phases. Compatibility of a reconstructed geological history with such mechanical models is encouraging. However, given the complexity of many geological processes, a mismatch is just as likely to suggest modification to the theoretical model as it is to the reconstruction.

1.7 The geological organization of Britain and Ireland

Chapter 2 in the introductory part of this book lays out the historical and spatial framework into which the geology of Britain and Ireland can be fitted. The remainder of the book is divided into five further parts, each describing a phase of the region's geological history. The relationship of these geological phases to the major tectonostratigraphic rock units of Britain and Ireland can be seen on a time-chart (Fig. 1.9a) and geological map (Fig. 1.9b). The boundaries between the units on these diagrams are regional unconformities, major gaps in the geological record each recording the crustal shortening, uplift and erosion associated with an orogenic event. Metamorphism and igneous activity are associated with each orogeny. The periods between the orogenic unconformities are times of net accumulation of rock successions, often because of episodes of crustal extension. Igneous activity and minor unconformities accompany some extensional phases.

The most significant orogenic unconformity is that associated with the Caledonian or Acadian Orogeny, culminating in Devonian time. This event finally welded the microcontinent of Eastern Avalonia — including England, Wales and south-east Ireland — to the large Laurentian continent — including Scotland and north-west Ireland. Before Silurian time, the two continents were separated by the wide Iapetus Ocean, and therefore had dissimilar geological histories. The northern Laurentian margin had a complex history (described in Part 2) involving Archaean, Proterozoic and Ordovician (Grampian) orogenies before the Caledonian climax. The tectonostratigraphic sequences on this margin differ between discrete fault-bounded terranes assembled late in the margin's history, and the chart simplifies this complex pattern. By comparison, the preserved pre-Caledonian history of the southern Iapetus margin (described in Part 3) is shorter and simpler. The Neoproterozoic rocks of the Cadomian orogenic belt are overlain by Cambrian to lower Devonian successions. Part 4 describes the culminating Caledonian events and the Devonian sedimentation that accompanied them.

Neither the Caledonian nor the Grampian orogenies deformed all the pre-existing rocks that are now exposed across Britain and Ireland. Consequently, relatively weakly deformed rocks, separately ornamented in Fig. 1.9, survive outside the Caledonian deformation front, both in the Welsh Borders to the south-east and in the Scottish Highlands to the north-west.

The Caledonian cycle was followed by the Variscan cycle of rock accumulation and deformation, described in Part 5. This cycle began in most areas with Late Devonian and then Carboniferous deposi-

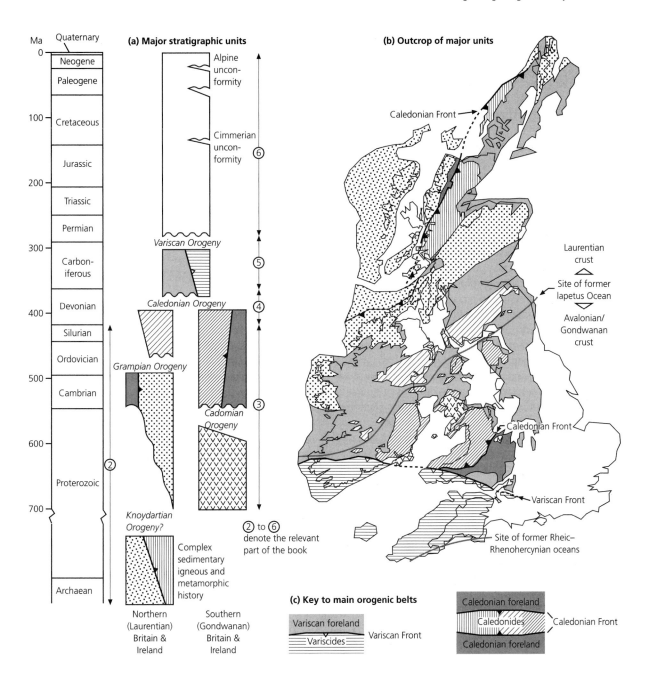

Fig. 1.9 Major tectonostratigraphic units in Britain and Ireland (a), with their distribution at outcrop (b) and a key to the main orogenic belts (c) (modified from Woodcock 1994, with permission from UCL Press (Routledge) 1999). The relevant parts of this book are labelled in (a).

tion over the Caledonian unconformity. The exception is in south-west England, where earlier Devonian sedimentation was probably continuous through the time of the Late Caledonian events. This was the area most affected by the Variscan Orogeny (Fig. 1.9b), whose culminating events resulted in an unconformity spanning Late Carboniferous and Early Permian time. A Variscan deformation front separates the strongly deformed zone from the area to the north, in which the post-Caledonian cover was relatively gently deformed (Fig. 1.9b).

By Early Permian time, the crust of Britain and Ireland had been assembled in more or less its present configuration. The post-Variscan cycle (Part 6) is dominated by sediment accumulation in basins formed by crustal extension, attributable to the marginal effects of the rifting and opening of the Atlantic Ocean. The pulsed nature of extension has produced gentle unconformities (such as the Cimmerian, Fig. 1.9a) and an episode of Jurassic magmatism. Palaeogene magmatism and unconformities resulted from the onset of a mantle plume, and there was gentle Neogene (Alpine) folding in southern England, but the post-Variscan cycle still awaits a culminating regional orogenic event.

References

Dewey, J.F. (1982) Plate tectonics and the evolution of the British Isles. *Journal of the Geological Society, London* **139**, 371–412.

Dickinson, W.R. & Suczek, C.A. (1979) Plate tectonics and sandstone compositions. *American Association of Petroleum Geologists Bulletin* **63**, 2164–2182.

Duff, D. (1993) *Holmes' Principles of Physical Geology*, 4th edn, pp. 1–791. Chapman & Hall, London.

Gradstein, F.M. & Ogg, J. (1996) A Phanerozoic time scale. *Episodes* **19**, 3–4.

Howell, D.G. (1989) *Tectonics of Suspect Terranes: mountain building and continental growth*, pp. 1–232. Chapman & Hall, London.

Scotese, C.R. & Barrett, S.F. (1990) Gondwana's movement over the South Pole during the Palaeozoic: evidence from lithological indicators of climate. In: *Palaeozoic Palaeogeography and Biogeography* (eds W. S. McKerrow & C. R. Scotese), Memoir 12, pp. 75–85. Geological Society, London.

Tucker, R.D., Bradley, D.C., Version Straeten, C.A. *et al.* (1998) New U–Pb zircon ages and the duration and division of Devonian time. *Earth and Planetary Science Letters* **158**, 175–186.

Wilson, M. (1989) *Igneous Petrogenesis*, pp. 1–466. Unwin Hyman, London.

Woodcock, N.H. (1994) *Geology and Environment in Britain and Ireland*, pp. 1–164. UCL Press, London.

Further reading

Allen, P.A. & Allen, J.R. (1990) *Basin Analysis. Principles and applications*, pp. 1–451. Blackwell Scientific, Oxford. [An excellent review of how sedimentary basins form and fill.]

Cocks, L.R.M. & Fortey, R.A. (1990) Biogeography of Ordovician and Silurian faunas. In: *Palaeozoic Palaeogeography and Biogeography* (eds W. S. McKerrow & C. R. Scotese), Memoir 12, pp. 97–104. Geological Society, London. [Illustrates the principles of global biogeography with respect to Early Palaeozoic examples.]

Doyle, P., Bennett, M.R. & Baxter, A.N. (1994) *The Key to Earth's History: an introduction to stratigraphy*, pp. 1–231. Wiley, New York. [An introduction to how stratigraphy is done, with helpful case studies.]

Faure, G. (1986) *The Principles of Isotope Geology*, 2nd edn, pp. 1–589. Wiley, New York. [The standard guide to isotope geology and radiometric dating.]

Frodeman, R. (1995) Geological reasoning: geology as interpretive and historical science. *Geological Society of America Bulletin* **107**, 960–968. [An incisive but readable analysis of how geologists think.]

Howell, D.G. (1989) *Tectonics of Suspect Terranes: mountain building and continental growth*, pp. 1–232. Chapman & Hall, London. [An excellent overview of terrane analysis, combining principles with case studies.]

Park, R.G. (1988) *Geological Structures and Moving Plates*, pp. 1–337. Blackie, Glasgow. [Reviews the main plate tectonic settings and their structural components.]

Reading, H.G. (ed.) (1996) *Sedimentary Environments: processes, facies and stratigraphy*, 3rd edn, pp. 1–688. Blackwell Science, Oxford. [An authoritative introduction to sedimentary environments.]

Snyder, D.B. & Barber, A.J. (1997) Australia–Banda Arc collision as an analogue for early stages in Iapetus closure. *Journal of the Geological Society, London* **154**, 589–592. [An example of the use of recent tectonic analogues to support a geological reconstruction.]

Whittaker, A., Cope, J.C.W., Cowie, J.W. *et al.* (1991) A guide to stratigraphical procedure. *Journal of the Geological Society, London* **148**, 813–824. [A concise statement of the rules of stratigraphy.]

2 Geological framework of Britain and Ireland

R. E. HOLDSWORTH, N. H. WOODCOCK AND R. A. STRACHAN

2.1 Global change through geological time

2.1.1 Introduction

The geological history of any region such as Britain and Ireland is the result of both local and large-scale processes. Individual chapters of this history, and of this book, will often seem to be dominated by the local complexities of climatic or tectonic setting, or of depositional or magmatic environment. Although every effort will be made to highlight the control by large-scale processes, these factors can be appreciated best when viewed on a long time-scale. This chapter provides this long-term overview of global evolution (see this section (2.1)), before focusing on the global palaeo-continental setting (see Section 2.2) and regional tectonic templates (see Section 2.3) of Britain and Ireland.

At least four types of global-scale process have an important influence on local geological history: plate tectonics, sea-level change, climatic change and biological evolution. Some of the parameters of each process can be charted for Phanerozoic time (Fig. 2.1) and will be described below (see Sections 2.1.2–2.1.7). However, all these processes are interrelated to some extent. Plate movements shift continents across latitudes and climatic zones. The development of spreading centres and hotspots affects sea levels, in turn influencing climate as the size of shallow seas changes. Climate changes too as oceans open and close, altering global circulation patterns in surface and deep waters. Climate is also affected by large-scale changes in surface topography, particularly the growth of mountain chains during continental collision. Global atmospheric temperatures are affected by the rate at which carbon dioxide is produced by volcanism.

All these factors were demonstrably important in influencing the geological evolution of Britain and Ireland from the Late Precambrian onwards. However, back through Precambrian time, the record of biological evolution becomes sparse, and that of sea-level and atmospheric change more difficult to detail (see Chapters 3 and 4). Plate tectonic processes are assumed to have operated back through the Proterozoic and into the Archaean. However, key evidence, such as ophiolites and blueschists, becomes harder to find in older rocks and the operation of less organized tectonic mechanisms in the earlier history of the Earth cannot be discounted.

2.1.2 Continental amalgamation and dispersal

On most integrated models of Earth evolution, the continual rearrangement of continents by plate tectonic processes plays a central role. The changing Neoproterozoic and Phanerozoic geometry of the continents and oceans will be detailed in a later section (see Section 2.2.2). However, at any time the continental arrangement can be simply assigned to one of two modes: a clustered supercontinent or a number of dispersed continents. A crude index of continental arrangement is therefore a graph of the number of continents through time (Fig. 2.1a). This graph shows the most recent supercontinent of Pangaea, formed during the Late Palaeozoic to Early Mesozoic. A Vendian supercontinent is similarly identified in the Late Neoproterozoic and a yet earlier supercontinent—Rodinia—in the Early Neoproterozoic. These times of clustered continents are separated by periods of continental dispersal

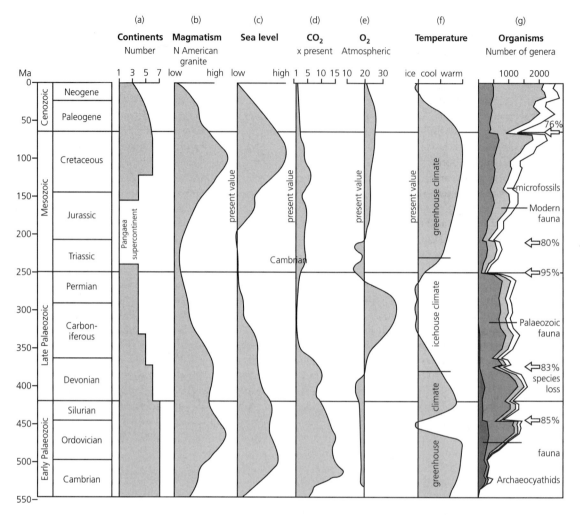

Fig. 2.1 Parameters of global change through Phanerozoic time: (a) number of major continents; (b) rate of intrusion of North American granites; (c) global sea level; (d) atmospheric concentration of carbon dioxide; (e) atmospheric concentration of oxygen; (f) global average surface temperature (g) number of genera in representative groups of organisms, subdivided into component faunas, and showing percentage species loss in major mass extinctions (data from Veevers 1990 (a,b,c,f); Berner 1991 (d); Berner & Canfield 1989 (e); Sepkoski 1998 (g)).

followed by coalescence. One such Phanerozoic phase lasted from Cambrian to Carboniferous time, and the second from the Cretaceous to the present day (Fig. 2.1a).

Later chapters will provide a number of examples of the influence of continental arrangement on the geological record of Britain and Ireland. However, these effects often acted indirectly through modifications of climate. Examples are the mega-monsoonal continental climate of the Permo–Triassic Pangaea, compared with the latitudinally zoned climate established in the Jurassic, when Pangaea was dispersing. More generally important is the observation that supercontinents coincide with periods of globally cool 'icehouse' climate and with low global sea level. However, the resulting hypothesis that continental arrangement is the primary control on global climate requires consideration of global magmatic activity before it can be explained or tested (see Section 2.1.3).

2.1.3 Magmatic activity

The Triassic world, when there was a single Pangaean continent and a single ocean (Panthalassa), must necessarily have had a relatively small length of both mid-ocean ridges and subduction zones. Such periods of low magmatic activity contrast with times of continental dispersal, when there were presumably more plate boundaries with higher spreading and subduction rates. There has been no adequate stocktaking of the global production of igneous rocks through time. However, an available curve of the volume of granites emplaced in the North American crust through the Phanerozoic confirms a correlation with the continental arrangements. Magma production at the subduction zones rimming North America was at a minimum during the time of the Vendian supercontinent, rose during continental dispersal and fell again during the continental re-amalgamation to form Pangaea. The same pattern was repeated during the Mesozoic dispersal of Pangaea and the continuing coalescence of the present continents.

Magmatic activity releases two particular products

of the Earth's interior that are of global significance: carbon dioxide and heat. These products provide a possible mechanistic link between continental patterns, climate and sea level. An Earth with a single supercontinent (Fig. 2.2a) should have a low magmatic output of carbon dioxide to the atmosphere, with a consequently weak greenhouse effect and a cool global climate. More heat should be stored in the Earth at such a time, particularly beneath the supercontinent. Consequently, the continent should stand high whereas the ocean should have slow-spreading ridges and a high ocean-water capacity. Global sea level should therefore tend to be low, possibly enhanced by storage of water in large continental ice sheets. Conversely, when the continents are dispersed (Fig. 2.2b), magmatic activity and carbon dioxide output are high. The consequent warm Earth and low ocean-basin capacity should result in high global sea levels.

These ideas can now be tested against the observed variations in sea level and atmospheric composition.

2.1.4 Global sea-level change

Relative changes in sea level can be estimated locally using the principles of sequence stratigraphy (see Section 1.3). If such local fluctuations in relative sea level can be correlated between regions with different tectonic histories, a case for their eustatic or global

Fig. 2.2 Postulated control on global climate and sea level by: (a) a single clustered supercontinent; (b) dispersed continents (modified from Veevers 1990, with permission from Elsevier Science 2000).

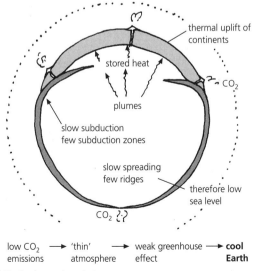

(a) Single clustered continent

low CO_2 → 'thin' → weak greenhouse → **cool**
emissions   atmosphere   effect   **Earth**

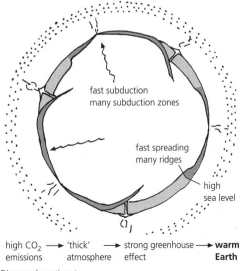

(b) Dispersed continents

high CO_2 → 'thick' → strong greenhouse → **warm**
emissions   atmosphere   effect   **Earth**

nature can be made. There is a vigorous debate about the evidence for global sea-level cycles on the order of millions or tens of millions of years, not least because appropriate driving mechanisms are not always apparent. However, there is wide acceptance of a first-order fluctuation on a period of several hundred million years (Fig. 2.1c). The first-order curve can be calibrated independently by estimating the proportion of continent covered by marine sediments through geological time. Sea-level changes on this scale are explained by the changing spreading rate and volume of mid-ocean ridges, and therefore the available capacity of the ocean basins to contain a fixed volume of sea water.

The first-order curve shows sea level rising during the Cambrian to a generally high level through early Palaeozoic time, a time of dispersed continents and presumed fast spreading. Sea level fell gently during the Late Palaeozoic to a lowstand during the Triassic amalgamation of Pangaea. A rapid rise through Jurassic and Early Cretaceous time accompanied the dispersal of Pangaea, culminating in a Phanerozoic highstand in the Late Cretaceous. Calibration of this curve is inexact, but sea level during the Late Cretaceous highstand is estimated to have been as much as 300 m above the present-day lowstand level. The post-Cretaceous fall in sea level was more rapid than might be expected on the hypothesis of impending continental coalescence, probably accelerated by the onset of global glaciation before the formation of the next supercontinent. Such a 'premature' glacial event also occurred midway through the Palaeozoic cycle, in Late Ordovician time.

The first-order sea-level curve is crudely but directly reflected in the rock record of Britain and Ireland. Highstands are recorded by the abundant marine deposits of the early Palaeozoic and late Mesozoic, whereas the intervening lowstand encouraged the widespread marginal marine and non-marine deposits of late Carboniferous to Triassic age. Only the predominantly continental deposits of the Devonian might be seen as anomalous, testifying to the effect of the Caledonian orogenic belt in raising most of Britain and Ireland above the level of even the high Devonian seas.

2.1.5 Atmospheric carbon dioxide and oxygen levels

Measurement of past atmospheric composition cannot be carried out directly, except from the air bubbles frozen in Quaternary ice. Further back in time, the stable isotopes of carbon and oxygen provide some constraints on the concentrations of atmospheric gases. However, estimates of changing atmospheric composition depend mainly on complex assessments of the carbon, oxygen and sulphur budgets as these elements are cycled through the various reservoirs of the Earth's surface and interior. The driving fluxes are physical, chemical and biological, and can rarely be accurately specified. The resulting curves (Fig. 2.1d,e) are necessarily inexact.

The carbon dioxide curve (Fig. 2.1d) correlates well with those for magmatic activity and continental arrangement (Fig. 2.1a,b), partly because these are factored into its derivation. Nevertheless, the broad features of the curve match independent geological evidence, not least from the geological record of Britain and Ireland. For instance, the low global carbon dioxide levels during the Carboniferous are balanced by the large amounts of carbon buried in limestone and coal deposits at that time. By contrast, global oxygen levels were exceptionally high at this time (Fig. 2.1e), due to the high rate of photosynthesis among Carboniferous plants. Oxygen levels were particularly low during the Triassic, perhaps due to erosion and oxidation of buried carbon and sulphur compounds from the uplifted Pangaean continent.

2.1.6 Global temperature

Earth surface temperatures can be estimated from oxygen isotope measurements and, more crudely, from the abundance and distribution of climatically sensitive sediments such as evaporites, coals and glacial tills. A first-order curve of average global temperature (Fig. 2.1f) shows a strong correlation with the atmospheric carbon dioxide concentration and in turn with magmatic activity. This similarity suggests that warm phases of the Earth's climate are due to the trapping of solar energy beneath the blanket of heat-absorbing carbon dioxide gas—the greenhouse effect (Fig. 2.2). Global 'greenhouse' states in the early Palaeozoic and Mesozoic–Cenozoic were separated by an 'icehouse' state during the Late Palaeozoic (Fig. 2.1f). The global icehouse period overlapped the time of the Pangaea supercontinent, low sea levels and low magmatic carbon dioxide output.

More problematical, on this model, are the two short glacial periods thought to fall within longer greenhouse states—in Late Ordovician and Late Neogene time. Explanations other than the simple clustering of continents are required for these events, such as the fortuitous existence of a large polar continent or of a land-locked polar ocean, either of which acts as a focus for ice accumulation.

The effect of the Phanerozoic cycling from greenhouse to icehouse states is only weakly recorded in the geological record of Britain and Ireland. A later section (see Section 2.2.3) will show that, by chance, this region lay in warm equatorial latitudes during the Late Palaeozoic global icehouse phase and cooler southern or northern latitudes during the earlier and later greenhouse periods, respectively.

2.1.7 Biological evolution

The final, and most complex, index of global change to be considered is the fluctuating diversity of organisms through Phanerozoic time (Fig. 2.1g). Problems of taxonomy and preservation complicate the construction of this curve, but plotting the number of genera of representative groups gives a robust estimate of diversity. The organisms can be subdivided into four distinctive faunas, each with a different pattern of abundance through time: a 'Cambrian fauna', dominated by trilobites and inarticulate brachiopods; a 'Palaeozoic fauna' including crinoids, corals, articulate brachiopods, graptolites and cephalopods; a 'Modern fauna' including bivalves, gastropods, echinoids and vertebrates; and a microfossil fauna.

The evident rise in organic diversity through Phanerozoic time has been spasmodic. Periods of stability or steady increase in diversity have been punctuated by rapid extinction events. The five most important events are in the Late Ordovician (when 85% of species were lost), in the Late Devonian (83%), end Permian (95%), end Triassic (80%) and end Cretaceous (76%). It has been argued that the clustered continents of Pangaea were to blame for the largest of these extinctions in the end Permian. The reduced area of shallow marine shelves, the smaller number of faunal provinces, the monsoonal climate and the cold polar temperatures have all been invoked as extinction mechanisms. However, other major extinction events fit less well or not at all

with this pattern, often coinciding with times of dispersed continents. Among alternative causes, the catastrophic influence of an asteroid or bolide impact on end-Cretaceous life is now strongly supported by observational evidence. However, it is likely that no single mechanism explains all the extinction events.

The changing compositions of fossil assemblages are apparent throughout the geological history described in this book. Most importantly, the available fossils and their evolutionary rates determine the potential for biostratigraphic dating at different periods. Examples are trilobites and graptolites in the Early Palaeozoic; corals, brachiopods and goniatite cephalopods in the Late Palaeozoic; and ammonite cephalopods through the Mesozoic. At times the organisms themselves built rock bodies, such as in the Devonian and Permian reefs, or their remains dominate the rock mass, such as in Carboniferous coals or Cretaceous chalks. Such rocks could only form after evolution of their constituent organisms. The appearance of abundant land vegetation during the Devonian was probably of fundamental sedimentological importance, stabilizing land surfaces for the first time and altering the sediment fluxes between continents and oceans.

2.2 Palaeocontinental framework of Britain and Ireland

2.2.1 Palaeocontinental setting

Earlier sections of this chapter have shown how the plate tectonic rearrangement of the continents has played a fundamentally important role in Earth history, affecting atmospheric compositions, sea level, climate and even biological evolution. However, the simple index of clustering or dispersal of continents (Fig. 2.1a) is too generalized to explain the many effects of changing continental configuration on the particular geological record of Britain and Ireland. The next section (2.2.2) presents global maps of continental positions. However, it is first necessary to clarify the position of Britain and Ireland with respect to these major continents.

When the continents are restored to their positions prior to the opening of the modern Atlantic Ocean, the continuity of basement rocks of different ages becomes apparent. Britain and Ireland are seen to straddle the Caledonian Orogen of Early to Mid-Palaeozoic age (Fig. 2.3). This orogenic system has three distinct arms: the North Atlantic Caledonian Belt and the Appalachian Belt—formed by closure of the Iapetus Ocean—and the Tornquist Belt—formed by closure of the Tornquist Sea. These belts separate areas of Archaean and Proterozoic crust, forming the palaeocontinents of Laurentia (North America and Greenland), Baltica (Scandinavia and the Baltic) and Eastern Avalonia (southern Ireland and Britain, with Belgium and the adjacent crust).

Eastern Avalonia is the northernmost of a group of continental fragments originally derived from the large Gondwana continent, and including Western Avalonia, Iberia and Armorica. The welding of these fragments through Mid- to Late Palaeozoic time resulted in the Variscan Orogen, the dominant component of the post-Caledonian orogens that now just impinge on the south of Britain and Ireland (Fig. 2.3).

2.2.2 Global palaeocontinental positions

Geological and geophysical information—particularly palaeomagnetic data—constrain the

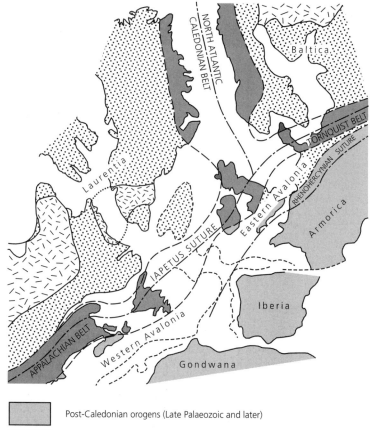

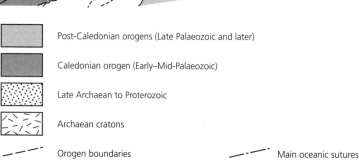

Post-Caledonian orogens (Late Palaeozoic and later)

Caledonian orogen (Early–Mid-Palaeozoic)

Late Archaean to Proterozoic

Archaean cratons

Orogen boundaries

Main oceanic sutures

Fig. 2.3 Map showing the palaeocontinental fragments around Britain and Ireland on a reconstruction before the opening of the Atlantic Ocean. The ages of the continental basement refer to the dominant period of orogeny, but younger belts are demonstrably floored by reworked older basement.

position of the palaeocontinents and associated marginal and oceanic terranes through time. Although the resulting maps change as new data become available, a consensus has emerged in recent years for the period from the Late Mesoproterozoic (*c.* 1100–1000 Ma) up until the present day. The configuration of the continents prior to this time is very uncertain.

Neoproterozoic supercontinents (1000–620 Ma)

Two supercontinents have been postulated during this period, based on palaeomagnetic data and on matching coeval, but now widely separated, orogenic belts recording major phases of continental amalgamation. The first supercontinent is referred to as Rodinia (Fig. 2.4a). It had Laurentia at its centre, surrounded by the cratons of East Gondwana (India,

Fig. 2.4 Global palaeocontinental reconstructions from the Neoproterozoic. (a) Rodinia at *c.* 750 Ma immediately prior to break-up. Arrows summarize plate motions leading to the Pan-African orogenesis. (b) Plate configuration as Laurentia began to rift away from the late Neoproterozoic (Vendian) supercontinent *c.* 610–580 Ma (modified from Dalziel 1992; Torsvik *et al.* 1996, with permission from Elsevier Science 2000).

Antarctica, Australia), Siberia, Baltica and the separated parts of West Gondwana (South America, Africa). The reconstruction is based on matching the Grenville orogenic belts (*c.* 1300–1000 Ma) thought to have resulted from the assembly of Rodinia. The south-eastern margin of Laurentia is placed against the proto-Andean margin of South America, while the south-western margin is linked to East Antarctica (the so-called SWEAT hypothesis). At about 770–750 Ma, Rodinia broke up with the opening of the proto-Pacific, separating East Gondwana from the western margin of Laurentia. Subduction of the Mozambique and Brazilide oceans led to the collision of East Gondwana and the several continental blocks forming West Gondwana, producing the Pan-African–Baikalian–Brasiliano orogens at about 620 Ma. This orogeny formed the second, late Neoproterozoic 'Vendian' supercontinent comprising Gondwana, Laurentia and Baltica (Fig. 2.4b).

Peripheral orogens and supercontinent break-up (620–540 Ma)

During and after the Late Neoproterozoic amalgama-

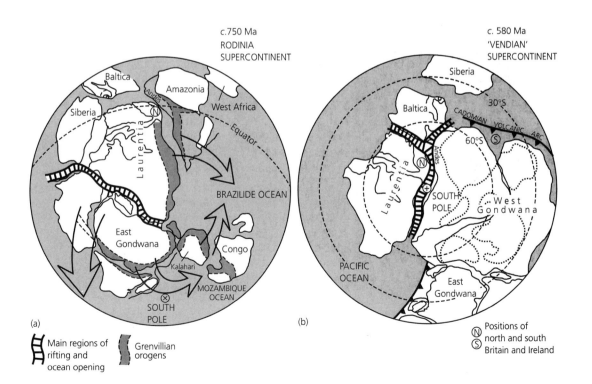

tion of the Vendian supercontinent, several Andean-type continental margins began to form around it. These peripheral orogens included the Cadomian, or Avalonian, belt of which southern Britain formed a part, probably in moderate southern latitudes (Fig. 2.4b). These margins were marked by widespread subduction of oceanic lithosphere, the growth of island arcs and the formation of associated accretionary prisms and back-arc basins, but with no continent–continent collisional episodes. Between 600 and 580 Ma, the Vendian supercontinent began to break up along the eastern margins of Laurentia. Baltica rifted from the Greenland margin, and Gondwana (western South America) rifted from the North American margin to initiate the Iapetus Ocean (Figs 2.4b, 2.5a).

Fig. 2.5 Global palaeocontinental reconstructions for the Palaeozoic: (a) 550 Ma; (b) 490 Ma; (c) 470 Ma; (d) 440 Ma; (e) 425 Ma; (f) 300 Ma. Note that not all continents are shown due to lack of data. Only selected subduction zones are shown (modified extensively from Torsvik *et al.* 1993, 1996, with permission from Elsevier Science 2000).

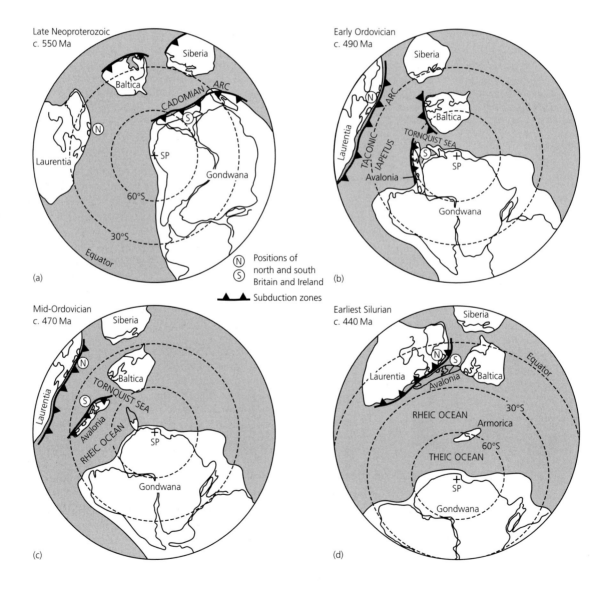

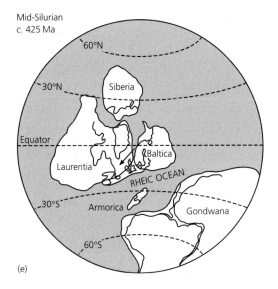

Mid-Silurian
c. 425 Ma

(e)

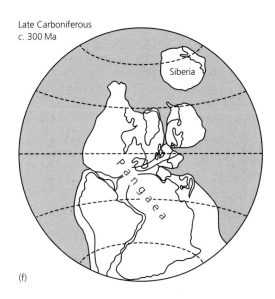

Late Carboniferous
c. 300 Ma

(f)

Fig. 2.5 *(Continued.)*

Arc–continent collisions and Gondwana break-up (540–460 Ma)

Localized orogenic events in Cambrian to Early Ordovician time are thought to reflect the collision of volcanic arcs with some of the continental margins of Iapetus (Fig. 2.5a,b). These events include the Finnmarkian Orogeny (540–490 Ma) on the Baltica margin and the Grampian Orogeny (470–460 Ma), which affected the Scottish and Irish sector of the Laurentian margin in low southern latitudes. The collision of volcanic arcs with continental margins resulted in closure of intervening marginal basins and in the obduction of ophiolites. Also during Early to Mid-Ordovician time, continental fragments were rifted away from the south polar latitudes of the Gondwana margin and migrated northwards, destined to accrete to the Laurentian margin in low southern latitudes (Fig. 2.5b,c). Avalonia, including southern Britain and Ireland, was rifted away in the Early Ordovician (*c.* 475 Ma), effectively starting the closure of Iapetus and the opening of a new ocean to the south known as the Rheic Ocean. Subduction was also occurring at this time along the western margin of Baltica, which was rotating anticlockwise and converging with Laurentia.

Iapetus closure (460–420 Ma)

The earliest linkage of Laurentian and Gondwanan marginal fragments occurred in Newfoundland in the Mid-Ordovician (*c.* 450 Ma). However, the main closure events appear to be Silurian (*c.* 425 Ma), with Baltica and parts of Avalonia converging obliquely with the Laurentian margin (Fig. 2.5c–e). Complex, collision-related deformation continued well into the Devonian in many areas (*c.* 390 Ma). The Iapetus margins that were finally sutured together were not the same segments that originally rifted apart. In addition, significant strike-slip faulting—mainly sinistral—associated with oblique closure *c.* 425–420 Ma disrupted the Laurentian continental margin and suture zone, especially in the UK and Newfoundland sectors of the orogen.

Late Caledonian events, the Variscan Orogeny and the formation of Pangaea (420–300 Ma)

Further collision events occurred as other slivers, rifted away from Gondwana during the Late Ordovician and Silurian, collided with the assembled northern continents (Fig. 2.5e–f). These slivers include Armorica (Acadian Orogeny; *c.* 400 Ma) and Iberia (Ligerian Orogeny; *c.* 390–370 Ma). The collision of the main Gondwana continent led to the Variscan Orogeny in north-west Europe (*c.* 370–

290 Ma) and to the equivalent Alleghanian orogenic episode in the USA. Thus, in many respects, the Caledonian and Variscan orogenic cycles are part of a continuum involving the break-up of one supercontinent in the Late Neoproterozoic and the reassembly of another, Pangaea, by the Late Palaeozoic.

Pangaea consolidation, rotation and rifting (300–180 Ma)

The Variscan collision continued to consolidate the Pangaea supercontinent into Permian time (Figs 2.5f, 2.6a). The nearest oceanic area to Britain was the Tethys seaway separating Eurasia and Africa. During Permian and Triassic time the consolidated continent rotated anticlockwise about a pole in the Gulf of Mexico. This rotation caused Britain and Ireland to move northward from an equatorial position during the late Carboniferous (Fig. 2.5f) to 25°N by the Late Permian (Fig. 2.6a) and to 40°N by the Jurassic (Fig. 2.6b). Although the supercontinent remained intact during this period, it was progressively cut by intracontinental rift zones that marked the future lines of Pangaea break-up.

Pangaea break-up (180–80 Ma)

During Jurassic time, the central part of the Atlantic Ocean began to spread, separating North America from South America and Africa (Fig. 2.6b). During the Cretaceous, spreading in the South Atlantic began to separate South America from Africa (Fig. 2.6c). The consequent anticlockwise rotation of Africa tended to narrow the Tethys Ocean, and to cause a complex belt of rifts and strike-slip faults to propagate westward towards the Atlantic. These rift systems demarcated the southern continents from Eurasia through the present Mediterranean region, but with the production of only limited areas of ocean crust.

North Atlantic separation (80–0 Ma)

By the Late Cretaceous, spreading of the North Atlantic was beginning to separate Greenland from Eurasia, and to propagate rifts into the Arctic Ocean (Fig. 2.6c). Continued rotation of Africa drove its northern edge into the Eurasian continent to produce the Alpine Orogeny (Fig. 2.6a). Britain and Ireland

continued to drift northward towards their present latitude between 50°N and 60°N (Fig. 2.6d).

2.2.3 Continental drift of Britain and Ireland

From a regional viewpoint, the Phanerozoic continental drift of Britain and Ireland records a generally northward shift of its crustal components. All these components started their Phanerozoic history in the middle latitudes of the southern hemisphere (Fig. 2.7a). However, during the Cambrian, Laurentian Britain drifted north whereas Gondwanan Britain drifted south. Not until Ordovician and Silurian time did Avalonia, followed by Armorica, drift northward to amalgamate with Laurentia. The assembled continental fragments crossed and hovered around the equator during the Carboniferous before heading, first rapidly and then more slowly, northward through the northern hemisphere. This northward drift is reflected directly in regional sedimentary facies, for instance in the arid Devonian and Permo-Triassic deposits bracketing the equatorial humid facies of the Carboniferous.

The global maps (Figs 2.5, 2.6) show that Northern Britain and Ireland were part of five successive Phanerozoic continents (Fig. 2.7b):
1 *Laurentia* until the collision with Avalonia during the Caledonian Orogen (Late Silurian–Devonian).
2 *Laurussia* until the final collision with Gondwana in the Variscan Orogeny (Carboniferous–Early Permian).
3 *Pangaea* until its break-up through the Central Atlantic and Tethys (Late Triassic–Early Jurassic).
4 *Laurasia* until the rifting of the North Atlantic (Late Cretaceous).
5 *Eurasia* since Late Cretaceous time.

Until the Caledonian Orogeny, Britain and Ireland south of the Iapetus Suture were part of first *Gondwana* and then, from Early Ordovician time, of *Avalonia* (Fig. 2.7b). The *Armorica* microcontinent, represented only on the southernmost tip of mainland Britain, amalgamated with Avalonia and the rest of Laurussia in Devonian time

2.3 Tectonic template of Britain and Ireland

2.3.1 Pre-Variscan tectonic template

The history of continental collisions and separations

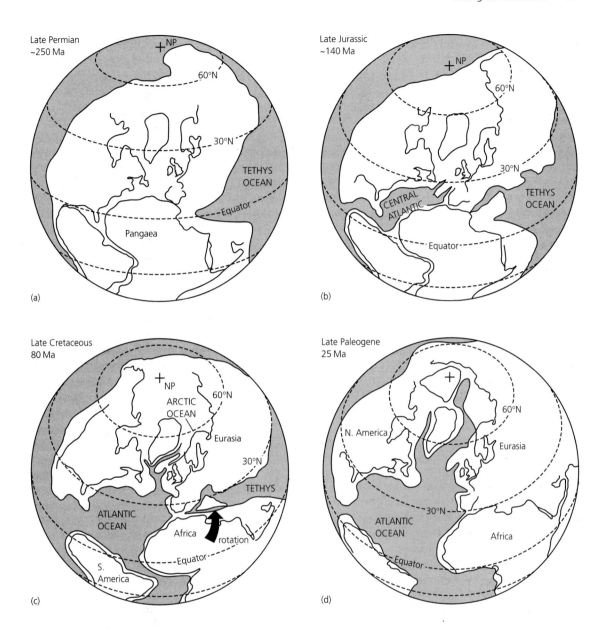

Fig. 2.6 Global palaeocontinental reconstructions for the late Palaeozoic, Mesozoic and Cenozoic: (a) 250 Ma; (b) 140 Ma; (c) 80 Ma; (d) 25 Ma (modified from Ziegler 1990).

documented in this chapter has left its permanent imprint on the crust of Britain and Ireland as a structural grain defined by faults, folds or tectonic fabrics. This grain has then acted as a template to influence, in some areas at least, the position and form of sedimentary basins and the location of magmatic activity. This section describes the Palaeozoic tectonic pattern, largely formed by the Grampian, Caledonian and Variscan orogenies.

The Precambrian and Palaeozoic rocks of Britain and Ireland are tectonically significant for four reasons:

1 They straddle the Caledonian orogenic belt and

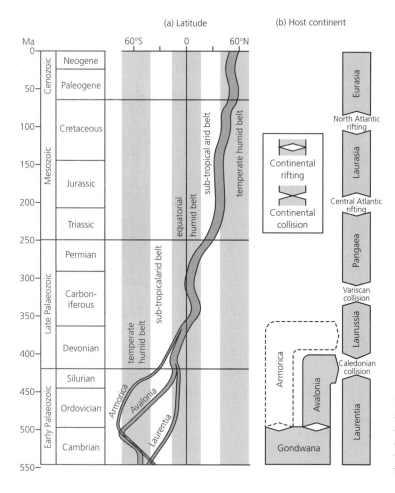

Fig. 2.7 Summary of the Phanerozoic continental drift of the British and Irish crust in terms of: (a) the latitude of its component fragments through time; (b) the successive continents of which these fragments formed a part.

Iapetus suture zone separating the original south-east margin of Laurentia from a marginal fragment of north-western Gondwana (Eastern Avalonia) (Fig. 2.3).

2 They lie close to the collisional triple junction between three palaeocontinents and to a sharp bend in the Laurentian continental margin (Fig. 2.3).

3 They nevertheless preserve evidence of pre-Caledonian and early Caledonian orogenic events.

4 They preserve part of the northern margin of the Variscan Orogen in north-west Europe, in particular part of the suture of the small Rhenohercynian Ocean (Fig. 2.3).

Inevitably, the earlier geological history of this region is the least well understood in terms of large-scale processes and plate tectonic reconstructions. The geological record is especially fragmentary in

Precambrian rocks, as they are often obscured by younger sequences, are mainly unfossiliferous and may display complex, heterogeneous patterns of tectonic overprinting that are difficult to unravel, even using isotopic dating techniques. These problems are compounded by the complexity of Late Precambrian to Early Palaeozoic events during the opening and subsequent closure of Iapetus. Three factors combine to make detailed reconstruction of Caledonian and pre-Caledonian geology problematical:

1 Caledonian and older rocks are often obscured by sequences of younger rocks or have been reworked during subsequent deformations, notably the Variscan Orogeny.

2 Following the opening of Iapetus, several fragments of Gondwana were rifted off, migrated across the ocean and collided diachronously with the

adjoining continents of Laurentia or Baltica (e.g. Fig. 2.5a–f). This complex Caledonian history varies along strike and significant evidence is often obscured or overprinted by later events related to final ocean closure.

3 Although early tectonic models of the Caledonian and Variscan belts considered only continental divergence and convergence normal to the orogenic strike, significant orogen-parallel displacements are now known to have occurred. These large-scale strike-slip displacements are notoriously difficult to unravel, even in modern orogenic systems, meaning that it is difficult to reconstruct the plate-boundary history using the preserved collisional architecture and rock record.

A practical approach to these problems has been to use the major faults of Britain and Ireland to delimit blocks with a coherent internal geological history (Fig. 2.8). Many of the block-bounding faults display evidence for long reactivation histories, but their last major displacement can typically be dated. For some faults this last displacement was in early Palaeozoic time, but for the majority, at and north of the Iapetus Suture, it was during the culmination of the Caledonian Orogeny in Devonian time. Only at and south of the diffuse Variscan Front did major displacement last through Carboniferous time.

Fault-bounded blocks with a distinct geological history have come to be known as *terranes* (see Section 1.4, p. 11), following studies of the displaced Mesozoic and Cenozoic blocks of western North America. In general, terranes may represent displaced fragments of continents, volcanic arcs or ocean basins accreted to continental margins by combinations of subduction, collision or strike-slip displacement. The terrane concept has been widely applied to the pre-Variscan geology of Britain and Ireland, and is correspondingly used in this book. However, the proven displacements on most 'terrane'-bounding faults in Britain and Ireland do not reach the hundreds of kilometres that characterize the type American region. The terrane concept must therefore be applied to Britain and Ireland with some caution.

The blocks or terranes of Britain and Ireland may be subdivided into three groups (Fig. 2.8):

1 Those north of the Fair Head–Clew Bay Line–Highland Boundary Fault, which are thought to have Laurentian affinities.

2 Those south of the Solway Line (thought to correspond to the main Iapetus Suture), which are thought to have Gondwanan affinities.

3 An intervening zone of continental margin, island arc and oceanic slivers, including at least one large displaced fragment of Laurentia (the Connemara Terrane). These rocks form a composite, complex suture zone separating Laurentia and Eastern Avalonia.

The geology of these terranes will be depicted in more detail in the final section of this chapter (Section 2.4, Fig. 2.11).

2.3.2 Post-Variscan tectonic template

The Variscan Orogeny was the last strong crustal shortening event to affect the north-west European crust. The Permian to Recent history of north-west Europe has been dominated by extensional tectonics, taken up on normal or strike-slip faults. These faults have small displacements by comparison with the supposed terrane-bounding faults of the Early Palaeozoic, and no dramatic reorganization of crustal blocks occurred. However, the larger normal faults defined the zones of active crustal subsidence that became the major Mesozoic rift basins (Fig. 2.9). Many of these basins persisted as zones of thermal subsidence, though not of crustal stretching, through Cenozoic time to the present day.

Analysis of Cenozoic subsidence patterns, vitrinite reflectance profiles and fission track data from both onshore and offshore areas show that northwestern parts of the region have undergone substantial amounts (locally up to 3 km) of epeirogenic uplift and exhumation. The cause of this Paleogene (Early Cenozoic) uplift is controversial, but is probably related to igneous underplating—addition and crystallization of magma at the base of the crust—associated with the opening of the North Atlantic above the Iceland plume. By contrast, some Mesozoic to Cenozoic basins in the southeastern part of the region have undergone Neogene (later Cenozoic) deformation of the basin-fill, with associated uplift and erosion to produce unconformities of local or regional extent. These inversion events are thought to reflect far-field effects of the Alpine collision in southern Europe or more localized, intrabasinal events such as the upward movement of salt.

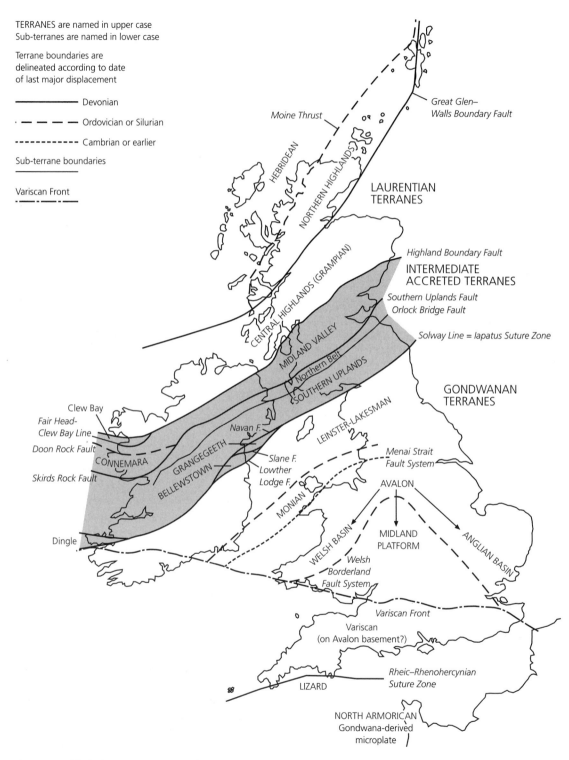

TERRANES are named in upper case
Sub-terranes are named in lower case

Terrane boundaries are
delineated according to date
of last major displacement

————————— Devonian

·— — — — Ordovician or Silurian

------------- Cambrian or earlier

Sub-terrane boundaries

Variscan Front
·—··—··—

Moine Thrust

Great Glen–
Walls Boundary Fault

HEBRIDEAN

NORTHERN HIGHLANDS

LAURENTIAN
TERRANES

CENTRAL HIGHLANDS (GRAMPIAN)

Highland Boundary Fault

INTERMEDIATE
ACCRETED TERRANES

Southern Uplands Fault
Orlock Bridge Fault

Solway Line = Iapatus Suture Zone

MIDLAND VALLEY

Northern Belt

SOUTHERN UPLANDS

GONDWANAN
TERRANES

Clew Bay

Fair Head-
Clew Bay Line

Doon Rock Fault

CONNEMARA

Skirds Rock Fault

Navan F.

GRANGEGEETH

BELLEWSTOWN

Slane F.
Lowther
Lodge F.

LEINSTER-LAKESMAN

Menai Strait
Fault System

MONIAN

AVALON

Dingle

WELSH BASIN

MIDLAND
PLATFORM

ANGLIAN BASIN

Welsh
Borderland
Fault System

Variscan Front

Variscan
(on Avalon basement?)

Rheic–Rhenohercynian
Suture Zone

LIZARD

NORTH ARMORICAN
Gondwana-derived
microplate

Fig. 2.8 Simplified Palaeozoic terrane map of Britain and Ireland
(after Bluck *et al.* 1992, with permission of the Geological
Society, London 1999).

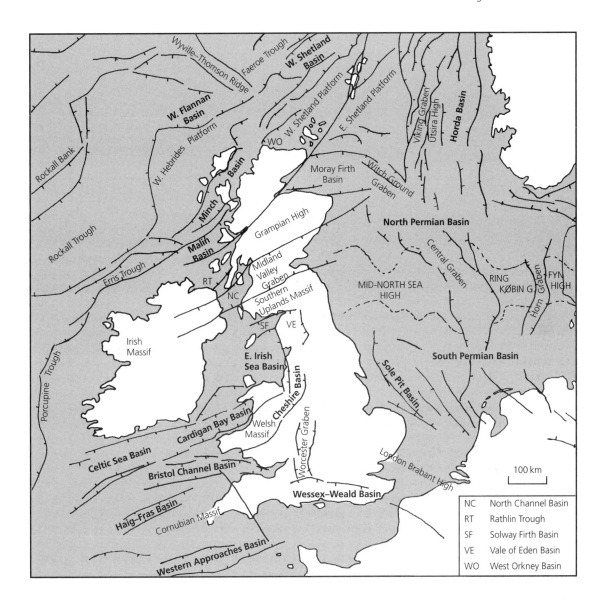

Fig. 2.9 A post-Variscan tectonic map of the area around Britain and Ireland, showing the main sedimentary basins and highs important during Permian to Recent time (modified from Ziegler 1990).

At first sight, the post-Variscan tectonic template (Fig. 2.9) seems significantly influenced by older structural trends. Many onshore and offshore basin-bounding faults appear to parallel adjacent Variscan or Caledonian structures, for example those around south-west Britain and north-west Scotland. There is

indeed direct geological and geophysical evidence for reactivation along some faults bounding basins such as the Minch Basin and Worcester Graben (Fig. 2.9). However, in other basins, such as the West Orkney Basin, offshore faults that appear to parallel older structures are, when traced onshore, completely discordant to basement fabrics. Thus, the hypothesis that the faulted architecture of the post-Variscan extensional basins in north-west Europe is substantially controlled by reactivated older crustal structures remains to be proved. In many cases, there is little or no evidence for reactivation other than a coincidence in trend or dip.

It is possible that some structural inheritance derives from reactivation of older fabrics not in the crust but the upper mantle, the strongest part of the lithosphere. Recent geophysical studies of the upper mantle beneath the British Isles suggest that, in some areas, the main mantle fabrics are parallel to Caledonian and Variscan structures in the overlying crust. Other basins, notably the Central Graben, Viking Graben and Moray Firth Basin of the North Sea (Fig. 2.9), appear to be strongly discordant to older structural trends. This mismatch suggests an independent control on Mesozoic rifting in this area. One hypothesis is that these basins formed as three arms of a trilete rift generated by Jurassic lithospheric doming above a mantle plume.

In conclusion, the post-Variscan tectonic template of Britain and Ireland probably results from a complex interaction between newly formed Mesozoic extensional faults and selective reactivation of some favourably orientated, especially weak structures in the crust and/or upper mantle. The present-day geography of the region is significantly influenced by Mesozoic extensional faulting combined with the effects of later Cenozoic uplift and exhumation related to underplating. For example, many of the islands on the north-west coast of Scotland, such as the Outer Hebrides, lie in the uplifted and eroded footwalls of Mesozoic normal faults (Fig. 2.10). Significantly, these uplifts form some important onshore areas where Precambrian basement rocks are exposed at the surface, so that the effects of Mesozoic faulting actually determine the present-day distribution of Precambrian rocks.

2.4 Synthesis: a chronostratigraphic chart of Britain and Ireland

The evolution of Britain and Ireland can be graphically summarized on a chart of geological events along a north to south transect (Fig. 2.11). Major depositional and igneous episodes are shown back into the Mesoproterozoic. Before this time, sedimentation events are difficult to date, and the record is defined more by igneous and tectono-metamorphic events. For graphical convenience, the scale of this chart changes at 700 Ma. Pre-Variscan and post-Variscan history can be followed in conjunction with the appropriate maps (Figs 2.8 and 2.9). Only major pre-Variscan terranes are shown.

The *Archaean to Palaeoproterozoic* basement to Gondwana is merely glimpsed in the North Armorican Terrane. By contrast, Laurentian rocks of this age are widely exposed in the Hebridean Terrane, with possible correlatives in inliers in the Northern Highlands Terrane. A rather different type of Laurentian basement is present east of the Great Glen Fault, on Islay and Colonsay, which probably underlies much of the Central Highlands Terrane.

The *Meso- to Neoproterozoic* record from Gondwana and Laurentia is more extensive than for previous eras. The Gondwanan terranes contain the remnants of the Late Neoproterozoic Avalonian arc sequences. By the end of the Cadomian Orogeny (early Cambrian) the Monian Terrane was probably already amalgamated with those bordering it. The three main Laurentian terranes each contain major sedimentary units of Late Meso- or Neoproterozoic age. Facies contrasts between the Hebridean and Northern Highlands terranes imply that they had not yet amalgamated, but a possible match across the Great Glen Fault implies proximity with the Grampian Terrane (Fig. 2.8). Here, the Dalradian unit spans the Neoproterozoic–Palaeozoic boundary.

Early Palaeozoic events are recorded in all terranes north of the Variscan Front (Fig. 2.11). Contrasts between most terranes are initially strong, but decline progressively through time. The Late Ordovician Shelvian events amalgamated the Midland Platform and Welsh Basin terranes, allowing a coherent Silurian to lower Devonian sequence to overstep the whole of Avalonia. On the Laurentian margin, the Mid-Ordovician Grampian Orogeny in the Highlands terranes preceded their amalgamation with the Hebridean Terrane in the Early Silurian Scandian Orogeny. By Late Silurian time, the Caledonian Orogeny was finally amalgamating all these terranes north of the Variscan Front, following the eradication of the Iapetus Ocean.

The *Late Palaeozoic* saw the end of the Caledonian Orogeny, and the overstep of the old terrane boundaries by the upper Devonian to Carboniferous sedimentary cover (Fig. 2.11). In the south, a number of microplates and their terrane assemblages continued to accrete to the southern margin of Laurussia during the protracted Variscan Orogeny. Armorica was juxtaposed against Avalonia, probably in Late Silurian

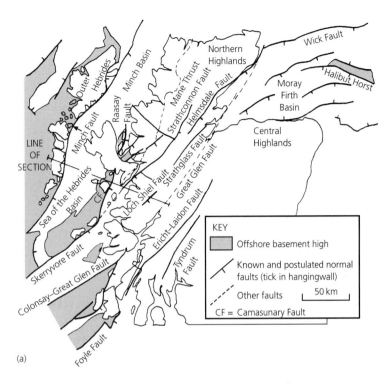

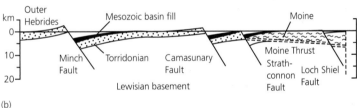

Fig. 2.10 (a) Map of Scotland showing the main Mesozoic normal faults that affect the distribution of Mesozoic basins and Precambrian basement. (b) Cross-section along the line marked on the map (modified from Roberts & Holdsworth 1999, with permission of the Geological Society, London 1999).

to Early Devonian time. This closure of the Rheic Ocean was followed by spreading of the small Rhenohercynian Ocean, and emplacement from it of the Lizard ophiolite. Continued Variscan compression produced widespread uplift and non-deposition across Britain and Ireland, recorded in the extensive unconformity below Permian sequences. The unconformity also marks the change from a tectonic template dominated by the north-east–south-west or east–west orientated terrane-bounding faults, to the more varied template of Mesozoic extensional tectonics.

The *Mesozoic* stratigraphic elements are necessarily generalized and simplified on the summary chart (Fig. 2.11), because many of them are best repre-

sented in offshore basins out of the line of transect. Representative post-Variscan basins near the transect line are named, but more extensive sequences typically occur in the North Sea basins further east. Nevertheless, the chart shows the widespread Permian to Jurassic fill to basins formed by crustal rifting. A major rift and thermal uplift in the Mid-Jurassic gave the Mid-Cimmerian unconformity, and an Early Cretaceous rifting the even more extensive Late Cimmerian unconformity. In each case, sedimentation in more rapidly subsiding basins, such as the Western Approaches, was relatively continuous.

The *Cenozoic* saw the igneous activity in north-west Britain and Ireland, during the rifting of the North Atlantic near the Iceland plume. The conse-

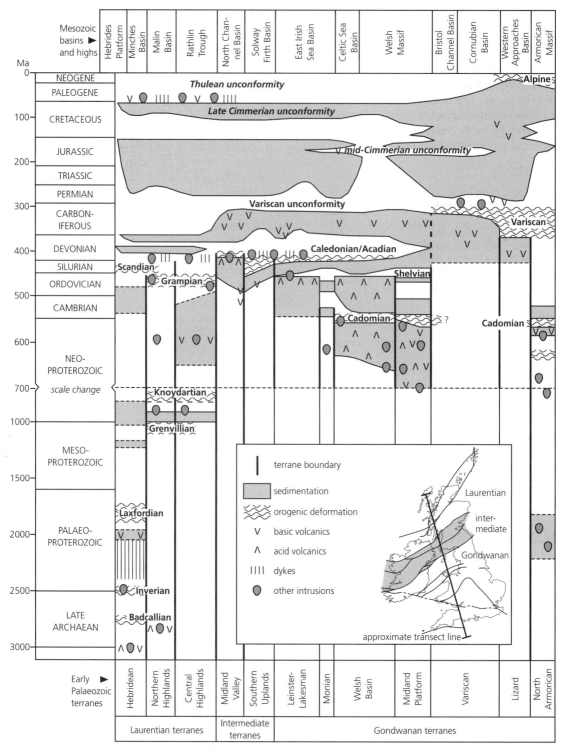

Fig. 2.11 A chronostratigraphic chart showing generalized geological relationships on a north to south transect through Britain. Only major pre-Variscan terranes are distinguished. Post-Variscan events are generalized from the western UK basins, some off the line of section. Note the time-scale change at 700 Ma.

quent Paleogene uplift of most of the area substantially produced its present-day physiography, augmented by the effects of earlier Mesozoic faulting and, in southern Britain, by the Neogene uplift associated with Alpine shortening.

References

Berner, R.A. (1991) A model for atmospheric CO_2 over Phanerozoic time. *American Journal of Science* **291**, 339–376.

Berner, R.A. & Canfield, D.E. (1989) A new model for atmospheric oxygen over Phanerozoic time. *American Journal of Science* **289**, 333–361.

Bluck, B.J., Gibbons, W. & Ingham, J.K. (1992) Terranes. In: *Atlas of Palaeogeography and Lithofacies* (eds J. C. W. Cope, J. K. Ingham & P. F. Rawson), Memoir 13, pp. 1–4. Geological Society, London.

Dalziel, I.W.D. (1992) Antarctica; a tale of two supercontinents? *Annual Review of Earth and Planetary Science* **20**, 501–526.

Roberts, A.M. & Holdsworth, R.E. (1999) Linking onshore and offshore structures: Mesozoic extension in the Scottish Highlands. *Journal of the Geological Society, London* **156**, 1061–1064.

Sepkoski, J.J., Jr (1998) Rates of speciation in the fossil record. *Philosophical Transactions of the Royal Society, London, B* **353**, 315–326.

Torsvik, T.H., Trench, A., Svensson, I. & Walderhaug, H.J. (1993) Palaeogeographic significance of mid-Silurian palaeomagnetic results from southern Britain—a major revision of the apparent polar wander path for Eastern Avalonia. *Geophysical Journal International* **113**, 651–668.

Torsvik, T.H., Smethurst, M.A., Meert, J.G. *et al.* (1996) Continental break-up and collision in the Neoproterozoic and Palaeozoic—a tale of Baltica and Laurentia. *Earth-Science Reviews* **40**, 229–258.

Veevers, J.J. (1990) Tectonic–climatic supercycle in the billion-year plate tectonic eon: Permian Pangaean icehouse alternates with Cretaceous dispersed-continents greenhouse. *Sedimentary Geology* **68**, 1–16.

Ziegler, P.A. (1990) *Geological Atlas of Western and Central Europe*. Shell International Petroleum Maatschappij BV, The Hague.

Further reading

van Andel, T.H. (1994) *New Views on an Old Planet—a history of global change*, 2nd edn. Cambridge University Press, Cambridge. [An entertainingly readable review of global mechanisms, including plate tectonics, climate change, mass extinctions and sea-level change.]

Bluck, B.J., Gibbons, W. & Ingham, J.K. (1992) Terranes. In: *Atlas of Palaeogeography and Lithofacies* (eds J. C. W. Cope, J. K. Ingham & P. F. Rawson), Memoir 13, pp. 1–4. Geological Society, London. [A concise review of the terranes of Britain and Ireland.]

Dalziel, I.W.D. (1992) Antarctica; a tale of two supercontinents? *Annual Review of Earth and Planetary Science* **20**, 501–526. [Review of the SWEAT hypothesis and the models of two Late Precambrian supercontinents.]

Dewey, J.F. (1969) Evolution of the Caledonian–Appalachian orogen. *Nature, London* **222**, 124–129. [Historically very significant paper as it is the first plate tectonic interpretation of the development and destruction of Iapetus.]

Glennie, K.W. (ed.) (1986) *Introduction to the Petroleum Geology of the North Sea*. Blackwell, Oxford. [An excellent synthesis of the post-Caledonian geology of the North Sea and related eastern areas of Britain.]

Hutton, D.H.W. (1987) Strike-slip terranes and a model for the evolution of the British and Irish Caledonides. *Geological Magazine* **124**, 405–425. [A useful summary of the terrane tectonics of the British Isles.]

Klemperer, S. & Hobbs, R. (1991) *The BIRPS Atlas: deep seismic reflection profiles around the British Isles*. Cambridge University Press, Cambridge. [An informative collection and interpretation of the deep seismic reflection profiles showing the deep structure of the crust around Britain and Ireland.]

Roberts, A.M. & Holdsworth, R.E. (1999) Linking onshore and offshore structures: Mesozoic extension in the Scottish Highlands. *Journal of the Geological Society, London* **156**, 1061–1064. [Proposes control of onshore structure and uplift by Mesozoic basin-bounding faults seen offshore.]

Torsvik, T.H., Smethurst, M.A., Meert, J.G. *et al.* (1996) Continental break-up and collision in the Neoproterozoic and Palaeozoic—a tale of Baltica and Laurentia. *Earth-Science Reviews* **40**, 229–258. [Comprehensive review of the palaeomagnetic and faunal evidence constraining global-scale palaeogeographic reconstructions from 800 Ma to the end of the Caledonian Orogeny.]

Veevers, J.J. (1990) Tectonic–climatic supercycle in the billion-year plate tectonic eon: Permian Pangaean icehouse alternates with Cretaceous dispersed-continents greenhouse. *Sedimentary Geology* **68**, 1–16. [A comprehensive review of the possible links between supposed cycles of climatic change and supercontinent assembly/destruction from *c.* 1100 Ma to the present day and into the future.]

Ziegler, P.A. (1990) *Geological Atlas of Western and Central Europe*. Shell International Petroleum Maatschappij BV, The Hague. [A masterly synthesis of the post-Caledonian history of the region, extensively illustrated with palaeogeographic maps and correlation charts.]

Part 2
The Northern Margin of the
Iapetus Ocean

Illustration overleaf: Metasedimentary gneisses of the Neoproterozoic Moine Supergroup near Invermoriston, Inverness-shire. The gneissic layering may have formed as a result of migmatization during the Neoproterozoic Knoydartian event; the upright folds formed during the Ordovician Grampian Orogeny.

3 Early Earth history and development of the Archaean crust

R. A. STRACHAN

3.1 Introduction

The earliest parts of Earth history are the most diffi-cult to understand because the oldest rocks have com-monly been either destroyed by erosion or modified strongly by later geological processes. In common with virtually all other areas on Earth, the British Isles contains no rocks formed during the earliest phase of Earth history, the *Hadean Eon* (4.6–4.0 Ga), during which time the Earth and a primitive atmosphere were formed. The next major phase is the *Archaean Eon* (4.0–2.5 Ga), characterized by the rapid and widespread development of crust which forms the nuclei of many continental blocks (Fig. 3.1). The Archaean atmosphere contained little or no oxygen, but none the less primitive bacterial life forms devel-oped in the Early Archaean oceans. The oldest rocks in the British Isles formed during the Late Archaean at about 3 Ga and belong to the Lewisian Complex of north-west Scotland. In order to understand fully the evolution of these rocks it is necessary to draw together information from other, more extensive areas of Archaean crust in regions such as Greenland, Canada, Australia and Africa. A question that has preoccupied geologists for many years is whether or not plate tectonics operated during the Archaean. As recently as the mid-1980s, many geologists argued that modern-style plate tectonics did not begin until about 2 Ga or later. However, there is now a substan-tial body of opinion that plate tectonics provides a firm basis for understanding Archaean crustal evolution.

3.2 The Hadean Eon: formation of the Earth and early atmosphere

The Hadean Eon comprises the first 600 million years of Earth history from initial formation of the planet at about 4.6 Ga to the boundary with the Archaean Eon at 4.0 Ga. There are virtually no rocks preserved from the Hadean Eon; our ideas about this early stage in the evolution of our planet come from theoretical models and evidence from other planets such as the Moon. It is commonly accepted that Earth, like the other inner planets of our solar system, formed by the amalgamation (*accretion*) of solid matter that condensed from a gaseous nebula centred on the Sun. The accretion of millions of small solid bodies, perhaps only a kilometre or so across, may have occurred relatively rapidly by geological standards, in less than 100 million years. At this stage, the young Earth was probably relatively homogeneous and also much hotter than at present: much of the outer part of the Earth was likely to have been covered in an ocean of magma. The Earth then underwent a process of *differentiation* to form the core, mantle and a primi-tive crust. Dense iron-based compounds gradually sank towards the interior of the planet to create a metallic core. The magma ocean crystallized to form a lower olivine-rich mantle and an outer crust formed of less dense basaltic material.

The Earth's early atmosphere was almost certainly very different from that at the present day. When the Earth first formed, a very thin *primary atmosphere* would have consisted of the gases most common in the early solar system. These were light gases such as hydrogen, helium, methane and ammonia. Almost all this light material (particularly hydrogen and helium) escaped into space. Subsequently, the Earth devel-

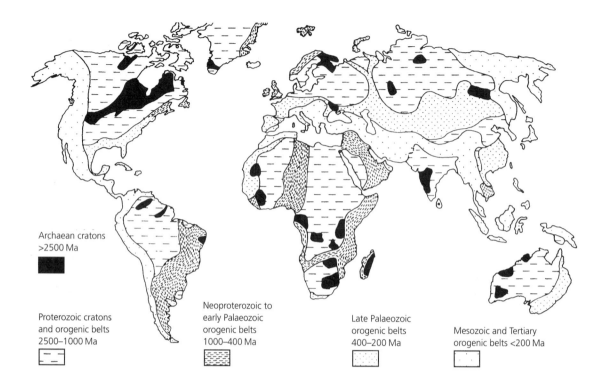

Archaean cratons
>2500 Ma

Proterozoic cratons
and orogenic belts
2500–1000 Ma

Neoproterozoic to
early Palaeozoic
orogenic belts
1000–400 Ma

Late Palaeozoic
orogenic belts
400–200 Ma

Mesozoic and Tertiary
orogenic belts <200 Ma

Fig. 3.1 Geological map of the world showing major tectonic provinces (redrawn from Anderton *et al.* 1979).

oped a *secondary atmosphere*, which was *outgassed* from the planet's interior during planetary differentiation. Gases were expelled from the interior of the planet mainly by volcanic processes resulting from the heat produced by accretion. The juvenile gases erupted from modern volcanoes consist largely of carbon dioxide and water vapour. These are thought to be the main components of the early atmosphere, together with minor quantities of nitrogen, carbon monoxide and hydrogen. There is hardly any molecular oxygen in modern volcanic exhalations, and it is therefore thought that the early atmosphere must also have contained very little oxygen, the free oxygen present in the atmosphere today having formed later.

The Earth's early crust was almost certainly very different from present-day continental crust, but no trace of this primitive crust has been preserved. The early lithosphere was probably thin and its interior

turbulent and vigorously convecting, so that any piece of solid crust that formed was rapidly subducted into the mantle. By analogy with the Moon, the juvenile Earth was probably subjected to intense bombardment by meteorites for much of the Hadean Eon. The heat released from meteorite impacts is likely to have penetrated the Earth's thin crust and promoted this recycling process. The surface of the young Earth was probably densely cratered and thus much like that of the Moon. Surface conditions would have been extremely hot and dry, and possibly not dissimilar to those on the surface of the planet Venus at the present day. It was far too hot for any water to exist on the surface and therefore no life could develop. The oldest preserved terrestrial rocks are the Acasta Gneisses in north-west Canada, which have yielded U–Pb zircon ages of 4.03–3.96 Ga. The gneisses comprise metamorphosed basalts, granites and sediments. Lithologically similar gneisses in Antarctica have yielded only slightly younger zircon ages of 3.92 Ga. The oldest known minerals are detrital zircons dated at 4.27 Ga within Archaean sandstones in Australia. The fate of the source rocks of these zircons is unknown; they were presumably destroyed either by meteorite impact or subduction. The oldest rocks found on Earth formed at about the same time as the rapid decline of the meteorite bombardment rate recorded on the Moon (at about 3.9 Ga). Before the decline in bombardment, presumably the Earth's early crust could not become stabilized at the surface.

3.3 Crustal evolution in the Archaean

Relatively small areas of Archaean crust outcrop at the surface of all the major continents, surrounded by belts of younger rocks; the most extensive areas of unmodified Archaean rocks occur in north-western Canada (Fig. 3.1). There is now widespread acceptance that Cenozoic-style plate tectonic processes account convincingly for the development of the Archaean crust. Furthermore, it is generally agreed that heat production in the Earth's interior as a result of the decay of radioactive isotopes was around three times higher during the Archaean than at present. This would have resulted in a higher-temperature Archaean mantle, a thicker continental lithosphere (150–200 km), a thicker oceanic crust (20–50 km) and sea-floor spreading rates two to three times greater than at present. At the present day, continental crust is formed mainly by the melting of oceanic crust at subduction zones and its subsequent differentiation to form the calc-alkaline igneous suite. Because plate movements were probably faster and subduction zones more numerous in the Archaean, then the rate of growth of continental crust should have been much faster than at present. Regions of Archaean crust comprise two contrasting but genetically related lithological assemblages: low-grade, volcanic-dominated *greenstone–granite* terranes that formed in the Archaean upper crust, and high-grade *granulite–gneiss* terranes that represent the Archaean mid–lower crust.

Greenstone–granite terranes are very important parts of the Archaean crust in areas such as the Superior and Slave provinces of North America, the Kaapvaal Craton of South Africa and the Pilbara Block in Australia. They comprise varied assemblages of igneous and sedimentary rock types associated with accretion at active plate margins. Volcanic rocks typically consist of calc-alkaline basalt–andesite–dacite–rhyolite associations which have the geochemical characteristics of igneous rocks formed at subduction zones (Fig. 3.2). Arc-type, intrusive diorite–granodiorite–granite plutons are coeval with the volcanic rocks. Thick piles of mafic intrusions and lavas identified in several belts are thought to represent Archaean ophiolites. Associated sedimentary rocks include oceanic pelagic sediments, arc-derived turbidites deposited in accretionary prisms (Fig. 3.2) and a range of continental and marine sediments deposited in fore-, inter- and back-arc basins. Although some greenstone belts are only weakly deformed, many are structurally highly complex as a result of thrusting.

Granulite–gneiss terranes form important parts of the Archaean crust in areas of India, Antarctica, West Greenland and the Limpopo belt of southern Africa. These rocks record metamorphism ranging from the amphibolite to granulite facies, which probably resulted from crustal thickening in a collision zone. The protoliths are mostly calc-alkaline tonalitic plutons typical of those inferred to form the bulk of the mid- to lower crust in Phanerozoic volcanic arcs such as the Andes. The Archaean calc-alkaline gneisses commonly intrude metavolcanic amphibolites and layered anorthosite–gabbro complexes which may represent pre-existing oceanic or marginal basin crust into which the arc plutons were intruded. Analogies may again be drawn with the Cretaceous tonalitic batholiths in the Andes, which intruded extensional marginal basins floored by oceanic-type crust on the active continental margin. It seems very likely that many Archaean granulite–gneiss terranes represent the deep crustal roots of high-level volcanic arcs such as those exposed in the greenstone belts.

The growth of continental crust in the Archaean was initiated by the global formation of innumerable island arcs now represented by the greenstone belts and the tonalitic protoliths of the granulite–gneiss terranes (Fig. 3.2). The progressive amalgamation of island arcs into cratons as a result of collision at subduction zones (also involving accretionary prisms and other sedimentary sequences as well as oceanic crust; Fig. 3.2) occurred in much the same manner as is occurring currently in south-east Asia, where Mesozoic–Cenozoic arcs are in various stages of amalgamation. Archaean arcs were probably amalgamated by thrusting, which led to considerable crustal thickening and high-grade metamorphism of the mid- to lower crustal levels of the arcs, and to the

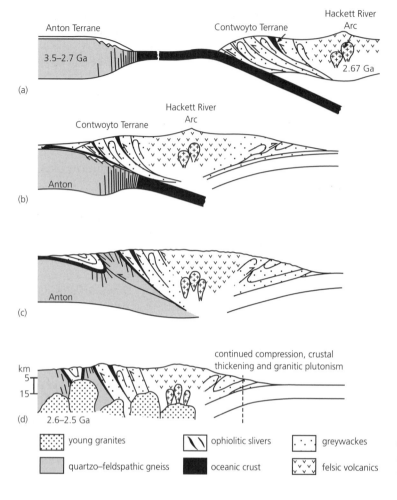

(a)

(b)

(c)

(d)

km
5
15

continued compression, crustal
thickening and granitic plutonism

2.6–2.5 Ga

	young granites		ophiolitic slivers		greywackes
	quartzo–feldspathic gneiss		oceanic crust		felsic volcanics

Fig. 3.2 Schematic illustration of the evolution of part of the Slave Province of Canada between 2.67 Ga and 2.5 Ga (redrawn from Kusky 1989). (a) A greenstone belt evolved above a subduction zone and comprised a volcanic arc (Hackett River Arc) and an accretionary prism with ophiolites (Contwoyto Terrane). (b–d) A high-grade gneiss microcontinent (Anton Terrane) collided with the subduction zone, resulting in crustal thickening, metamorphism and interthrusting and interfolding of the various crustal units. A deep-level section through the deformed and metamorphosed greenstone belt would reveal high-grade calc-alkaline gneisses, very similar to the older gneisses of the Anton Terrane.

formation of the granulite–gneiss terranes. Small gneissic microcontinents had developed by 3.5–3.2 Ga; the evolution of stable crustal blocks is indicated by the gradual appearance of shelf-type sediments such as quartzites and marbles, which had hitherto been largely absent. These microcontinents were progressively enlarged by the addition of other gneissic terranes (older arcs) and younger marginal island arcs, as a result of continued subduction and collision around their margins. A number of substantial cratons had developed by the end of the Archaean, by which time it has been estimated that 75–80% of the present-day continental crust had been formed. It seems likely that the Archaean crust must have been approximately its present thickness and strength.

Many granulite–gneiss terranes contain evidence for metamorphism at depths of over 30 km. It follows that because these granulites are today underlain by 30–35 km of continental crust, Archaean orogenic belts must have attained crustal thicknesses of *c.* 60–70 km, comparable with that of the modern Himalayas and Andes.

3.4 Development of the hydrosphere

As the Earth's surface and atmosphere gradually cooled through the later part of the Hadean and the early Archaean, most of the water vapour condensed as liquid and fell as rain. The condensation of water vapour would have greatly reduced the atmospheric pressure. Major depressions in the crust (probably impact craters) filled with water to form the earliest oceans, and a hydrologic system of rivers and oceans gradually developed. The oldest known sedimentary rocks are present in the *Isua supracrustal belt* of southern West Greenland (Fig. 3.3). They contain banded iron formations (see below), a lithology thought to have been precipitated in an aqueous environment. These metasediments are cut by igneous rocks dated by U–Pb zircon methods at *c*. 3.85 Ga, which is therefore a minimum age for the accumulation of water on the Earth's surface. Surface temperatures on the Earth have therefore been more or less constant from 3.8 Ga to the present. Early Archaean sedimentary rocks show evidence of deposition in fluvial and marine environments and it is generally believed that the physical processes controlling sedimentation were much the same as they are today. However, because the composition of the atmosphere and oceans was different, so too were the chemical processes by which some sedimentary rocks formed.

Highly distinctive banded iron formations formed during the Archaean and Palaeoproterozoic. These consist of alternating centimetre-thick layers of chert and iron minerals (siderite, magnetite, haematite, iron silicates and pyrite), which were chemically or biochemically precipitated. This requires that the oceans built up huge dissolved concentrations of ferrous iron, some of which must have come from the weathering of continents, but most were probably produced by submarine volcanic activity. Such sedimentary rocks could not form at the present day. In the presence of an atmosphere free of molecular oxygen such as existed in the Archaean, iron is relatively soluble in water. Deposition of the banded iron formations occurred as the ferrous iron reacted with dissolved oxygen to form insoluble ferric compounds. The dissolved oxygen was probably produced by early photosynthetic life forms which evolved in the Archaean oceans. Archaean and Palaeoproterozoic fluvial sediments are also notable because they contain detrital pyrite and uraninite, minerals that oxidize rapidly under present-day surface conditions—a further indication of the lack of oxygen in the early atmosphere.

3.5 The evolution of life

Metasedimentary rocks from the Isua supracrustal belt contain chemical evidence for the existence of life forms that probably evolved at about 4.0–3.8 Ga

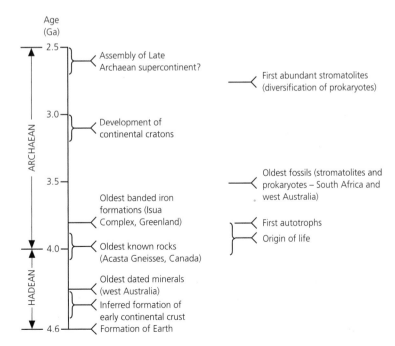

Fig. 3.3 Summary of major biotic and related geological events during the Hadean and Archaean.

(Fig. 3.3). No organic remains are preserved, but it seems likely that the earliest life forms were probably very simple, one-celled organisms known as heterotrophs. These had poorly developed metabolic systems and survived by absorbing nutrients from their surroundings. By about 3.8 Ga (Fig. 3.3), these had evolved into autotrophs, which were cells capable of manufacturing their own food, probably by photosynthesis. Such early organisms most likely developed from proteins and DNA that gradually amalgamated in the Early Archaean seas. Energy sources such as ultraviolet radiation, electrical discharges and hydrothermal vents on the sea floor may have initiated the chemical reactions that produced complex organic compounds from mixtures of simple gases such as hydrogen, carbon dioxide, water vapour and ammonia. These compounds collected together to form cells, and life began when they acquired the ability to reproduce.

The oldest described structures that are unambiguously of organic origin are 3.5-Ga-old stromatolites from the Pilbara region of western Australia (Fig. 3.3). Microscopic spheroids found within cherts in the 3.5-Ga-old Barberton greenstone belt in South Africa are also of probable organic origin. The widespread development of bacterial communities by Middle Archaean times (3.5–3.0 Ga) is indicated by abundant organic carbon in cherts, and by laminated carbonaceous layers that appear to be stromatolites. Additionally, other microfossils, including rod-shaped bodies, filamentous structures and spheroidal bodies that range in size from <1 mm to ~20 mm, have been described from the Barberton and other Archaean greenstone belts. All these life forms were formed of prokaryotic cells—primitive cells that lack a cell wall around the nucleus and are not capable of cell division. These very simple life forms typify the limited extent of biological activity in the Archaean. Early life must have been anaerobic; the oxygen produced by photosynthetic organisms was taken up by the reservoir of reduced elements in the oceans, in particular iron, resulting in its local precipitation as banded iron formations.

3.6 The Archaean in the British Isles: the Lewisian Complex

There are no Early or Mid-Archaean rocks exposed in the British Isles. The oldest exposed rocks are of Late Archaean age and outcrop extensively in the Hebridean Terrane along the north-west coast of the Scottish mainland and in the Outer and Inner Hebrides (Fig. 3.4). These rocks are mostly high-grade orthogneisses known collectively as the *Lewisian Complex*. The Lewisian Complex is a typical Archaean high-grade granulite–gneiss terrane and one of the most intensively studied areas of Precambrian basement geology in the world. However, various aspects of Lewisian geology remain poorly understood, mainly because of the extremely complex nature of these rocks. Normal methods used to establish a stratigraphy based upon sedimentary successions cannot be applied to such basement complexes. Analysis depends mainly upon recognition of sequences of deformation, metamorphism and igneous intrusion, and radiometric dating of these events (Fig. 3.5).

The Archaean history of the Lewisian Complex is difficult to unravel because large parts were affected by intense deformation and metamorphism during the Palaeoproterozoic. The gneiss complex is divided into the Southern, Central and Northern regions by Palaeoproterozoic shear zones (Fig. 3.4). The magnitude of displacements across these shear zones, and hence the nature of the original relationships between different blocks of Late Archaean crust, are the subject of debate. The oldest components of the Lewisian Complex are the gneisses of the Central and Southern regions (Fig. 3.4). These gneisses mainly originated as plutonic igneous rocks, which were affected by Late Archaean granulite facies metamorphism and polyphase deformation. The strong similarities between the gneisses of the Central and Southern region implies that they formed part of the same block of crust in the Late Archaean. Although a Palaeoproterozoic subduction suture now separates the two regions in the region of Loch Maree (see Chapter 4), it is not thought that the oceanic tract which once existed was necessarily very wide. For these reasons the gneisses of the Central and Southern regions are grouped together as the *Scourian gneisses*. However it is now established that the gneisses that outcrop to the north of Scourie (informally termed here the *Laxford gneisses*) are a slightly different and younger suite of Late Archaean plutonic igneous rocks. The timing of early deformation and metamorphism of the Laxford gneisses is uncertain, but it seems likely that they were also affected by the Late Archaean events recorded by the Scourian gneisses.

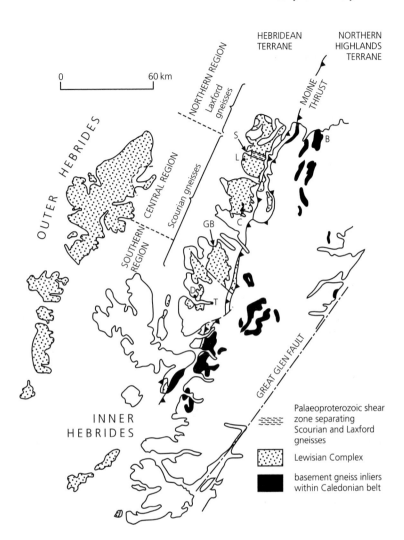

Fig. 3.4 Location of the Archaean rocks of the Lewisian Complex in north-west Scotland. B, Borgie; C, Canisp; D, Diabaig; GB, Gruinard Bay; L, Laxford; S, Scourie.

3.6.1 Early history of the Scourian gneisses

As is the case in most high-grade gneiss complexes, elucidation of the parent rock types ('protoliths') is difficult because original igneous and/or sedimentary textures and minerals have been largely obliterated during deformation and high-grade metamorphism. Most of the Scourian gneisses (75–80%) are banded, acid to intermediate 'grey' gneisses, thought to have originated as plutonic igneous rocks. These gneisses have geochemistries comparable with tonalites, granodiorites and trondhjemites. Minor occurrences of interbanded quartzite and schist are probably metamorphosed sediments. Units of fine-grained, banded amphibolite are thought to represent metavolcanics.

The metasediments and metavolcanics are commonly closely associated with strips of basic, ultrabasic and anorthositic rocks, which may show a relict mineral banding characteristic of layered igneous intrusions.

Recognition of original relationships between the protoliths of the various gneiss types is difficult because most lithological contacts are concordant. Originally cross-cutting contacts, such as intrusive boundaries and unconformities, are likely to have been sheared into parallelism during deformation. Intrusive contacts are occasionally preserved in modified form, and show that the basic and ultrabasic rocks were intruded by the plutonic protoliths of the grey gneisses.

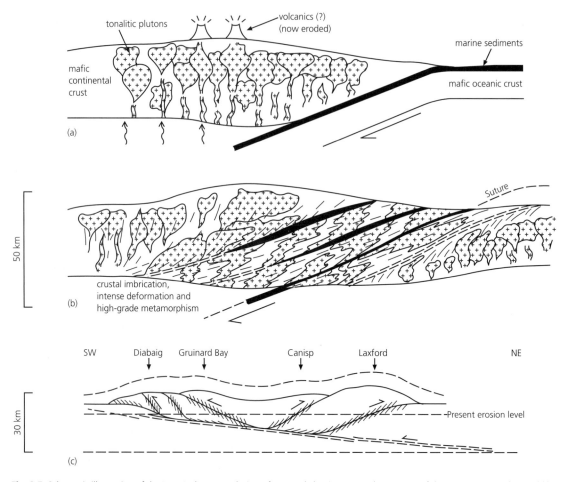

Fig. 3.5 Schematic illustration of the Late Archaean evolution of the Scourian gneisses (modified from Park & Tarney 1987). (a) Generation of the tonalitic protoliths at a subduction zone. (b) Widespread Badcallian deformation and high-grade metamorphism as a result of collision of island arcs at the subduction zone. The outcrop of the Lewisian Complex could lie entirely within the central deformation zone located between the two colliding blocks. (c) Renewed Inverian deformation concentrated along discrete ductile shear zones which have oblique dextral senses of movement.

Integrated field and geochemical studies provide clues to the tectonic environment in which the Scourian gneisses were formed. The ultramafic and mafic rocks have geochemistries similar to those of modern tholeiitic basalts. Their association within the Scourian gneisses with metasediments suggests that some of these rock units may represent fragments of oceanic-type crust. The grey gneisses have geochemistries comparable with magmas generated at subduction zones by partial melting of oceanic crust and/or upper mantle. In common with many other Archaean granulite–gneiss terranes, the igneous pro-

toliths of the Scourian gneisses may therefore have evolved along an active plate margin as an island arc constructed on oceanic crust (Fig. 3.5a). An appropriate modern analogue may be the Tertiary-Recent island arcs of the south-west Pacific.

Dating the protoliths of the Scourian gneisses is difficult, because isotope systematics have been modified during later high-grade metamorphism. Early attempts using Sm–Nd and U–Pb techniques yielded a range of Late Archaean ages (c. 3.0–2.5 Ga). However, some of these ages are of uncertain significance because of the now outdated analytical tech-

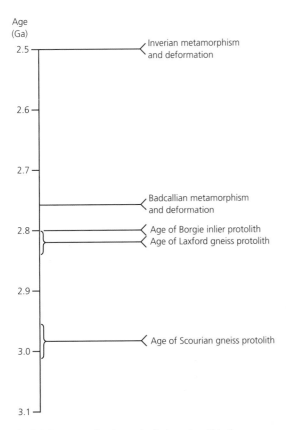

Fig. 3.6 Summary of major geological events within the Lewisian Complex during the Archaean.

niques used, and others have unacceptably large errors. Recent U–Pb dating of the igneous cores of zircons within Scourian tonalitic gneisses has yielded a much more restricted range of ages between *c.* 2.96 and *c.* 3.03 Ga, which confirms a Late Archaean age for these gneisses (Fig. 3.6).

3.6.2 Deformation and high-grade metamorphism of the Scourian gneisses

The earliest period of deformation and metamorphism is known as the *Badcallian* event. Widespread isoclinal folding of the protoliths of the Scourian gneisses resulted in formation of a flat-lying gneissic banding. In the area between Scourie and Gruinard Bay, deformation was associated with granulite facies metamorphism and formation of anhydrous orthopyroxene–quartz–feldspar assemblages. Metamor-

phism occurred at temperatures of 950–1000°C and pressures of 11–15 kbar, corresponding to crustal depths of 35–50 km. Pb–Pb dating of monazite inclusions in early formed garnets suggests that high-grade metamorphism occurred at *c.* 2.76 Ga (Fig. 3.6). The gneisses contain very low quantities of uranium, rubidium, thorium and potassium. These elements were probably removed during widespread melting at deep crustal levels to result in the 'depleted' geochemistry characteristic of many basement complexes. Granulite facies mineralogies are only locally present to the south of Gruinard Bay. It is possible that originally extensive granulite facies mineralogies were obliterated by subsequent retrogression, or alternatively these regions may have been metamorphosed within the amphibolite facies at higher crustal levels. The tectonic setting of the Badcallian event is uncertain, partly because of the relatively small area where the structures and metamorphic assemblages are preserved. One possibility is that it resulted from collision with another volcanic arc terrane at a subduction zone (Fig. 3.5b). The effects of such a collision are usually widespread on a scale of hundreds of kilometres and the suture might therefore lie beyond the present area of exposure of the Lewisian Complex.

Later deformation during an *Inverian* event was associated with development of wide, steep zones of dextral shearing (Fig. 3.5c), perhaps the result of renewed, oblique collision between crustal blocks. The shear zones are characterized by retrogression within the amphibolite facies of the older high-grade granulite facies assemblages. The influx of water along shear zones resulted in replacement of pyroxene by hydrous mineral phases such as biotite and hornblende. U–Pb dating of zircons provides evidence for a major metamorphism at *c.* 2.50 Ga, which probably corresponds to the Inverian event (Fig. 3.6).

3.6.3 The age and history of the Laxford gneisses

The Laxford gneisses are banded, acid to intermediate orthogneisses, which are lithologically very similar to the Scourian gneisses. However, there is no evidence that the Laxford gneisses have ever been affected by granulite facies metamorphism. It was thought for many years that the Laxford gneisses were essentially Scourian gneisses that had undergone intense reworking during the Palaeoproterozoic. The absence of any traces of granulite facies metamorphism was thought to indicate that the Laxford gneisses had been located at a higher crustal level than the Scourian gneisses during the Badcallian metamorphism. However, later geochemical studies showed that the Laxford gneisses are quite dissimilar to the Scourian gneisses. U–Pb dating of the igneous cores of zircons within Laxford gneisses has yielded a range of protolith ages between *c.* 2.80 and *c.* 2.84 Ga, which is distinctively younger than that obtained for the Scourian gneisses (Fig. 3.6). In combination, these studies suggest that the Laxford and Scourian gneisses are two quite different segments of Late Archaean crust. None the less, very similar origins seem likely, as the overall assemblage of primary igneous rock types which make up the Laxford gneisses is consistent with generation at an active plate margin. The Laxford gneisses could therefore represent a volcanic arc, similar to but younger and separate from the arc rocks that make up the Scourian gneisses. The Laxford gneisses certainly underwent deformation and metamorphism prior to juxtaposition with the Scourian gneisses in the Palaeoproterozoic. The precise timing of these events is uncertain. A Late Archaean age for these events is *assumed* but needs to be proven by isotopic dating.

3.6.4 Basement inliers within the Caledonian orogenic belt

Inliers of basement gneisses outcrop extensively east of the Moine Thrust within the Lower Palaeozoic Caledonian orogenic belt of northern Scotland (Fig. 3.4). The inliers are dominated by basic to acidic orthogneisses with occasional strips of metasediment (e.g. marbles and semipelitic schists) and are thought to represent fragments of the high-grade basement on which the early Neoproterozoic sedimentary rocks of the Moine Supergroup were deposited (see Chapter 4). This basement complex probably extends at depth at least as far east as the Great Glen Fault. Correlation with the Lewisian Complex has been generally *assumed* on the basis of general lithological and geochemical similarities. A U–Pb zircon age of *c.* 2.8 Ga has been obtained from the Borgie basement inlier (Fig. 3.4) in north Sutherland. This indicates a Late Archaean age for the igneous protolith and might permit correlation with the Laxford gneisses.

However, it is also possible that these basement inliers represent an entirely different Archaean gneiss complex, which was thrust into its present position relative to the Lewisian Complex during the Caledonian Orogeny. Correlation with the Lewisian Complex remains to be proven and further isotopic data are required to resolve this issue.

3.6 Summary

Models for the evolution of the Archaean crust, atmosphere and early life are inevitably tentative because of the fragmentary nature of the preserved rock record. However, understanding of this important period in Earth history has advanced tremendously since the mid-1980s as a result of multidisciplinary field and laboratory studies and further insights into the evolution of other planets in our solar system. There is now a consensus that present-day plate tectonic models can be used successfully to interpret crustal evolution during the Archaean. This does not mean, however, that geological processes have operated in an *identical* manner through time. Many features of the Archaean crust are the result of the greater heat production at that time. The importance of Archaean high-grade gneiss terranes such as the Lewisian Complex lies in their use as analogues for middle- to lower-crustal behaviour in younger mountain belts where these crustal levels are generally unexposed and the only structures seen are those inferred from deep seismic studies. The Scourian gneisses have therefore yielded valuable information on deep crustal metamorphic processes. It is for these reasons that the Lewisian Complex continues to be the focus of research that is providing valuable insights into orogenic processes within the middle to lower crust.

References

Anderton, R., Bridges, P.H., Leeder, M.R. & Sellwood, B.W. (1979) *A Dynamic Stratigraphy of the British Isles*, pp. 1–301. Unwin Hyman, London.

Kusky, T.M. (1989) Accretion of the Slave Province. *Geology* **17**, 63–87.

Park, R.G. & Tarney, J. (1987) The Lewisian complex: a typical Precambrian high-grade terrain? In: *Evolution of the Lewisian and Comparable Precambrian High Grade Terrains*, Special Publication 27, pp. 13–25. Geological Society, London.

Further reading

Condie, K.C. (1997) *Plate Tectonics and Crustal Evolution*, 4th edn. Butterworth Heinemann, Oxford, pp. 1–282. [An overview of Earth history with major sections on the accretion of the Earth and the Archaean.]

Park, R.G. (1991) The Lewisian Complex. In: *Geology of Scotland* (ed. G. Y. Craig), 3rd edn, pp. 25–64. Geological Society, London. [A review of the Lewisian Complex.]

Windley, B.F. (1993) Uniformitarianism today: plate tectonics is the key to the past. *Journal of the Geological Society, London* 150, 7–19. [A brief overview of global tectonics through time.]

Windley, B.F. (1995) *The Evolving Continents*, 3rd edn, pp. 1–526. John Wiley, Chichester. [A comprehensive overview of Earth history with major sections on the Archaean.]

4 Proterozoic sedimentation, orogenesis and magmatism on the Laurentian Craton (2500–750 Ma)

R. A. STRACHAN AND R. E. HOLDSWORTH

4.1 Introduction

The Palaeoproterozoic records a gradual transition, over several hundred million years, from the Archaean, when continents were growing, to the Mesoproterozoic, when large, stable continents had formed and were close to their present extent. In terms of tectonic processes, the Archaean–Proterozoic boundary heralds the change from the widespread crustal mobility characteristic of the Archaean to a two-fold division into stable continental blocks and approximately linear mountain belts that persists to the present day. Proterozoic orogens appear to have resulted from the collision of large continental blocks and the closure of intervening oceans; they contain discrete sutures, ophiolites, calc-alkaline igneous rocks and evidence for high-pressure metamorphism. In most respects these orogens therefore match Phanerozoic examples formed by plate tectonic processes.

The formation of large, stable continents resulted in the development of extensive passive margins characterized by thick, mature quartzite–carbonate sedimentary successions, which therefore form a significantly higher proportion of the total volume of sedimentary rock than in the Archaean. There are few other significant changes in the rock record at or near the Archaean–Proterozoic boundary. The percentage of free oxygen in the atmosphere increased only slowly through most of the Proterozoic. The gradual change, through the Palaeoproterozoic, from a reducing to an oxidizing environment led to the end of deposition of most banded iron formations after 1.9 Ga. Palaeoproterozoic terrestial sediments younger than 1.9 Ga include red-beds, a sediment type largely absent in the Archaean. The presence of ferric oxide in these rocks as grain coatings and/or pore fillings indicates an oxidizing atmosphere. Additional evidence is provided by the first appearance of oxidized fossil soil profiles beneath major unconformities in Palaeoproterozoic volcano-sedimentary sequences. Primitive algae remained the dominant life form, although they became more diverse and complex and evolved into the multicellular *eukaryotes*, which were capable of utilizing oxygen.

For most of the Proterozoic, northern Britain formed part of a continental mass known as *Laurentia*. This chapter examines the geological evolution of northern Britain between 2.5 Ga (the start of the Proterozoic) and 0.75 Ga (the onset of rifting related to the Iapetus Ocean and the initiation of the Caledonian orogenic cycle).

4.2 Early Palaeoproterozoic rifting and sedimentation: the Scourie dykes and the Loch Maree Group

The Lewisian Complex of north-west Scotland (Fig. 4.1) preserves a more or less complete record of the various Palaeoproterozoic tectonic events that affected this part of Laurentia. There is very little evidence worldwide for orogenic activity around the Archaean–Proterozoic boundary. By contrast, extensive, stable cratons were intruded by basic dykes and overlain by widespread platform sediments on passive continental margins. This evidence suggests that an Archaean supercontinent broke up into smaller continental blocks. The break-up is dated by the emplacement of mafic dyke swarms at *c.* 2.4–2.0 Ga in many continental blocks. In north-west Scotland, continental rifting was associated with emplacement of the *Scourie dykes*, which intrude the

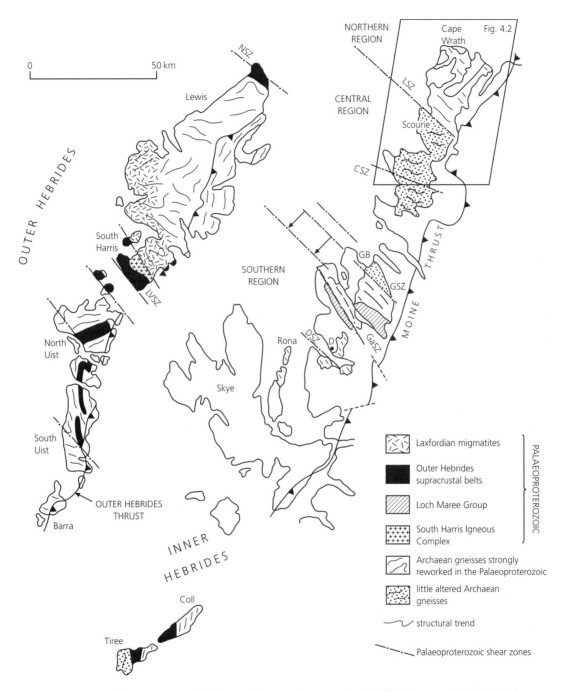

Fig. 4.1 Simplified map of the Lewisian Complex showing the main rock units and the large-scale subdivision of the mainland into three regions (=terranes?) by Palaeoproterozoic shear zones (modified from Park 1991). D, Diabaig; GB, Gruinard Bay; LSZ, Laxford shear zone; CSZ, Canisp shear zone; GSZ, Gruinard shear zone; GaSZ, Gairloch shear zone; DSZ, Diabaig shear zone; NSZ, Ness shear zone; LVSZ, Langavat shear zone. The senses of shear on these structures are not depicted because they characteristically display evidence for displacements during both the Inverian and Laxfordian events.

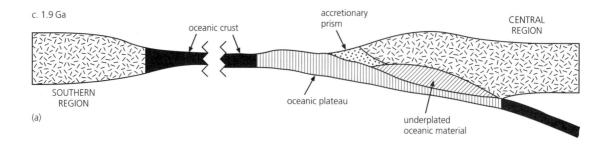

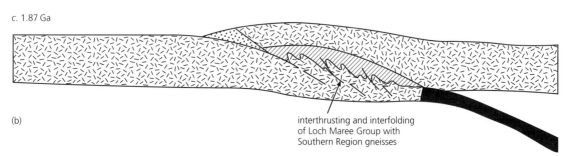

Fig. 4.2 Schematic model for the evolution of the Loch Maree Group following rifting and separation of the Scourian gneisses of the Central and Southern regions at c. 2.0 Ga and consequent development of oceanic crust (modified from Park et al. 2000). (a) Oceanic closure are a result of subduction beneath the Central region resulted in the amalgamation at c. 1.9 Ga of a number of disparate units which together comprise the Loch Maree Group. These include oceanic plateau basalts and sediments which were scraped off the subducting plate, tectonically mixed with accretionary prism clastics derived from the upper plate, and underplated beneath the Central region continental crust. (b) Widespread Laxfordian deformation and metamorphism at c. 1.87 Ga resulted from the collision at the subduction zone of the Southern and Central regions.

various Archaean gneisses of the Lewisian Complex, cross-cutting folds formed during Badcallian and Inverian deformation. Sm–Nd mineral and U–Pb baddeleyite ages suggest that dyke emplacement spanned a considerable period of time from c. 2.4 to c. 2.0 Ga; the Scourie dyke swarm may therefore actually comprise at least two intrusive suites. The anorthosites and metagabbros of the South Harris igneous complex in the Outer Hebrides (Fig. 4.1) were intruded at c. 2.2 Ga and may be part of the same phase of igneous activity. Supercontinent break-up was also recorded in many continental blocks by the widespread development of extensional rift basins and passive margins, and the accumulation of extensive sedimentary and extrusive volcanic sequences (known collectively as 'supracrustals'). The metavolcanic and metasedimentary rocks of the *Loch Maree Group* of the Lewisian Complex, and possibly also the metasedimentary belts of the Outer Hebrides, are considered to result from the continental break-up at c. 2.0 Ga. The Loch Maree Group (Figs 4.1, 4.2a) comprises both oceanic and continental components. The oceanic sequence includes plateau basalts, ferruginous hydrothermal deposits (banded iron formation) and platform carbonates, while the continental component consists of metagreywackes interpreted as deltaic sediments. Detrital zircons from the metagreywackes are derived from both Archaean and Palaeoproterozoic (c. 2.2–2.0 Ga) sources. These supracrusted rocks are intruded by calc-alkaline granodioritic gneisses with a U–Pb zircon age of c. 1903 Ma. Contacts with adjacent Archaean gneisses are now concordant and highly deformed. The Loch Maree Group is considered to have formed as a result of the accretion of oceanic and volcanic arc components to the upper continental plate of a subduction zone formerly located along the present southwest margin of the Central region.

4.3 Palaeoproterozoic orogeny: Laxfordian deformation and metamorphism of the Lewisian Complex

The period *c.* 2.0 Ga to *c.* 1.6 Ga was characterized worldwide by convergence of continental blocks and development of active plate margins. This later stage of the Palaeoproterozoic is therefore associated with widespread orogeny and continental growth. In north-west Scotland, Palaeoproterozoic orogenic activity is represented by the *Laxfordian event*. Widespread deformation and metamorphism of the Lewisian Complex resulted from the final amalgamation of the Archaean gneisses of the Southern, Central and Northern regions (Fig. 4.1). In the Southern and Northern regions, pervasive ductile deformation has largely obliterated original discordant contacts between the host gneisses and the Scourie dykes. By contrast, the gneisses and Scourie dykes of the intervening Central region (Fig. 4.1) mostly escaped Laxfordian deformation and metamorphism. As a result, Archaean (Badcallian) structures, their associated granulite facies mineral assemblages and intrusive relations between the dykes and the gneisses, are often preserved intact. Laxfordian deformation and metamorphism is widespread in the Outer Hebrides (Fig. 4.1).

Laxfordian crustal thickening was associated with regional folding and the development of major ductile shear zones (Figs 4.1 & 4.3). The dextral strike-slip component of movement along many of the shear zones on the mainland indicates an overall transpressive tectonic regime. Many of these structures, such as the Laxford and Diabaig shear zones, are apparently located along steeply dipping zones of pre-existing Inverian deformation which possessed a strong structural anisotropy and were consequently reactivated during the Laxfordian event. The Laxford shear zone represents the boundary between the Northern and Central regions. It comprises a zone of intense north-vergent folding associated with oblique northward overthrusting, under amphibolite facies conditions, of the granulites of the Central region (Fig. 4.3). Granite sheets formed by melting of gneisses at deeper structural levels were emplaced syn-tectonically along the shear zone. The boundary between the Central and Southern regions is less well defined, but a subduction suture separating the two blocks is presumed to lie somewhere in the Loch Maree–Gairloch area (Figs 4.1

& 4.2b). Peak metamorphism in the Outer Hebrides was within the granulite facies and associated with the formation of the extensive migmatite and granite complexes of Harris and Lewis. The Laxfordian shear zones in this area are mainly flat lying and their relationship to those of the mainland is uncertain. Sm–Nd mineral and U–Pb zircon dating of metamorphic assemblages and granites suggests that the main phase of Laxfordian deformation and metamorphism occurred at *c.* 1.9–1.75 Ga.

Reconstruction of the continental blocks in the North Atlantic region suggests that the Lewisian Complex forms part of a continuous Palaeoproterozoic orogenic belt linking the Labrador belt of the Canadian Shield with the Nagssugtoqidian and Ammassalik belts of Greenland and the Lapland–Kola belt of Scandinavia (Fig. 4.4). This orogenic belt mainly comprises Archaean rocks that were variably deformed and metamorphosed at *c.* 1.9–1.8 Ga. The evolution of the Ammassalik belt in particular is very similar to that of the Lewisian Complex. The Ammassalik belt comprises Archaean gneisses, mafic dykes correlated with the Scourie dykes and abundant sediments which may be of the same age as the Loch Maree Group. The presence of calc-alkaline igneous rocks interpreted as subduction-related in several parts of this belt, including the Loch Maree Group, has led to the interpretation that it results from oceanic closure followed by collision of the various continental blocks surrounding the belt (Fig. 4.4). This process resulted in crustal thickening and high-grade metamorphism of the rocks within the belt and would account for the Laxfordian events of north-west Scotland.

4.4 Palaeoproterozoic subduction: evolution of the Rhinns and Annagh Gneiss complexes

The convergence of continental blocks between *c.* 2.0 Ga and *c.* 1.6 Ga resulted in the formation of a broadly continuous supercontinent that included Laurentia and Baltica. A major northward-dipping (present reference frame) subduction zone and Andean-type active plate margin developed between *c.* 1.9 Ga and *c.* 1.6 Ga along the southern margin of this supercontinent and is represented by the *Ketilidian belt* of southern Greenland and the *Svecofennian belt* of Scandinavia (Fig. 4.4). Both belts are charac-

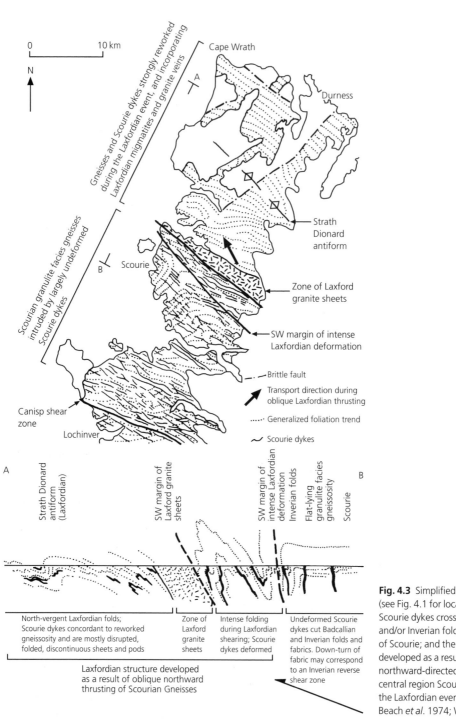

Fig. 4.3 Simplified map and cross-section (see Fig. 4.1 for location) showing: Scourie dykes cross-cutting Badcallian and/or Inverian folds in the region south of Scourie; and the principal structures developed as a result of the oblique northward-directed thrusting of the central region Scourian gneisses during the Laxfordian event (modified from Beach et al. 1974; Watson 1983).

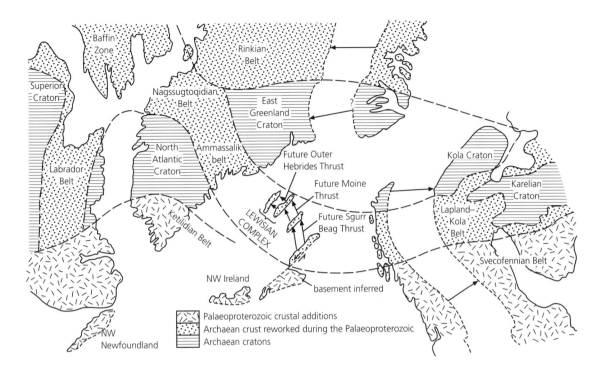

Fig. 4.4 Reconstruction of the Palaeoproterozoic orogenic belts and Archaean cratons of the North Atlantic region (modified from Park 1991). The various crustal blocks which make up Scotland, East Greenland and Scandinavia are depicted in their likely relative positions prior to Caledonian thrusting (arrows). In the case of Scotland, it is also necessary to remove late Caledonian sinistral strike-slip displacements along the Great Glen Fault (GGF).

terized by high-temperature, low-pressure metamorphism and abundant calc-alkaline magmatism, features consistent with a volcanic arc setting. The majority of these igneous rocks are newly differentiated mantle-derived material, with only a small proportion of older Archaean crust. Two small areas of basement in Scotland and Ireland, the *Rhinns and Annagh Gneiss complexes*, may expose parts of this active plate margin.

The Rhinns Complex of the Inner Hebrides (Fig. 4.5) mainly comprises a metamorphosed and weakly deformed alkaline igneous association of syenite and subordinate mafic material intruded by gabbro sheets. These (meta)igneous rocks are characterized by major and trace element patterns similar to igne-

ous rocks generated in a subduction-related magmatic arc. U–Pb zircon ages obtained from syenites on Islay (1782 ± 5 Ma) and Inishtrahull (1779 ± 3 Ma) are interpreted as igneous crystallization ages. Isotopic evidence indicates that the syenite was juvenile mantle-derived material and not reworked Archaean crust. A $^{40}Ar/^{39}Ar$ hornblende age of *c.* 1710 Ma obtained from a metagabbro at Inishtrahull dates cooling following metamorphism. The Annagh Gneiss Complex of north-west Ireland (Fig. 4.5) consists mainly of orthogneisses cut by granites and metabasic dykes. U–Pb and Sm–Nd data indicate that the igneous protoliths of some of the gneisses formed at *c.* 1900 Ma and these rocks may therefore have affinities with the Rhinns Complex.

Although the Rhinns Complex is at present exposed only in a very small area, several lines of evidence indicate that it may underlie much of the Central Highlands Terrane, south-east of the Great Glen Fault (Figs 4.4 & 4.5). The isotopic signatures of Caledonian granites in this area are consistent with derivation, at least in part, from juvenile Proterozoic crust at depth. Locally derived granitic boulders within Late Neoproterozoic glacial tillites within the

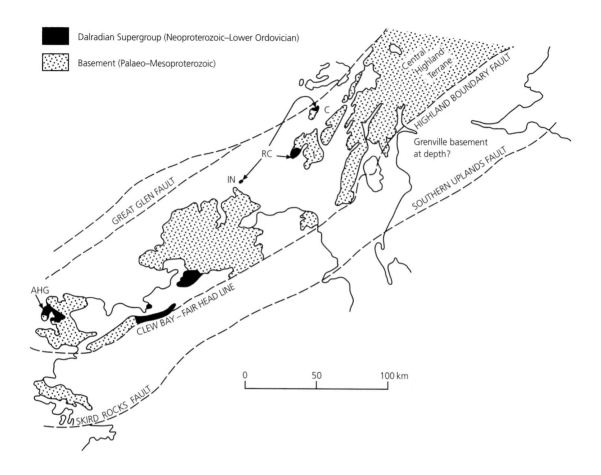

Dalradian Supergroup (Neoproterozoic–Lower Ordovician)

Basement (Palaeo–Mesoproterozoic)

Fig. 4.5 Location of Palaeo- and Mesoproterozoic basement inliers within north-west Ireland and the Inner Hebrides. RC, Rhinns Complex (C, Colonsay; I, Islay; IN, Inishtrahull); AHG, Annagh Gneiss Complex; SD, Slishwood Division.

Dalradian Supergroup (see Chapter 5) are petrographically and chemically very similar to components of the Rhinns Complex, suggesting that these basement rocks were more extensively exposed in the region at the time of glaciation.

4.5 Mesoproterozoic rifting: deposition of the Stoer Group

The supercontinent that formed during the Late Palaeoproterozoic is thought to have existed until c. 1.4–1.3 Ga when a prolonged period of rifting and continental dispersal was marked by intrusion of regional basic dyke swarms in Canada, Greenland

and Scandinavia. Palaeomagnetic evidence indicates that north-east Laurentia and Baltica were separated by rifting at c. 1.3–1.1 Ga (Fig. 4.6a,b). In north-west Scotland, it seems likely that this episode of rifting was associated with deposition of the continental sedimentary rocks of the *Stoer Group*, which rest unconformably on the Lewisian Complex between Stoer and Gairloch (Fig. 4.7). The Stoer Group has traditionally been thought to belong to a 'Torridonian' succession, which included the superficially similar continental sedimentary rocks of the overlying Sleat and Torridon groups (Fig. 4.7; see Section 4.7). However, it now seems likely that the 'Torridonian' comprises two entirely separate sedimentary sequences of which the Stoer Group is the older.

The Stoer Group comprises c. 2 km of fluvial, aeolian and lacustrine red-bed deposits (Fig. 4.8). A hot, arid climate is indicated by the generally red colouration of the sediments and the local presence of

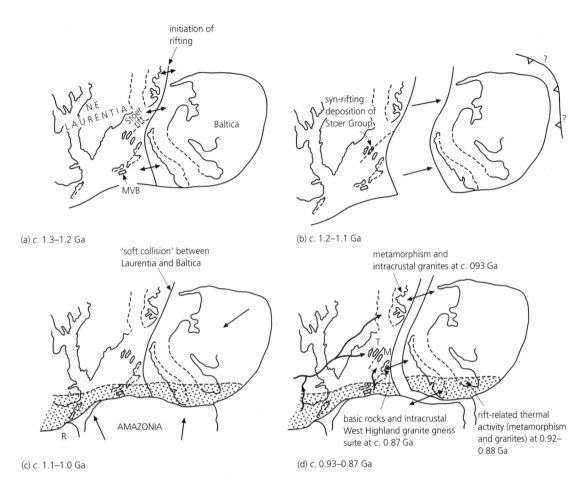

Fig. 4.6 Generalized palaeocontinental reconstructions for the period c. 1.3–0.9 Ga. As in Fig. 4.5, East Greenland, Baltica and Scotland are depicted in their likely configurations prior to Caledonian thrusting. (a) Onset of rifting (bivergent arrows) between Laurentia and Baltica at c. 1.3–1.2 Ga is accompanied by the development of a failed rift (aulacogen) striking at a high angle into Laurentia. MVB, Midland Valley basement. (b) Separation of the two continents and syn-rifting deposition of the Stoer Group at c. 1.2–1.1 Ga. Width of oceanic tract notional because the timing of rifting is poorly constrained. (c) Collision of Rio de Plata microcontinent (R) and Amazonia with Laurentia and Baltica to result in the Grenville–Sveconorwegian orogeny at c. 1.1–1.0 Ga (dotted region). This collision also involved the basement of north-west Ireland and the Midland Valley. Note the 'soft collision' and lack of major crustal thickening along the majority of the Laurentia–Baltica suture. (d) Renewed rifting (bivergent arrows) of Baltica from Laurentia and Amazonia at c. 0.93–0.87 Ga, accompanied by a protracted period of magmatism, metamorphism and emplacement of granites. Heavy wavy arrows represent likely transport directions of sediment into the Sleat/Torridon (T) and Moine (M) Basins. These basins are thought to have been separated by a basement high (dashed-dotted line) which could have formed during rifting in either (b) or (d).

algal limestones. Palaeomagnetic evidence indicates deposition at palaeolatitudes of between 10° and 20°N. The lowest Stoer Group rocks are basal fan-glomerate breccias, which infill an irregular landscape, with a relief of up to 300 m cut in the Lewisian Complex (Fig. 4.8). Clasts were derived mainly from adjacent Lewisian units. The breccias pass upwards into cross-bedded sandstones, which were deposited by braided rivers (Fig. 4.8). These are intercalated with rare red lacustrine shales and aeolian sandstones. The petrography and geochemistry of the sandstones suggests that they were derived mainly from erosion of Scourian gneisses. Quartzose sandstone and metaquartzite clasts, unlike any lithology at present exposed in the Lewisian Complex, may have been derived from supracrustal units which formerly either

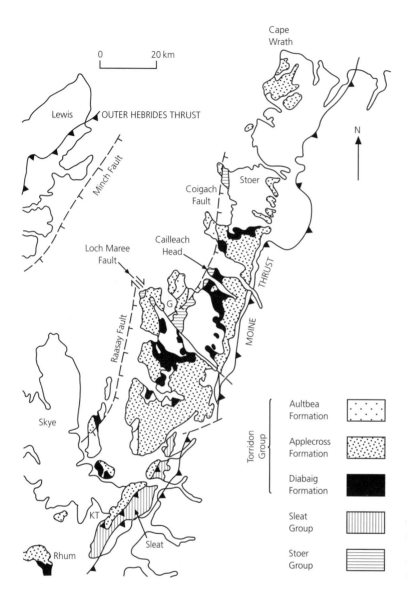

Fig. 4.7 Distribution of the 'Torridonian' sediments in north-west Scotland (modified from Anderton *et al.* 1979), together with the location of important steep, normal faults which were probably active during sedimentation (a), and the younger Moine, Outer Hebrides and Kishorn (KT) thrusts which formed during the Caledonian Orogeny (b). G, Gairloch.

overlay or were infolded within the basement. The volcanic Stac Fada Member forms a key marker horizon, which can be traced along almost the entire north–south outcrop of the Stoer Group. It comprises muddy sandstone with up to 30% devitrified glass, and has been interpreted as either an ashflow or a volcanic mudflow. The geochemistry of some other Stoer Group siltstones is also consistent with them representing reworked volcanic tuffs.

The dating of Precambrian sedimentary sequences presents many problems because of the lack of bio-

stratigraphic control. Proterozoic microfossils and algal remains occur in shales of the Stoer Group, but do not provide accurate age constraints. Furthermore, the Stoer Group does not contain precisely datable volcanic strata. It is therefore necessary to apply less reliable isotopic and palaeomagnetic techniques and compare the results. A Pb–Pb isochron of 1199 ± 70 Ma obtained from a Stoer Group limestone has been interpreted as the time of early diagenesis, which is probably close to the time of deposition. The isotopic results are in broad agreement with palaeo-

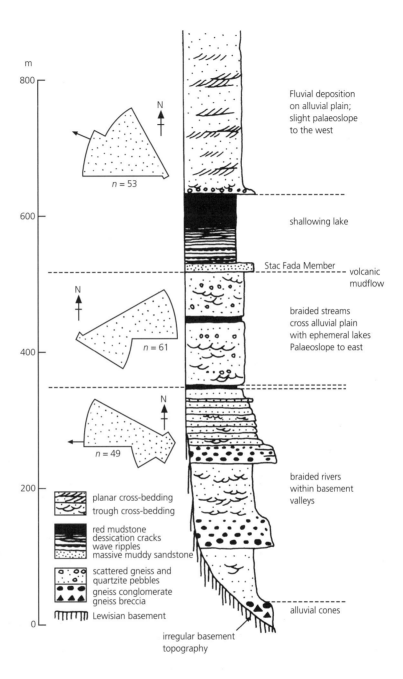

Fig. 4.8 Schematic stratigraphic profile through the Stoer Group at Stoer (redrawn from Stewart 1982).

Within the figure:

m
800

600

400

200

0

N
n = 53

N
n = 61

N
n = 49

Fluvial deposition
on alluvial plain;
slight palaeoslope
to the west

shallowing lake

Stac Fada Member — volcanic
mudflow

braided streams
cross alluvial plain
with ephemeral lakes
Palaeoslope to east

braided rivers
within basement
valleys

alluvial cones

irregular basement
topography

Legend:
planar cross-bedding
trough cross-bedding
red mudstone
dessication cracks
wave ripples
massive muddy sandstone
scattered gneiss and quartzite pebbles
gneiss conglomerate
gneiss breccia
Lewisian basement

magnetic data, which imply an age of *c.* 1.2 Ga for the Stoer Group.

Palaeocurrent data indicate both westerly and easterly directed current flows (Fig. 4.8). The limits of the Stoer Group basin are uncertain, but it may have been bounded to the west by the Minch Fault (Fig. 4.7) and to the east by a normal fault now unexposed and buried beneath the Moine Nappe. Rapid palaeocurrent reversals through 180° at two levels in the sequence suggest active faulting during deposition. Small-scale prelithification faults are a common feature of Stoer Group sediments within a kilometre of the Coigach Fault (Fig. 4.7), which may therefore have been active during deposition. In the same area, fractures in the Lewisian basement are infilled with Stoer Group clastic material that was deformed prior to lithification by continued fault movements. Similarly orientated brittle fracture arrays are recognized regionally in the Lewisian basement along the north-west coast of Scotland, pre-dating deposition of the younger Torridon Group sediments (see Section 4.8). Structural analysis of these fractures indicates that the Stoer Group basin resulted from ENE–WSW-directed crustal stretching coupled with a component of dextral transtension. Evidence for syn-depositional faulting and active volcanism during basin development, when taken in conjunction with the available isotopic and palaeomagnetic evidence, supports the interpretation that the Stoer Group is a syn-rift sequence deposited during continental break-up at *c.* 1.2 Ga (Fig. 4.7a,b).

4.6 Late Mesoproterozoic Grenville orogenesis: reworking of basement

Renewed continental convergence at *c.* 1.1–1.0 Ga resulted in the formation of the supercontinent *Rodinia*, which included Laurentia, Baltica, most of Gondwana and Siberia. The North American segment of Laurentia is thought to have collided with the Rio de Plata microcontinent and Amazonia (South America) to form the *Grenville* orogenic belt (Fig. 4.6c). Baltica collided with Amazonia to form the *Sveconorwegian* orogenic belt (Fig. 4.6c). Various basement rocks within the Caledonian belt of the British Isles and Ireland record Grenville deformation and high-grade metamorphism. U–Pb zircon ages of *c.* 1.1–1.0 Ga obtained from syn-tectonic granites emplaced during metamorphism and ductile deformation of the Annagh Gneiss Complex (Fig. 4.5) show that these basement rocks were located within the Grenville belt. Metasedimentary basement rocks of the *Slishwood Division* in north-west Ireland (Fig. 4.5) record high-pressure granulite facies metamorphism that is imprecisely dated but probably also Grenville in age. The northern margin of the Grenville belt must lie between these basement rocks and the Rhinns Complex (Fig. 4.5), because the latter are apparently unaffected by Grenville deformation and metamorphism. Unexposed Grenville basement may also be present at depth south-east of the High-

land Boundary Fault in Scotland (Fig. 4.5), because blocks of granulite facies gneisses contained within Carboniferous volcanic vents in the Midland Valley have yielded U–Pb and Sm–Nd mineral ages of *c.* 1.1–1.0 Ga.

The extent of Grenville orogenic activity within the basement rocks that underlie the Scottish Highlands is uncertain, and for this reason these rocks are considered to lie to the north of the mountain belt (Fig. 4.6c,d). However, the basement gneiss inlier of Glenelg in north-west Scotland (Fig. 4.11) is a notable exception. These gneisses are in tectonic contact with overlying Neoproterozoic Moine sediments (see Section 4.8) and contain eclogites, which are diagnostic of metamorphism at very high pressures. The eclogites here formed at temperatures of 750±25°C and pressures of *c.* 16 kbar, corresponding to crustal depths of *c.* 45 km. Sm–Nd whole-rock and mineral isochrons of 1082±24 Ma and 1010±13 Ma date peak metamorphism. The present location of these gneisses is apparently anomalous: one solution is that they were faulted into their present position some time after the Grenville Orogeny but prior to the Caledonian Orogeny. Further isotopic investigations of the basement rocks of Scotland are clearly needed to refine existing palaeocontinental reconstructions for this period.

4.7 Early Neoproterozoic continental sedimentation: the Sleat and Torridon groups

In the North Atlantic region, the period between *c.* 1000 Ma and *c.* 800 Ma was dominated by the accumulation of thick sedimentary sequences derived from the erosion of the Grenville mountain chains. An Early Neoproterozoic continental sedimentary sequence comprising the *Sleat* and *Torridon* groups outcrops extensively along the north-western seaboard of Scotland and on the islands of Skye, Raasay and Rhum (Figs 4.7, 4.9). It also exists at subcrop as far west as the Minch Fault, and to the south-west of Rhum it forms the sea bed for *c.* 125 km as far south as the Great Glen Fault.

The sedimentary rocks of the Sleat Group occur within the Caledonian Kishorn thrust nappe of Skye and are weakly deformed and metamorphosed. They comprise *c.* 3.5 km of coarse, grey fluvio-deltaic sand-

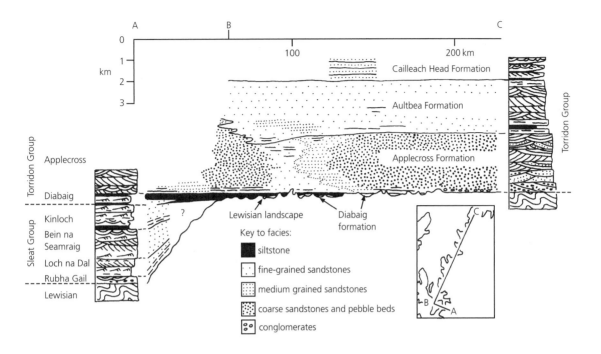

Fig. 4.9 Stratigraphic profile along the line shown on the inset map, and logs for the Sleat and Torridon groups (redrawn from Anderton *et al.* 1979).

stones with subordinate grey lacustrine or marine shales. Palaeocurrents are mainly from the west. Most clasts are of porphyritic rhyolite or rhyodacite. The lower boundary of the Sleat Group is unexposed and is presumed to be an unconformity with the Lewisian Complex; the upper boundary is a conformable transition into the overlying Torridon Group.

The coarse red fluvial arkosic sandstones of the 5–6 km thick Torridon Group rest with strong angular unconformity on the Lewisian Complex and the Stoer Group. They cover an old land surface, which in the Loch Maree area has a relief of up to 600 m. An overall wet, possibly subtropical depositional environment is implied by the geochemistry of the sediments, the dominance of fluvial deposits and the total absence of caliche, evaporites or aeolian deposits. This is consistent with palaeomagnetic evidence that Torridon Group sedimentation occurred at palaeolatitudes of between 30° and 40°N.

The lowest *Diabaig Formation* (0.6 km thick) comprises red breccias, grey sandstones and shales. The breccias and sandstones are interpreted as fan deposits that accumulated in palaeovalleys, with the grey shales recording ephemeral lakes in valley bottoms. Diabaig sedimentation was terminated abruptly by the influx from the west of the coarse fluvial clastics of the *Applecross* (3–3.5 km thick) and *Aultbea* (1.5–2 km thick) formations. Diabaig lakes were replaced by fluvial conditions as a major alluvial braidplain prograded eastwards across the basin, infilling valleys and burying remaining hills. Fluvial sedimentation resulted in a major fining-upwards succession, 5–6 km thick, from coarse pebbly Applecross arkoses through to more homogeneous medium-grained Aultbea sandstones. Large-scale bar structures, up to 9 m thick, within the Applecross Formation indicate the occurrence of major rivers. Soft-sediment deformation structures are ubiquitous and were probably triggered by turbulent current activity. Palaeocurrents flow consistently eastwards. Geochemical evidence suggests that in much of the Applecross Formation quartzofeldspathic detritus was derived from erosion of Lewisian gneisses. 'Exotic' clast types which cannot be matched with parts of the Lewisian Complex exposed at present include lithologies such as muscovite schist, porphyritic rhyolite and dacite, chert and jasper, banded iron formation, and tourmalinized quartzite, which were presumably derived from erosion of supracrustal units. The *Cailleach Head Formation* is about 800 m thick and comprises numerous grey shale-to-sandstone cyclothems that probably result from the repeated progradation of deltas into freshwater lakes.

Proterozoic microfossils and algal remains occur in Torridon Group shales but do not provide accurate age constraints. Rb–Sr whole-rock regressions from siltstones of the Diabaig and Applecross formations produce ages of 994±48 Ma and 977±39 Ma, respectively. These dates are interpreted as the time of early diagenesis, which is probably close to the time of deposition. The isotopic results are in broad agreement with palaeomagnetic data, which imply

an age of *c.* 1040 Ma for sedimentation of the Torridon Group. U–Pb ages as young as 1088 Ma have been obtained from individual detrital zircons within the Applecross Formation and confirm that deposition of the Torridon Group must post-date *c.* 1100 Ma.

The Sleat and Torridon groups were probably deposited shortly after the Grenville Orogeny (Fig. 4.6d). The Applecross Formation's stratigraphic thickness, outcrop belt length (200 km length) and overall uniformity of sedimentary facies, together with evidence for major rivers, indicates an enormous depositional basin (Fig. 4.10). The Torridon Group basin is therefore most likely to represent a second stage extensional basin formed by thermal relaxation processes. The Sleat Group may represent the initial rifting stage. This is supported by the Sleat Group's

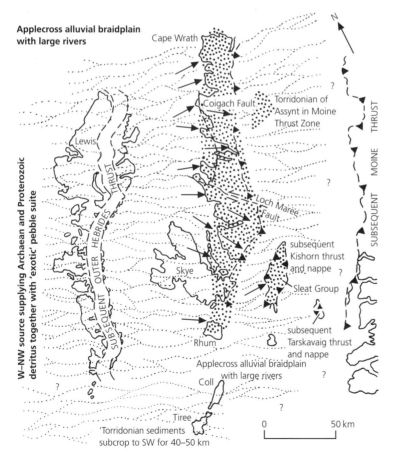

Fig. 4.10 Generalized palaeogeography during deposition of the Torridon Group (redrawn from Nicholson 1993).

more diverse sedimentary facies, less mature petrography, including large amounts of volcanic detritus, and its conformable relationship with the overlying Torridon Group. Modelling of the Applecross river system suggests a basin comparable in size with those of the present-day South Saskatchewan and Platte rivers with river lengths of up to 500 km and drainage basin areas of the order of 10 million km². This implies a distal rather than proximal source region for some of the 'exotic' clast lithologies of the Torridon Group. The U–Pb ages obtained from detrital zircon grains within the Applecross Formation indicate that the source region(s) comprised Late Archaean to Mesoproterozoic basement rocks that had been affected by Grenville high-grade metamorphism and/or igneous activity. Likely source areas are the basement rocks of north-east Canada and southern Greenland (Fig. 4.6d).

4.8 Early Neoproterozoic marine sedimentation: the Moine Supergroup

The Moine Supergroup is a thick succession of strongly deformed and metamorphosed sedimentary rocks, which crops out extensively within the Caledonian orogenic belt between the Moine Thrust Zone and the Great Glen Fault—the Northern Highlands Terrane (Fig. 4.11). Inliers of Late Archaean basement orthogneisses (see Chapter 3) are thought to represent the continental basement on which the Moine sediments were deposited. Moine sedimentation occurred after *c.* 1000 Ma (the age of the youngest detrital zircon so far dated) but before *c.* 873 Ma (the age of the oldest igneous rocks that intrude the sediments). The tectonic history and affinities of the Moine Supergroup and associated basement gneisses continue to be the subject of much vigorous debate. Were they actually part of Laurentia in the Early Neoproterozoic, or do they form an 'exotic' terrane derived from another continental block and later accreted to the Laurentian margin?

The sedimentary precursors of the Moine Supergroup were deposited as fine-grained sands, silts and muds. The sandstones were later metamorphosed into psammites and the silts and muds into pelites. The Moine Supergroup has been subdivided into the *Morar*, *Glenfinnan* and *Loch Eil groups* (Fig. 4.11). Each is characterized by differing proportions of psammite and pelite together with striped units that comprise rapid alternations of psammite and pelite. The Moine rocks of Shetland (Fig. 4.11) are very similar, but precise correlation with the groups established on the mainland is uncertain. The top of the Moine succession is not exposed. Minimum thicknesses of *c.* 5–6 km have been proposed for both the Morar and Loch Eil groups in West Inverness. Sedimentary structures are widely preserved within the thick psammites of the Morar and Loch Eil groups. A shallow marine environment of deposition is indicated by bipolar cross-stratification, wave ripples, possible lenticular and flaser bedding, and the rarity of channelling. Upper parts of the Morar Group include complex sand waves that compare with those found in present-day shelf environments. Palaeocurrents deduced from cross-bedding show a flow towards the north and north-east.

The Moine sediments were probably deposited in NNE–SSW trending half-grabens that were bounded to the west by east-dipping normal faults (Fig. 4.12). In Morar and Ardnamurchan, the Upper Morar Group psammite displays a marked westward thickening consistent with deposition in a half-graben. Soft-sediment deformation structures, probably seismically induced, are common in the west outcrop of the unit (proximal to the inferred normal fault) and absent in the east. The unit appears to become progressively more distal eastwards and the striped and pelitic rocks of the Glenfinnan Group may represent a distal facies of the Morar Group (Fig. 4.12a). Both are thought to rest unconformably on basement gneisses. An upward transition from Morar to Glenfinnan lithologies preserved on south-west Mull suggests that the mixed and muddy deposits of Glenfinnan type prograded across the tidal sand shelf represented by the Upper Morar Psammite. Early rift-related sedimentation (Morar Group) may thus have been followed by transgressive, more quiescent deposition (Glenfinnan Group), perhaps reflecting a decline in active rifting and greater control by thermal subsidence. The Loch Eil Group may represent the effects of renewed extensional rifting. An asymmetrical facies distribution and westward thickening of the sequence is again consistent with deposition in a half-graben bounded by an east-dipping normal fault (Fig. 4.12b). This rifting phase must have occurred later than the Morar phase, because the Loch Eil

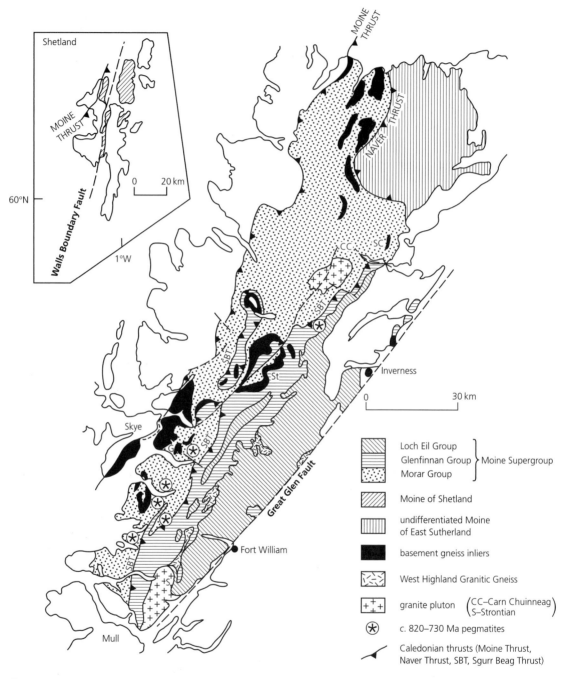

Fig. 4.11 Location of the Moine rocks of north-west Scotland and Shetland. Minor outcrops of Moine rocks on the islands of Mull and Mainland Orkney are not shown. G, Location of Glenelg basement inlier which contains Grenvillian eclogites. Locations of Moine basal conglomerates: S, Strathan; St, Strathfarrar; SC, Strath Carnaig.

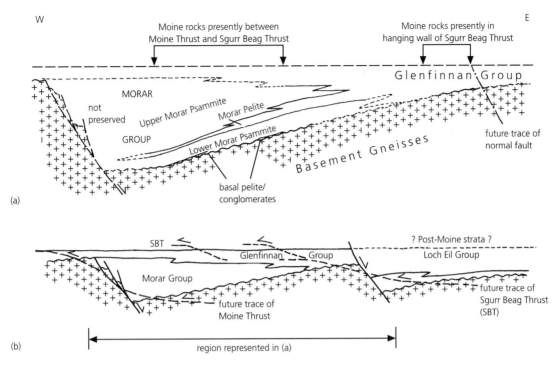

Fig. 4.12 (a) Schematic west–east restored section across the Moine outcrop between Skye and Fort William to show the original stratigraphic relationships of the Morar and Glenfinnan groups, and likely geometry of the Morar rift basin. (b) Inferred original stratigraphic relationships of the three Moine groups and the basins in which they were deposited, showing the position of the major Caledonian thrusts that subsequently disrupted their basin fills (modified from Soper et al. 1998).

Group conformably overlies the Glenfinnan. It seems likely that deposition occurred in a separate sub-basin to the east in the southeast part of the Moine group. Amphibolites are common in the Glenfinnan and Loch Eil groups, but largely absent from the Morar Group, and are thought to represent basic igneous sheets that were emplaced during rifting. The amphibolites that intrude the Glenfinnan and Loch Eil groups have a tholeiitic chemistry, which implies that the rift basins developed on thinned continental crust.

Several lines of evidence indicate that the Moine Supergroup was deposited unconformably upon the basement gneisses, which at present occur as a series of infolds and thrust sheets interleaved with the Moine rocks (Fig. 4.11). Many of the basement inliers within the Morar Group are located in the cores of isoclinal folds and at the base of local Moine successions, which invariably young *away* from the basement. Contacts between the basement gneisses and the Moine rocks are commonly highly strained, but in places the level of strain is low and cross-bedding is preserved within a few metres of the contact, implying that it represents a tectonized unconformity. Basal Moine conglomerates occur adjacent to basement gneiss inliers at Strathfarrar, Strath Carnaig and Strathan (Fig. 4.11). The mineralogical and textural differences between the high-grade (mid–upper amphibolite facies), basement gneisses and the medium-grade (low amphibolite facies) Morar Group metasediments imply that the former were already gneissic prior to the deposition of the Moine Supergroup.

U–Pb dating of detrital zircons within the Moine Supergroup (as well as granitic gneisses formed as a result of the melting of these rocks) has yielded ages almost entirely between c. 1.9 Ga and c. 1.0 Ga. The scarcity of Archaean zircons means that the Moine rocks were not derived from the erosion of the Lewisian Complex of the Hebridean Terrane. The Palaeo- to Mesoproterozoic basement rocks of the Rhinns Complex and the Annagh Gneiss Complex represent one possible source for the Moine sediments (Fig. 4.6d). A southerly provenance area in the Inner Hebrides and north-west Ireland is consistent with the northerly sediment dispersal patterns indicated by cross-bedding in the Moine sediments. Alternatively, the Moine sediments may have been derived from the erosion of more distal Proterozoic basement blocks in the Grenville mountain belt and/or South America (Fig. 4.6d).

There has been much debate over the past 100 years or so as to whether or not the Torridonian and Moine sediments are, respectively, the continental and marine parts of a single sedimentary sequence deposited in one major basin, and later juxtaposed by Caledonian thrusting. It now seems likely that the Stoer Group is much older than the Moine Supergroup and the two sequences cannot therefore be correlated. The available isotopic constraints suggest that the Torridon Group and the Moine Supergroup were probably deposited at about the same time, following the Grenville orogeny, and the possibility that they might represent lateral correlatives deserves consideration. However, the rather different source areas indicated by the dating of detrital zircons tends to rule out the possibility that the shallow marine Moine sediments originated from the Applecross Formation rivers. The two sequences may have been deposited in different sedimentary basins that received detritus from contrasting source regions (Fig. 4.6d).

4.9 The Central Highland Migmatite Complex: Moine rocks south-east of the Great Glen Fault?

A sequence of high-grade, migmatized psammitic and pelitic metasedimentary rocks, known informally as the *Central Highland Migmatite Complex*, outcrops south-east of the Great Glen Fault in the Central Highlands Terrane (Fig. 4.13). It strongly resembles parts of the Moine Supergroup with which it has been correlated by some workers. However, sedimentary structures are almost entirely absent and no formal lithostratigraphic subdivision exists. U–Pb dating of detrital zircons has provided a range of Palaeo- to Mesoproterozoic ages similar to that obtained from the Moine Supergroup, which may indicate derivation from the same source area. The Central High-

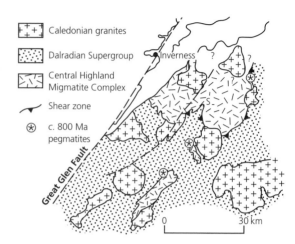

Fig. 4.13 Simplified geological map of the Central Highlands, showing the location of the Central Highland Migmatite Complex. S, Slochd.

land Migmatite Complex is overlain by Neoproterozoic metasedimentary rocks of the Dalradian Supergroup (see Chapter 5). In some places, apparent onlap of Dalradian units onto the migmatite complex may indicate that the contact is an unconformity, although any angular discordance which might have once existed has been obliterated by strong ductile deformation. In others areas, the contact is marked by a ductile shear zone (Fig. 4.13).

4.10 Neoproterozoic Knoydartian orogenesis in the Scottish Highlands?

The Moine Supergroup and the Central Highland Migmatite Complex both contain evidence for polyphase deformation and amphibolite facies metamorphism. Isotopic dating has shown that widespread deformation and metamorphism occurred during the Middle Ordovician Grampian Orogeny (see Chapter 6). Several lines of evidence indicate, however, that some deformation, metamorphism and igneous activity occurred during an important and controversial Neoproterozoic tectonothermal event for which two contrasting models have been proposed. The traditional model is that this event is compressional and represents an orogeny; the alternative is that it is extensional with metamorphism and igneous activity occurring during rifting at the base of a thick sedimentary pile.

4.10.1 The ages of granitic gneiss formation and migmatization

The West Highland Granitic Gneiss which intrudes the Glenfinnan and Loch Eil groups (Fig. 4.11) has traditionally been assumed to have formed during orogenesis as a result of the melting of Moine rocks. U–Pb zircon dating has established an age of 873 ± 3 Ma for this event. However, the granitic gneisses are cut by a suite of gabbros and tholeiitic basic rocks which have yielded identical zircon ages. This strongly suggests that the granitic and the basic rocks might together form a bimodal igneous suite emplaced during rifting. The heat necessary to melt the Moine rocks and form the granitic gneisses may have been provided in part by the basic magmas. An extensional setting therefore seems possible for magmatism and crustal melting at $c.$ 870 Ma. Crustal extension at this time is most plausibly related to the rifting post-930 Ma of Baltica from Laurentia and Amazonia (Fig. 4.6d)

4.10.2 The ages of regional metamorphism and pegmatite generation

The Moine Supergroup and Central Highland Migmatite Complex are widely veined by pegmatites, and most of these undoubtedly formed during Caledonian metamorphism. However, the results of early isotopic studies in the 1960s identified a distinctive suite of older, usually foliated pegmatites that formed in the Neoproterozoic. The most recent and accurate isotopic dating of these pegmatites has yielded U–Pb zircon and monazite ages in the range $c.$ 820–780 Ma, although some ages range as young as $c.$ 730 Ma (Figs 4.11, 4.13). Field relationships indicate that the pegmatites were emplaced during ductile deformation and metamorphism of their country rocks. In the south-west part of the Morar Group, garnets which grew after early isoclinal folding have yielded Sm–Nd ages between 823 ± 5 Ma and 788 ± 4 Ma. Metamorphic analysis shows that these garnets record peak temperatures and pressures of 575–625°C and 10–12 kbar. These P–T conditions could not be achieved during extension but, instead, are entirely consistent with crustal thickening and orogeny. East of the Great Glen Fault at Slochd (Fig. 4.13), kyanite-bearing migmatites of the Central Highland Migmatite Complex have yielded a U–Pb zircon age

of 840 ± 11 Ma, which is thought to date melting during crustal thickening.

It is concluded that the balance of evidence indicates that the Moine Supergroup and the Central Highland Migmatite Complex were deformed and metamorphosed during a Neoproterozoic orogeny at $c.$ 820–780 Ma. This event is referred to widely as the *Knoydartian* (or *Morarian*) event. The geotectonic significance of the event is uncertain at present. One possibility is that following limited separation of Baltica by $c.$ 870 Ma (Fig. 4.7d), it rotated clockwise to result in transpressive orogenesis of the Scottish segment of Laurentia. A similar-aged event to the Knoydartian has been recorded in northern Norway (the Porsanger orogeny). It is necessary to involve such a rotation of Baltica because it is widely thought that Norway was opposite East Greenland by $c.$ 750 Ma (see Chapter 2). In this interpretation, the boundary between the Central Highland Migmatite Complex and overlying Dalradian strata probably represents the unconformity resulting from crustal uplift during the orogeny. However, at the time of writing the long-lived controversy as to the nature of the Knoydartian event is far from settled! Some workers are of the opinion that there is no major structural or metamorphic difference across the Central Highland Migmatite Complex–Dalradian contact, and *if* an orogenic unconformity exists at all (which in the view of some is by no means proven) then it must be located elsewhere, perhaps higher in the Dalradian succession. Further detailed fieldwork and isotopic dating are needed to finally resolve the continuing debate.

4.11 Do the Moine Supergroup and Central Highland Migmatite Complex represent an exotic terrane?

The $c.$ 1100–800 Ma histories of the Northern Highlands and Central Highlands terranes are remarkably similar (Fig. 4.14). This suggests that the Great Glen Fault does not separate fundamentally different crustal blocks. The apparent difference in the age of Late Archaean to Palaeoproterozoic basement either side of the fault could simply result from late Caledonian sinistral strike-slip displacements of no more than several hundred kilometres that cut discordantly across age provinces in the basement (Fig. 4.4). However, the $c.$ 1100–800 Ma histories of the areas west and east of the Moine Thrust Zone could be

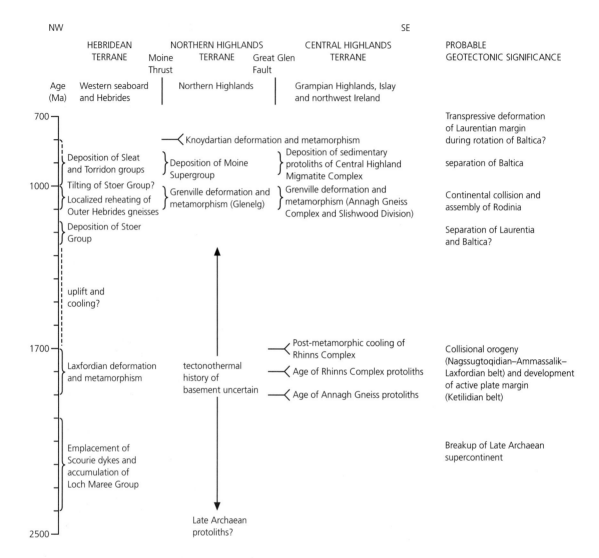

Fig. 4.14 Summary of Palaeoproterozoic to Early Neoproterozoic histories of the Hebridean, Northern Highlands and Central Highlands terranes in Scotland and Ireland. Note also the likely presence of unexposed Grenville basement southeast of the Highland Boundary Fault in Scotland.

interpreted as being very different (Fig. 4.14). This brings us to a rather more contentious issue — were the Moine Supergroup and underlying basement gneisses demonstrably part of Laurentia during the Early Neoproterozoic, as boldly implied by the title of this chapter? The traditional interpretation is that the basement to the Moine sediments is part of the Lewisian Complex of the Hebridean Terrane, and that

the Moine sediments accumulated in rift basins along the margin of Laurentia (Fig. 4.6d). However, this need not necessarily be the case. There is no single line of evidence that demonstrates conclusively correlation of the sub-Moine basement with the Lewisian Complex. Notably, there is no evidence elsewhere along the Laurentian margin (e.g. East Greenland, north-east Canada) for any orogeny at *c.* 800 Ma. Accordingly, some workers have argued that the Moine rocks and underlying basement are instead a fragment of either Baltica or South America that was accreted to the Laurentian margin, perhaps as a result of strike-slip faulting during the rotation of Baltica between *c.* 870 and *c.* 750 Ma. According to this interpretation, accretion of this exotic terrane to Laurentia must have occurred after the Knoydartian Orogeny at c. 800 Ma, but before the deposition of the Late Neoproterozoic–Cambrian Dalradian Supergroup, which is tied firmly to Laurentia on the basis of faunal evidence (see Chapter 5). Evidence for the c. 800 Ma Porsanger Orogeny in northern Norway opens up the possibility that the Moine rocks and associated basement may have been derived from Baltica.

4.12 Summary

There is strong evidence that, during the Proterozoic, continents moved about relative to each other with repeated assembly, dispersal and reassembly of supercontinents in various configurations. Supercontinents appear to have existed at approximately 2.5, 1.8 and 1.1–1.0 Ga, showing a periodicity of assembly and major orogenic activity of c. 0.7–0.8 Ga. Even though the northern British Isles represents a very small crustal block and there are substantial gaps in the geological record, the various basement complexes and sedimentary sequences together provide a more or less complete picture of the global reorganizations of continents that occurred between c. 2.5 Ga and c. 0.75 Ga.

References

Anderton, R., Bridges, P.H., Leeder, M.R. & Sellwood, B.W. (1979) *A Dynamic Stratigraphy of the British Isles*, pp. 1–301. Unwin Hyman, London.

Beach, A., Coward, M.P. & Graham, R.H. (1974) An interpretation of the structural evolution of the Laxford Front, north-west Scotland. *Scottish Journal of Geology* 9, 297–308.

Nicholson, P.G. (1993) A basin reappraisal of the Proterozoic Torridon Group, northwest Scotland. In: *Tectonic Controls and Signatures in Sedimentary Successions* (eds L. E. Frostick & R. J. Steel), pp. 183–202. Blackwell Scientific Publications, Oxford.

Park, R.G. (1991) The Lewisian. In: *Geology of Scotland* (ed. G. Y. Craig), 3rd edn, pp. 25–64. Geological Society, London.

Park, R.G. & Tarney, J. (1987) The Lewisian Complex: a typical Precambrian high-grade terrain? In: *Evolution of the Lewisian and Comparable Precambrian High Grade Terrains* (eds R. G. Park & J. Tarney), Special Publication 27, pp. 13–25. Geological Society, London.

Soper, N.J., Harris, A.L. & Strachan, R.A. (1998) Tectonostratigraphy of the Moine Supergroup: a synthesis. *Journal of the Geological Society, London* 155, 13–24.

Stewart, A.D. (1982) Late Proterozoic rifting in NW Scotland: the genesis of the 'Torridonian'. *Journal of the Geological Society, London* 139, 413–420.

Watson, J.V. (1983) Lewisian. In: *Geology of Scotland* (ed. G. Y. Craig), 2nd edn, pp. 23–47. Geological Society, London.

Further reading

Bluck, B.J., Dempster, T.J. & Rogers, G. (1997) Allochthonous metamorphic blocks on the Hebridean passive margin, Scotland. *Journal of the Geological Society, London*, 154, 921–924. [Questions the Laurentian affinities of the Moine Supergroup and basement inliers.]

Holdsworth, R.E., Strachan, R.A. & Harris, A.L. (1994) Precambrian rocks in northern Scotland east of the Moine Thrust: the Moine Supergroup. In: *A Revised Correlation of Precambrian Rocks in the British Isles* (eds W. Gibbons & A. L. Harris), Special Report 22, pp. 23–32. Geological Society, London. [Overview of the Moine Supergroup and associated basement inliers.]

Muir, R.J., Fitches, W.R., Maltman, A.J. & Bentley, M.R. (1994) Precambrian rocks of the southern Inner Hebrides–Malin Sea region: Colonsay, West Islay, Inishtrahull and Iona. In: *A Revised Correlation of Precambrian Rocks in the British Isles* (eds W. Gibbons & A. L. Harris), Special Report 22, pp. 54–58. Geological Society, London. [Overview of the Rhinns Complex basement.]

Nicholson, P.G. (1993) A basin reappraisal of the Proterozoic Torridon Group, northwest Scotland. In: *Tectonic Controls and Signatures in Sedimentary Successions* (eds L. E. Frostick & R. J. Steel), pp. 183–202. Blackwell Scientific Publications, Oxford. [New ideas on the sedimentation of the Torridon Group.]

Park, R.G. (1991) The Lewisian. In: *Geology of Scotland* (ed. G. Y. Craig), 3rd edn, pp. 25–64. Geological Society, London. [Overview of the Lewisian Complex.]

Park, R.G. (1992) Plate kinematic history of Baltica during the Middle to Late Proterozoic: a model. *Geology* 20, 725–728. [Plate tectonic model mainly for the Mesoproterozoic evolution of Baltica and Laurentia.]

Park, R.G. (1994) Early Proterozoic tectonic overview of the northern British Isles and neighbouring terrains in Laurentia and Baltica. *Precambrian Research* 68, 65–79. [Overview of Palaeoproterozoic tectonics in the British Isles and adjacent regions.]

Park, R.G., Tarney, J. & Connelly, J.N. (2000) The Loch Maree Group: Palaeoproterozoic subduction–accretion complex in the Lewisian of NW Scotland. *Precambrian Research*.

Smith, M., Robertson, S. & Rollin, K.E. (1999) Rift basin architecture and stratigraphical implications for basement–cover relationships in the Neoproterozoic Grampian Group of the Scottish Caledonides. *Journal of the Geological Society, London* **156**, 1163–1173. [Discussion of geological relationships in the Central Highlands.]

Soper, N.J., Harris, A.L. & Strachan, R.A. (1998) Tectonostratigraphy of the Moine Supergroup: a synthesis. *Journal of the Geological Society, London* **155**, 13–24. [Basin model for the Moine Supergroup.]

Stewart, A.D. (1991) Torridonian. In: *Geology of Scotland* (ed. G. Y. Craig), 3rd edn, pp. 66–85.

Geological Society, London. [Overview of the Torridonian, slightly superseded by Nicholson 1993 above.]

Vance, D., Strachan, R.A. & Jones, K.A. (1998) Extensional versus compressional settings for metamorphism: Garnet chronometry and pressure–temperature–time histories in the Moine Supergroup, northwest Scotland. *Geology* **26**, 927–930. [Isotopic evidence for a Knoydartian orogeny.]

Windley, B.F. (1995) *The Evolving Continents*, 3rd edn, pp. 1–526. John Wiley, Chichester. [A comprehensive overview of Earth history with major sections on the Proterozoic.]

5

Late Neoproterozoic (<750 Ma) to Early Ordovician passive margin sedimentation along the Laurentian margin of Iapetus

R. A. STRACHAN AND R. E. HOLDSWORTH

5.1 Introduction

The Late Neoproterozoic was a period of major tectonic, climatic, biogeochemical and biological change. The previous chapter detailed how the collision of Laurentia, Baltica and Amazonia at *c.* 1.1–1.0 Ga resulted in formation of the supercontinent *Rodinia*. The Late Neoproterozoic break-up of Rodinia resulted from the separation of East Antarctica from western Laurentia and was followed by the amalgamation of East and West Gondwana blocks to form the Vendian supercontinent (see Chapter 2). Other important Late Neoproterozoic events include extensive and repeated glaciations and a marked increase in the oxygen concentration of the hydrosphere and atmosphere. It seems increasingly likely (although not understood in detail) that all these phenomena are interrelated and that plate tectonics and continental movements influenced major climatic and biogeochemical cycles (see Chapter 2). The increased rapidity of build-up of oxygen led to the evolution of the first *metazoans* in the form of the *Ediacara* fauna, which appeared by *c.* 570 Ma. These were complex, multicellular, soft-bodied organisms that required oxygen for their growth; the first hard-bodied organisms appeared a little later at approximately the Precambrian–Cambrian boundary (544 Ma).

The Late Neoproterozoic break-up of the Vendian supercontinent resulted in the separation of Laurentia, Baltica and Amazonia to form the Iapetus Ocean (Fig. 5.1). Extensive passive margin sedimentary sequences were deposited along the eastern side of Laurentia during a prolonged period of continental rifting and ocean widening, which lasted from *c.* 750 Ma until the early Ordovician. In the British Isles, this rifting history is recorded in two separate rock sequences in Scotland and Ireland (Fig. 5.2). On the Caledonian foreland of north-west Scotland, a *Cambrian–Ordovician shelf succession* (Fig. 5.2b) can be correlated with very similar sequences in East Greenland and north-west Newfoundland (Fig. 5.2c). To the south-east, within the Caledonian orogenic belt and exposed widely between Shetland and Ireland, is a thick sequence of deformed and metamorphosed Late Neoproterozoic to Early Ordovician shallow- and deep-water sediments and volcanics, the *Dalradian Supergroup*. This is very similar in age to the Fleur de Lys Supergroup in Newfoundland and the Eleonore Bay Supergroup in East Greenland (Fig. 5.2c). The Cambrian–Ordovician and Dalradian rocks and their correlatives formed two subparallel sedimentary belts located along the eastern margin of Laurentia, with the shelf Cambrian–Ordovician rocks lying on the landward side of the generally deeper water Dalradian-type lithologies. These two sedimentary belts are now in much closer proximity due to the effects of crustal shortening during the Grampian and Caledonian orogenies. Oceanic crust is presumed to have existed to the south-east of the present Dalradian outcrop.

5.2 Basin history of the Dalradian Supergroup

5.2.1 Lithostratigraphy

The Dalradian Supergroup is located mainly in the Central Highlands Terrane between the Great Glen Fault and the Fair Head–Clew Bay Line–Highland Boundary Fault (Fig. 5.2). The exception to this is the Dalradian outlier of the Connemara Terrane in west

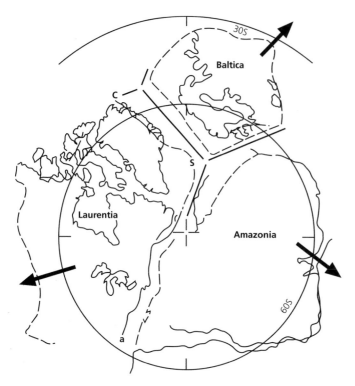

Fig. 5.1 Reconstruction of Baltica, Amazonia and Laurentia (B, Az, L) in the late Neoproterozoic; a–c, western margin of Appalachian–Caledonian orogen; S, Scotland; EG, East Greenland; N, Newfoundland; arrows indicate directions of relative movement of continental blocks during rifting and opening of the Iapetus Ocean (redrawn from Soper 1994).

Ireland (Fig. 5.2). These rocks are thought to have been part of the main Dalradian outcrop, which was detached and moved into its present position during strike-slip faulting at a late stage of the Ordovician Grampian Orogeny (see Chapter 6).

The Dalradian sediments were deposited as a sequence of marine sands, silts, muds and limestones, which are subdivided into the *Grampian, Appin, Argyll* and *Southern Highland groups* (Fig. 5.3). The apparent total thickness of the Dalradian succession is at least ~25 km, although it is unlikely that the complete vertical thickness was ever deposited at one place. It is more likely that the depocentre migrated south-eastwards with time. Although the groups are mainly lithostratigraphic divisions, in part they may also represent chronostratigraphic units. The base of the Argyll Group is defined by the Port Askaig Tillite (Fig. 5.3), and the top of the Argyll Group by large quantities of basic volcanics (Fig. 5.3). These horizons are thought to correspond, respectively, to important periods of glaciation and rift-related magmatism, and probably represent events that were broadly synchronous on the scale of the Dalradian Basin. Large parts of the Dalradian (e.g. south-west Scottish Highlands) have only been affected by low-grade metamorphism and are not strongly deformed, with the result that the sedimentological history and basin evolution are well understood. Lithological variations indicate periods of basin deepening and shallowing which, by analogy with Phanerozoic basins, may correspond to multiple periods of rifting and thermal subsidence (Fig. 5.3). It seems likely that in north-west Ireland and south-west Scotland, the succession was deposited on basement similar to the Rhinns and Annagh Gneiss complexes and the Slishwood Division, although all contacts are now faulted. In the north-east Grampian Highlands of Scotland, the Dalradian may have been deposited unconformably on the Central Highland Migmatite Complex (but see discussion in Chapter 4).

5.2.2 Age of the Dalradian Supergroup

The age of the middle to upper part of the Dalradian succession is constrained by isotopic dating and occasional fossils. A latest Neoproterozoic (Vendian) age for a large part of the succession is indicated by correlation of the Port Askaig Tillite with the Varanger Tillite in Norway, the age of which has been estimated at *c.* 650 Ma. Volcanics occurring at the top of

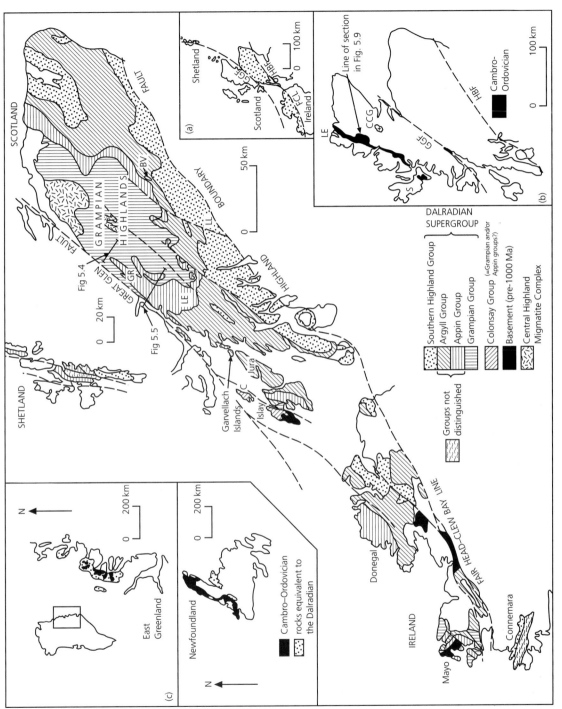

Fig. 5.2 Simplified outcrop map of the Dalradian rocks of Shetland, Scotland and Ireland (various younger rocks omitted for clarity). BV, Ben Vurich granite; LL, location of Leny Limestone. Insets are as follows: (a) Relative locations of Shetland, Scotland and Ireland with Dalradian rocks shown in dotted ornament; FCL, Fair Head–Clew Bay Line; GGF, Great Glen Fault; HBF, Highland Boundary Fault; LE, Loch Eriboll; S, Skye; LE, Loch Eriboll; CCG, Carn Chainneag Granite. (b) Location of Cambrian–Ordovician rocks in north-west Scotland. The rocks equivalent to the Dalradian in East Greenland and north-west Newfoundland. (c) Location of equivalent rocks in East Greenland and north-west Newfoundland. The rocks equivalent to the Dalradian in East Greenland are the Eleonore Bay Supergroup and the Tillite Group and in Newfoundland the Fleurs de Lys Supergroup.

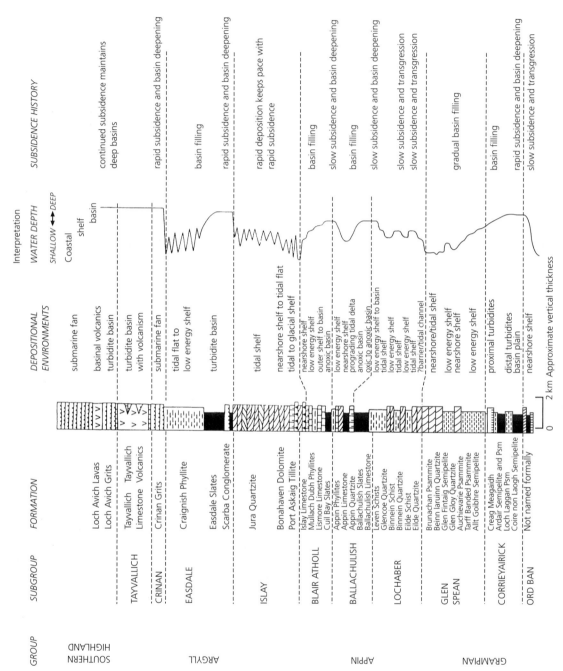

Fig. 5.3 Dalradian stratigraphy and sedimentary evolution (modified from Anderton 1985). Note that the thickness of the Ord Ban Group has been exaggerated in the interests of clarity.

the Argyll Group have been dated at 594 ± 4 Ma, thus constraining deposition of the Argyll Group to a maximum time-span of *c*. 55 Ma. The uppermost part of the Southern Highland Group is Lower Palaeozoic, as shown by the presence of late Lower Cambrian trilobites in the Leny Limestone (see Fig. 5.2 for location) and Lower Ordovician acritarchs in the MacDuff Slates of north-east Scotland. The age of the Grampian and Appin groups is less well constrained, but the general consensus is that the base of the Grampian Group was probably deposited no earlier than *c*. 750 Ma.

5.2.3 The Grampian Group: early rifting and basin development

The main outcrop of the group is in the Grampian Highlands of Scotland (Fig. 5.2); it has also been recognized on Islay and in western Ireland, but is absent in Shetland. The fault-bounded metasedimentary rocks of the Colonsay Group exposed on the islands of Islay and Colonsay (Fig. 5.2) may also, at least in part, belong to the Grampian Group. The Grampian Group is composed mainly of micaceous to quartzose psammites and semipelites, which may total *c*. 7–8 km in thickness. The group is lithologically more uniform overall than the well-differentiated metasedimentary rocks of the Appin and Argyll groups, and regional subdivision has only recently been attempted. Sedimentary structures are common locally, despite widespread ductile deformation and greenschist to amphibolite facies metamorphism; they include grading, cross-bedding, loading and water escape structures, scouring, mudflake breccias and internal laminations and ripples.

The Grampian Group in Scotland has been subdivided into three units, the Ord Ban, Corrieyairick and Glen Spean subgroups. The Ord Ban Subgroup comprises a thin (<100 m) assemblage of meta-limestone and pelite (and intrusive amphibolite) immediately overlying parts of the Central Highland Migmatite Complex. Sedimentological detail is lacking, but the assemblage has the overall lithological characteristics of a shallow marine shelf succession. A major period of rifting and basin deepening is indicated by deposition of the overlying Corrieyairick Subgroup, which has been interpreted as a sequence of turbiditic sandstones. Thickness variations indicate the existence of a major syn-sedimentary normal fault within the Cor-

rieyairick basin (Fig. 5.4a), which probably trended between north and north-east, approximately parallel to the present orogenic trend. The upper Glen Spean Subgroup is a mixed sequence of semipelite and psammite with quartzite thought to represent deltaic and shallow marine tidal sediments, which gradually prograded south-eastwards to infill the former turbidite-dominated basin (Fig. 5.4b). The Grampian Group succession thus preserves a sequence very similar to that observed in Phanerozoic basins where basin shallowing and the diminution of syn-sedimentary faulting accompany the transition from rift to thermally induced subsidence.

5.2.4 The Appin Group: stable platform sedimentation

The Appin Group is dominated by quartzites, lime-

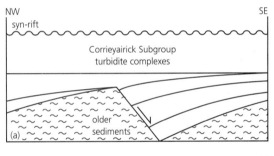

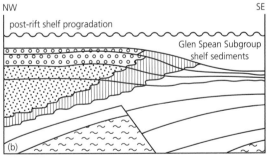

Fig. 5.4 Schematic summary of the basin evolution of the Grampian Group in the Grampian Highlands (see Fig. 5.2 for location; modified from Glover *et al.* 1995). (a) Syn-rift turbidite sedimentation produced major thickness variations in the Corrieyairick Subgroup (unornamented). (b) Latest Grampian Group times characterized by south-eastward progradation of the shelf sediments of the Glen Spean Subgroup. In (b), open circles = Brunachan Psammite (coastal sands and muds); dots = Beinn Iaruinn Quartzite (nearshore sands); dashed lines = Glen Fintaig Semipelite (offshore muds).

stones, semipelites and pelites, which were deposited on a marine shelf. The group has been subdivided into three subgroups; lateral facies changes occur locally, but many units can be traced over large distances. Correlations between local successions have therefore been made throughout Ireland and the Grampian Highlands, mainly as a result of detailed mapping, but also using the distinctive chemical signatures of certain units.

The complex interdigitation of uppermost Grampian and lowermost Appin Group (Lochaber Subgroup) units indicates that the boundary between the two groups is transitional and diachronous. Basal Lochaber Subgroup sedimentation appears to represent a renewed marine transgression following upper Grampian Group shoaling. Large-scale, cyclic alternations of quartzites and muddy sediments are thought to correspond to episodes of rifting and thermal subsidence (Fig. 5.3), although in all cases the rate of change of water depth and the rate of stretching were probably gradual. Thick sand wedges such as the Binnein and Glencoe quartzites are characterized by cross-stratification (commonly herring-bone), ripple marks and erosional features, and were probably deposited on a tidal shelf. Occasional mud cracks and rain pits indicate intermittent emergence. Palaeocurrents flowed north-east–south-west and the quartzites all fine and become less feldspathic towards the north-east. By analogy with modern tidal shelves, these facies variations and palaeocurrents suggest deposition in a gulf which opened to the north-east and had a major sediment input from the south-west.

Upper parts of the Appin Group (Ballachulish and Blair Atholl subgroups) are dominated by extensive offshore carbonate and anoxic mudstone sequences deposited during a period of low sediment supply, possibly during a widespread phase of thermal subsidence and basin widening. The Islay Limestone, which defines the top of the Appin Group, marks a return to shallow to intertidal conditions. It contains various features indicative of deposition in a warm, arid coastal setting, including stromatolites, flake breccias, ooids and possible pseudomorphs after an evaporite mineral. Facies and thickness variations in the south-west Scottish Highlands suggest that the Appin Group was deposited in a series of half-graben basins, very similar to that already proposed for the Grampian Group basin. It is thought that major south-east-dipping normal faults located in the Great Glen and Loch Etive–Glen Roy areas bounded basins during Ballachulish and Blair Atholl subgroup times (Fig. 5.5).

5.2.5 The Port Askaig Tillite: evidence for late Precambrian glaciation

The tillite at the base of the Argyll Group (Islay Subgroup) is thought to have formed during the late Neoproterozoic Varangian glaciation. The Port Askaig Tillite can be traced from west Ireland to north-east Scotland, but is not found in Shetland. It is *c.* 750 m thick in its type area and consists of a sequence of marine sandstones, siltstones, conglomerates and dolostones with numerous boulder beds (diamictites). These latter beds range from 0.5 m to 65 m in thickness and contain boulders up to 2 m in diameter. Spectacular outcrops on the Garvellach

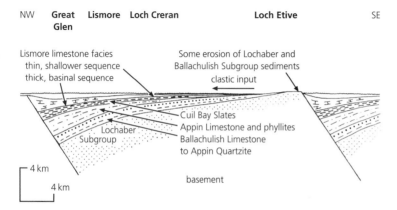

Fig. 5.5 A tectonic cross-section for the Lochaber to Lismore area during deposition of the upper Appin Group (see Fig. 5.2 for location; after Anderton 1985).

Islands (Fig. 5.2) include the 'Great Breccia', which contains dolostone rafts up to 320 m long. The lower boulder beds contain clasts of dolostone, presumably eroded from underlying Dalradian carbonate units. The upper boulder beds contain abundant granitic clasts, which are isotopically and chemically similar to the nearby basement of the Rhinns Complex.

All workers are agreed that the boulder beds are glacial in origin, although whether they were deposited by grounded ice sheets or floating ice has been controversial; quite different interpretations have been proposed for the same glacio-sedimentary sequence (Fig. 5.6). One interpretation views these sequences as resulting from repeated glacioeustatic sea-level changes, which led to numerous periods of direct glaciation by grounded ice, deposition of the boulder beds as tillites and subsequent glacial weathering. An alternative view is that the boulder beds were deposited as greatly increased concentrations of suspended sediment from floating ice rafts derived from continental ice sheets in a nearby coastal region. Polygonal structures developed on the surface of some boulder beds and interpreted initially as subaerial, periglacial features may in fact result from subaqueous soft-sediment deformation.

Overlying the Port Askaig Tillite, the shallow water to intertidal dolomitic shales, siltstones and sandstones of the Bonahaven Dolomite indicate a rapid return to a warm, arid climate. Domal stromatolites include quartz–calcite nodules which may be pseudomorphs after anhydrite. The abundant chemically precipitated carbonate within the dolomite is also an important indicator of aridity. The evidence for warm, arid conditions both above and below the Port Askaig Tillite has been taken to indicate that the Dalradian Basin was located at low latitudes during the Late Neoproterozoic and that the tillite was the result of a unique, almost worldwide, glaciation. Glacial deposits of similar age are found in North America, East Greenland, Russia, China and Africa, where there is also evidence that glaciation interrupted the deposition of warm-water sediments.

In order to place the Varangian glaciation in context, it is necessary to understand that the Neo-

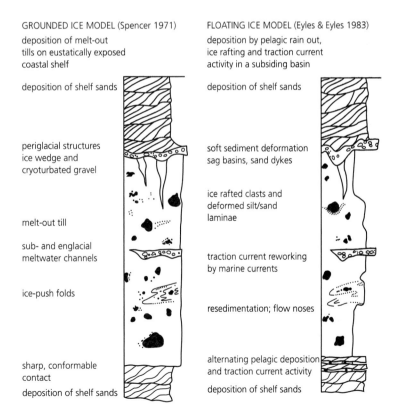

GROUNDED ICE MODEL (Spencer 1971)
deposition of melt-out
tills on eustatically exposed
coastal shelf

deposition of shelf sands

periglacial structures
ice wedge and
cryoturbated gravel

melt-out till

sub- and englacial
meltwater channels

ice-push folds

sharp, conformable
contact
deposition of shelf sands

FLOATING ICE MODEL (Eyles & Eyles 1983)
deposition by pelagic rain out,
ice rafting and traction current
activity in a subsiding basin

deposition of shelf sands

soft sediment deformation
sag basins, sand dykes

ice rafted clasts and
deformed silt/sand
laminae

traction current reworking
by marine currents

resedimentation; flow noses

alternating pelagic deposition
and traction current activity

deposition of shelf sands

Fig. 5.6 Contrasting models proposed by Spencer (1971) and Eyles & Eyles (1983) for the deposition of the Port Askaig Tillite in the Garvellach Islands (see Fig. 5.2 for location; redrawn from Eyles & Eyles 1983).

proterozoic was characterized by major atmospheric and climatic fluctuations. The Varangian glaciation was one, albeit the most extensive, of six major glaciations that occurred between *c.* 970 Ma and *c.* 570 Ma. These represent the most extensive period of glaciation(s) in Earth history, whose effects were much more widespread than subsequent Permo-Carboniferous or Quaternary glaciations. The precise cause of these glaciations, and the near worldwide Varangian glaciation in particular, is poorly understood although some clues are provided by studies of carbon isotope ratios in Neoproterozoic sediments. These indicate that each glaciation appears to correspond to a rapid decline in the amount of atmospheric carbon dioxide. As this gas is very effective in absorbing solar radiation, such a fall could have resulted in global cooling and glaciation. Large-scale tectonic cycles involving the amalgamation of supercontinents in the Neoproterozoic and the consequent reduction in magmatic activity (and hence output of carbon dioxide into the atmosphere) are likely to be the ultimate cause of the glaciations (see discussion in Chapter 2, pp. 22–23).

5.2.6 The Argyll Group: widespread fault-controlled sedimentation

Following deposition of the Bonahaven Dolomite, a marine transgression resulted in deposition of the thick tidal shelf sands of the Jura Quartzite. North-north-east-directed palaeocurrents probably flowed parallel to a contemporary coastline to the north-west. Exotic pebbles, including haematitic quartzites and cherts, are unlike any in the Port Askaig Tillite and were most probably derived from Archaean or Proterozoic banded iron formations on the Caledonian foreland. The uniform nature of the sequence, which is up to 5 km thick, implies a delicate balance between subsidence rate and deposition.

Thickness and facies variations in the Dalradian rocks below the Jura Quartzite are generally gradual and can be accounted for by differential subsidence of a relatively stable continental shelf. By contrast, above the Jura Quartzite, the Easdale and Crinan subgroups are characterized by variations that are much more marked and probably the result of the onset of active syn-depositional faulting. Following deposition of the Jura Quartzite, a major pulse of rapid stretching and subsidence resulted in formation of a series of fault-bounded, deep-water turbidite basins. The Scarba Conglomerate and Easdale Slates represent, respectively, submarine fan turbidites and deep-water muds deposited during this event. Marked thickness variations within all sedimentary units provide evidence for syn-depositional faults that trended both parallel and transverse to basin margins (Fig. 5.7). The interbasinal highs are marked by local thinning of units, facies changes, erosion and non-deposition. Syngenetic barium and base metal mineralization in the Easdale Slates in parts of the Grampian Highlands is probably related to the regional crustal extension. This process provided deep basins, in which exaled brines could be ponded, and also reactivated faults that permitted sea water to leach metals from the underlying sediments via hydrothermal convection cells.

The deep marine basins gradually filled up as shown by the succeeding Craignish Phyllite, which includes shallow marine and tidal flat facies with gypsum pseudomorphs. Earthquake-induced liquefaction structures are frequent within upper parts of the Craignish Phyllite and herald the onset of another rapid rifting event. The shallow-water sediments are overlain directly by submarine fan turbidites, the Crinan Grits, indicating renewed rapid subsidence, presumably as a result of faulting. Palaeocurrents indicate that a major deep-water basin extended north-eastwards from north-west Ireland. The basin was fed axially from both ends, although the overall fining towards the north-east suggests a major input from the south-west. The mineralogical maturity of most of the sandstones of the Appin and Argyll groups probably reflects their derivation from a largely (meta)sedimentary source area.

The uppermost part of the Argyll Group comprises the Tayvallich Subgroup. The base of the unit is defined by the Tayvallich Limestone, which has been traced from north-west Ireland to north-east Scotland. The limestones are mainly turbiditic and were probably deposited in pre-existing Crinan Subgroup basins following a reduction in supply of clastic sediment to the adjacent shelves. The overlying Tayvallich Volcanic Formation is a sequence of basaltic pillow lavas and hyaloclastite and airfall tuffs with interbedded deep-water turbiditic limestones and black pelites. An eruption age of 595 ± 4 Ma has been obtained from U–Pb dating of zircons of a keratophyre, which forms part of the volcanic suite. The

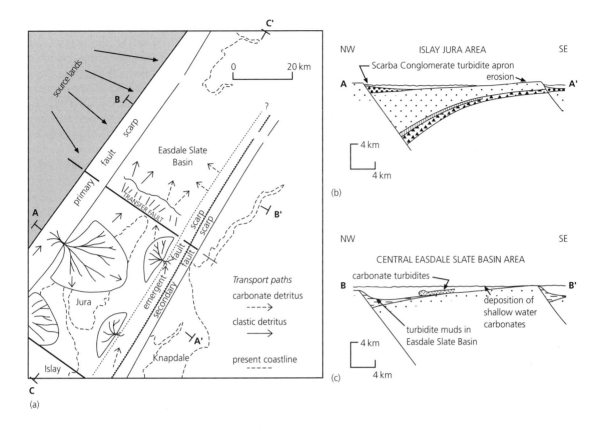

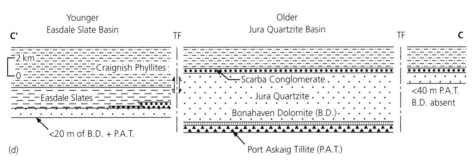

Fig. 5.7 (a) Suggested palaeogeography for the Argyll seaboard area in the vicinity of Jura during Scarba Conglomerate to Easdale Slate times, showing location of submarine fans and directions of sediment transport; (b,c) tectonic cross-sections for early Easdale Subgroup times (see (a) for locations of sections); (d) north-east–south-west longitudinal section of the Islay and Easdale subgroups in the area of map (a) showing how movement along transfer faults can account for thickness variations (the latter shown schematically) (modified from Anderton *et al.* 1979; Anderton 1985).

Tayvallich volcanism began with deposition of thin tuffs and then by the injection of up to 3 km cumulative thickness of basaltic sills and dykes. These were injected progressively from bottom to top of the sedimentary pile at shallow depths into wet sediment.

The intrusions have a tholeiitic chemistry and most likely resulted from mantle melting and magma migration upwards along deep-seated faults during crustal rifting.

A prominent north-west–south-east trending geophysical feature, the *Cruachan Lineament* (Fig. 5.8), coincides with the north-east limit of abundant basic magmatism and may correspond to a major structure within the Dalradian basin, transverse to the regional north-east–south-west structural trend. Many of the tholeiitic feeder dykes on Jura have a north-west–south-east trend when the effects of later deformation are removed. It is possible that these transverse trends may reflect a transtensional component to the rifting history, with the Tayvallich Vol-

canics confined to a pull-apart basin bounded by oblique-slip rifts and the Cruachan Lineament (Fig. 5.8). A similar deep-seated fault is recognized in north-west Ireland where the Donegal Lineament (Fig. 5.8) apparently controlled the emplacement of basic magmas and the distribution of sedimentary facies within the Appin and Argyll groups. Other transverse structures in Scotland, such as the Ossian, Rothes, Glenlivet and Deeside lineaments (Fig. 5.8), also coincide with facies changes in the Dalradian

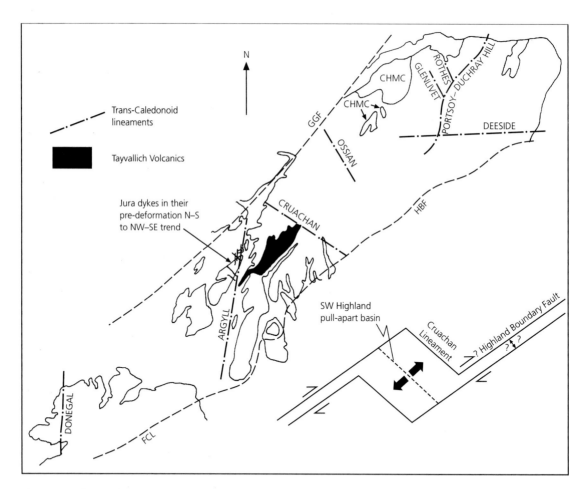

Fig. 5.8 Map showing the main trans-Caledonoid lineaments in Scotland and Ireland that are thought to have been active as faults during Dalradian sedimentation (data from Fettes *et al.* 1986; Hutton & Alsop 1996). Map also shows the present extent of the Tayvallich Volcanics and the predeformation trend of basic dykes in the Jura area (modified from Soper 1994). CHMC,

Central Highland Migmatite Complex; GGF, Great Glen Fault; FCL, Fair Head–Clew Bay Line; HBF, Highland Boundary Fault. Diagram below right shows the geometry of a transtensional, pull-apart basin which may have controlled the location of the Tayvallich Volcanics (redrawn from Graham 1986).

succession and therefore may also have originated as transtensional faults that were active during sedimentation.

5.2.7 The Southern Highland Group: deep-water turbidite basins

The Southern Highland Group has been traced from Mayo in west Ireland to north-east Scotland. The sedimentology of the muds, sands and gravels of the group is poorly known, partly because of the strongly deformed nature of these rocks. The finer sediments are generally strongly cleaved and reveal little about their depositional environment. By contrast, the sands and gravels often show the characteristics of submarine fan facies, with thick (up to 15 m), variably graded, laterally impersistent beds with channelled bases. Palaeocurrents flowed mainly to the south-east, indicating a north-westerly source region. Minor currents which flowed north-east–south-west are probably axial flows trending along the length of the basin. These sediments are characterized by a marked increase in the proportion of feldspar. This need not indicate that these were derived from a different source area to underlying Dalradian sediments. It is possible that by Southern Highland Group times, the source area had been stripped of its sedimentary cover to expose a gneissic or igneous basement. Sedimentation was periodically accompanied by basic volcanism represented by pillow lavas. There is little evidence for active faulting and sedimentation probably accompanied thermal subsidence. During Cambrian to Lower Ordovician times, the Southern Highland Group was most likely located on a sediment-starved outer continental margin, which was slowly accumulating hemipelagic muds and deep water, trilobite-bearing carbonates such as the Leny Limestone. Evidence for a Late Cambrian–Early Ordovician glacial episode is provided by dropstones and diamictites preserved in deep-water Southern Highland Group rocks (MacDuff Slates) in north-east Scotland.

In Scotland, Argyll Group rocks were intruded by the Ben Vurich granite (Fig. 5.2), which has yielded a U–Pb zircon age of 590 ± 2 Ma. This is similar to the *c.* 580 Ma U–Pb zircon age obtained for emplacement of the Carn Chuinneag granite into the Moine rocks of Northern Scotland (Fig. 5.2b). The close correspondence of these ages to that obtained for the Tayvallich

Volcanics suggests that these granites may also have been produced during mantle melting and rifting. Granites form important parts of rift-related igneous suites in contemporary zones of crustal extension such as the Red Sea and the East African Rift Valley.

5.3 Cambrian–Ordovician shelf sedimentation in north-west Scotland

Cambrian–Ordovician marine sediments crop out on the Caledonian foreland and within the Moine Thrust Zone between Loch Eriboll and Skye (Fig. 5.2b). On the foreland, these sediments overlie a remarkably planar unconformity, which truncates the Lewisian Complex and the Torridon Group. West of the Moine Thrust Zone in the Assynt area, Cambrian sediments progressively overstep the underlying Torridon Group eastwards, to rest on the Lewisian Complex (Fig. 5.9a). Cambrian sediments also overlie unconformably Lewisian gneisses and the Torridon Group further east in the Ben More thrust sheet. A restored section across the area indicates that the Torridon Group was probably disposed in two half-grabens prior to deposition of the Cambrian succession (Fig. 5.9b). The easterly dipping normal faults that are thought to have bounded the half-grabens probably resulted from crustal extension across the Laurentian margin as the Dalradian basin developed further to the south-east.

The Cambrian–Ordovician succession consists of a basal clastic sequence (*Eriboll Sandstone* and *An t-Sron* formations), up to 250 m thick, which is overlain by over 1000 m of Cambrian–Ordovician carbonates (*Durness Formation*) (Fig. 5.10). The present north-east–south-west outcrop pattern is inferred to be broadly parallel to a contemporary shoreline to the north-west. The succession overall records a marine transgression (Fig. 5.11). The Eriboll Sandstone mainly comprises cross-bedded quartz arenites, which show an upward transition from a barrier island tidal inlet and shoreface sequence to a tidal shelf sequence. Bimodal and polymodal palaeocurrent patterns reflect complex interactions between tidal currents and wave activity. Low-angle erosion surfaces and winnowed granule lags may represent the effects of shelf erosion during major storms. The upper part of the unit (Pipe Rock Member) contains abundant vertical burrows (*Skolithos* and *Monocraterion*). A general reduction

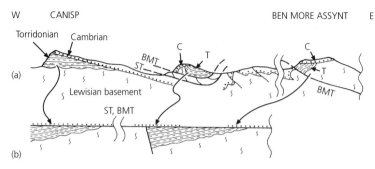

Fig. 5.9 (a) Schematic cross-section of the Caledonian foreland and Moine Thrust Zone in the Assynt area (see Fig. 5.2b for location) showing the overstep relationships discussed in the text; (b) a restored section following removal of the effects of Caledonian thrusting, and showing the inferred late Neoproterozoic faults that are thought to have developed prior to the basal Cambrian transgression (modified from Soper & England 1995). C, Cambrian; T, Torridonian; ST, Sole Thrust; BMT, Ben More Thrust.

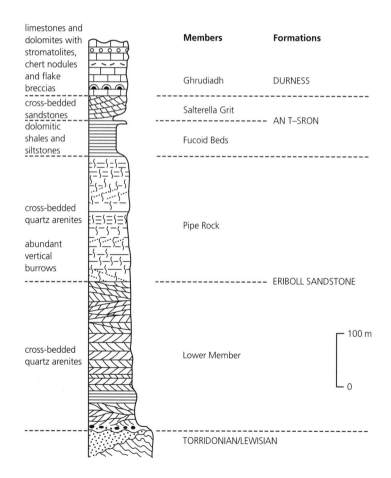

Fig. 5.10 Generalized sedimentary log through the Cambrian–Ordovician rocks of north-west Scotland (redrawn from McKie 1990).

NW   SE

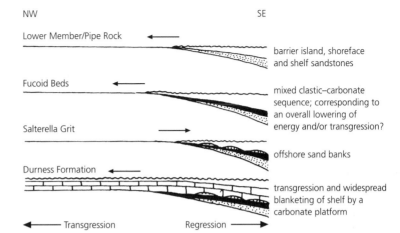

Lower Member/Pipe Rock ← ───

barrier island, shoreface
and shelf sandstones

Fucoid Beds ← ───

mixed clastic–carbonate
sequence; corresponding to
an overall lowering of
energy and/or transgression?

Salterella Grit ───→

offshore sand banks

Durness Formation ← ───

transgression and widespread
blanketing of shelf by a
carbonate platform

Fig. 5.11 Simplified sedimentary
evolution of the Cambrian–Ordovician
succession of north-west Scotland
(redrawn from McKie 1990).

← ── Transgression   Regression ──→

in tidal strength through the Pipe Rock is indicated by
progressive thinning of bed sets, suggesting a gradual
transgression and deposition of more distal sands.
This transgressive event culminated in deposition of
the lower part of the overlying An t-Sron Formation,
which consists of mixed carbonate and clastic storm-
dominated sediments (Fucoid Beds). Olenellid trilobites within this unit indicate a lower Cambrian age.
The upper part of the An t-Sron Formation (Salterella
Grit) comprises cross-bedded and bioturbated sandstones, which were probably deposited as offshore
sandbanks during a regression. Clastic deposition
ceased and the shelf was blanketed by a widespread
transgressive carbonate platform represented by the
limestones and dolomites of the Durness Formation.
Well-preserved flake breccias, stromatolites, oolites
and pseudomorphs of evaporite minerals suggest arid
conditions and deposition in tidal flat, sabkha and
shallow subtidal environments. A sparse fauna of gastropods, cephalopods and conodonts indicates a
lower Ordovician age (Arenig?–Llanvirn) for upper
parts of the sequence. The negligible clastic input
during carbonate deposition is consistent with low-
lying relief within interior parts of Laurentia and the
sedimentological evidence for an arid climate. The
presence of carbonates is itself an important indicator
of low-latitude deposition, and contrasts strongly
with the carbonate-free Cambrian–Ordovician successions being deposited on the high-latitude margin
of Gondwana (see Chapter 9).

The upward passage from basal shallow marine
clastics into intertidal to subtidal carbonates displayed by the Cambrian–Ordovician rocks of north-
west Scotland is almost identical to that seen in
correlative Laurentian margin sequences in East
Greenland and eastern North America. This remarkable similarity in these sequences suggests similar,
widespread controls on sediment deposition along
this margin of the developing Iapetus Ocean. A combination of several factors, including flexural downloading of the crust (due to the accumulated thickness
of Dalradian and equivalent successions), thermal
subsidence following rifting, and a eustatic sea-level
rise probably accounts for the overall transgressive
nature of the sequences.

5.4 Summary

The sedimentological and basin history of the Dalradian Supergroup is consistent with a prolonged
phase of lithospheric stretching leading to continental
rifting and eventually break-up to form the Iapetus
Ocean. The Grampian and Appin groups were
deposited on the relatively passive continental shelf
on the north-west side of the rift (Fig. 5.12a). Continued crustal extension widened and deepened the
rift during deposition of the Argyll Group (Fig.
5.12b). The appearance of large amounts of basic volcanic material towards the top of the Argyll Group
resulted from rupturing of the continental crust (Fig.
5.12c). Dalradian sedimentation was controlled
principally by the development of a series of normal
fault-bounded basins, which appear to have trended
subparallel to the present north-east–south-west oro-

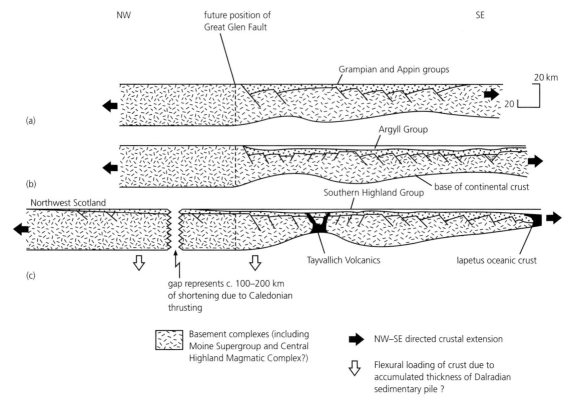

Fig. 5.12 Schematic cross-sections to show the progressive development of the rifted Laurentian margin in Scotland in: (a) late Appin Group times; (b) Crinan Subgroup (Argyll Group) times; (c) late Argyll Group to Southern Highland Group times (after Anderton 1985). The passive margin is thought to have widened progressively through time, mainly as a result of northwest–south-east-directed crustal extension (black arrows) combined with flexural downloading of the crust (open arrows), thermal subsidence and a eustatic sea-level rise. The inferred lateral passage from the shallow-water platform Cambrian–Ordovician sediments of the foreland to the deepwater Southern Highland Group depicted in (c) is no longer exposed, but is presumed to have formerly overlain the Moine Supergroup of north-west Scotland.

genic trend. Transverse basin structures such as the Cruachan and Deeside lineaments may also have exerted an important control on sedimentation.

The shallow marine Cambrian–Ordovician succession of north-west Scotland was contemporaneous with the deep-water hemipelagic muds of the Southern Highland Group. The stable shallow shelf conditions that characterized the early stages of the Dalradian Basin transgressed north-westwards through time as a result of continued rifting, subsidence due partly to flexural downloading, and a eustatic sea-level change. The Moine Supergroup of north-west Scotland is therefore presumed to once have been overlain by Cambrian–Ordovician marine rocks, which were formerly continuous to the south-east with the Dalradian basins (Fig. 5.12c). However, all traces of this Cambrian–Ordovician succession have been removed as a result of uplift and erosion during and after the Caledonian orogeny.

References

Anderton, R. (1985) Sedimentation and tectonics in the Scottish Dalradian. *Scottish Journal of Geology* **21**, 407–426.

Anderton, R., Bridges, P.H., Leeder, M.R. & Sellwood, B.W. (1979) *A Dynamic Stratigraphy of the British Isles*, pp. 1–301. Unwin Hyman, London.

Eyles, C.H. & Eyles, N. (1983) Glaciomarine model for upper Precambrian diamictites of the Port Askaig Formation, Scotland. *Geology* **11**, 692–696.

Fettes, D.J., Graham, C.M., Harte, B. & Plant, J.A. (1986) Lineaments and basement domains: an alternative view of Dalradian evolution. *Journal of the Geological Society, London* **143**, 453–464.

Glover, B.W., Key, R.M., May, F., Clark, G.C., Phillips, E.R. & Chacksfield, B.C. (1995) A Neoproterozoic multiphase rift sequence: the Grampian and Appin groups of the southwestern Monadhliath Mountains of Scotland. *Journal of the Geological Society, London* **152**, 391–406.

Graham, C.M. (1986) The role of the Cruachan Lineament during Dalradian evolution. *Scottish Journal of Geology* **22**, 257–270.

Hutton, D.H.W. & Alsop, G.I. (1996) The Caledonian strike-swing and associated lineaments in NW Ireland and adjacent areas: sedimentation, deformation and igneous intrusion. *Journal of the Geological Society, London* **153**, 345–360.

McKie, T. (1990) Tidal and storm influenced sedimentation from a Cambrian transgressive passive margin sequence. *Journal of the Geological Society, London* **147**, 785–794.

Soper, N.J. (1994) Neoproterozoic sedimentation on the NE margin of Laurentia and the opening of Iapetus. *Geological Magazine* **131**, 291–299.

Soper, N.J. & England, R.W. (1995) Vendian and Riphean rifting in NW Scotland. *Journal of the Geological Society, London* **152**, 11–14.

Spencer, A.M. (1971) *Late Precambrian Glaciation in Scotland*. Memoir 6. Geological Society, London.

Further reading

Anderton, R. (1985) Sedimentation and tectonics in the Scottish Dalradian. *Scottish Journal of Geology* **21**, 407–426. [An authoritative review of Dalradian sedimentation.]

McKie, T. (1990) Tidal and storm influenced sedimentation from a Cambrian transgressive passive margin sequence. *Journal of the Geological Society, London* **147**, 785–794. [Sedimentological analysis of the lower clastic part of the Cambrian–Ordovician succession of north-west Scotland.]

Soper, N.J. (1994) Neoproterozoic sedimentation on the NE margin of Laurentia and the opening of Iapetus. *Geological Magazine* **131**, 291–299. [Synthesis of late Proterozoic sedimentation in the North Atlantic region in relation to continental reconstructions.]

Stephenson, D. & Gould, D. (1995) *British Regional Geology: the Grampian Highlands*, 4th edn. HMSO (for the British Geological Survey), London. [Synthesis of Dalradian stratigraphy and sedimentation.]

Walton, E.K. & Oliver, G.J.H. (1991) Lower Palaeozoic—stratigraphy. In: *Geology of Scotland* (ed. G. Y. Craig), 3rd edn, pp. 161–193. Geological Society, London. [Includes a summary of the Cambrian–Ordovician succession of north-west Scotland.]

6 The Grampian Orogeny: Mid-Ordovician arc–continent collision along the Laurentian margin of Iapetus

R. A. STRACHAN

6.1 Introduction

In the Early Ordovician, north-west Britain consisted of a shallow carbonate shelf which passed south-eastwards into deep marine turbidite basins. Sedimentation was halted in the Mid-Ordovician by the Grampian Orogeny. This resulted in regional deformation and metamorphism of the Moine and Dalradian rocks of Scotland and Ireland to form a mountain chain that has supplied detritus for surrounding areas ever since. The orogeny marks the initial stages of narrowing of the Iapetus Ocean and probably results from the collision of the passive Laurentian margin with a volcanic arc developed above an intraoceanic subduction zone. The southern margin of autochthonous Laurentian crust is thought to correspond broadly to the Highland Boundary Fault–Fair Head–Clew Bay Line. The lower Palaeozoic rocks that lie south-east of this lineament within the Midland Valley Terrane include ophiolites, calc-alkaline igneous rocks and an accretionary prism, which together record the development of a subduction zone within the Iapetus Ocean between the Cambrian and Mid-Ordovician.

6.2 Grampian regional deformation and metamorphism

The Grampian Orogeny resulted in the dominant structures and metamorphic assemblages present in the Moine and Dalradian rocks of Scotland and Ireland (Fig. 6.1). Stratigraphic constraints on the timing of the Grampian Orogeny are good. The youngest rocks affected by the orogeny are upper-most Dalradian sediments, which may be as young as

Early Ordovician (Arenig). The oldest metamorphic detritus eroded from the developing Grampian mountain chain was deposited in the Mid-Llanvirn. This represents a time gap of $c.$ 15 Ma, and thus the Grampian Orogeny is a relatively short-lived event. It is convenient for the purposes of description of Grampian events in Scotland to use the Great Glen Fault as a dividing line, although it is not thought that the structure separates fundamentally different rock units.

6.2.1 Grampian events in Scotland north of the Great Glen Fault

The Moine Supergroup was affected by widespread ductile thrusting, folding and development of foliations and lineations under amphibolite facies conditions. The most significant regional structure is the Sgurr Beag–Naver ductile thrust, which divides the Moine Supergroup into two major thrust nappes (Moine and Sgurr Beag–Naver nappes; Figs 6.2, 6.3). Both nappes also contain subordinate ductile thrusts such as the Knoydart and Ben Hope thrusts (Figs 6.2, 6.3). Because thrusting occurred during regional medium- to high-grade metamorphism, the ductile thrusts are defined by broad belts, up to hundreds of metres wide, of recrystallized mylonites. The total displacement on the Sgurr Beag–Naver thrust is unknown but could be over a hundred kilometres and thus the structure may be at least as significant as the younger Moine Thrust. Ductile thrusting was accompanied by intense foliation development and tight to isoclinal, recumbent folding of the Moine rocks on all scales. A strong lineation is commonly defined by the alignment of minerals such as micas and elongated and recrystallized grains of feldspar and quartz. The

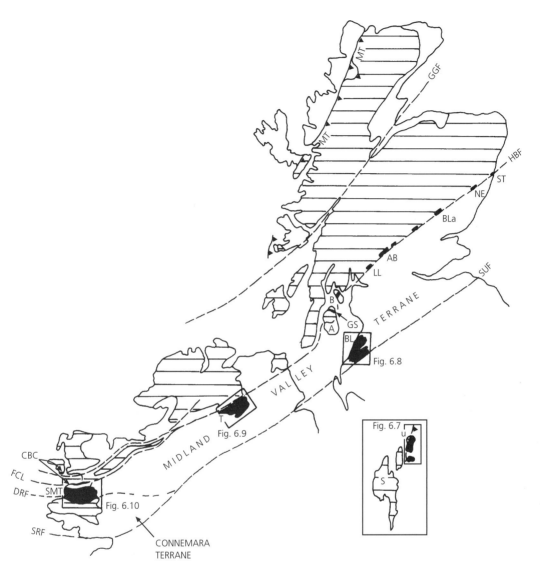

Fig. 6.1 Map of Scotland and Ireland showing the area affected by the Grampian Orogeny (horizontal lines), dominated by metasedimentary rocks of the Moine and Dalradian supergroups, with associated basement inliers; the location of Cambrian–Ordovician rocks that were accreted onto the Laurentian margin during the Grampian Orogeny (in black); regional scale faults. Note that many of these faults are younger than the Grampian Orogeny; see text for discussion. S, Shetland; U, Unst; MT, Moine Thrust; GGF, Great Glen Fault; HBF, Highland Boundary Fault; SUF, Southern Uplands Fault; ST, Stonehaven; NE, North Esk; BLa, Blairgowrie; AB, Aberfoyle; LL, Loch Lomond; B, Bute; A, Arran; GS, Glen Sannox; BL, Ballantrae; AGL, Antrim–Galway Line; T, Tyrone; SRF, Skerd Rocks Fault; SMT, South Mayo Trough; CBC, Clew Bay Complex; FCL, Fair Head–Clew Bay Line; DRF, Doon Rock Fault.

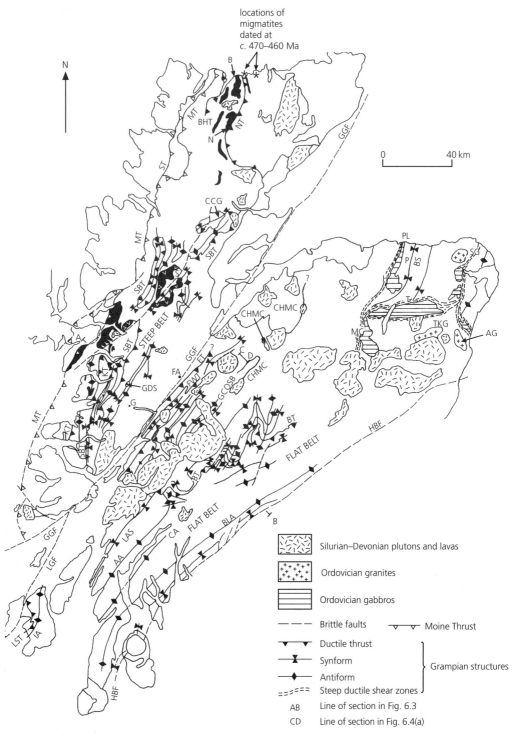

Fig. 6.2 Grampian structures and associated Ordovician intrusions in Scotland: BHT, Ben Hope Thrust; NT, Naver Thrust; SBT, Sgurr Beag Thrust; KT, Knoydart Thrust; GDS, Glen Dessary syenite; ET, Eilrig Thrust; FWT, Fort William Thrust; BT, Boundary Thrust; LST, Loch Skerrols Thrust; IA, Islay Anticline; LAS, Loch Awe Synform; AA, Ardrishaig Antiform; GCOSB, Geal Charn–Ossian Steep Belt; CA, Cowal Antiform; BLA, Ben Ledi Antiform; PL, Portsoy Lineament; AG, Aberdeen granite; SG, Strichen granite; TKG, Tillyfourie–Kenmay granite; H, P, Huntly, Portsoy

gabbro; I, Insch gabbro; MC, Morven–Cabrach gabbro; BS, Boyndie Syncline. Note that the Moine (MT) and Sole (ST) thrusts, the Great Glen (GGF), Highland Boundary (HBF) and Loch Gruinart (LGF) faults are significantly younger structures. Other abbreviations: N (Naver), B (Borgie) and S (Scardroy) basement inliers; CHMC, Central Highland Migmatite Complex; FA, Fort Augustus; G, Glenfinnan; CCG, Carn Chuinneag granite.

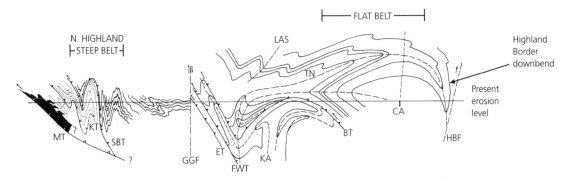

Fig. 6.3 General cross-section across the Grampian structures of Scotland (see Fig. 6.2 for line of section AB). MT, Moine Thrust; KT, Knoydart Thrust; SBT, Sgurr Beag Thrust; GGF, Great Glen Fault; ET, Eilrig Thrust; FWT, Fort William Thrust; KA, Kinlochleven Anticline; LAS, Loch Awe Synform; TN, Tay Nappe; BT, Boundary Thrust; CA, Cowal Antiform; HBF, Highland Boundary Fault (sources: section west of GGF after Powell 1988; section east of GGF after Roberts & Treagus 1977).

lineations lie parallel to the main directions of thrust transport. Evidence from kinematic indicators, together with the regional lineation trajectories, indicates that thrusts moved broadly north-westwards. There is clear evidence for at least two major periods of deformation, because ductile thrusts and associated recumbent folds are commonly deformed by the later open to tight upright folds which form the Northern Highlands Steep Belt (Fig. 6.3).

The numerous basement units that occur between the Moine Thrust and the Great Glen Fault are thought to represent tectonically emplaced inliers of the gneissic substrate upon which the Moine Supergroup sediments were originally deposited (see Chapter 4). Some inliers lie in the cores of anticlines (e.g. Borgie and Naver inliers, Fig. 6.2), whereas others were carried as allochthonous slices along Grampian thrusts such as the Sgurr Beag Thrust (e.g. Scardroy, Fig. 6.2). The basement gneisses are mostly strongly recrystallized and retrogressed, so that few traces remain of any pre-existing Archaean or Palaeoproterozoic mineral assemblages. The intense nature of Grampian ductile deformation (combined with the effects of ductile shearing during the Neoproterozoic Knoydartian event) has obliterated any obvious traces of sedimentary or structural discordances across the original basement–Moine unconformity.

Grampian metamorphic grade within the Moine rocks is everywhere in the amphibolite facies, as indicated by the widespread syn-tectonic crystallization of garnet and hornblende. Detailed evaluation of metamorphic grade is difficult because the Moine rocks are characteristically aluminium poor, which means that Barrovian index minerals such as kyanite and staurolite are relatively rare as compared with the Dalradian Supergroup. However, kyanite of probable Grampian age is present at various localities scattered through the Moine outcrop, suggesting that the regional metamorphic grade is at least within the middle amphibolite facies. The Moine rocks above the Sgurr Beag–Naver thrust (Glenfinnan and Loch Eil groups) are commonly gneissic; although some migmatization may date from the Neoproterozoic Knoydartian event, there is isotopic evidence for widespread melting of Moine rocks in East Sutherland during the Grampian Orogeny.

A Mid-Ordovician age (c. 470–460 Ma) for early Grampian metamorphism and accompanying deformation is indicated by U–Pb zircon ages of 461 ± 13 Ma and 467 ± 10 Ma obtained from Moine migmatites in East Sutherland (Fig. 6.2), and a U–Pb titanite age of 470 ± 2 Ma obtained from the West Highland Granitic Gneiss near Fort Augustus (Fig. 6.2). The Glen Dessary syenite (Fig. 6.2) was intruded into Moine rocks at 456 ± 5 Ma after early Grampian deformation but prior to the phase of intense upright folding that formed the Northern Highlands Steep Belt (Fig. 6.3). This latter deformation phase probably occurred at 450 ± 10 Ma, the age of monazite in syn-metamorphic pegmatites that cut the Moine gneisses at Glenfinnan (Fig. 6.2).

6.2.2 Grampian events in Scotland south-east of the Great Glen Fault

The Dalradian Supergroup was deformed into a complex primary regional fold pattern defined by north-east–south-west trending, often tight to isoclinal primary folds with amplitudes of up to tens of kilometres (Fig. 6.2). These folds and their associated axial–planar cleavage 'fan' across the region, from north-west facing in the north-west to south-east facing in the south-east (Fig. 6.3). These contrasting sectors are separated by a central zone of upright folding represented by the Loch Awe Syncline and Geal Charn–Ossian Steep Belt. The north-west-facing folds developed in association with north-west-directed ductile thrusts, such as the Loch Skerrols, Eilrig, Fort William and Boundary thrusts (Figs 6.2, 6.3). The total displacements along these ductile thrusts are unknown, but seem likely to total at least tens of kilometres. To the south-east of the central zone, much of the Dalradian is inverted, lying in the upside-down limb of a major recumbent fold known as the Tay Nappe (Fig. 6.3). A secondary phase of deformation resulted in refolding of these primary folds, also along north-east–south-west axes, to produce large-scale interference patterns as well as secondary foliations and crenulation cleavages. Later folds and steeply dipping shear zones are developed locally, particularly in north-east Scotland (Fig. 6.2).

It is important to emphasize that in places the pre-existing structural architecture of the Dalradian extensional rift-basins (see Chapter 5) exerted a strong influence on the geometry of the primary folds and thrusts that formed during the Grampian Orogeny (Fig. 6.4). A particularly good example of this control is found in the northern part of the Geal Charn–Ossian Steep Belt (Fig. 6.2). In this area, the steep belt appears to be coincident with the eastern margin of a major Dalradian basin. Although the metasedimentary rocks are intensely folded (Fig. 6.4a), it is clear once the effects of upright folding are removed (Fig. 6.4b) that Grampian and Appin Dalradian strata originally onlapped onto rocks of Central Highland Migmatite Complex to the east. The intense upright Grampian folding characteristic of the steep belt is thought to have resulted from the compression of the thick sequence of Grampian Group sediments against a rigid, upstanding fault-

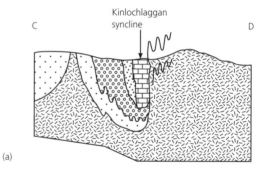

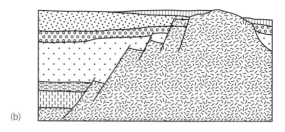

Fig. 6.4 (a) Cross-section across the Geal Charn–Ossian Steep Belt (see Fig. 6.2 for line of section CD). Random dashes are Central Highland Migmatite Complex, other ornaments represent Grampian and Appin group strata. (b) Basin reconstruction along the same line of section (modified from Robertson & Smith 1999). The steep belt is thought to have formed as a result of compression of the Grampian and Appin strata against the upstanding block of Central Highland Migmatite Complex.

block formed of the Central Highland Migmatite Complex.

The Dalradian outcrop is a classic area for the study of metamorphism. The Grampian Highlands were the location of the early studies of metamorphic isograds carried out by Barrow in 1912. Growth of peak metamorphic index minerals such as staurolite, kyanite and sillimanite overlapped and slightly post-dated deformation in many areas. The zonal metamorphic pattern (Fig. 6.5) mainly reflects the variation in temperature at the peak of Grampian metamorphism. Low-temperature greenschist facies prevails in the south-west, and high-temperature middle amphibolite facies occurs in the north-eastern part of the Dalradian where local migmatites are associated with sillimanite-bearing rocks. Pressures

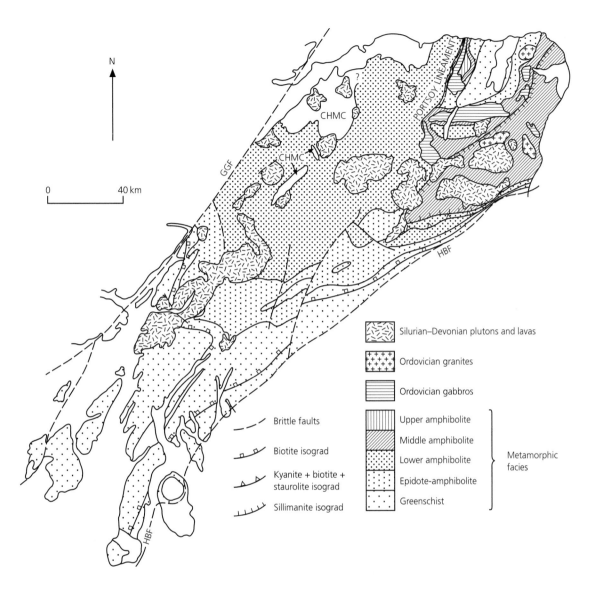

Fig. 6.5 Metamorphic zones within the Dalradian rocks of the Grampian Highlands (modified from Stephenson & Gould 1995). CHMC, Central Highland Migmatite Complex; GGF, Great Glen Fault; HBF, Highland Boundary Fault.

between 9 and 12 kbar have been recorded over large parts of the south-west and central Highlands, corresponding to depths of burial of *c.* 25–35 km. Pressures are significantly less in the north-eastern part of the Dalradian, which may thus correspond to a higher structural level.

The Dalradian rocks of north-east Scotland were intruded by a suite of gabbros, some of which display good igneous layering. The suite includes the gabbros at Portsoy, Huntly, Insch and Morvern–Cabrach (Fig. 6.2). The gabbros were intruded at about the peak of regional metamorphism and produced, in places, a sillimanite overprint of the regional metamorphic pattern, prior to themselves being involved in late-stage regional folding. Closely associated with the gabbros are a series of granites, including the Aberdeen, Strichen and Tillyfourie–Kenmay plutons (Fig. 6.2). These granites commonly contain foliations formed as a result of both magmatic flow during emplacement and subsequent solid-state deformation after crystallization. Geochemical evidence suggests that the granites formed as a result of the melting of Dalradian sedimentary rocks, not far below the present exposure level, presumably during peak regional metamorphism. Structural evidence indicates that the granites were emplaced during late-stage regional folding. A U–Pb zircon age of 468 ±8 Ma for the Insch gabbro and a U–Pb monazite age of 470±1 Ma for the Aberdeen granite provide tight constraints on the timing of the Grampian Orogeny in this part of Scotland (Fig. 6.2). It seems reasonable to assume that regional deformation and metamorphism elsewhere across the Dalradian outcrop also occurred at this time. Rb–Sr biotite ages of *c.* 460–440 Ma obtained from various parts of the Dalradian outcrop correspond to the time of regional uplift and cooling following the close of the Grampian Orogeny.

6.2.3 Grampian events in Ireland

The Dalradian rocks of Donegal and Mayo (Fig. 6.6) display a structural history very similar to that outlined above in Scotland. Early north-west-directed compression resulted in the formation of large-scale recumbent, tight to isoclinal primary fold nappes in association with major ductile thrusts such as the Ardsberg, Horn Head, Knockateen, Central Donegal and Central Achill thrusts (Fig. 6.6, section AB). The cumulative displacement on these structures may amount to many tens of kilometres. This early episode of crustal thickening was followed by the development of major, south-east-vergent folds, such as the Ballybofey Nappe, synchronous with ductile thrusting along the Lough Derg slide and the Omagh Thrust, which resulted in south-eastward displacement of the Dalradian over the Lough Derg basement gneisses and the Tyrone Igneous Complex (Fig. 6.6, section AB). The major Dalradian folds thus have a fan-shaped regional geometry very similar to that in Scotland. Barrovian amphibolite to greenschist facies regional metamorphism accompanied crustal thick-

ening. On Achill Island, regional *P–T* conditions were 8±2 kbar and 620±30°C.

The Dalradian rocks of the Connemara Terrane (Figs 6.1, 6.6) occupy an apparently anomalous position relative to the remainder of the Dalradian rocks in north-west Ireland, because they are located to the *south* of the Fair Head–Clew Bay Line. This is thought to be the result of strike-slip faulting during the Grampian Orogeny (see Section 6.6). The structural and metamorphic evolution of the Connemara Dalradian is very similar to that already outlined from Mayo and Donegal. Two major phases of north-directed, recumbent tight to isoclinal folding accompanied regional Barrovian amphibolite facies metamorphism, which produced garnet–staurolite ± kyanite assemblages in pelitic rocks. This event was partly overprinted by contact metamorphism associated with emplacement of a suite of mafic intrusions, the Connemara Gabbros, which are commonly strongly foliated and have an amphibolite facies mineralogy, together with calc-alkaline dioritic gneisses. There are two groups of metagabbro bodies, the Dawros–Currywongaun–Doughruagh Complex in the north, and the Cashel–Lough Wheelaun intrusion to the south. The gabbros were intruded during the first phase of recumbent folding. U–Pb dating of zircons from these intrusions has yielded ages of 475 ±1 Ma (Currywongaun intrusion) and 470±1 Ma (Cashel Gabbro) which are thus thought to correspond closely to the time of folding. Recumbent folding was followed by the formation of a regional open, upright fold, the Connemara Antiform (Fig. 6.6, section CD). This late phase of folding accompanied southward-directed translation of the Dalradian metasediments, the metagabbros and the dioritic gneisses along the Mannin Thrust. The late to post-tectonic Oughterard granite has yielded a U–Pb xenotime age of 463±3 Ma, which dates the close of the Grampian Orogeny in Connemara. The rocks in the footwall to the Mannin Thrust are a series of metamorphosed and deformed rhyolites of unknown, but probable Ordovician, age.

6.3 What caused the Grampian Orogeny?

Isotopic and other evidence indicates that Grampian orogenic activity occurred in the Mid-Ordovician at *c.* 475–460 Ma, with deformation continuing until *c.* 450 Ma in certain sectors (e.g. Northern Highlands

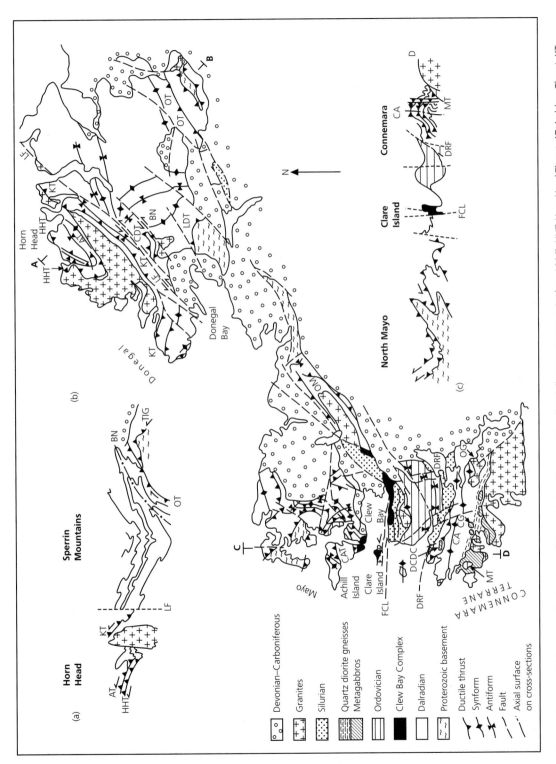

Fig. 6.6 Grampian folds and ductile thrusts in north-west Ireland (sources: Powell & Phillips 1985; Hutton & Alsop 1996). HHT, Horn Head Thrust; AT, Ardsbeg Thrust; KT, Knockateen Thrust; CDT, Central Donegal Thrust; LDT, Lough Derg Thrust; OT, Omagh Thrust; CAT, Central Achill Thrust; MT, Mannin Thrust; BN, Balleybofey Nappe; FCL, Fair Head–Clew Bay Line; DRF, Doon Rock Fault; TIG, Tyrone Igneous Complex; CA, Connemara Antiform; DCDC, Dawros–Currywongaun Complex; CG, Cashel Gabbro; OG, Oughterard granite; OM, Ox Mountains. Note that the synform developed within the Silurian rocks south of Clew Bay (section CD) formed later than the Grampian Orogeny, during the final closure of the Iapetus Ocean in the Late Silurian–Early Devonian (see Chapter 12).

Steep Belt). It is clear that the Grampian Orogeny cannot have resulted from the hard collision of Laurentia with southern Britain (Eastern Avalonia), as this did not begin until the Silurian (see Chapter 12). Critical clues as to the cause of the Grampian Orogeny are obtained by examination of Cambrian to Mid-Ordovician rocks, which occur either along or to the south-east of the Highland Boundary Fault–Fair Head–Clew Bay Line within the Midland Valley Terrane (Fig. 6.1). These rock sequences have oceanic affinities and include ophiolites, a possible accretionary prism and fore-arc basin, and a volcanic arc, which together are thought to record the development of a subduction zone within the Iapetus Ocean. Attempts to develop a model for the Grampian Orogeny are complicated by the likelihood that many of these units have undergone potentially very large lateral displacements relative to Laurentia as a result of late Caledonian strike-slip faulting (see Chapter 12). The importance of such displacements is emphasized by the anomalous position of the Dalradian rocks of the Connemara Terrane, which have been structurally interleaved with the oceanic terranes. Furthermore, there is considerable controversy as to the magnitude of these displacements and the original absolute and relative palaeogeographic settings of the various oceanic terranes. None the less, there is a broad consensus that the Grampian Orogeny resulted from the thrusting (obduction) of ophiolites onto continental crust and the collision of Laurentia with a volcanic arc.

6.4 Evidence for obduction of ophiolites during the Lower–Middle Ordovician

6.4.1 Shetland ophiolite

In north-eastern Shetland, on the islands of Unst and Fetlar, metasedimentary rocks correlated with the Dalradian of the mainland of Scotland are overlain by two thrust nappes mainly comprising basic and ultrabasic igneous rocks (Fig. 6.7). The lower nappe exhibits a layered sequence with olivine-rich lherzolite–harzburgite at the base overlain by (in ascending order) dunite, banded gabbro, wehrlite–clinopyroxenite and a sheeted dyke complex composed mainly of microgabbros. This sequence is identical to that characteristic of the lower parts of oceanic crust and thus it is termed an ophiolite; the sequence would normally be capped by the basaltic pillow lavas typical of the ocean floor, but these are absent here. The upper nappe is composed entirely of olivine-rich lherzolite–harzburgite. These ophiolitic rocks are mainly undeformed but have been affected by low-grade metamorphism, including serpentinization prior to thrusting. The total reassembled thickness of the ophiolite is c. 8 km.

Melange zones between the nappes, below the lower nappe and above the upper nappe contain blocks of various igneous and metamorphic rocks. These were produced as a result of the erosion of the ophiolite nappes as they were thrust up onto the surface. Continued thrusting resulted in these erosion products being overriden, metamorphosed and deformed. Fragments include serpentinized peridotite and gabbro clearly derived from erosion of the ophiolite, as well as slices of migmatitic rocks sheared off the footwall over which the nappes travelled. Pebbles of meta-volcanic rocks such as spilite, quartz–albite–porphyry and trondhjemite were probably derived from erosion of a volcaniclastic cover to the ophiolite. Low-grade pelites and gritty sandstones within the melanges are presumably remnants of the marine sedimentary cover. Atypically, the Funzie Conglomerate on Fetlar is composed entirely of highly deformed quartzite pebbles; the origin of this unit is uncertain, but it may represent a submarine fan debris flow derived from erosion of the Laurentian continental margin to the north-west. Plagiogranites formed during crystallization of the complex have yielded an early Arenig U–Pb zircon age of 492 ± 3 Ma.

6.4.2 Ballantrae Igneous Complex

The Ballantrae Igneous Complex occurs near Girvan, in the southern part of the Midland Valley (Fig. 6.8). The complex consists of two main rock associations. The first comprises spilitic pillow lavas overlain by agglomerates, tuffs and cherts. Black shales within the tuffs yield the brachiopods *Acotretra* and *Lingula* together with planktonic graptolites *Tetragraptus* and *Didymograptus*. The second association consists of serpentinized peridotite and harzburgite with minor gabbro and trondhjemite. The rocks that underlie the complex are not exposed. Although it seems likely that the complex forms part of an ophiolite, original stratigraphic relationships are not pre-

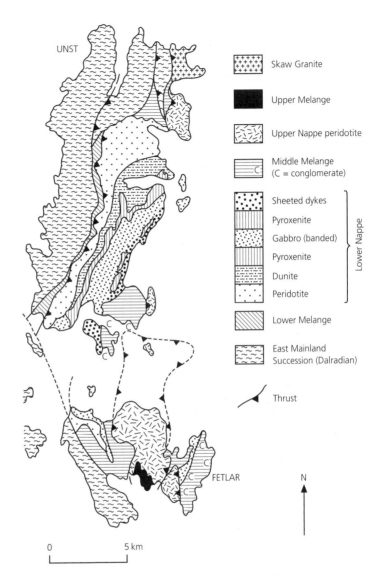

UNST

FETLAR

N

+++	Skaw Granite
■	Upper Melange
	Upper Nappe peridotite
C	Middle Melange (C = conglomerate)
	Sheeted dykes
	Pyroxenite
	Gabbro (banded)
	Pyroxenite
	Dunite
	Peridotite
	Lower Melange
	East Mainland Succession (Dalradian)
	Thrust

Lower Nappe

0   5 km

Fig. 6.7 Geology of the Shetland ophiolite (for location see Fig. 6.1; modified from Flinn 1985).

served because internal contacts are mostly tectonic. Olisthostromes within the complex include fragments of serpentinite, gabbro, dolerite, spilite, amphibolite, chert and shale, all presumably derived from the erosion of the ophiolite during obduction. Amphibolites and schists beneath the serpentinites resemble the metamorphic soles found in ophiolites where obduction of hot oceanic lithosphere over pre-existing rocks has caused metamorphism. A K–Ar age of $478 \pm 8\,\mathrm{Ma}$ (Llanvirn) obtained from amphibolites in the sole probably dates cooling following obduc-

tion. These isotopic ages are broadly consistent with the Low–Mid-Arenig age of graptolite faunas within black shales in the complex. Obduction was followed by intrusion of gabbros and microgabbros, which cut both the serpentinite and the metamorphic sole.

6.4.3 Highland Border Complex

A series of generally low-grade metamorphic rocks known collectively as the Highland Border Complex occur as discontinuously exposed and fault-bounded

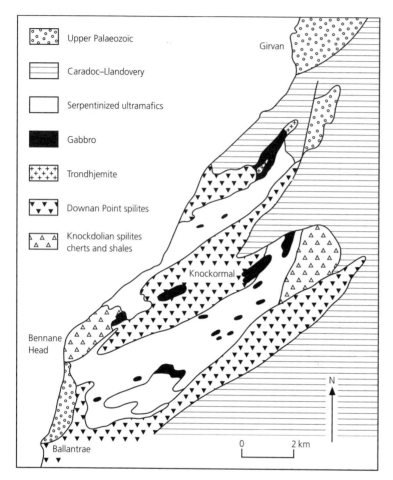

Fig. 6.8 Geology of the Ballantrae Igneous Complex and adjacent rock units (for location see Fig. 6.1; modified from Brown 1991).

slivers along the Highland Border Fault between Stonehaven in north-east Scotland and the island of Arran (Fig. 6.1). The complex has been subdivided into four distinct assemblages, although the entire sequence is not present at any single locality. The oldest assemblage mainly comprises serpentinites; on Arran and Bute they occur in association with hornblende schists and metamorphosed pillow lavas. These rock types have been interpreted as remnants of a tectonically dismembered ophiolite. Ophiolitic rocks are overlain unconformably by the second assemblage, conglomerates and shallow-water dolomitic limestones, which contain a middle Arenig trilobite and brachiopod fauna. A third assemblage is an oceanic sequence of spilitic pillow lavas, tuffs and breccias, black shales and cherts, and sandy turbidites. The sedimentary rocks contain

brachiopods and chitinozoans of late Arenig to Llandeilo age. The fourth assemblage is a sequence of limestones and mature sandstones; chitinozoans have been interpreted as indicating a Caradoc age. These latter sedimentary rocks are thought by some workers to have been deposited uncon-formably on older parts of the complex after they had been deformed, presumably during the Grampian Orogeny.

The tectonic history and regional significance of the Highland Border Complex are highly controver-sial. One viewpoint interprets the four assemblages together as a 'suspect terrane' derived from an ocean basin that once existed to the south of Laurentia. Because the fourth assemblage was apparently deposited *after* the Grampian Orogeny, but contains none of the coarse debris which might be expected if it was located close to the eroding mountain belt, it has been suggested that the Highland Border Complex and the Dalradian were not juxtaposed until some time in the Silurian or Devonian. An alternative view, adopted here, is that the complex comprises only two main elements: sedimentary rocks, which form the upper part of the Dalradian Supergroup, and an ophiolite, which was obducted onto the Laurentian margin during the Grampian Orogeny. This view is based on the observation that sedimentary rocks of the complex (including the fourth assemblage) and the ophiolitic rocks apparently *all* carry the same Grampian cleavage as adjacent Dalradian rocks. The palaeontological evidence for the age of the fourth assemblage has been questioned, and this second hypothesis requires that this assemblage is older than thought previously, possibly Arenig in age.

6.4.4 Tyrone Igneous Complex

The Tyrone Igneous Complex (Fig. 6.9) consists of metamorphosed gabbros, sheeted microgabbro dykes and locally preserved pillow lavas, overlain by a sequence of Arenig–Llanvirn volcanics and black shales. Although strongly deformed and incomplete, the presence of a sheeted dyke complex permits interpretation of the igneous rocks as an ophiolite. The gabbros commonly incorporate small bodies of tonalite, quartz diorite and trondhjemite, which are thought to be silicic differentiates from the developing oceanic crust. The outcrop of the ophiolite surrounds basement schists and gneisses of unknown age. Structural evidence suggests that the ophiolite was thrust north-westwards onto this basement. The ophiolite is intruded by the Craigballyharky tonalite; evidence for mingling and mixing of the tonalite and basic rocks of the ophiolite means that the latter must have been at a high temperature at the time of intrusion. As there is no evidence for a separate, post-thrusting, pre-tonalite heating event, this high temperature must have been a relic of the ophiolite ocean-floor metamorphism. The tonalite is not an integral part of the ophiolite as it contains inherited Proterozoic zircons, possibly derived from the underlying basement rocks; it must have been intruded into the ophiolite shortly after thrusting onto the basement. A U–Pb zircon age of $471+2/-4$ Ma obtained from the tonalite is thus interpreted as dating obduction.

6.4.5 Clew Bay Complex

The Clew Bay Complex of western Ireland lies immediately to the south of the Fair Head–Clew Bay Line (Fig. 6.1). Three main fault-bounded units are

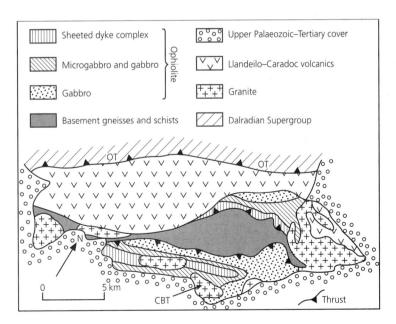

Fig. 6.9 Geology of the Tyrone Igneous Complex and adjacent rock units (for location see Fig. 6.1; modified from Hutton *et al.* 1985). CBT, Craigballyharky tonalite; OT, Omagh Thrust.

Sheeted dyke complex

Microgabbro and gabbro

Gabbro

Ophiolite

Basement gneisses and schists

Upper Palaeozoic–Tertiary cover

Llandeilo–Caradoc volcanics

Granite

Dalradian Supergroup

Thrust

0 5 km

present. The Deer Park Complex (Fig. 6.10) comprises serpentinized ultramafic rocks, amphibolites, meta-gabbros, a sheeted dyke complex of microgabbros, and meta-pelites set in a sheared talc–carbonate matrix. The assemblage has been interpreted as a highly deformed ophiolite. To the north these are in fault contact with the highly sheared, low-grade sediments of the Killadangen Formation (Fig. 6.10). These are mainly grits locally interbedded with cherts, shales and basaltic pillow lavas. Occasional

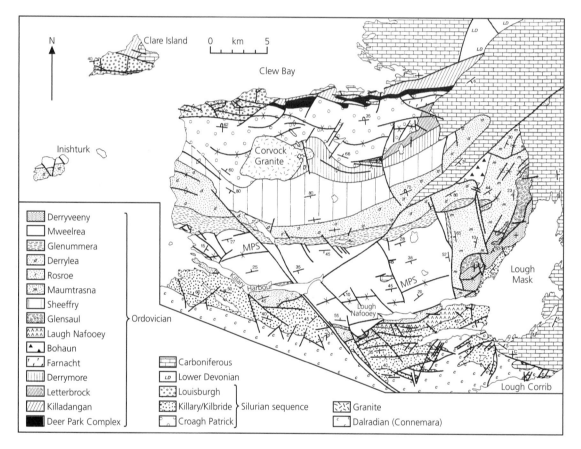

Fig. 6.10 Simplified geological map of the South Mayo Trough (for location see Fig. 6.1). Modified from Dewey and Mange (1999). PMS, Mweelrea–Party Mountains Syncline.

sponges and conodonts indicate an Early to Mid(?)-Ordovician age for much of this complex. The third sequence occurs further north on Achill Island (Fig. 6.6) and contains retrogressed blueschists with blocks of mafic volcanics. The presence of blueschists is particularly important, because they are indicative of the high-pressure, low-temperature metamorphic regime typical of subduction zones. On Achill Island, the main lineation formed during the Grampian Orogeny can be traced from the Dalradian into the blueschists, and is also present in the Westport and Deer Park complexes.

6.4.6 Synthesis

Field and isotopic evidence in combination indicate that ophiolites in Clew Bay, Tyrone and Shetland were obducted onto the Laurentian continental margin during the Early to Mid-Ordovician. The Ballantrae and Highland Border ophiolites are presumed to have been obducted onto the underlying rocks of the Midland Valley also at this time. Whether the ophiolites originally formed oceanic crust *sensu stricto* or marginal basin crust is difficult to assess, partly because geochemical studies have yielded ambiguous results. In the case of the Ballantrae Complex, however, the close association of the ophiolite with volcaniclastic sediments implies a marginal basin setting near an active volcanic arc. In order to understand the likely mechanism by which these ophiolites were obducted and how this might relate to the Grampian Orogeny, it is now necessary to review the evidence for the development of Early Ordovician volcanic arcs in the Iapetus Ocean.

6.5 Evidence for Ordovician volcanic arc complexes

6.5.1 South Mayo Trough

The South Mayo Trough of western Ireland is an early Mid-Ordovician sedimentary basin which lies immediately to the south of the Clew Bay Complex (Fig. 6.10). It is separated from the Dalradian of Connemara by the Doon Rock Fault (Fig. 6.6) beneath the Silurian sedimentary sequence south of Killary Harbour (Fig. 6.10). The basin contains an extremely thick sequence of Tremadoc–Llanvirn volcaniclastic sediments that occupies a large east–west trending

fold, the Mweelrea–Partry Mountains Syncline. The north limb comprises a 10-km-thick conformable succession of volcaniclastic sediments. The base is unexposed, but it probably rests upon mafic volcanic rocks (Fig. 6.10). The Letterbrock, Derrymore, Sheeffry and Derrylea formations are dominated by volcaniclastic turbidites that contain 17 andesitic to rhyolitic tuff bands. Occasional conglomerates in the Letterbrock Formation contain clasts of mafic, intermediate and acidic volcanics, quartzite, granite gneiss, dolerite, gabbro and low-grade schists. The Derrylea Formation contains mafic and ultramafic detritus (including chromite) and detrital garnet and staurolite, which were derived from a northerly source area thought to have included an ophiolite and a Barrovian metamorphic complex. The incoming of metamorphic detritus occurs within the *Didymograptus artus* Zone at c. 470–467 Ma. These lower units are succeeded by the Glenummera Formation, comprising siltstones and turbiditic sandstones, then overlain by the easterly derived fluvio-deltaic sandstones and conglomerates of the Mweelrea Formation, which include clasts of rhyolitic and dacitic porphyries and granites. The proportion of metamorphic detritus increases up the succession.

The southern limb succession mainly comprises arc volcanics and proximal deep-sea sediments. The Arenig Lough Nafooey Group is a 2.6-km sequence of mafic to intermediate volcanics, the chemistry and petrography of which are consistent with eruption during the early evolution of an oceanic arc. The Rosroe Formation contains conglomerates, coarse sandstones and andesitic pyroclastic rocks, which are interpreted as deep-sea fan deposits laterally equivalent to the Derrylea Formation. As on the northern limb, the top of the southern limb succession is occupied by the Glenummera and Mweelrea formations. In the east, the Maumtrasma and Derryveeny formations comprise fluvio-deltaic sandstones and conglomerates with abundant igneous and metamorphic clasts, some derived from the south. The exact stratigraphic positions of these units are uncertain but they are presumed to lie towards the top of the Ordovician succession.

The South Mayo Trough is interpreted as a basin that developed adjacent to an active oceanic volcanic arc represented by the Lough Nafooey Group. The progressive shallowing of the sequence from turbidites upwards into fluvio-deltaic sediments is thought to reflect collision with the margin of Laurentia to the north. The incoming of northerly derived ophiolitic and metamorphic detritus in the earliest Llanvirn corresponds to the time of ophiolite obduction and of the Grampian Orogeny, as indicated elsewhere by isotopic dating. The presence of southerly derived metamorphic detritus in the Maumtrasma and Derryveeny formations has been taken to date the arrival of the Connemara Terrane to the south. It is presumed that this resulted from strike-slip faulting during the Grampian Orogeny, which detached a segment of the Laurentian margin and moved it along and across strike. As a result, this Dalradian segment now occupies an apparently anomalous position south of the South Mayo Trough. The Mweelrea–Partry Mountains Syncline is thought to have developed initially during the Grampian Orogeny, prior to deposition of Silurian sediments that overlie unconformably the Ordovician succession (Fig. 6.10).

6.5.2 Midland Valley

Devonian and Carboniferous rock units cover all pre-Mid-Ordovician rock units in the Midland Valley apart from the Ballantrae Igneous Complex. Circumstantial evidence that an Ordovician volcanic arc lies beneath the Upper Palaeozoic cover is provided by provenance studies of Mid-Ordovician sediments at Girvan and Mid-Ordovician–Mid Silurian rocks south of the Southern Uplands Fault (see Chapter 7). These sediments were deposited after the Grampian Orogeny and mostly derived from a proximal north-westerly source, which is presumed to be within the Midland Valley. They include acidic and intermediate igneous clasts, some of which have yielded Ordovician isotopic ages and are thought to have been eroded from a volcanic arc analogous to the Lough Nafooey arc in western Ireland.

6.6 Summary: the Grampian Orogeny—a result of arc–continent collision and ophiolite obduction?

Models for the Grampian Orogeny have been influenced greatly by understanding of the geology of western Ireland, where original relationships between major rock units are most completely preserved (although still modified by later faulting). Critical observations include:

1 The continuity of Grampian structures from the Dalradian into the Clew Bay Complex, indicating a common structural evolution.
2 The presence of blueschists on Achill Island, indicative of a former subduction zone in this area.
3 The existence of the Deer Park Complex ophiolite.
4 Provenance studies in the South Mayo Trough, which show that Early Ordovician sedimentation occurred at the same time as erosion of an ophiolite and a metamorphic complex to the north (presumably the obducted Deer Park Complex and underlying Dalradian rocks) and eruption of a volcanic arc to the south.

These observations allow development of an integrated model for the Grampian Orogeny in western Ireland (Fig. 6.11). This involves development during the Latest Cambrian–Early Ordovician of an intra-oceanic subduction zone dipping southwards (present reference frame) (Fig. 6.11a). The Westport Complex has been interpreted as an accretionary prism developed above the subduction zone, and the blueschists of Achill Island as segments of this sequence metamorphosed within the subduction zone. The oceanic crust situated above the subduction zone may be correlated with the Deer Park Complex and the Tyrone Igneous Complex. An island arc (Lough Nafooey Group) and associated fore-arc basin (South Mayo Trough) are located immediately to the south. When the Laurentian continental margin collided with the subduction zone, Dalradian sediments were regionally deformed and metamorphosed beneath a large ophiolite thrust nappe, resulting in several phases of widespread north-west-directed, recumbent, tight to isoclinal folding and regional Barrovian metamorphism (Fig. 6.11b). The gabbros and dioritic gneisses of Connemara represent subduction-related magmas that became incorporated into the developing orogenic belt. The timing of ophiolite obduction and orogeny is constrained tightly by the influx of northerly derived ultramafic detritus into the South Mayo Trough at around the Arenig–Llanvirn boundary. The Lough Nafooey Arc was accreted onto the Laurentian margin at this time. Sinistral movement along the Doon Rock Fault at a late stage in the Grampian Orogeny detached the Dalradian rocks of the Connemara Terrane from the continental margin and displaced them to their present location south of the Lough Nafooey arc (Fig. 6.12).

A very similar model may account for Grampian

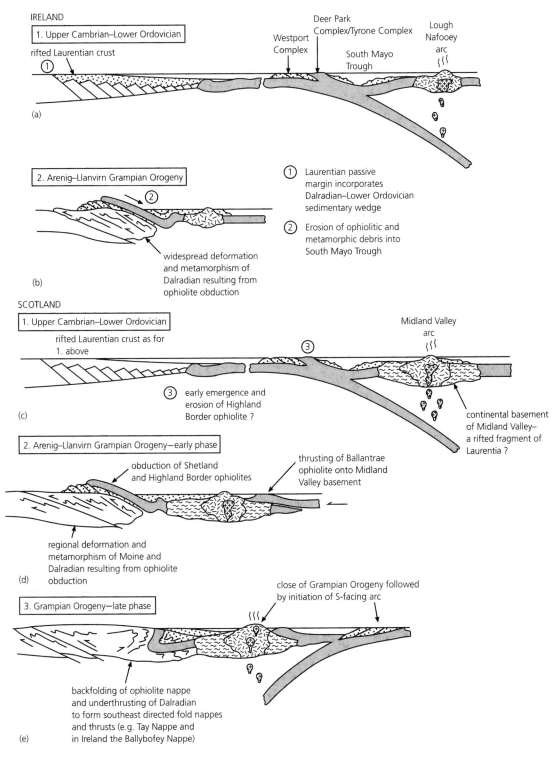

Fig. 6.11 A possible model for the Grampian Orogeny in Ireland (a,b) (modified from Ryan & Dewey 1991) and Scotland (c–e). See text for explanation.

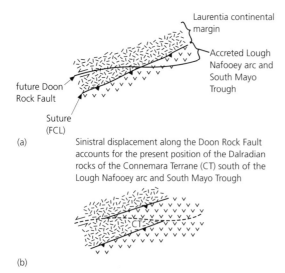

Fig. 6.12 Generalized model explaining how the Dalradian rocks of the Connemara Terrane were displaced to their present position south of the Lough Nafooey Arc and the South Mayo Trough. (a) Juxtaposition of the arc with the Laurentian margin; the dashed line represents the future course of the Doon Rock Fault; (b) sinistral displacement along the Doon Rock Fault interleaves a segment of the Laurentian margin (Connemara) with the accreted arc and associated fore-arc basin. FCL, Fair Head–Clew Bay Line.

events in Scotland (Fig. 6.11c–e). In this case, the volcanic arc was probably developed on the continental basement of the Midland Valley (Fig. 6.11c). This basement block may represent a fragment of Laurentia that was detached during continental rifting in the late Neoproterozoic–Early Cambrian. Overthrusting of an ophiolite nappe resulted in northwest-directed ductile thrusting and folding and Barrovian metamorphism of the Dalradian and Moine rocks (Fig. 6.11d). The Shetland and Highland Border ophiolites are intepreted as remnants of this nappe and are thus broadly correlated with the Deer Park and Tyrone ophiolites in Ireland. The Ballantrae Igneous Complex was also obducted onto the Midland Valley basement at this time (Fig. 6.11d). Continued crustal thickening resulted in upright folding of ductile thrusts and early recumbent folds in association with peak regional metamorphism and emplacement of the gabbros and granites of northeast Scotland. In Scotland and Ireland, south-east-vergent folds and thrusts such as the Tay and

Ballyboffey nappes and Omagh Thrust probably resulted from underthrusting of the volcanic arc and/or Midland Valley basement along the location of the present Highland Boundary Fault–Fair Head–Clew Bay Line (Fig. 6.11e). Continued underthrusting is also likely to have resulted in downfolding of the Tay Nappe (Highland border downbend, Fig. 6.3), steepening and overturning of the Clew Bay Complex, which now dips moderately to the northwest, and formation of the Mweelrea–Partry Mountains Syncline in the South Mayo Trough.

The model outlined above is similar to that proposed for early middle Ordovician orogeny in the Appalachians of eastern North America, where the Taconic Orogeny resulted from the thrusting of major ophiolite nappes onto the Laurentian margin during arc–continent collision. In North America (e.g. Newfoundland), the ophiolites are preserved as thrust klippe which overlie Cambro-Ordovician shelf carbonates equivalent to those exposed on the Caledonian foreland in north-west Scotland (see Chapter 5). Accordingly, it seems likely that ophiolites must have originally overlain much of the orogenic belt but have now been largely removed by erosion. The only difference between the Taconic and Grampian orogenies is that the former occurred over a slightly longer period (Arenig–Caradoc), possibly the result of several phases of arc–continent collision.

A currently active modern analogue for the Grampian Orogeny may be found in the Australia–Banda Arc collision zone (Fig. 6.13). Over the past 40 Myr, the northern margin of the Australian continent has been subducted beneath the oceanic crust of the Banda Sea. The collision of Australia with the Banda volcanic arc developed above the northward-dipping subduction zone resulted in continent-ward thrusting of ophiolites onto Timor and in kyanite-grade Barrovian metamorphism. Collision and 'jamming' of this subduction zone was followed by a change in the polarity of subduction as the plate boundary moved to the opposite, Banda Sea side of the volcanic arc. Such a 'flip' in the direction of subduction is precisely what appears to have followed the Grampian Orogeny in Scotland and Ireland in order to accommodate continued closure of the Iapetus Ocean and is the subject of the next chapter.

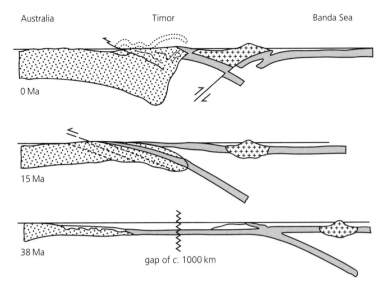

Australia   Timor   Banda Sea

0 Ma

15 Ma

38 Ma   gap of c. 1000 km

Fig. 6.13 Generalized sequence of events in the Australia–Banda Arc collision zone, which may represent a modern analogue for the Grampian Orogeny (modified from Snyder & Barber 1997). At 38 Ma, oceanic closure resulted from consumption of oceanic crust along a north-dipping subduction zone. Note the volcanic arc developed above the subduction zone (analogous to the Lough Nafooey arc?) and early intraoceanic obduction of ophiolite (analogous to the Highland Border ophiolite?). At 15 Ma, collision of rifted Australian continental crust with the subduction zone resulted in widespread obduction of ophiolites, deformation and Barrovian metamorphism in present-day Timor (equivalent to obduction of the Shetland, Ballantrae and Tyrone ophiolites and the Grampian orogeny?). At the present day, following a 'flip' in subduction polarity, Timor comprises an exhumed metamorphic belt bounded to the north by a volcanic arc and a south-dipping subduction zone (analogous to Scotland in the Middle Ordovician?).

References

Brown, P.E. (1991) Caledonian and earlier magmatism. In: *Geology of Scotland* (ed. G. Y. Craig), pp. 229–295. Geological Society, London.

Dewey, J. F. & Mange, M. (1999) Petrography of Ordovician and Silurian sediments in the western Irish Caledonides: tracers of a short-lived Ordovician continent–arc collision orogeny and the evolution of the Laurentian Appalachian–Caledonian margin. In: *Continental Tectonics* (eds C. MacNiocaill & P. D. Ryan), Geological Society, London, Special Publication **164**, 55–107.

Flinn, D. (1985) The Caledonides of Shetland. In: *The Caledonide Orogen—Scandinavia and related areas* (eds D. G. Gee & B. A. Sturt), pp. 1159–1171. John Wiley, Chichester.

Hutton, D.H.W., Aftalion, M. & Halliday, A.N. (1985) An Ordovician ophiolite in County Tyrone, Ireland. *Nature (Physical Sciences)* **315**, 210–212.

Powell, D. (1988) Excursion 4: Glenfinnan to Morar. In: *An Excursion Guide to the Moine Geology of the Scottish Highlands* (eds I. Allison, F. May & R. A. Strachan), pp. 80–102. Scottish Academic Press, Edinburgh.

Powell, D. & Phillips, W.E.A. (1985) Time of deformation in the Caledonide orogen of Britain and Ireland. In: *The Nature and Timing of Orogenic Activity in the Caledonian Rocks of the British Isles* (ed. A. L. Harris), Memoir 9, pp. 18–30. Geological Society, London.

Roberts, J.L. & Treagus, J.E. (1977) Polyphase generation of nappe structures in the Dalradian rocks of the southwest Highlands of Scotland. *Scottish Journal of Geology* **13**, 237–254.

Robertson, S. & Smith, M. (1999) The significance of the Geal Charn–Ossian steep belt in basin development in the Central Scottish Highlands. *Journal of the Geological Society, London* **156**, 1175–1182.

Ryan, P.D. & Dewey, J.F. (1991) A geological and tectonic cross-section of the Caledonides of western Ireland. *Journal of the Geological Society, London* **148**, 173–180.

Snyder, D.B. & Barber, A.J. (1997) Australia–Banda Arc collision as an analogue for early stages in Iapetus closure. *Journal of the Geological Society, London* **154**, 589–592.

Stephenson, D. & Gould, D. (1995) *British Regional Geology: the Grampian Highlands*, 4th edn. HMSO for the British Geological Survey, London.

Further reading

Bevins, R.E., Bluck, B.J., Brenchley, P.J. *et al.* (1992) Ordovician. In: *Atlas of Palaeogeography and Lithofacies* (eds J. C. W. Cope, J. K. Ingham & P. F. Rawson), pp. 19–36. Geological Society, London. [Palaeogeographic maps and summary of regional stratigraphy during the Ordovician.]

Bluck, B.J. (1985) The Scottish paratectonic Caledonides. *Scottish Journal of Geology* **21**, 437–464. [Synthesis which includes discussion of the tectonic setting of the Highland Border Complex and the Ballantrae Igneous Complex.]

Dewey, J.F. & Ryan, P.D. (1990) The Ordovician evolution of the South Mayo Trough, western Ireland. *Tectonics* **9**, 887–901. [Synthesis of the stratigraphic and tectonic evolution of a critical part of the Grampian orogen.]

Dewey, J. F. & Mange, M. (1999) Petrography of Ordovician and Silurian sediments in the western Irish Caledonides: tracers of a short-lived Ordovician continent–arc collision orogeny and the evolution of the Laurentian Appalachian–Caledonian margin. In: *Continental Tectonics* (eds C. MacNiocaill & P. D. Ryan), Geological Society, London, Special Publication **164**, 55–107. [Shows how detailed analysis of detrital heavy minerals within the South Mayo Trough sequences can be used to elucidate the history of the Grampian orogeny in western Ireland.]

Flinn, D. (1985) The Caledonides of Shetland. In: *The Caledonide Orogen—Scandinavia and related areas* (eds D. G. Gee & B. A. Sturt), pp. 1159–1171. John Wiley, Chichester. [Discussion of the Shetland ophiolite.]

Friedrich, A.M., Bowring, S.A., Martin, M.W. & Hodges, K.V. (1999) Short-lived continental magmatic arc at Connemara, western Irish Caledonides: implications for the age of the Grampian orogeny. *Geology* **27**, 27–30. [Isotopic evidence for the timing of the Grampian Orogeny in Connemara.]

Harris, A.L. & Johnson, M.R.W. (1991) Moine. In: *Geology of Scotland* (ed. G. Y. Craig), pp. 87–123. Geological Society, London. [Includes a synthesis of the Grampian orogenic evolution of the Moine.]

Phillips, W.E.A. (1981) The Orthotectonic Caledonides. In: *A Geology of Ireland* (ed. C. H. Holland), pp. 17–40.

Scottish Academic Press, Edinburgh. [Regional overview of the Irish Caledonides.]

Ryan, P.D. & Dewey, J.F. (1991) A geological and tectonic cross-section of the Caledonides of western Ireland. *Journal of the Geological Society, London* **148**, 173–180. [Synthesis of the Grampian Orogeny in western Ireland.]

Snyder, D.B. & Barber, A.J. (1997) Australia–Banda Arc collision as an analogue for early stages in Iapetus closure. *Journal of the Geological Society, London* **154**, 589–592. [A modern analogue for the Grampian Orogeny.]

Soper, N.J., Ryan, P.D. & Dewey, J.F. (1999) Age of the Grampian Orogeny in Scotland and Ireland. *Journal of the Geological Society, London* **156**, 1231–1236. [Summarizes the various lines of evidence relating to the timing of the Grampian Orogeny.]

Stephenson, D. & Gould, D. (1995) *British Regional Geology: the Grampian Highlands*, 4th edn. HMSO for the British Geological Survey, London. [Detailed description of the regional structure and metamorphism of the Dalradian Supergroup.]

Tanner, P.W.G. (1995) New evidence that the Lower Cambrian Leny Limestone at Callander, Perthshire, belongs to the Dalradian Supergroup, and a reassessment of the 'exotic' status of the Highland Border Complex. *Geological Magazine* **132**, 473–483. [Detailed structural analysis of parts of the Highland Border Complex, arguing against an exotic origin for these rocks; see also discussion of this paper by B. J. Bluck and J. K. Ingham and reply by P. W. G. Tanner in *Geological Magazine* **134**, 563–570.]

Van Staal, C. R., Dewey, J. F., McKerrow, W. S. & MacNiocaill, C. (1999) The Cambrian–Silurian tectonic evolution of the northern Appalachians and British Caledonides: history of a complex, southwest Pacific-type segment of Iapetus. In: *Lyell: the Present is in the Past* (eds D. Blundell & A. C. Scott), Geological Society, Special Publications. [Synthesis of the Grampian orogeny and related events in Scotland, Ireland and the northern Appalachions.]

Walton, E.K. & Oliver, G.J.H. (1991) Lower Palaeozoic—stratigraphy. In: *Geology of Scotland* (ed. G. Y. Craig), pp. 161–193. Geological Society, London. [Synthesis, including detailed discussion of the Highland Border Complex and the Ballantrae Igneous Complex.]

7 Mid-Ordovician to Silurian sedimentation and tectonics on the northern active margin of Iapetus

R. A. STRACHAN

7.1 Introduction

In the last chapter it was shown how the collision of a volcanic arc with the margin of Laurentia in the Early–Mid-Ordovician resulted in the Grampian Orogeny and the formation of a major mountain belt located to the north of the Highland Boundary Fault–Fair Head–Clew Bay Line. This region is thought to have remained uplifted throughout the remainder of the Ordovician and all of the Silurian, shedding sedimentary rocks into adjacent basins. Continued closure of the Iapetus Ocean after the Grampian Orogeny was achieved apparently by a 'flip' in the direction of subduction to northerly directed. The Mid-Ordovician to Silurian rocks of the Midland Valley and Southern Uplands terranes of Scotland, and their lateral equivalents in Ireland (Fig. 7.1), record the development of this active plate margin south of the Grampian mountain belt. The tectonic and relative palaeogeographic settings of many of these sedimentary sequences have been the subject of much controversy. This has arisen partly because geological relationships have been modified substantially by major sinistral strike-slip faulting and thrusting during the final collision of Eastern Avalonia with Laurentia, which resulted in the Caledonian Orogeny (see Chapter 12).

7.2 Mid-Ordovician to Silurian fore-arc and inter-arc sedimentation in the Midland Valley Terrane of Scotland and Ireland

Mid-Ordovician to Silurian sedimentary rocks occur as a series of small inliers in the Midland Valley of Scotland and Ireland (Fig. 7.1). The sedimentary record of this important period following the Grampian Orogeny is rather more fragmented than in the Southern Uplands. However, a broad picture of the evolution of the Midland Valley Basin can be pieced together by linking information from the various inliers.

7.2.1 Mid- to Upper Ordovician of the Girvan area

A sequence of sedimentary rocks, c. 2.6 km thick, was deposited unconformably upon the Ballantrae Igneous Complex in Llanvirn–Caradoc times (Figs 7.1, 7.2). The succession comprises conglomerates, sandstones, mudstones and occasional carbonates, deposited in a range of fluvio-deltaic and marine environments. Successively younger strata overlap northwards onto various parts of the ophiolitic basement (Fig. 7.2). The abundant fossil fauna includes brachiopods, trilobites and graptolites. In the Stinchar Valley, the lowermost unit is the Kirkland Conglomerate, deposited in fluvial and marine fan-delta environments. The majority of clasts are of basic and ultrabasic igneous rocks derived from erosion of the underlying Ballantrae Igneous Complex (Fig. 7.2). Palaeocurrents flowed towards the south-east. A marine transgression is indicated by deposition of the succeeding Stinchar Limestone, deposited in an inshore environment. This is overlain by an extensive unit, the Benan Conglomerate, which reaches a thickness of c. 640 m (Fig. 7.2). Lower parts of the conglomerate were deposited in a submarine fan (Fig. 7.3a); a progressive upward shallowing of water depth and eventual emergence is indicated as upper parts of the unit display evidence for fluvial deposition (Fig. 7.3b). Palaeocurrents again indicate flow towards the south-east.

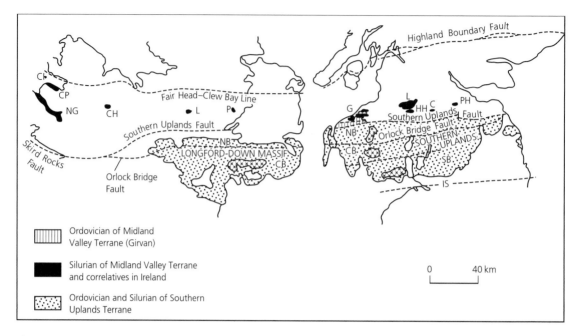

Fig. 7.1 General map showing the locations of the Ordovician and Silurian rocks of the Midland Valley and Southern Uplands terranes of Scotland and their lateral equivalents in Ireland. Midland Valley inliers: G, Girvan; L, Lesmahagow; HH, Hagshaw Hills; C, Carmichael; PH, Pentland Hills; P, Pomeroy; L, Lisbellaw; CH, Charlestown; NG, North Galway; CP, Croagh Patrick; CI, Clare Island. Southern Uplands units: NB, Northern Belt; CB, Central Belt; SB, Southern Belt.

The clasts in the Benan Conglomerate are of particular interest. The conglomerates contain some basic and ultrabasic clasts, derived most probably from the Ballantrae Igneous Complex; also present is a distinctive suite of fine-grained acidic to intermediate igneous clasts, including hornblende–biotite granite (Fig. 7.2). Two lines of evidence rule out the Ordovician granites of the Grampian mountain belt as a source. Firstly, the igneous clasts have a calc-alkaline chemistry which suggests derivation from a volcanic arc; they are compositionally very different, for example, from the crustally derived, two-mica granites of north-east Scotland. Secondly, some of the clasts are over a metre long and clearly from a proximal source. Granite clasts have yielded Rb–Sr ages that range from 559 ± 20 Ma to 451 ± 8 Ma. This spectrum of isotopic ages (although imprecise) overlaps with the age of the basin in which they accumulated, implying that the sediments were derived from the erosion of an active volcanic arc located immediately to the north and/or north-west.

It is presumed that this arc lies within the Midland Valley Terrane but is now obscured by Upper Palaeozoic sediments.

The Benan Conglomerate is overlain by the outer shelf mudstones of the Balclatchie Formation, which probably passed laterally northwards into the shallow-water Kilranny Conglomerate (Fig. 7.2). The Ardwell, Whitehouse and Shalloch formations are mainly deep-water mudstones and sandstones with occasional conglomerates. Some display evidence for deposition by turbidity currents; pebbly clasts comprise ophiolitic material, granite and occasional mica schist. These deep-water sediments also passed northwards into shallow-water environments as recorded by the contemporaneous mudstones, sandstones and limestones at Craighead, north-east of Girvan (Fig. 7.2). These rocks are particularly notable for a prolific fauna of echinoderms, brachiopods, corals, trilobites and gastropods. The lateral variations in thickness and sedimentary facies within the Girvan succession

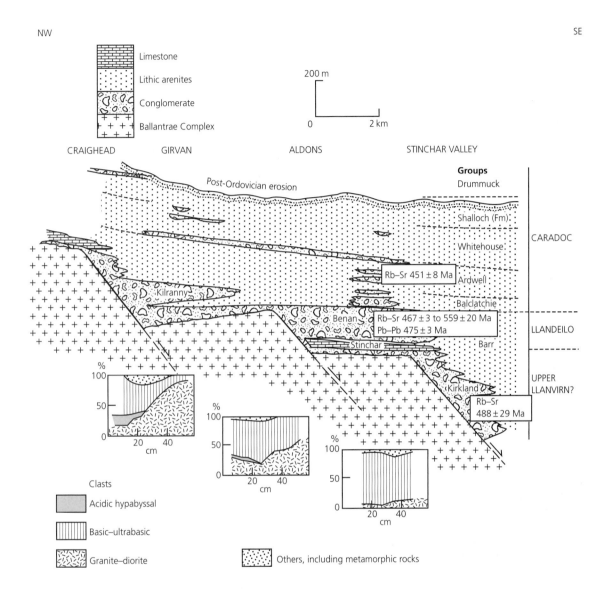

Fig. 7.2 Ordovician sedimentary successions in the Girvan area, also depicting the variation in clast composition in conglomerates, and isotopic ages derived from igneous clasts (modified from Anderton *et al.* 1979 and Walton & Oliver 1991). Younger conglomerates show increasing granite–diorite clasts.

have been interpreted as indicating that deposition was controlled by three faults, each with downthrow to the south-east and active during sedimentation (Figs 7.2, 7.3).

7.2.2 Silurian sedimentation in the Midland Valley Terrane of Scotland

Silurian rocks occur in a series of inliers in the southern part of the Midland Valley of Scotland (Figs 7.1, 7.4). In the Girvan Inlier, the Silurian rocks rest unconformably on the Mid–Upper Ordovician sediments described above. Elsewhere, the stratigraphic bases of the local Silurian successions are not exposed. A rich fossil fauna permits correlation of Silurian successions between the inliers and recognition

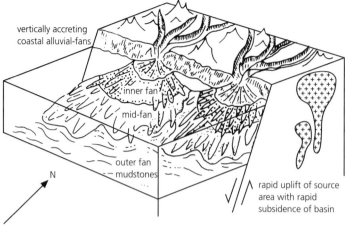

(a) LOWER HORIZONS OF FORMATION

Not to scale

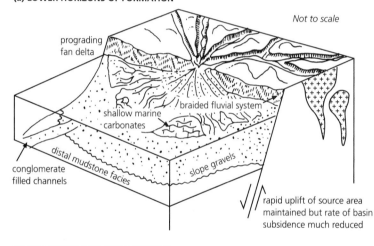

(b) UPPER HORIZONS OF FORMATION

Fig. 7.3 Palaeogeographic reconstruction of the Girvan area during the deposition of lower and upper parts of the Benan Conglomerate (after Ince 1984).

of a general regression through the Llandovery into the Wenlock. The succession in the Pentlands Inlier may extend up into the Ludlow. In all the inliers, early marine turbidite sequences pass up into shallow-water sediments. The Girvan Inlier comprises entirely marine sediments. The other major inliers (Lesmahagow, Hagshaw Hills, Carmichael and Pentland Hills) all record an upward passage from marine turbidites into fluvial sequences. The Llandovery is everywhere mainly composed of shales and mudstones with occasional graptolites. Thin interlayered sandy beds were deposited as turbidites, which may contain a resedimented shallow-water fauna, including bryozoans, bivalves and crinoids. Palaeocurrents in the Girvan Inlier indicate that some turbidites were derived from a northerly direction.

The change from marine to terrestrial sedimentation probably occurs at the base of the Wenlock and is marked by the deposition of major alluvial fan conglomerates in the Pentlands, Carmichael and Hagshaw Hills inliers (Fig. 7.4). The conglomerates thin towards the north-west over *c.* 8 km from 50 m to 0 m near Lesmahagow. Cross-stratification indicates derivation mainly from a southerly direction. At least three different conglomerates have been identified on the basis of clast composition. In the Hagshaw Hills, Carmichael and Pentland Hills inliers, the lowermost conglomerates contain mainly igneous clasts, which has led to these being informally termed the 'igneous conglomerates' (Fig. 7.4). Spilites, keratophyres, andesites and quartz porphyries frequently make up 70–80% of the total, with chert, sedimen-

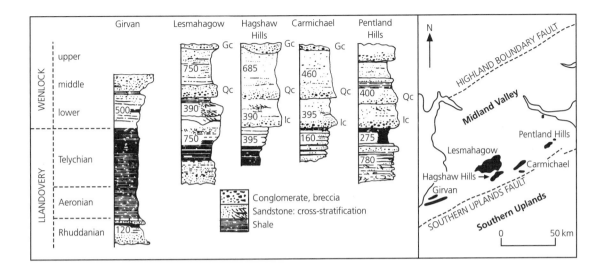

Fig. 7.4 Silurian successions of the Midland Valley (modified from Anderton *et al.* 1979). Logs in metres. Ic, igneous conglomerate; Qc, quartzite conglomerate; Gc, greywacke conglomerate.

tary quartz and quartzite. These conglomerates are overlain by fluvial sandstones and interbedded lacustrine mudstones. The latter may contain a rich fossil fish fauna with arthropods and plant fragments. A higher series of conglomerates known informally as the 'quartzite conglomerates' occurs in all the inliers except for Girvan (Fig. 7.4). These show similar alluvial fan features to the igneous conglomerates with the beds apparently derived from a southerly direction and thinning north-westwards. Clasts are similar to those before, but the proportion of quartz and quartzite is often over 50%. An upper 'greywacke conglomerate' composed mainly of greywacke clasts is present in the Carmichael, Hagshaw Hills and Lesmahagow inliers (Fig. 7.4).

7.2.3 Silurian sedimentation in the Midland Valley Terrane of Ireland

In contrast to the Scottish Midland Valley, the Silurian sedimentary rocks of western Ireland (Fig. 7.5) clearly rest unconformably on a range of older rocks that had been exhumed following the Grampian Orogeny. The Silurian successions of western Ireland are broadly similar in age and sedimentary facies, but detailed correlation is difficult, probably as a result of

Silurian–Devonian sinistral strike-slip displacements of up to several tens of kilometres along the various faults in the region. The Clare Island–Louisburgh succession is a *c.* 1.7-km-thick sequence of fluvial to marginal marine siliciclastic sediments of Wenlock–Ludlow age (Fig. 7.5). On Clare Island, the succession rests unconformably on the ophiolite of the Deer Park Complex. This succession is in fault contact to the south with strongly cleaved, greenschist facies Silurian sediments of Croagh Patrick, which are distinctively older (Llandovery–Wenlock). The two successions were possibly juxtaposed by Caledonian displacements along the Fair Head–Clew Bay Line (see Chapter 12).

The most complete Silurian succession in western Ireland is that preserved in North Galway (Fig. 7.5), which unconformably overlies the Dalradian of Connemara and the Ordovician sediments and volcanics of the South Mayo Trough. The *c.* 3-km-thick succession commences in the Middle Llandovery with the Lough Mask Formation, an unfossiliferous sequence of red cross-bedded sandstones with minor red mudstones, conglomerates and breccias. Palaeocurrents flowed southwards, and the formation has been interpreted as a fluvial sequence deposited by braided rivers. The overlying siltstones and sandstones of the Kilbride Formation are clearly marine, because they contain brachiopods, trilobites, corals, crinoids and *Skolithos* burrows. These are succeeded by red shales of the Tonalee Formation, thought to have been deposited in an offshore shelf setting. The

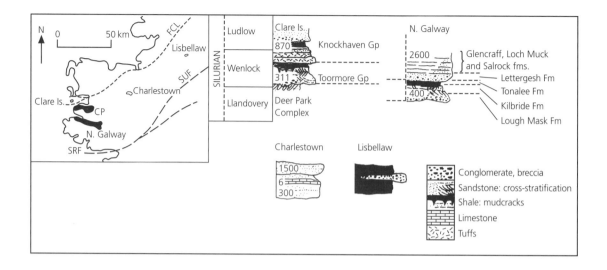

Fig. 7.5 Silurian successions of western Ireland (modified from Anderton *et al.* 1979). Logs in metres. Inset: CP, Croagh Patrick succession; SRF, Skird Rocks Fault; SUF, Southern Uplands Fault; FCL, Fair Head–Clew Bay Line.

Llandovery–Wenlock transition is marked by the incoming of a 1500-m-thick regressive sequence of conglomerates and sandstones with occasional graptolitic mudstones, the Lettergesh Formation, for which a fan-delta environment of deposition has been suggested. There is strong evidence for contemporaneous volcanism. Lithic clasts in the sandstones include porphyritic volcanics with fresh (andesine) feldspars and embayed quartz. The formation also incorporates lapilli tuffs, reworked tuffs and devitrified glass shards. Overlying Wenlock units (Glencraff, Lough Muck and Salrock formations) are dominated by shallow marine sandstones, siltstones and shales. Palaeocurrent data indicate a general north-easterly trending shoreline. Other inliers to the north-east comprise thinner sequences, which are equivalent to different levels of the North Galway succession (Fig. 7.5). The petrology of conglomerate clasts indicates some common features of source type for many of the Irish and Scottish Silurian sediments. Conglomerates in the Galway, Croagh Patrick, Lisbellaw and Pomeroy inliers (Figs 7.1, 7.5) are dominated by metaquartzite clasts, very similar to the 'quartzite conglomerates' of the Scottish inliers.

7.2.4 Synthesis

The Mid- to Upper Ordovician succession of the Girvan area has generally been interpreted as providing evidence for a southward transition from shallow- to deep-water sedimentation in a basin located to the south of an active volcanic arc. A proximal fore-arc or interarc setting is thus likely for the Girvan succession. If the Girvan succession can be taken to be representative of conditions during the Mid–Upper Ordovician, the generally shallow-water nature of Silurian sedimentation in Scotland and Ireland indicates a progressive infilling of the Midland Valley Basin. Evidence (although limited) for contemporaneous volcanic activity and source areas located to the north and the south suggests that the Silurian successions accumulated in a series of possibly separate inter-arc basins.

7.3 Development of a Mid-Ordovician to Silurian accretionary prism in the Southern Uplands Terrane

The Southern Uplands Terrane is characterized by thick sequences of Ordovician and Silurian sedimentary rocks that were deposited in deep marine environments. In Scotland, the terrane has been divided into three major fault-bounded tectono-stratigraphic units: the Northern, Central and Southern Belts (Fig.

7.1); the Northern and Central Belts probably extend laterally into Ireland. The Northern Belt comprises Ordovician sediments and rare volcanics; the Central Belt is formed of Ordovician and Silurian rocks, and the Southern Belt is entirely Silurian in age. Each belt is divided internally into a series of tracts, each at least several kilometres in width and bounded by major reverse faults. Although the sediments are all steeply dipping and strongly folded, the overall younging direction within each tract is towards the north-west.

7.3.1 The Northern Belt

This is bounded to the north by the Southern Uplands Fault and to the south by the Orlock Bridge Fault (Fig. 7.6). The Ordovician succession ranges in age from Arenig to Ashgill. It consists of a basal sequence of lava and chert (Crawford Group) and black shale (Moffat Group), which was gradually buried under advancing clastic turbidites (Leadhills Supergroup) deposited as large-scale submarine fans (Fig. 7.7).

Mixed volcanic and chert successions known collectively as the Crawford Group form the oldest components of many of the fault-bounded tracts. The oldest rocks are of Arenig age and found only near Abington (Fig. 7.7 tract 4), where the local succession consists of interbedded basaltic pillow lavas, blue-grey radiolarian cherts and brown mudstones, with dolerite intru-

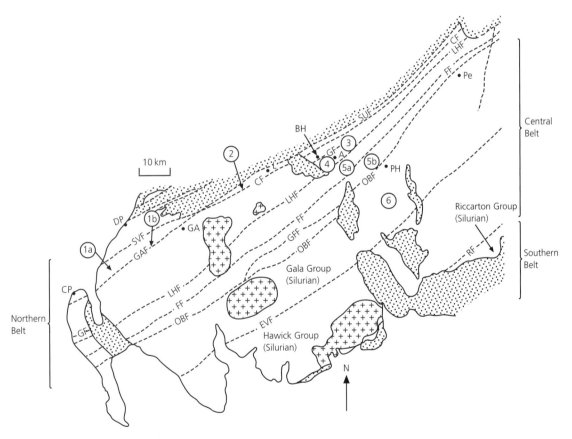

Fig. 7.6 Geological map of the Southern Uplands (modified from Owen *et al.* 1999). Major reverse faults separate structural units with different stratigraphic sequences; beds within units young predominantly to the north-west. The numbers refer to the structural tracts depicted in Fig. 7.7. Faults: SUF, Southern Uplands Fault; SVF, Stinchar Valley Fault; GAF, Glen App Fault; CF, Carcow Fault; GF, Grassfield Fault; FF, Fardingmullach Fault; GFF, Glen Fumart Fault; LHF, Leadhills Fault; OBF, Orlock Bridge Fault; EVF, Ettrick Valley Fault; RL, Riccarton Fault. Place names: A, Abington; BH, Bail Hill; CP, Corsewall Point; DP, Downan Point; GA, Glen App; L, Leadburn; Pe, Peebles; PH, Pinstane Hill. 1a–6, structural units (see text). Crosses represent Siluro–Devonian granites; stipple denotes post-Silurian strata.

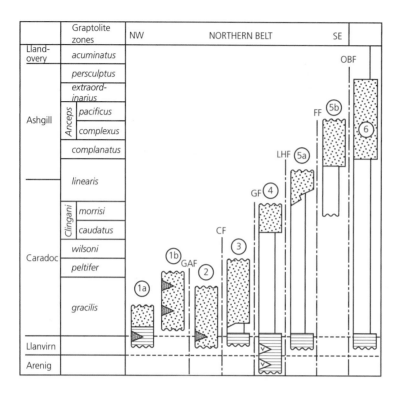

Fig. 7.7 Time–stratigraphic diagram showing positions of coarse clastic sediments and pelagic mudstones and shales (modified from Owen *et al.* 1999). Numbers correspond to the structural units identified in Fig. 7.6. Abbreviations of faults as described in Fig. 7.6. Horizontal lines = Crawford Group (cherts and mudstones); 'v' ornament = volcanics; blank = Moffat Group (shales); dots = Leadhills Supergroup (sandstone-dominated turbidites).

sions or lava flows. These are overlain by a sequence of red and grey cherts and siliceous mudstones of Llan-virn–lowermost Caradoc age, which is correlatable with similar units of this age in other tracts (Fig. 7.7 tracts 3, 5 and 6). Basaltic pillow lavas of similar age occur at Downan Point and in the Leadburn area (Figs 7.6, 7.7 tracts 1 and 2). The Moffat Group comprises mainly dark grey and black graptolitic shales. Traced towards the north-west, the Moffat Group shales are replaced by sandstone turbidites at progressively older stratigraphic levels in successive fault-bounded tracts (Fig. 7.7). In the most northerly tract (Fig. 7.7 tract 1), no shales are present and turbidites of the Leadhills Supergroup rest directly on red cherts and lavas of the Crawford Group.

The Leadhills Supergroup comprises thick sequences of sandstone-dominated turbidites, which overlie the Moffat Group in most tracts (Fig. 7.7). Of particular interest are major conglomerate horizons, which occur at Corsewall Point and Glen App (Fig. 7.6). These contain a wide variety of clast types yielding valuable information on the nature of the source terrain. The Corsewall Point Conglomerate is *c.*

500 m thick; the clast assemblage is dominated by granites, but acid volcanics, basalts and microgab-bros are also common. The most common granite types are unfoliated hornblende–biotite granites (very similar to that identified in the Benan Conglomerate near Girvan) and muscovite–biotite granites. Palaeocurrent data indicate derivation from the north-west. The movement of boulder-sized gravels requires steep slopes and the flows which deposit them are not easily diverted during their movement. Palaeoflow directions obtained from the Corsewall Conglomerate are thus thought to be a reliable indication of a major source region located to the north-west. The Glen App Conglomerate is *c.* 180 m thick; individual beds are up to 12 m thick with erosive channelized bases. Clasts up to 1.5 m in diameter are present, although most are between 0.5 and 1 m. The clasts are dominated by granites similar to those at Corsewall Point, but also include granitic gneisses, felsic volcanics, acid porphyries, microgabbros, basalts, gabbros and quartzites. Palaeoflow directions obtained from the Glen App Conglomerate again indicate derivation from the north-west.

The sandstone turbidites in the north-western and central parts of the Northern Belt are typically grey, quartz rich and include small clasts of quartzite, phyllite, schist, gneiss and hornfels, implying derivation from a metamorphic source region. $^{40}Ar/^{39}Ar$ dating of detrital muscovite grains has yielded ages mostly in the range c. 480–460 Ma. These are very similar to the regional cooling ages characteristic of the deformed and metamorphosed Dalradian Supergroup, which was being exhumed to the north-west following the Grampian Orogeny. Palaeoflows are more variable than in the conglomerates. Some are from the north-west, but the majority are from the north-east and are interpreted as axial flows directed along the strike of the basin. Interestingly, some sandstones record palaeoflows from the south-east and south. These sandstones incorporate hornblende and pyroxene grains and fragments of andesite. $^{40}Ar/^{39}Ar$ dating of two volcanic clasts has yielded ages of 560 ± 50 Ma and 530 ± 10 Ma, implying derivation from a Cambrian arc or seamount. Remnants of a possible volcanic seamount developed *in situ* are preserved at Bail Hill (Fig. 7.6), where sediments are overlain by andesite lavas and pyroclastic rocks. The sandstone turbidites in the south-east part of the Northern Belt are characteristically dark because of their basic composition. They have high proportions of hornblende, pyroxene and andesitic fragments, and blueschist clasts have also been identified. In the Tweed Valley, near Peebles (Fig. 7.6), the mid-Caradoc age sediments contain transported blocks up to several metres in diameter of alkaline volcanics and shallow-water Llanvirn limestones. Both the volcanics and the limestones are interpreted as an olistostrome which brought a Llanvirn sequence from a seamount down into deep waters in Mid-Caradoc time.

7.3.2 The Central and Southern Belts

The pattern of sedimentation established in the Northern Belt continues through the Central and Southern Belts with only minor differences. In north-western parts of the Central Belt, the oldest sediments are Ordovician graptolitic mudstones and carbonaceous shales of Caradoc–Ashgill age (Fig. 7.7), which have been correlated with the Moffat Group. These are only very minor components of the Central Belt, which mostly comprises thick sequences of Silurian

sandstone greywackes and minor conglomerates deposited in submarine fans. These have been subdivided into the Gala Group (north-west of the Ettrick Valley Fault) and the Hawick Group (to the south-east). The influx of sandstones occurred generally diachronously southwards through the Llandovery and possibly into the Wenlock (Fig. 7.7). Palaeocurrents are commonly south-west directed and again interpreted as axial flows along the strike of the basin. The greywackes of the various structural units can be distinguished on the basis of clast suites and the presence or absence of detrital minerals such as pyroxene and garnet. Sandstones and associated conglomerates in the north-westernmost part of the Gala Group outcrop, for example, commonly include detrital pyroxene and hornblende. The Pinstane Hill submarine fan conglomerate (Fig. 7.6) resembles the Corsewall and Glen Afton conglomerates of the Northern Belt. However, the maximum clast size here is 30 cm and granite clasts form a much smaller proportion of the coarse detritus. Volcanic rocks, cherts and quartzite are the most common types, and basalt, vein quartz and schist are also present. By contrast, Hawick Group greywackes lack pyroxene, are typically more siliceous and may contain garnet. The greywackes in the southernmost part of the Hawick Group outcrop are distinguished by common carbonate cement, the presence of laminae and beds of reddish mudstone, and by haematite staining of detrital white mica grains. The Southern Belt lies south of the Riccarton Fault (Fig. 7.6) and is composed entirely of a 4-km-thick succession of Wenlock greywacke sandstones known as the Riccarton Group, which are characterized by numerous thin silty graptolite horizons (Fig. 7.7).

7.3.3 Structural evolution

A similar structural evolution is recorded throughout the Southern Uplands, but has been studied in most detail in the Northern and Central belts that are well exposed on the coast. Initial deformation probably occurred while the sediments were still wet and unlithified, and is characterized by the development of slump sheets and slump folds. The main stage of deformation is associated with south-east-directed thrusting and widespread south-east-vergent folding. Folding may have initiated in partially lithified sediments and progressed until deformation was proceeding under low-grade metamorphic conditions (prehnite–pumpellyite grade), as indicated by the development of cleavage. In the Northern Belt, cleavage is axial-planar to folds, indicating approximately coaxial deformation. By contrast, in the Central Belt, the clockwise obliquity between cleavage and fold hinges indicates that compression was accompanied by a component of sinistral strike-slip movement (transpression). In both the Northern and Central belts, early thrusts were later reactivated as sinistral strike-slip faults. A final stage of deformation common to both belts resulted in the development of low-angle, north-west-directed thrusts and a crenulation cleavage in parts of the Central Belt.

7.3.4 Synthesis

A model for the sedimentological and tectonic evolution of the Ordovician and Silurian rocks of the Southern Uplands Terrane must account for the following features in particular:

1 The dominance of deep marine sediments deposited in a variety of submarine fan and pelagic settings.
2 Their location north of the Iapetus Suture (see Chapter 12) and south of an active volcanic arc in the Midland Valley region.
3 The occurrence of numerous major reverse faults, spaced several kilometres apart.
4 Differences in detailed stratigraphy between fault-bounded units.
5 The occurrence to the south-east of progressively younger rocks, and in particular the diachronous south-eastward spread of submarine fan deposition.

One solution to this problem is provided by studies of modern active plate margins. Thick piles of deformed sediment known as accretionary prisms develop in the fore-arc regions of modern convergent plate boundaries because the sedimentary cover of the oceanic floor is generally not subducted along with the ocean-floor basalt (Fig. 7.8). Instead, packets of sediment are sequentially scraped off the downgoing oceanic plate. Modern accretionary prisms are typically divided up into a series of fault-bounded slices. The faults are initially low-angle thrusts, which are rotated progressively into a steep orientation as a result of continuous underthrusting and build-up of the accretionary prism. Within each slice, sediments young continentward, yet the absolute ages of the sediments in the slices will be younger towards the ocean (Fig. 7.8). Thrusting is

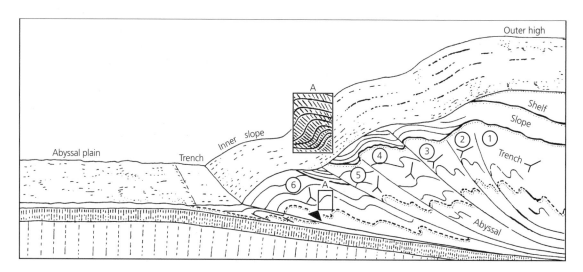

Fig. 7.8 A model for the development of an accretionary prism at an oceanic trench (modified from Anderton *et al.* 1979). Note the different fault-bounded packages of deformed sediments that were accreted in the order 1–6. The faults originate as low-angle thrusts (e.g. base of unit 6), which are passively rotated into a steep orientation as younger units are continually underthrust. Thrusting is accompanied by oceanward-vergent folding and cleavage development (see inset A), but an overall upward younging direction is characteristic of each unit.

typically accompanied by widespread folding at all stages prior to, during and following lithification. Folding in the latter stage is associated with the development of cleavage under low-grade metamorphic conditions resulting from progressive thickening of the accretionary prism.

The generally accepted interpretation of the Ordovician and Silurian rocks of the Southern Uplands is that they represent an accretionary prism developed above a northerly dipping subduction zone. This accounts satisfactorily for all their intrinsic sedimentological and structural features and is consistent with the inferred overall tectonic setting. The proposed Southern Uplands accretionary prism compares well in terms of scale with modern accretionary prisms such as the Oaxaca prism off Mexico. Differences in stratigraphy between each fault-bounded unit can be explained by their having been initially deposited on widely separated areas of the ocean floor, and progressively juxtaposed by accretion as they were transported in conveyor belt fashion towards the subduction zone. The Arenig basalts of the Leadhills area (Fig. 7.7 tract 4) may represent the uppermost layer of oceanic crust, which apparently was not subducted. Other occurrences of basalt *within* turbidite successions (e.g. Fig. 7.7 tract 1b) presumably represent submarine volcanics erupted onto the oceanic floor. Convincing evidence that south-east-directed thrusting and folding and accompanying low-grade metamorphism of older parts of the prism occurred synchronously with sedimentation of younger parts is provided by the presence of recycled grains of prehnite and pumpellyite in turbidites. The change from early coaxial deformation in the Northern Belt to sinistral transpression in the Central Belt may reflect a change in the angle of subduction relative to the continental margin.

7.4 The Southern Uplands controversy — alternatives to the accretionary prism model

The accretionary prism model for the Southern Uplands was first published in the mid-1970s and generally accepted for the following 10 years. However, it was always evident that the structural geometry of the Southern Uplands successions is not *by itself* diagnostic of an accretionary prism. A similar geometry might be expected in any thrust belt

whether developed in a foreland basin, a back-arc basin or an accretionary prism. If further data emerged that indicated a different palaeogeographic setting for the Southern Uplands successions, then the accretionary prism model might require modification. Further research, particularly in the Northern Belt, has indeed led to the development of alternative tectonic models and much controversy as to the regional palaeogeographic setting of the Southern Uplands.

7.4.1 The Northern Belt of the Southern Uplands as a back-arc basin?

Several models published in the mid- to late 1980s envisage that the Northern Belt sediments were deposited in a back-arc basin located to the north of a major volcanic arc (Fig. 7.9a). Such models hinge critically on the presence in the Northern Belt of volcaniclastic-rich turbidites apparently derived from the south. The mixture of volcaniclastic and metamorphic detritus in these turbidites suggests that any such arc was developed on a microcontinent. The location of this now unexposed arc would be either at the margin of the Northern Belt or in the Central Belt. There are various explanations to account for the subsequent removal of the arc from southern Scotland. Some workers have argued that it is buried at depth as a result of thrusting. It has been proposed that the northern basin closed by the end of the Ordovician when thrusting carried deformed sediments over the remnant arc (Fig. 7.9b). The Central and Southern belts are thought to represent recycled sediments derived from the rising thrust stack to the north and which were progressively accreted into the thrust stack as it migrated southwards (Fig. 7.9b). Alternatively, it has been suggested that the arc was removed by strike-slip faulting along the Orlock Bridge Fault. The presence of numerous late to post-tectonic calc-alkaline lamprophyre dykes in the Southern Uplands has also been cited in support of a back-arc basin model on the basis that such igneous rocks might not normally be expected to occur in a fore-arc region so close to the subduction zone. Various counter-arguments have been put forward by supporters of the accretionary prism model against the back-arc basin interpretation. These may be summarized briefly as follows:

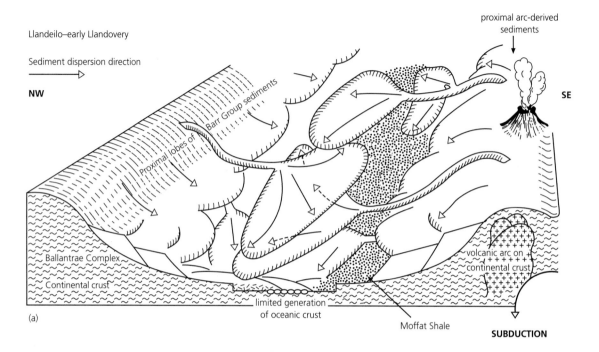

Llandeilo–early Llandovery

Sediment dispersion direction
→

NW

SE

proximal arc-derived
sediments

Proximal lobes of Barr Group sediments

Ballantrae Complex

Continental crust

limited generation
of oceanic crust

Moffat Shale

volcanic arc on
continental crust

SUBDUCTION

(a)

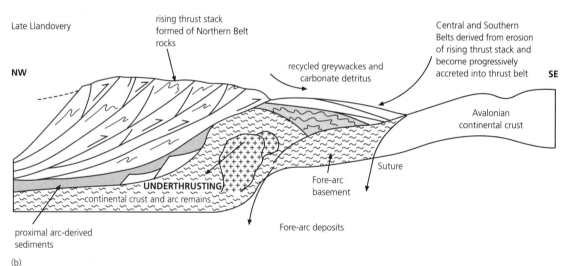

Late Llandovery

NW

SE

rising thrust stack
formed of Northern Belt
rocks

recycled greywackes and
carbonate detritus

Central and Southern
Belts derived from erosion
of rising thrust stack and
become progressively
accreted into thrust belt

Avalonian
continental crust

UNDERTHRUSTING
continental crust and arc remains

Suture

Fore-arc
basement

Fore-arc deposits

proximal arc-derived
sediments

(b)

Fig. 7.9 A back-arc interpretation for the Northern Belt of the Southern Uplands. A volcanic arc on continental crust supplied sediment into a back-arc basin until Llandovery times (a), when thrusting carried deformed sediments over the remnant arc (b) (modified from Stone *et al.* 1987).

1 Experience in modern trench systems (e.g. Nankai trough, Japan) shows that axial turbidite channels can meander appreciably and in some circumstances may direct sediment towards an accretionary prism. The southerly derived palaeocurrents in the Northern Belt should not therefore be taken as firm evidence for the location of the source area; the volcaniclastic turbidites could have come from an axial volcanic source, not an outboard volcanic source.

2 Even if volcaniclastic turbidites were derived from the south, they could have resulted from the erosion of seamounts on the oceanic plate. Examples of such volcanics which might be interpreted as seamounts are preserved *in situ* at Bail Hill and as blocks in an olistostrome in the Tweed Valley.

3 The 'southerly derived' volcanic detritus has so far only yielded latest Precambrian–early Cambrian isotopic ages that are considerably older than any sediments within the Southern Uplands itself, thus ruling out derivation from a contemporaneous arc.

4 Modern back-arc basins (e.g. Sea of Japan) contain enormous volumes of arc-derived volcanic detritus in comparison with the very modest amount of such material in the Northern Belt.

5 It seems highly unlikely that an entire microcontinental arc (and its fore-arc) could be overthrust by back-arc basin sediments, especially given the buoyancy of the arc crust and its underlying continental basement, and the fact that no trace of the arc has been exposed by subsequent uplift and erosion.

6 The widespread lamprophyre dykes have yielded Early Devonian isotopic ages (*c.* 418–395 Ma) and are therefore appreciably younger than any sediments

in the Southern Uplands. These dykes were probably intruded during final Caledonian collision and underthrusting of the Southern Uplands by the Avalonian continental crust of the Lake District (see Chapter 12). Accordingly, irrespective of the petrogenetic origin of the dykes, they may be of no relevance to an understanding of the palaeogeographic setting of the Southern Uplands through the Ordovician and Silurian.

7.4.2 The Southern Uplands as a rifted continental margin?

A further model published in the mid- to late 1990s considers the Northern Belt of the Southern Uplands to have been deposited on rifted continental crust in a fore-arc basin (Fig. 7.10). Advocates of this model argue for only minor displacements across the Southern Uplands Fault on the basis that there are considerable sedimentological and faunal similarities linking the Mid- to Late Ordovician successions of Girvan and the Northern Belt. These successions together could be viewed as recording progressive rifting of a continental margin and outbuilding of submarine fans onto a subsiding shelf. This model is based in part on the chemistry of the lower Caradoc basalt lavas within the Northern Belt. If the accretionary prism model is correct, then these lavas should have a

Fig. 7.10 A fore-arc interpretation for the Northern Belt.

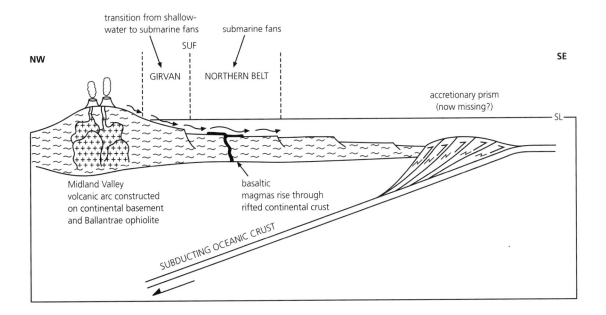

geochemical signature comparable with modern-day oceanic basalts. However, the lavas are more comparable on geochemical grounds, with basalts erupted into rifted continental crust (Fig. 7.10). Additional supporting evidence is provided by the rare earth element (REE) signatures of Caradoc and Ashgill radiolarian cherts and siliceous mudstones from the Northern Belt. Determination of the REE signatures of such rocks has become a powerful tool in determining the environmental setting of Mesozoic and Cenozoic oceanic and ocean margin sedimentary successions. The cherts and siliceous mudstones of the Northern Belt provide REE signatures that are comparable with those of more recent deposits from continental margin settings, and different from those characteristic of the open oceanic setting required in the accretionary prism model. Whether or not a fore-arc basin model can also be applied to the Central and Southern belts is at present unclear and awaits further research.

7.5 Summary: regional tectonic framework of the Midland Valley and Southern Uplands terranes

Although the Southern Uplands controversy is currently unresolved, the general consensus in recent years is that the tectonic evolution of the Mid-Ordovician to Silurian rocks of the Midland Valley and Southern Uplands can be reconstructed as follows. Various lines of evidence indicate that an active volcanic arc was located in the region of the Midland Valley, to the south of the Grampian mountain belt (Fig. 7.11). The volcanic arc was presumably constructed on the Midland Valley basement and a remnant Ordovician arc, and developed in response to north-west-directed subduction as indicated by the geometry of the Southern Uplands accretionary prism. Substantial amounts of metamorphic and igneous detritus derived from erosion of the Dalradian rocks and the volcanic arc were shed southeastwards into fore-arc basins (e.g. Girvan) and ultimately distributed via submarine fans onto oceanic crust (Southern Uplands accretionary prism). The present location of a proximal fore-arc basin at Girvan adjacent to an accretionary prism is, however, puzzling when the scale of modern active margins is considered. In present-day examples, there is normally a gap between the arc and the trench of c. 90 km that is occupied by a fore-arc basin, now apparently largely absent in Scotland and Ireland. The Silurian conglomerates of the Midland Valley of

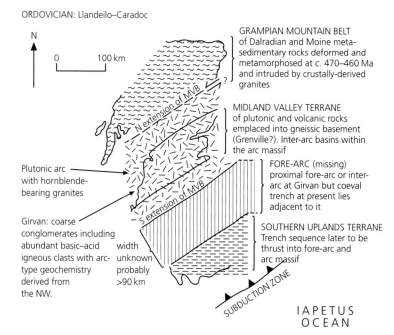

Fig. 7.11 A tectonic reconstruction of the region between the Grampian mountain belt and the Iapetus Ocean during Ordovician (Llandeilo–Caradoc) times (after Bluck 1984). MVB, Midland Valley Block.

Scotland provide additional evidence for a 'lost' area of crust to the south. Calculations of drainage length and basin dimensions for the Wenlock conglomerates suggest that they were deposited in a large alluvial fan system, which had a source possibly as far south as the present-day Solway Firth. Clast compositions indicate a source area composed of acidic and basic volcanic rocks with a pre-existing metaquartzite cover. Such rocks do not now exist in the Southern Uplands, and it is unlikely that they were derived from erosion and re-sedimentation of older conglomerates in this area.

A solution to the problem posed by these areas of 'lost' crust is that a Midland Valley arc terrane comprising basement and associated volcanic rocks once occupied a very much wider area than at present, and was bordered to the south by a fore-arc basin of unknown width, but probably >90 km, and the

Fig. 7.12 Comparison between the tectonic elements of the Caledonides and of Sumatra and Java. (a) British Isles superimposed on Sumatra and Java with no palinspastic reconstruction for the Caledonian orogenic belt. (b) Java superimposed on a palinspastic reconstruction of the Caledonides of northern Britain (modified from Bluck 1984).

Southern Uplands accretionary prism (Fig. 7.11). A palinspastic reconstruction of the Midland Valley and Southern Uplands terranes along these lines bears close comparison in terms of the overall scale of tectonic elements with currently active plate margins in south-east Asia (Fig. 7.12). It has been suggested that the 'lost' metamorphic–igneous source for the Silurian conglomerates of the Midland Valley is at present located at depth *beneath* the Southern Uplands accretionary prism, which is thought to be allochthonous (Fig. 7.13). Two lines of evidence support this interpretation. Firstly, deep seismic reflection profiling has shown that the Southern Uplands is underlain by a major southward-dipping reflector, which is interpreted as a major northward-directed thrust formed during the 'late' Caledonian collision of Eastern Avalonia with Laurentia (see Chapter 12). The crust beneath this reflector has the same geophysical characteristics as the continental crust of the Midland Valley. Secondly, Carboniferous volcanic vents in the Southern Uplands incorporate xenoliths of granulite facies gneisses derived presumably from an underlying continental basement. The late, low-angle north-west-directed thrusts within the Northern and Central belts are likely to be related to

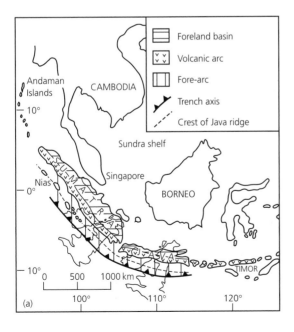

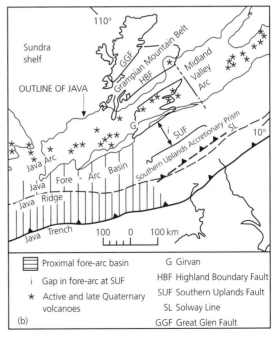

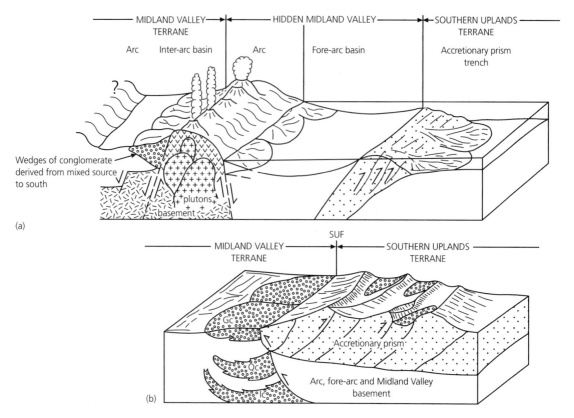

Fig. 7.13 Tectonic reconstruction of the Midland Valley and Southern Uplands terranes during Silurian times: (a) showing derivation of conglomerates from a southerly continental–volcanic arc source; (b) explaining the present juxtaposition across the Southern Uplands Fault of these conglomerates with the accretionary prism by northward overthrusting of the accretionary prism onto various components of the arc, fore-arc and basement (modified from Bluck 1984). Ic, igneous conglomerate; Qc, quartzite conglomerate; Gc, greywacke conglomerate.

large-scale thrusting of the Southern Uplands Terrane over the Midland Valley Terrane (Fig. 7.13).

References

Anderton, R., Bridges, P.H., Leeder, M.R. & Sellwood, B.W. (1979) *A Dynamic Stratigraphy of the British Isles*, pp. 1–301. Unwin Hyman, London.

Bluck, B.J. (1984) Pre-Carboniferous history of the Midland Valley of Scotland. *Transactions of the Royal Society of Edinburgh: Earth Sciences* 75, 275–295.

Ince, D. (1984) Sedimentation and tectonism in the middle Ordovician of the Girvan district, SW Scotland. *Transactions of the Royal Society of Edinburgh: Earth Sciences* 75, 225–237.

Owen, A.W., Armstrong, H.A. & Floyd, J.D. (1999) Rare earth element geochemistry of upper Ordovician cherts from the Southern Uplands of Scotland. *Journal of the Geological Society, London* 156, 191–204.

Stone, P., Floyd, J.D., Barnes, R.P. & Lintern, B.C. (1987) A sequential back-arc and foreland basin duplex model for the Southern Uplands of Scotland. *Journal of the Geological Society, London* 144, 753–764.

Walton, E.K. & Oliver, G.J.H. (1991) Lower Palaeozoic — structure and palaeogeography. In: *Geology of Scotland* (ed. G. Y. Craig), pp. 195–228. Geological Society, London.

Further reading

Armstrong, H.A., Owen, A.W., Scrutton, C.T., Clarkson, E.N.K. & Taylor, C.M. (1996) Evolution of the Northern Belt, Southern Uplands: implications for the Southern Uplands controversy. *Journal of the Geological Society, London* 153, 197–205. [Argues for a rifted continental margin setting for the Northern Belt, based on the chemistry of the Caradoc lavas.]

Bassett, M.G., Bluck, B.J., Cave, R., Holland, C.H. & Lawson, J.D. (1992) Silurian. In: *Atlas of Palaeogeography and Lithofacies* (eds J. C. W. Cope, J. K. Ingham & P. F. Rawson), pp. 37–56. Geological Society, London. [Palaeogeographic maps and summary of regional stratigraphy during the Silurian.]

Bevins, R.E., Bluck, B.J., Brenchley, P.J. *et al.* (1992) Ordovician. In: *Atlas of Palaeogeography and Lithofacies* (eds J. C. W. Cope, J. K. Ingham & P. F. Rawson), pp. 19–36. Geological Society, London. [Palaeogeographic maps and summary of regional stratigraphy during the Ordovician.]

Bluck, B.J. (1984) Pre-Carboniferous history of the Midland Valley of Scotland. *Transactions of the Royal Society of Edinburgh: Earth Sciences* 75, 275–295. [Major synthesis of regional tectonics in the Midland Valley and the Southern Uplands.]

Holland, C.H. (1981) Silurian. In: *A Geology of Ireland* (ed. C. H. Holland), pp. 65–81. Scottish Academic Press, Edinburgh. [Regional overview of the stratigraphy of the Silurian in Ireland.]

Hutton, D.H.W. & Murphy, F.C. (1987) The Silurian of the Southern Uplands and Ireland as a successor basin to the end-Ordovician closure of Iapetus. *Journal of the Geological Society, London* 144, 765–772. [An alternative model to the accretionary prism interpretation.]

Leggett, J.K. (1987) The Southern Uplands as an accretionary prism: the importance of analogues in reconstructing palaeogeography. *Journal of the Geological Society, London* 144, 737–752. [Thoughtful reappraisal of the accretionary prism model and discussion of alternative models.]

Leggett, J.K., McKerrow, W.S. & Eales, M.H. (1979) The Southern Uplands of Scotland: a Lower Palaeozoic accretionary prism. *Journal of the Geological Society, London* 136, 755–770. [First detailed paper proposing the accretionary prism model.]

Morris, J.H. (1987) The Northern Belt of the Longford–Down inlier, Ireland and Southern Uplands, Scotland: an Ordovician back-arc basin. *Journal of the Geological Society, London* 144, 773–786. [An alternative to the accretionary prism model.]

Needham, D.T. (1993) The structure of the western part of the Southern Uplands of Scotland. *Journal of the Geological Society, London* 150, 341–354. [Structural analysis and discussion of regional tectonic models.]

Owen, A.W., Armstrong, H.A. & Floyd, J.D. (1999) Rare earth element geochemistry of upper Ordovician cherts from the Southern Uplands of Scotland. *Journal of the Geological Society, London* 156, 191–204. [Argues for a rifted continental margin setting for the Northern Belt based on the geochemistry of radiolarian cherts and siliceous mudstones.]

Stone, P., Floyd, J.D., Barnes, R.P. & Lintern, B.C. (1987) A sequential back-arc and foreland basin duplex model for the Southern Uplands of Scotland. *Journal of the Geological Society, London* 144, 753–764. [A back-arc interpretation for the Northern Belt of the Southern Uplands.]

Walton, E.K. & Oliver, G.J.H. (1991) Lower Palaeozoic — stratigraphy. In: *Geology of Scotland* (ed. G. Y. Craig), pp. 161–193. Geological Society, London. [Synthesis of the Ordovician and Silurian stratigraphy of the Midland Valley and the Southern Uplands.]

Walton, E.K. & Oliver, G.J.H. (1991) Lower Palaeozoic — structure and palaeogeography. In: *Geology of Scotland* (ed. G. Y. Craig), pp. 195–228. Geological Society, London. [Overview of the regional structure of the Midland Valley and Southern Uplands and contrasting tectonic models.]

Williams, D.M. & Harper, D.A.T. (1988) A basin model for the Silurian of the Midland Valley of Scotland and Ireland. *Journal of the Geological Society, London* 145, 741–748.

Part 3
The Southern Margin of the Iapetus Ocean

Illustration overleaf: Pentamerid brachiopods within a Middle Llandovery (Aeronian Stage) sandstone (Derwyddon Formation) from the type area of the Llandovery series in Crychan Forest, near Llandovery, Mid-Wales. The brachiopods record the pulsed Llandovery (Early Silurian) transgression over the Midland Platform as a result of sea-level rise after the end-Ordovician ice age.

8 Late Neoproterozoic to Cambrian accretionary history of Eastern Avalonia and Armorica on the active margin of Gondwana

R. A. STRACHAN

8.1 Introduction

Palaeomagnetic data summarized in Chapter 2 indicate that, during the late Neoproterozoic, southern Britain and south-east Ireland formed part of a crustal block, *Eastern Avalonia*, which was located along the edge of Gondwana (Fig. 8.1). The Channel Islands formed part of another block, *Armorica*, also thought to have been situated along the Gondwanan margin (Fig. 8.1). The late Neoproterozoic rocks of Eastern Avalonia and Armorica record the development of volcanic arcs and marginal basins produced by oceanic plate subduction (Fig. 8.1). A similar geological history is preserved in related Neoproterozoic rocks of Maritime Canada, Carolina and Florida (Fig. 8.1). This contrasts markedly with events of the same age in northern Britain, which were dominated by rifting and passive margin sedimentation associated with the break-up of Laurentia, Baltica and Amazonia and the formation of the Iapetus Ocean (see Chapter 5). Armorica and Eastern Avalonia were largely stable crustal blocks by the end of the Precambrian, although accretionary tectonism continued along the north-west margin of Eastern Avalonia into the Cambrian, thus partly overlapping the history of Cambrian sedimentation described in Chapter 9 from the interior parts of Eastern Avalonia.

8.2 Precambrian evolution of Armorica: the Cadomian Orogeny

The Channel Islands and north-west France together form the type area of the late Neoproterozoic *Cadomian* orogenic belt, which extends eastwards into central Europe as far as Bohemia (Fig. 8.1). The Cadomian belt in the Channel Islands and north-west France is dominated by calc-alkaline plutons and volcanics, and deformed and metamorphosed volcano-sedimentary sequences of the *Brioverian Supergroup* (Fig. 8.2). The belt is transected by steep strike-slip shear zones and faults that divide it into tectonic blocks with contrasting tectonothermal histories, and which represent dismembered parts of the orogen juxtaposed in latest Neoproterozoic time.

The Palaeoproterozoic *Icartian* granite gneisses exposed on Guernsey and on the French mainland at La Hague and in the Tregor peninsular (Fig. 8.2) represent the only ancient basement known to outcrop between the Iapetus Suture and north-west Africa. The Icartian gneisses of Guernsey have yielded a precise U–Pb zircon age of 2061 ± 2 Ma. Similar, although less reliable, zircon ages ranging between *c.* 2.2 Ga and *c.* 1.8 Ga have been obtained from the Icartian granite gneisses on the French mainland. These ages are thought to date crystallization of the igneous protoliths. Small areas of metasedimentary schist and gneiss exposed on Guernsey, La Hague and possibly Sark probably represent the country rock into which the granites were intruded. Isotopic and geochemical studies of Cadomian igneous rocks imply the existence of a similar-aged basement at depth throughout much of the Cadomian belt in France and also in Spain. The Icartian granite gneisses have been correlated with the granitoids of the *c.* 2.1-Ga Eburnian belt of north-west Africa (Fig. 8.1). Whether or not the Icartian basement was continuous with that of north-west Africa during the late Neoproterozoic is uncertain; it is possible that Armorica may already have rifted away from Africa to form a separate microcontinent.

The onset of Neoproterozoic subduction along the margin of Gondwana may have occurred as early as

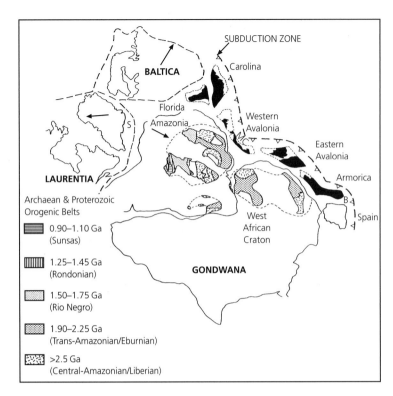

Fig. 8.1 Distribution of the various components of the Avalonian–Cadomian orogenic belt (present-day land in black) in the late Neoproterozoic (redrawn from Nance & Murphy 1996). B, Bohemia; MC, Maritime Canada; S, Scotland. Arrows indicate relative divergence of Laurentia, Baltica and Amazonia to form the Iapetus Ocean at the same time as oceanic plate subduction along the margin of Gondwana.

c. 750 Ma, the U–Pb zircon age of calc-alkaline gneisses in the Penthièvre Complex on the French mainland (Fig. 8.2). Clasts of foliated granodiorite within a Brioverian conglomerate in the Baie de St Brieuc have yielded U–Pb zircon ages of *c.* 670–655 Ma, although the early Cadomian arc from which these clasts were derived has not yet been identified. Subsequent evolution of the Cadomian belt can be divided into three broad phases: (i) early calc-alkaline plutonism (*c.* 615–600 Ma); (ii) rifting and accumulation of the Brioverian Supergroup (*c.* 600–570 Ma); and (iii) regional deformation and metamorphism, and continued magmatism (*c.* 570–540 Ma). A major phase of calc-alkaline plutonism at *c.* 615–605 Ma was associated with the emplacement of quartz diorites on Guernsey and Sark, and at La Hague and in the Tregor Peninsula. On Guernsey and Sark, the oldest plutons, dated at *c.* 615–611 Ma, are foliated and were emplaced during regional deformation and metamorphism. The Icartian granite gneisses probably acquired their dominant deformation fabrics at this time. The youngest plutons are undeformed and were emplaced at *c.* 610–608 Ma.

Rifting of the Icartian continental basement and early Cadomian magmatic arcs at *c.* 600 Ma resulted in development of one or more marginal basins and accumulation of the metavolcanic and metasedimentary rocks of the *Brioverian Supergroup* (Fig. 8.2). The precise age span of the Brioverian is at present unclear, but it is likely to have accumulated between *c.* 600 Ma and *c.* 570 Ma. A stratigraphic succession is difficult to erect because of the tectonically disrupted nature of the belt, but an original thickness of *c.* 4–5 km seems likely. In the Baie de St Brieuc, the lowest levels of the Brioverian are dominated by a *c.* 2-km succession of submarine basaltic volcanics. An acidic member of the volcanic suite has yielded a U–Pb zircon age of 588 ± 11 Ma, which is interpreted to date eruption. The basalts have a subalkaline to tholeiitic chemistry and occur as pillowed and massive flows, interdigitated with intrusive sheets. Pillow breccias, hyaloclastites and peperites are locally common. Similar volcanics form the lowest parts of the Brioverian succession in Normandy. The volcanics are overlain by thick turbidite sandstone sequences with minor volcaniclastic horizons such as

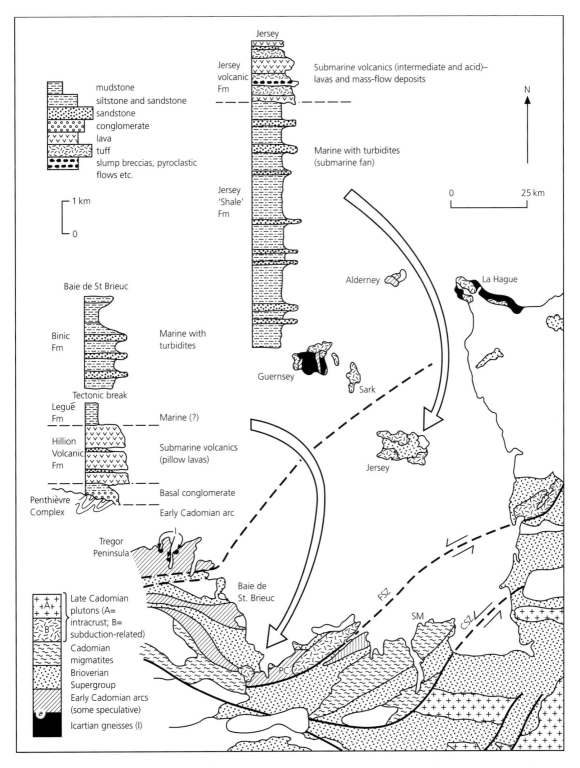

Fig. 8.2 Regional map of the Cadomian belt of Northern France and the Channel Islands, with schematic lithological logs of parts of the Brioverian succession. FSZ, Fresnaye shear zone; CSZ, Cancale shear zone; SM, St Malo; PC, Penthièvre Complex.

those exposed on Jersey (Fig. 8.2). These were probably deposited in various submarine fan environments. Clast types include intermediate–acid volcanic rocks, shales, sandstones, schists, gneisses, black cherts and plutonic rocks. These indicate several source regions, including volcanic arcs, metamorphic basement and reworked contemporary shelf sediments. On Jersey, the Brioverian is overlain conformably by over 2 km of intermediate–acidic volcanic rocks composed of lava flows, pyroclastic ash-flow deposits and mass flow volcaniclastic deposits. Similar volcanics occur in upper parts of the Brioverian in Normandy and together they imply that calc-alkaline volcanism (and hence subduction) accompanied Brioverian sedimentation.

Regional transpressive deformation and greenschist to amphibolite facies metamorphism of the Brioverian Supergroup occurred between *c.* 580 Ma and *c.* 540 Ma and corresponds to the Cadomian orogeny *sensu stricto*. The final stages of regional deformation at *c.* 540 Ma were associated with mainly sinistral displacements along steep, strike-slip shear zones (Fresnaye and Cancale shear zones; Fig. 8.2) and lateral translation of regional tectonic units. The cumulative lateral displacement is unknown but may be at least tens of kilometres. Emplacement of subduction-related plutons continued until *c.* 570 Ma. Spectacular calc-alkaline plutonic suites exposed on Guernsey display abundant evidence for mixing between gabbro, diorite and granite, indicating that these magmas were emplaced contemporaneously. On the French mainland, the Fresnaye shear zone (Fig. 8.2) separates contrasting Cadomian igneous provinces: subduction-related calc-alkaline igneous complexes occur to the northwest and crustally derived melts to the south-east. The migmatite belts of the St Malo region (Fig. 8.2) resulted from the partial melting and migmatization of Brioverian rocks during regional deformation at *c.* 550–540 Ma. The migmatite belts are interpreted as the mid-crustal sources for the Cadomian granites exposed at higher structural levels east of the Cancale shear zone (Fig. 8.2).

In summary, the Cadomian belt resulted from the onset during the late Neoproterozoic of oceanic plate subduction. Calc-alkaline magmatic arcs and marginal basins developed on continental crust were deformed episodically and juxtaposed as a result of sinistral strike-slip displacements at a late stage in the Cadomian Orogeny. Despite the importance of strike-slip displacements within the belt, there is no evidence for the existence of any 'exotic' terranes. A plausible geological history can be constructed without recourse to *major* displacements along any faults. The various tectonic units within the Cadomian belt apparently represent a series of 'proximal' magmatic arc and marginal basin terranes that evolved along the same active continental margin.

8.3 Precambrian evolution of Eastern Avalonia: the Avalon and Monian–Rosslare terranes

Eastern Avalonia comprises two late Precambrian crustal blocks: the Avalon and Monian–Rosslare terranes (Fig. 8.3). The Avalon Terrane underlies central England and much of Wales; the Monian–Rosslare Terrane is exposed in north-west Wales and southeast Ireland and incorporates rather different rock units from those of the Avalon Terrane. The two terranes are separated by a major tectonic boundary, the *Menai Strait Line*.

8.3.1 The Avalon Terrane

In southern Britain, the Avalon Terrane is largely concealed by Phanerozoic rocks. Neoproterozoic rocks outcrop as inliers, many of which are structurally controlled as a result of displacements along long-lived crustal lineaments such as the *Malvern Lineament* and the *Welsh Borderlands Fault System* (Fig. 8.3). Isotopic dating indicates that these rocks formed between *c.* 700 Ma and *c.* 560 Ma. In contrast to Armorica, there are no exposures of any significantly older basement. Regional metamorphic rocks are confined to poorly exposed inliers in east Shropshire (Fig. 8.3). The gar-

Fig. 8.3 (*Opposite.*) Regional map of the Avalon Terrane of southern Britain with schematic lithological logs of the Longmyndian and Charnian supergroups. Squares refer to U–Pb zircon ages, which are interpreted to date the crystallization of plutonic or volcanic units: 1, Malverns Complex granodiorite; 2, Warren House basaltic volcanics; 3, Johnstone complex diorite; 4, Sarn Complex diorite; 5, Arfon Group tuff; 6, diorite intrusive into Charnian Supergroup; 7, Ercall Granophyre; 8, Uriconian rhyolite; 9 and 10, felsic tuffs in the Orton (O) and Glinton (G) boreholes; MSL, Menai Strait Line.

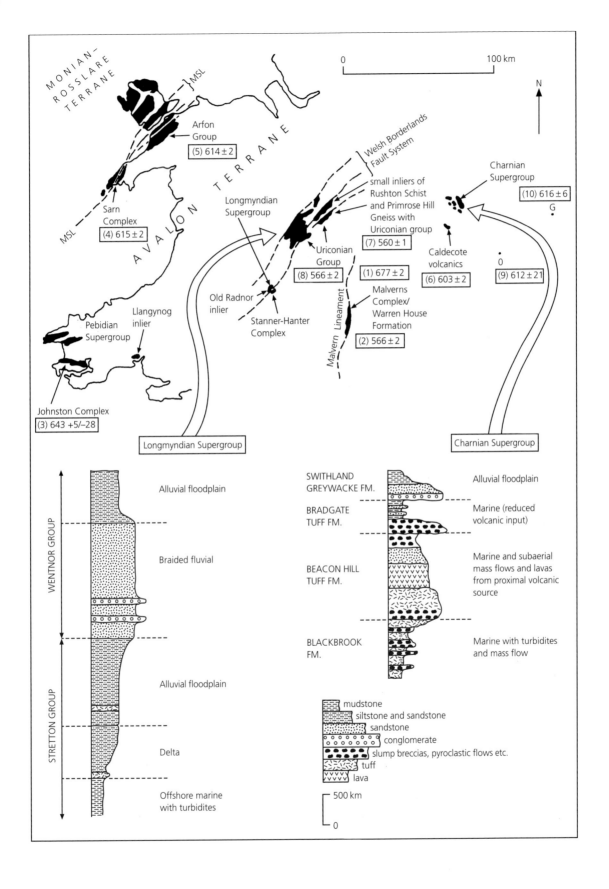

MONIAN–ROSSLARE TERRANE

MSL

Arfon Group
(5) 614±2

AVALON TERRANE

Sarn Complex
(4) 615±2

MSL

Longmyndian Supergroup

Welsh Borderlands Fault System

small inliers of Rushton Schist and Primrose Hill Gneiss with Uriconian group
(7) 560±1

Charnian Supergroup
(10) 616±6
G

Uriconian Group
(8) 566±2

(1) 677±2

Caldecote volcanics
(6) 603±2

0

(9) 612±21

Pebidian Supergroup

Llangynog inlier

Old Radnor inlier

Stanner-Hanter Complex

Malverns Complex/ Warren House Formation
(2) 566±2

Malvern Lineament

Johnston Complex
(3) 643 +5/−28

0   100 km

N

Longmyndian Supergroup

Charnian Supergroup

WENTNOR GROUP

Alluvial floodplain

Braided fluvial

STRETTON GROUP

Alluvial floodplain

Delta

Offshore marine with turbidites

SWITHLAND GREYWACKE FM.

Alluvial floodplain

BRADGATE TUFF FM.

Marine (reduced volcanic input)

BEACON HILL TUFF FM.

Marine and subaerial mass flows and lavas from proximal volcanic source

BLACKBROOK FM.

Marine with turbidites and mass flow

mudstone
siltstone and sandstone
sandstone
conglomerate
slump breccias, pyroclastic flows etc.
tuff
lava

500 km

0

netiferous *Rushton schist* has yielded a Rb–Sr whole-rock isochron of 667±20 Ma, which probably dates metamorphism. The nearby Primrose Hill 'gneiss' incorporates banded mafic units which might represent metamorphosed basic volcanic or plutonic rocks.

Calc-alkaline igneous complexes, which include gabbro, diorite, granodiorite and granite and/or their extrusive volcanic equivalents, form important components of the Avalon Terrane. The oldest rocks are possibly those of the Stanner–Hanter Complex, which lies along the Welsh Borderlands Fault System (Fig. 8.3). The complex comprises sheared and altered gabbro and microgabbro intruded by granite. A Rb–Sr whole-rock isochron of 702±8 Ma obtained from a granite requires confirmation by high-precision U–Pb zircon dating. The oldest reliably dated rocks of the Avalon Terrane are those of the *Malverns Complex*, which forms a series of fault-bounded blocks located along the Malvern Lineament (Fig. 8.3). The complex was intruded at *c.* 680–670 Ma (U–Pb zircon, monazite) and is mainly composed of diorite, tonalite and granite, which are intruded by microgabbro, microdiorite and pegmatite. Metasedimentary(?) gneisses exposed at the south end of the Malvern Hills may represent the country rock into which the plutonic rocks were emplaced. The Malverns Complex was variably deformed and metamorphosed within the low amphibolite facies, and as a result many of the plutonic rocks are banded and schistose. $^{40}Ar/^{39}Ar$ mineral cooling ages place a lower limit of *c.* 650 Ma on this event, with greenschist facies metamorphism and localized shearing occurring between *c.* 650 Ma and *c.* 600 Ma. Other Avalon plutons are generally less deformed and metamorphosed. These include the diorite–granodiorite–granite suite of the Johnston Complex of South Wales, the gabbroic to granitic rocks of the Sarn Complex in north-west Wales, the Charnwood and Nuneaton diorites, and the Ercall Granophyre of Shropshire (Fig. 8.3). The latter is of international importance with regard to calibration of the timescale, because it is overlain disconformably by lower Cambrian quartzites.

The most important volcanics are those of the *Uriconian Group* of Shropshire (Fig. 8.3). These are a thick pile of pyroclastic breccias, agglomerates, tuffs and lavas, which range from basalt through andesite and dacite to rhyolite. The common occurrence of welded tuffs, combined with a lack of well-developed

pillow structures in the basalts, suggests subaerial eruption. Isotopic dating indicates ages of *c.* 566 Ma for these lavas and the similar basalt and basaltic andesite lavas of the Warren House Formation in the Malvern Hills (Fig. 8.3).

Major volcano-sedimentary sequences crop out in Wales, Shropshire and central England (Fig. 8.3). In north-west Wales, the *c.* 4 km *Arfon Group* comprises mainly welded ash-flow tuffs that pass upwards into interbedded sandstones and tuffs. A Precambrian age for the lower part of the Arfon Group is indicated by a U–Pb zircon age of 614±2 Ma obtained from a rhyolitic ash-flow tuff. Unconformity between the Arfon Group and undisputed Cambrian strata in this part of Wales has not been proven, and it is therefore possible that upper parts of the Arfon Group could be Cambrian in age. In south-west Wales, the *Pebidian Supergroup* comprises *c.* 2 km of weakly metamorphosed and cleaved trachytic, andesitic and felsic tuffs with lavas and thin conglomerate horizons. At Llangynog in central South Wales (Fig. 8.3), a small inlier of Precambrian rocks includes rhyolitic and andesitic lavas and tuffs intruded by dolerite, and siltstones and sandstones that are interbedded with the extrusive igneous rocks. The sediments contain a late Neoproterozoic Ediacaran fauna, which includes *Cyclomedusa* sp., *Medusinites* sp., shallow branching burrow systems and feeding trails.

In Shropshire, the *Longmyndian Supergroup* is a *c.* 6 km succession of sandstones and mudstones, with occasional conglomerates and tuffs, mainly exposed within the Welsh Borderlands Fault System (Fig. 8.3). The lower part (Stretton Group) records a regressive sequence from basin-plain muds, through turbidites and shallow marine-deltaic sands to alluvial flood-plain sandstones and siltstones. Possible biogenic structures are present within some shallow marine, low-energy deposits. The upper part (Wentnor Group) is dominated by braided alluvial sandstones and conglomerates. Palaeocurrent data indicate that the basin was probably orientated north-east–south-west and sourced in part from the south-east. Minor pyroclastic deposits (lapilli and ash tuffs) occur in the lower part of the succession. Rare clasts of schist and fragments of detrital garnet indicate a metamorphic source area similar to the Rushton schist. Some of the sediments contain volcanic debris, probably derived from the nearby Uriconian Group. The succession was tightly folded during sinistral transpression, probably during the late Precambrian, although this cannot be demonstrated unequivocally. The supergroup was probably deposited in a basin either within or close to the Uriconian volcanic arc.

In Central England, the *Charnian Supergroup* consists of *c*. 3 km of tuffs, pelites and greywackes with subordinate slump breccias, volcanic breccias, quartz arenites and conglomerates (Fig. 8.3). The large volume of volcaniclastic material contrasts with the Longmyndian Supergroup. Lower parts of the succession (Blackbrook Formation and Beacon Hill Tuff and Bradgate Tuff formations) contain extensive pyroclastics. These were deposited by submarine pyroclastic flows and turbidity currents, and probably developed close to active volcanic centres. Some of the fine-grained volcaniclastic rocks contain an Ediacaran fauna, including impressions of the frond-like *Charnia masoni* and the disc-like *Charniodiscus concentricus*. Examples of contemporaneous volcanic centres may be represented by the Whitwick and Bardon Hill Complexes, which comprise volcanic breccias and tuffs intruded by acid–intermediate quartz–feldspar porphyries. The upper part of the succession (Swithland Greywacke Formation) records an upward transition into fluvial conglomerates and sandstones. The Charnian Supergroup apparently records the rapid degradation of a volcanic arc eruptive centre associated with inter-arc basin sedimentation. A minimum age for deposition of the supergroup is provided by a U–Pb zircon age of 603 ± 2 Ma obtained for a diorite, which intrudes the sequence at Nuneaton (Fig. 8.3). Deformation and metamorphism of the sequence occurred prior to deposition of basal Cambrian quartzites exposed at Nuneaton. The Charnian Supergroup was tightly folded and cleaved during dextral transpression. Accompanying metamorphism was at low greenschist facies. Felsic tuffs similar in age to parts of the Charnian Supergroup occur at depth in nearby boreholes in the east Midlands (Fig. 8.3).

Geochemical studies of the Uriconian, Charnwood and Warren House volcanics demonstrate that all three suites show evidence for subduction-related arc magmatism. The geochemistry of the Warren House lavas is consistent with a primitive marginal basin or oceanic arc. The Charnwood lavas were probably erupted in a primitive arc located on oceanic or strongly attenuated immature continental crust. By contrast, the Uriconian volcanics show a strong within-plate component in their geochemical signature and may have been erupted in a marginal basin floored by continental crust. The geochemical differences between these three volcanic sequences, although subtle, may indicate that initially they evolved as separate terranes that were later amalgamated along the Malvern Lineament.

The nature of the concealed basement of the Avalon Terrane is largely conjectural. Data from numerous boreholes drilled mostly in central England indicate Precambrian rocks broadly similar to those exposed at surface. The poor acoustic reflectivity of the middle and lower crust of the English Midlands is compatible with a deep basement composed largely of arc-related plutonic igneous complexes. However, isotopic analysis of the plutons of the Malverns Complex suggests that they have been produced by the melting at depth of Mesoproterozoic basement aged *c*. 1.2–1.0 Ga. It is therefore likely that the Avalon Terrane is at least partly underlain by continental basement, but the areal extent of this is uncertain.

In summary, the Avalon Terrane of England and Wales comprises a series of island arcs and associated marginal basin sedimentary sequences. Isotopic dating is consistent with a broad continuum of magmatism, sedimentation and localized structural–metamorphic events in the period *c*. 700–560 Ma. In a wider context, Neoproterozoic rocks that belong to the probable western extension of the Avalon terrane occur in Western Avalonia (Maritime Canada) (Fig. 8.1). In all these areas, Neoproterozoic calc-alkaline magmatic arc assemblages associated with thick volcano-sedimentary successions display a very similar evolutionary history to those of the Avalon Terrane in southern Britain.

8.3.2 The Monian–Rosslare Terrane

The geological evolution of the Monian–Rosslare Terrane of north-west Wales and south-east Ireland (Fig. 8.4) is controversial, partly because the ages of many rock units are poorly constrained. Proven Precambrian rocks include high-grade gneisses, a calc-alkaline granite pluton and a belt of blueschist facies metamorphic rocks. Thick low-grade metasedimentary sequences and the complex deformation events that affect them have been viewed as Precam-

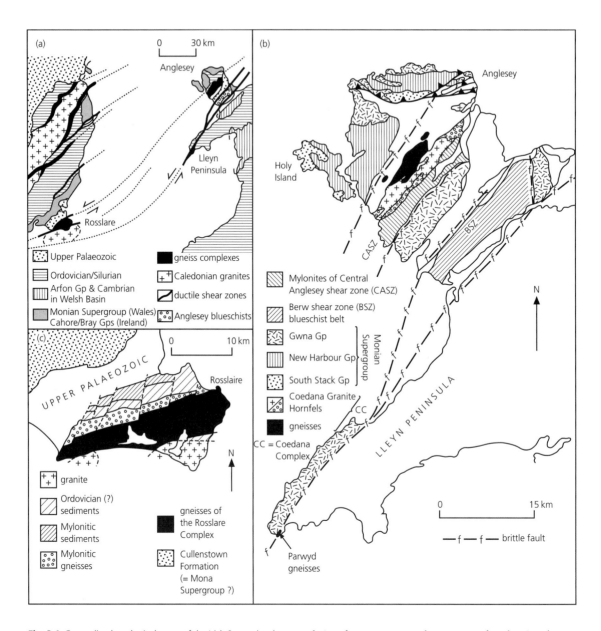

Fig. 8.4 Generalized geological maps of the Irish Sea region (a, modified from Gibbons *et al.* 1994), Anglesey (b, modified from Gibbons & Horak 1990) and Rosslare (c), showing the distribution of the various Precambrian rock units and younger sedimentary sequences and regionally important ductile shear zones and faults.

brian for many years but may in fact be Cambrian. The junction between the Monian–Rosslare and Avalon terranes is defined by a network of steep sinistral shear zones and brittle faults, the Menai Strait Line (Fig. 8.4a,b).

The Monian rocks of Anglesey and the Lleyn Peninsula can be divided into three tectono-stratigraphic units (Fig. 8.4b):

1 High-grade gneisses and a granite, collectively known as the *Coedana Complex*.

2 A belt of *blueschist* facies rocks.

3 A thick sequence of low-grade metasediments, the *Monian Supergroup*.

The contacts between all these units are defined by faults. In most cases these are ductile dip-slip and strike-slip shear zones which have often been reactivated as brittle faults. It is therefore difficult on the basis of field evidence alone to decipher the relative ages of these rock units.

The oldest rocks of the Coedana Complex are amphibolites and migmatitic, sillimanite-bearing paragneisses of unknown age. The gneisses are intruded by the undeformed and unmetamorphosed calc-alkaline Coedana granite. An intrusion age of 614±4 Ma (U–Pb zircon) for the granite places a minimum age on the deformation and metamorphism of the gneisses.

The blueschists of south-east Anglesey (Fig. 8.4b) are very important for regional tectonic models, because such rocks are normally indicative of the high-pressure/low-temperature metamorphic conditions associated with subduction zones. In Anglesey they occur within a belt of metabasic igneous rocks characterized by crossite and glaucophane. Deformation is generally intense, although some relic gabbroic igneous textures are preserved in the low strain centres of some of the larger metabasic masses. These rock units have been interpreted as representing highly tectonized oceanic basaltic crust. Phengite schists probably represent associated deep-water sediments. An early generation of actinolitic amphibole has yielded a ^{40}Ar/^{39}Ar cooling age of *c.* 580 Ma, which may correspond to sub-ocean-floor metamorphism. This was followed by subduction and high-pressure/low-temperature metamorphism, which produced blue amphibole. Final uplift is dated by a ^{40}Ar/^{39}Ar age of *c.* 550 Ma. The most likely interpretation of the blueschist belt is that it represents a tectonic slice of an accretionary prism formed above a late Neoproterozoic subduction zone.

The Monian Supergroup is a *c.* 5–7-km-thick sequence of polydeformed metasediments that range in metamorphic grade from anchizone to low greenschist facies. Lowest parts are dominated by orthoquartzites and psammites with minor pelites (*South Stack Group*). The arenaceous rocks contain graded bedding and flute casts and were deposited in a sand-rich turbidite fan system (Fig. 8.5). The orthoquartzites form major channel-fill sequences that transported mature quartz-arenites from shallow marine environments into the deep basin. Palaeocurrent data suggest deposition in a north-east–south-west-orientated basin, with a primary (continental) source of detritus from the south-east (Fig. 8.5). These are overlain by phyllites (*New Harbour Group*) with metabasalts, serpentinites and metagabbros. The sediments were probably deposited as turbidites and contain quartz–feldspar and granite clasts, probably derived from erosion of a pre-existing andesitic arc complex. The metabasalts

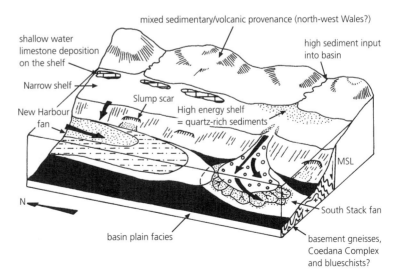

Fig. 8.5 Schematic representation of the depositional basin of the South Stack, New Harbour and Gwna Groups of the Mona Supergroup (redrawn from Phillips 1991a). MSL, Menai Strait Line.

mixed sedimentary/volcanic provenance (north-west Wales?)

shallow water limestone deposition on the shelf

high sediment input into basin

Narrow shelf

Slump scar

High energy shelf = quartz-rich sediments

New Harbour fan

MSL

N

South Stack fan

basin plain facies

basement gneisses, Coedana Complex and blueschists?

have a volcanic arc geochemical signature. Upper parts of the Monian Supergroup are formed of the *Gwna Group*, which mainly comprises a chaotic melange. Clasts vary in size from a few centimetres to (rarely) several kilometres and include shallow-water lithologies such as stromatolitic limestone and orthoquartzite, mixed with deep-water turbiditic greywacke sediments and metabasalts (commonly pillowed). The petrographic, isotopic and geochemical similarity of some granite clasts to the Coedana granite suggests that the Coedana Complex may have formed part of the source region for the Gwna Group. If this assumption is correct, the Gwna Group must have been deposited after *c.* 614 Ma. The melange is probably an *olistostrome*, produced by a large-scale submarine debris flow. Modern examples are invariably associated with tectonic instability and have been recorded from passive continental margins and collision zones. The Monian Supergroup thus preserves a history of deep-water sedimentation, which culminates in tectonic instability associated with melange formation. Several lines of palaeontological evidence have been proposed in support of a Cambrian age for the Monian Supergroup. These include the presence of Lower to Middle Cambrian acritarchs in the Gwna Group, possible Cambrian *Skolithus* burrows in the South Stack Group, and Neoproterozoic or Lower Cambrian stromatolites in a limestone clast within the Gwna Group melange.

The Monian Supergroup displays a polyphase structural history characterized by multiple fold and foliation development, although the style and intensity of deformation is very variable. The ages of these deformation events are uncertain. The supergroup is overlain unconformably by Ordovician (Arenig) sediments, which are mostly significantly less deformed and metamorphosed. This would suggest that deformation occurred during either the Neoproterozoic or Cambrian. Boulders of deformed New Harbour Group rocks occur within Mid-Ordovician (Caradoc) sediments in Anglesey, which places a minimum age on the earliest Monian deformation.

The *Rosslare Complex* of south-east Ireland is interpreted as the along-strike continuation of the Monian rocks (Fig. 8.4a,c). The oldest components of the complex are interbanded paragneisses and amphibolites of unknown age and display a polyphase

tectonothermal history. The paragneisses are locally migmatitic and contain relic kyanite. Interbanded amphibolites have a geochemistry comparable with continental tholeiitic basalts. A ^{40}Ar/^{39}Ar hornblende age of 626±6 Ma dates cooling following amphibolite facies metamorphism. The gneisses are cut by the calc-alkaline St Helens gabbro; a ^{40}Ar/^{39}Ar hornblende age of 618±6 Ma is thought to closely date the intrusion. Other units within the complex include variably deformed and metamorphosed diorites and granites, some of which may be Palaeozoic in age. The complex is bordered to the north by a wide belt of mylonitic sediments and gneisses belonging to the Menai Strait Line and which display evidence for sinistral strike-slip displacements.

A sequence of deformed, low-grade quartzose turbidites known as the *Cullenstown Formation* outcrops to the north-west of the Rosslare Complex (Fig. 8.4c). For many years this was thought to be Precambrian in age and equated on lithological and structural grounds with lower parts of the South Stack Group in Anglesey (at that time also thought to be Precambrian). It is now clear that the Cullenstown Formation is part of a thick series of Mid- to Late Cambrian sediments, collectively termed the *Cahore Group*. This reassessment of the age of the Cullenstown Formation is obviously consistent with palaeontological evidence that part or all of the Monian Supergroup is also Cambrian.

8.3.3 Avalon/Monian–Rosslare relationships

The relationship of the Monian–Rosslare and Avalon terranes has been the subject of much debate. The presence of blueschist facies rocks in Anglesey led some workers in the 1970s to suggest that these lay along the surface trace of a late Neoproterozoic subduction zone that dipped south-eastwards beneath the Avalon Terrane. The deep-water sediments of the Monian Supergroup, and in particular the Gwna Group melange, were thought to have been deposited either close to or within the ocean trench. The later recognition that a series of strike-slip shear zones and faults separates the rocks of Anglesey and the north-west Lleyn Peninsula from the Avalonian rocks of north-west Wales prompted suggestions that the two crustal blocks represent Cordilleran-type 'suspect' terranes that were formerly widely separated prior to final amalgamation. Various lines of evidence force a reassessment of these views and support a more conservative interpretation whereby the two terranes may have evolved in relative proximity:

1 The likelihood that the Monian Supergroup is Cambrian dissociates it from the blueschists, which, on the basis of isotopic evidence, formed during the latest Precambrian.

2 Although the blueschists probably formed within an accretionary prism, the recognition of strike-slip shear zones along the Menai Strait Line implies that these rocks could have been translated some distance from their origin. The Menai Strait Line does not therefore necessarily correspond to the subduction zone where the blueschists were formed.

3 Isotopic evidence demonstrates that exactly coeval calc-alkaline igneous complexes occur on either side

of the Menai Strait Line (Sarn Complex and Coedana granite).

The present consensus is that the Monian–Rosslare and Avalon terranes are part of essentially the same unit, the *Avalon Superterrane*. The early evolution of the Monian–Rosslare Terrane appears to have involved the amalgamation of basement gneisses, calc-alkaline igneous complexes and an accretionary prism during the latest Neoproterozoic in a very similar tectonic setting to that already invoked for the Avalon Terrane of central England and Wales.

8.4 Summary: Neoproterozoic tectonics of Eastern Avalonia and Armorica

Eastern Avalonia and Armorica are characterized by

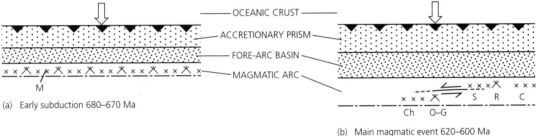

(a) Early subduction 680–670 Ma

(b) Main magmatic event 620–600 Ma

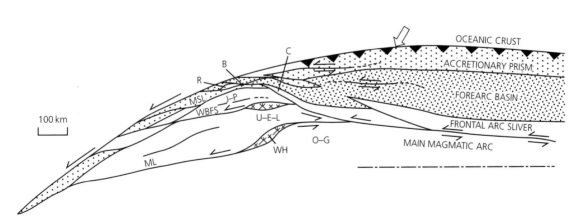

(c) Oblique convergence and terrane dispersal 570–550 Ma

Fig. 8.6 (a–c) Model for the evolution of the Avalonian subduction system in England and Wales (redrawn from Gibbons & Horak 1996). B, Anglesey blueschists; C, Coedana Complex; Ch, Charnian plutonism; J–P, Johnston plutonic complex and Pebidian volcanics; M, Malverns; ML, Malverns Lineament;

MSL, Menai Strait Line; O–G, Orton and Glinton boreholes; R, Rosslare Complex; S, Sarn Complex; U–E–L, Uriconian volcanics, Ercall Granophyre, Longmyndian Supergroup; WBFS, Welsh Borderlands Fault System; WH, Warren House volcanics.

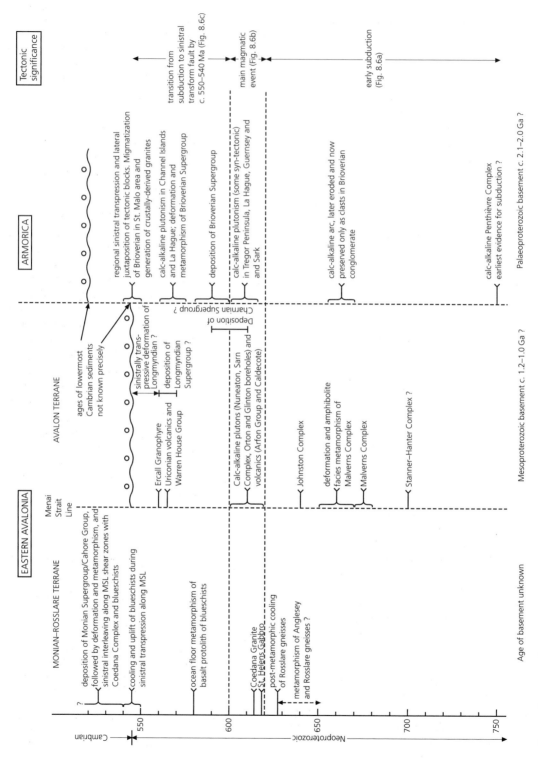

Fig. 8.7 Summary of late Neoproterozoic events in Eastern Avalonia and Armorica.

very similar late Precambrian–Early Cambrian accretionary histories. There is no evidence that deformation was associated with major crustal thickening and high-grade metamorphism. It is therefore unlikely that accretion was associated with continent–continent collision. Mesozoic–Cenozoic active plate margins such as the Andes and East Indies, which are characterized by prolonged periods of ocean–continent convergence, may therefore provide useful analogies for the Neoproterozoic evolution of Eastern Avalonia and Armorica.

Neoproterozoic tectonic events in Eastern Avalonia can be summarized in a generalized, three-stage model (Fig. 8.6). An early period of arc magmatism, represented by the Malverns Complex, and possibly also the Stanner–Hanter Complex, was initiated by c. 680 Ma (Fig. 8.6a). Most Avalonian exposures in England and Wales formed during a main magmatic event at c. 620–600 Ma (Fig. 8.6b). These include the Charnian plutons and associated volcano-sedimentary sequences, the Sarn Complex, Coedana granite and the St Helens gabbro. The final stages of Avalonian arc evolution at c. 570–550 Ma are thought to have been dominated by the effects of oblique convergence (Fig. 8.6c). This resulted in the detachment from the arc of elongate frontal arc slivers that were transported at low angles across the strike of the subduction system. This process can account for the interleaving of the Rosslare and Coedana Complexes with the Anglesey blueschist belt. The latter represents a thin slice of an accretionary prism interleaved with slivers of the magmatic arc. The development of major intra-arc, strike-slip fault systems may have controlled contemporaneous volcanism and sedimentation. By analogy with present-day obliquely convergent plate margins, the Warren House and Uriconian volcanics may have been erupted within releasing bends along strike-slip faults. This same mechanism could account for the development of the narrow, fault-bounded basin in which the Longmyndian Supergroup accumulated. The final result was a tectonic collage of various components of the subduction system. It has been suggested that a modern analogy is provided by the effects of increasing obliquity along the Indonesian subduction system from Java through the active-arc strike-slip deformation of Sumatra into the tectonic collage of Burma.

A very similar general tectonic model could be applied to the Cadomian orogeny in Armorica. Particular points of similarity include a major period of calc-alkaline magmatism at c. 615–600 Ma, and the dominance of oblique convergence and strike-slip tectonics at a late stage in the orogeny at c. 550–540 Ma (Fig. 8.7). In both Eastern Avalonia and Armorica, the termination of calc-alkaline magmatic activity has been attributed to the replacement of a subduction zone by a sinistral transform fault. These similarities are entirely consistent with palinspastic reconstructions, which depict Armorica and Eastern Avalonia in adjacent positions along a margin of the Gondwanan craton (Fig. 8.1).

Compressive deformation within interior parts of the Avalon Terrane had ceased by the start of the Cambrian, but tectonic activity continued along the Menai Strait Line. Cambrian dip-slip and/or transtensional displacements along this fault resulted in rifting of the Avalonian tectonic collage and development of the deep-water Monian–Cahore sedimentary basin to the north-west. During the Late Cambrian, transpressive deformation was localized along and to the north-west of the Menai Strait Line. In Anglesey this resulted in widespread folding and low-grade metamorphism of the Monian Supergroup and interleaving of these metasediments with the Coedana Complex and the blueschists along strike-slip shear zones and faults. This represents the final stage in the late Precambrian–Cambrian accretionary history of Eastern Avalonia, which thereafter formed a generally coherent crustal block until collision with Laurentia in the Silurian.

References

Gibbons, W. & Horak, J. (1990) Contrasting metamorphic terranes in northwest Wales. In: *The Cadomian Orogeny* (eds R. S. D'Lemos, R. A. Strachan & C. G. Topley), Special Publication 51, pp. 315–327. Geological Society, London.

Gibbons, W. & Horak, J.M. (1996) The evolution of the Neoproterozoic Avalonian subduction system: evidence from the British Isles. In: *Avalonian and Related Peri-Gondwanan Terranes of the Circum-North Atlantic* (eds R. D. Nance & M. D. Thompson), Special Paper 304, pp. 269–280. Geological Society of America, Boulder, Co.

Gibbons, W., Tietzsch-Tyler, D., Horak, J.M. & Murphy, F.C. (1994) Precambrian rocks in Anglesey, southwest Llyn and southeast Ireland. In: *A Revised Correlation of Precambrian Rocks in the British Isles* (eds W. Gibbons

& A. L. Harris), Special Report 22, pp. 75–84. Geological Society, London.

Nance, R.D. & Murphy, J.B. (1996) Basement isotopic signatures and Neoproterozoic palaeogeography of Avalonian–Cadomian and related terranes in the Circum-North Atlantic. In: *Avalonian and Related Peri-Gondwanan Terranes of the Circum-Atlantic* (eds R. D. Nance & M. D. Thompson), Special Paper 304, pp. 333–346. Geological Society of America, Boulder, Co.

Phillips, E. (1991) The lithostratigraphy, sedimentology and tectonic setting of the Monian Supergroup, western Anglesey, North Wales. *Journal of the Geological Society, London* **148**, 1079–1090.

Further reading

Egal, E., Guerrot, C., Le Goff, E., Thieblemont, D. & Chantraine, J. (1996) The Cadomian Orogeny revisited in North Brittany. In: *Avalonian and Related Peri-Gondwanan Terranes of the Circum-Atlantic* (eds R. D. Nance & M. D. Thompson), Special Paper 304, pp. 281–318. Geological Society of America, Boulder, Co. [Review of the Cadomian belt in North Brittany.]

Gibbons, W. & Horak, J.M. (1996) The evolution of the Neoproterozoic Avalonian subduction system: evidence from the British Isles. In: *Avalonian and Related Peri-Gondwanan Terranes of the Circum-North Atlantic* (eds R. D. Nance & M. D. Thompson), Special Paper 304, pp. 269–280. Geological Society of America, Boulder, Co. [Review and tectonic model for Avalonian magmatism.]

Gibbons, W., Tietzsch-Tyler, D., Horak, J.M. & Murphy, F.C. (1994) Precambrian rocks in Anglesey, southwest Llyn and southeast Ireland. In: *A Revised Correlation of Precambrian Rocks in the British Isles* (eds W. Gibbons & A. L. Harris), Special Report 22, pp. 75–84. Geological Society, London. [Review of the Monian–Rosslare Terrane.]

Murphy, F.C. (1990) Basement–cover relationships of a reactivated Cadomian mylonite zone: Rosslare Complex, SE Ireland. In: *The Cadomian Orogeny* (eds R. S. D'Lemos, R. A. Strachan & C. G. Topley), Special Publication 51, pp. 329–339. Geological Society, London. [Detailed structural analysis of part of the Menai Strait Line.]

Nance, R.D., Murphy, J.B., Strachan, R.A., D'Lemos, R.S. & Taylor, G.K. (1991) Late Proterozoic tectonostratigraphic evolution of the Avalonian and Cadomian terranes. *Precambrian Research* **53**, 41–78. [Regional overview of the Avalonian and Cadomian terranes of the northern Appalachians and western Europe.]

Pauley, J. (1990) Sedimentology, structural evolution and tectonic setting of the Late Precambrian Longmyndian Supergroup of the Welsh Borderland, UK. In: *The Cadomian Orogeny* (eds R. S. D'Lemos, R. A. Strachan & C. G. Topley), Special Publication 51, pp. 341–351. Geological Society, London. [Detailed summary of the geology of the Longmyndian Supergroup.]

Pharaoh, T.C. & Gibbons, W. (1994) Precambrian rocks in England and Wales south of the Menai Strait Fault System. In: *A Revised Correlation of Precambrian Rocks in the British Isles* (eds W. Gibbons & A. L. Harris), Special Report 22, pp. 85–97. Geological Society, London. [Review of the Avalon Terrane.]

Pharaoh, T.C., Webb, P.C., Thorpe, R.S. & Beckinsale, R.D. (1987) Geochemical evidence for the tectonic setting of late Proterozoic volcanic suites in central England. In: *Geochemistry and Mineralization of Proterozoic Volcanic Suites* (eds T. C. Pharaoh, R. D. Beckinsale & D. Rickard), Special Publication 33, pp. 541–552. Geological Society, London. [Review of the geochemistry and tectonic setting of Avalon Terrane igneous suites.]

Phillips, E. (1991) The lithostratigraphy, sedimentology and tectonic setting of the Monian Supergroup, western Anglesey, North Wales. *Journal of the Geological Society, London* **148**, 1079–1090. [Detailed account of the sedimentology of the South Stack Group and New Harbour Group of the Monian Supergroup.]

Phillips, E. (1991) Progressive deformation of the South Stack and New Harbour Groups, Holy Island, western Anglesey, North Wales. *Journal of the Geological Society, London* **148**, 1091–1100. [Structural analysis of regional deformation within the Monian Supergroup.]

Strachan, R.A., D'Lemos, R.S. & Dallmeyer, R.D. (1996) Neoproterozoic evolution of an active plate margin: North Armorican Massif, France. In: *Avalonian and Related Peri-Gondwanan Terranes of the Circum-Atlantic* (eds R. D. Nance & M. D. Thompson), Special Paper 304, pp. 319–332. Geological Society of America, Boulder, Co. [Review and tectonic model for the evolution of the Cadomian belt of NW France and the Channel Islands.]

9 The Cambrian and earliest Ordovician quiescent margin of Gondwana

N. H. WOODCOCK

9.1 Palaeocontinental setting: a quiescent margin of the Gondwana continent

The evidence has already been reviewed (see Chapter 2) for the existence of the wide Iapetus Ocean separating the northern and southern parts of Britain and Ireland during much of Early Palaeozoic time. The most important lines of evidence are:

1 Major differences in the age and character of the Precambrian basement: no older than 700 Ma in most of southern Britain and Ireland, but including components as old as 2900 Ma north of the Iapetus suture.

2 Contrasts in sedimentary facies, particularly between the warm-water Cambro-Ordovician carbonate-rich successions of Scotland and the carbonate-poor sediments deposited in southern Britain and Ireland until Late Ordovician time.

3 Faunal contrasts, particularly between provinces of Cambrian trilobites and Ordovician brachiopods and trilobites.

4 Palaeomagnetic results implying low southern hemisphere latitudes for northern Britain but moderate to high southern latitudes for southern Britain until earliest Silurian time.

These four types of geological data have been used to position the two main halves of Britain and Ireland on the Cambrian globe and to match them against possible neighbours among the larger continents. Cambrian data, particularly from palaeomagnetism, give poorer positional constraints than for Ordovician and later time (see also Fig. 2.5, pp. 26–27). However, most available palaeogeographic maps for this time show the same essential features, exemplified by one drawn at about the Neoproterozoic–Cambrian boundary (Fig. 9.1):

1 Scotland and north-western Ireland lay at between 40°S and 25°S latitude, on the southern edge of a Laurentian continent that straddled the equator throughout Early Palaeozoic time.

2 England, Wales and south-eastern Ireland lay further south — across an arm of the Iapetus Ocean — at a latitude varying between 60°S, in Early and Late Cambrian time, to 40°S in Mid-Cambrian time.

3 These parts of southern Britain and Ireland, later to become the Avalonian microcontinent, were attached to the edge of the Gondwana continent, probably opposite what is now north-west Africa.

4 Baltica, comprising Scandinavia and much of north-eastern Europe, was separated from Laurentia by the northern Iapetus Ocean and from Gondwana by the Tornquist Sea.

In contrast to its Neoproterozoic history as an active plate margin (see Chapter 8), the Avalonian margin of Gondwana was almost quiescent throughout Cambrian time. This sedimentary margin, founded on the dismembered, eroded and subsided roots of the Neoproterozoic volcanic arc, is the subject of the present chapter. Its demise was to come during Tremadoc time, the first epoch of the Ordovician period, when a volcanic arc was re-established along the Gondwana margin. This volcanism marked the onset of subduction of Iapetus ocean crust beneath the Avalonian segment of Gondwana, a phase that was to last through much of Ordovician time (see Chapter 10).

9.2 The problems of the Cambrian time-scale

Geological history is difficult to chart without an internationally agreed chronostratigraphic frame-

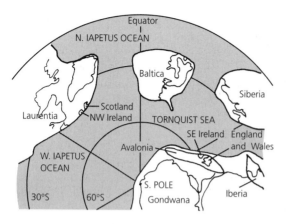

Fig. 9.1 Palaeocontinental reconstructions for Late Cambrian time (modified from Torsvik *et al.* 1996, with permission from Elsevier Science (2000)).

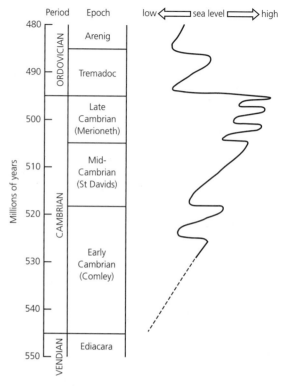

Fig. 9.2 An estimate of global sea-level change through Cambrian and earliest Ordovician time (data from Leggett *et al.* 1981; Conway Morris & Rushton 1988; and Fortey 1984; time-scale from Tucker & McKerrow 1995).

work and its adequate radiometric calibration (see Chapter 1). Both these stratigraphic components are particularly problematical for Cambrian time.

The historical status of the Cambrian as the first system of 'fossiliferous' rocks has necessarily meant difficulties in defining its base and zoning its lower parts by conventional biostratigraphy, dependent mainly on trilobites. Progress is now being made by using small shelly fossils, but a consensus on an international scale is yet to be reached. An informal subdivision into Early, Mid- and Later Cambrian is widely used, approximately corresponding to the Comley, St Davids and Merioneth Series in Britain (Fig. 9.2). Fossil dating is suggesting that 'Cambrian' successions in areas such as Siberia and China range below the stratigraphic level of the traditional basal Lower Cambrian in much of the southern British Isles (Fig. 9.3). There is a clear unconformity below Cambrian successions in England and southern Wales, but in northern Wales there may be a conformable boundary with Neoproterozoic rocks, which will need redefining.

Radiometric calibration of the Cambrian time-scale is sparse, because of the lack of suitable lithologies in fossiliferous successions, and because of resetting of isotope systems during later Phanerozoic time. Past estimates of the age of the base of the Cambrian have ranged from about 600 Ma to 530 Ma, with 570 Ma as a widely used compromise. However, U–Pb dates from igneous zircon grains have shown

that this value is too high and that an age around 545 Ma is preferable. Such dates also suggest that Early Cambrian time may be longer than the Mid- and Late Cambrian combined (Fig. 9.2). Doubts have also been raised about estimates of the basal Ordovician age of between 505 Ma and 510 Ma, which makes the Tremadoc (earliest Ordovician) appear long relative to its preceding Cambrian epochs.

The time-scale used in this chapter (Figs 9.2, 9.3, see also Fig. 9.8) will undoubtedly be modified as international subdivisions are standardized and as new radiometric age data become available. It is important to remember the poorly calibrated time-scale when trying to quantify factors such as the synchroneity of sea-level changes (see Section 9.3) and the subsidence rates of sedimentary basins (see Section 9.5).

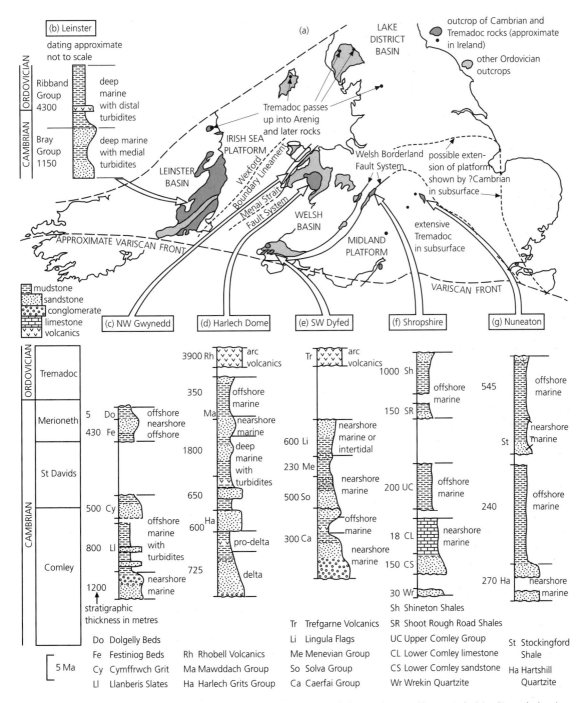

Fig. 9.3 Representative stratigraphic columns from southern Britain and Ireland for Cambrian and lowest Ordovician (Tremadoc) rocks.

9.3 Changes in global sea level and climate

Variations in global sea level can be expected to exert a first-order control on sedimentary sequences on a passive continental margin facing a large ocean (Fig. 9.1). However, the establishment of a eustatic sea-level curve requires good dating and correlation of stratigraphic sections on different continents, just the procedures that are most difficult for Cambrian and Lower Ordovician rocks (see Section 9.2). The curve in Fig. 9.2 is based preferentially on sequences from southern Britain, and must be regarded only as a first approximation to a global curve. However, some components of it seem to be recognized on other palaeocontinents:

1 A rise in sea level from late Neoproterozoic into Early Cambrian time.
2 A continued pulsed rise throughout Cambrian time to a Late Cambrian highstand.
3 Three rise–fall cycles in Late Cambrian time, marked by major faunal changes.
4 A strong regression and sea-level fall close to the beginning of Ordovician (Tremadoc) time.
5 A rise during Early Tremadoc time.

The steady rise in sea level during Cambrian time was probably driven partly by increase in volume of the oceanic ridge system as the late Neoproterozoic supercontinent began to fragment (see Chapter 2). However, a major contributary factor was the climatic change from a cold Vendian (Neoproterozoic) to a warmer Cambrian period. The nearly worldwide Varangian glacial epoch (about 600 Ma) marks the acme of the Vendian cold period, and may have been followed by a later weaker glacial event close to the beginning of the Cambrian (see Chapter 5). However, indicators of warmer global climates become increasingly common from late Vendian time through the Early Cambrian, particularly evaporites, and carbonaceous shales and phosphorites formed as nutrient-rich waters flooded continental shelves. The switch from global 'icehouse' to 'greenhouse' climates probably also promoted the marked diversification in marine invertebrates at this time.

Sea-level changes through later Cambrian and Early Ordovician time may have been caused by the fluctuating volume of a persistent southern hemisphere ice sheet, which was briefly to expand again to give the latest Ordovician glacial event (see Section 11.7, p. 174).

9.4 Large-scale stratigraphic signatures

Cambrian and lowest Ordovician rocks are exposed or penetrated by boreholes in a number of isolated areas between the Variscan Front and the Iapetus Suture (Fig. 9.3a). Rocks of this age may have accumulated south of the Variscan Front, but their metamorphosed equivalents would now underlie southernmost England and Ireland. The Cambro-Ordovician rocks north of the Iapetus Suture were formed during the separate history of the Laurentian margin (see Chapter 5), only to be united with the Gondwana terranes during Silurian time. Large post-Cambrian displacements are possible between different areas within the southern British Isles, so that the present-day relationships of Cambrian outcrops may be misleading. Three fault zones in particular could separate distinct Cambrian terranes (Fig. 9.3a):

1 The *Wexford Boundary Line* and its possible continuation to the north-west of Anglesey had substantial Cambrian or Early Ordovician displacements, and marked the south-eastern edge of the Leinster Basin.
2 The *Menai Strait Fault System* and its possible continuation off south-east Ireland had large late Neoproterozoic displacements (see Chapter 8) and controlled the north-west margin of the Welsh Basin through early Palaeozoic time.
3 The *Welsh Borderland Fault System* had significant strike-slip displacement in Late Ordovician time—possibly up to tens of kilometres—and influenced the south-eastern margin of the Welsh Basin through early Palaeozoic time.

Although the possibility of large (tens of kilometres) post-Cambrian displacements on any of these zones cannot be discounted, a plausible geological evolution can be constructed without such movements (see Sections 9.6, 9.9). Nevertheless, the fault systems separate four areas of distinctive Cambro-Ordovician stratigraphy, depositional environment and tectonic setting (Fig. 9.3a).

The *Midland Platform* is the area south-east of the Welsh Borderland Fault System. Successions here begin with a marked unconformity on Neoproterozoic rocks and pass up into about 1 km of mudstones with subordinate sandstones and limestones (Fig. 9.3f,g). The sediments were apparently deposited in a shallow marine environment. There are some thickness variations against faults active during Cambrian time, but these reflect only modest subsidence of a relatively stable segment of continental crust. Tremadoc (Early Ordovician) successions vary more markedly in thickness, from zero to over 2000 m. These variations record pronounced local subsidence in fault-controlled rifts. Subsidence rates must have approached those in the Welsh Basin. However, the term Midland Platform is still useful, because this area was restabilized by Mid-Ordovician time, and most of it acted as a shallow marine or emergent platform for the rest of Early Palaeozoic time. Quartzites penetrated by boreholes in the north-east Midlands have been assigned a Cambrian age on lithological grounds only, a conclusion that would imply a more extensive area of platform than shown by the inner boundary on the map (Fig. 9.3a).

The *Welsh Basin* is the area between the Welsh Borderland and Menai Strait Fault systems. It is characterized by higher subsidence rates and consequently thicker successions than the Midland Platform, at least during Cambrian time. Sequences lie unconformably or, in north-west Gwynedd, apparently conformably on volcanic successions, parts of which give late Neoproterozoic ages (see Chapter 8). The basin crust subsided rapidly enough for shallow marine environments in North Wales to be replaced by deeper marine turbidites in late Early and Mid-Cambrian time (Fig. 9.3d). Shallow marine conditions were re-established in the Late Cambrian. The succession in south-west Dyfed is intermediate between the Welsh Basin and its bounding platform to the south-east (Fig. 9.3e). The Dyfed succession has over 1600 m of Cambrian sediments, thicker than on the platform. However, it deepens only to offshore shelf depths rather than to those dominated by turbidite deposition. In Dyfed and on the south-east side of the Harlech Dome, major volcanic successions of probable late Tremadoc age record a new subduction-related arc and mark the onset of a volcanically dominated phase of Eastern Avalonian history (see Chapter 10).

The *Irish Sea Platform* lies between the Menai Strait Fault System and the Wexford Boundary Line. This crustal sliver contains Precambrian gneisses both at Rosslare and on Anglesey (see Section 8.2, p. 133). The platform hosts no undoubted Cambrian rocks, and may therefore have acted as an emergent

massif separating Cambrian basins in Wales and Leinster. Alternatively, the Mona Complex of Anglesey may correlate with the Cambrian deepwater successions of the Cahore Group in Leinster (see Section 8.2, p. 136), now exposed to the north-west of the Wexford Boundary Lineament. In this case, the lineament may run through, rather than to the north-west of, Anglesey, and the Cambrian Irish Sea Platform would be reduced in its size and significance.

The *Leinster and Lake District Basins* lie northwest of the Wexford Boundary Lineament. Extensive areas of probable Cambrian and Early Ordovician sediments outcrop in south-east Ireland, but they are sparsely fossiliferous and therefore only imprecisely dated (Fig. 9.3b). Exposed rocks in the Lake District and the Isle of Man apparently range down into the Tremadoc but not into the Cambrian. Despite these uncertainties over the stratigraphic successions in the Leinster and Lake District basins, they contrast in three main ways with those to the south-east of the Menai Strait Line. Firstly, they comprise only deep marine sediments, dominated by turbidites. Secondly, their basal contacts are unexposed. Thirdly, they show continuous sedimentation from Tremadoc into Arenig times, lacking the major unconformity that occurs below the Arenig over the Welsh Basin and Midland Platform. These contrasts are compatible with the Leinster–Lakes belt occupying a relatively outboard position on the Gondwana continental margin, possibly on thin continental or even marginal oceanic crust. The Ribband Group contains basic pillow lavas that have been interpreted as thrust slivers of ocean floor. More acidic rocks in the same group are the Tremadoc or possibly latest Cambrian products of renewed subduction under the Gondwana margin.

Despite the variable detailed character of the Cambrian and Early Tremadoc rocks of the southern British Isles, they have one important feature in common: the predominance of sedimentary over volcanic rocks. Minor tuffs occur in places, particularly in the Lower Cambrian of the Welsh Basin, and some of the Leinster Basin volcanics may be Cambrian in age. However, these occurrences are minor compared with the large volumes of volcanics that dominate late Tremadoc to Caradoc successions during Ordovician subduction of Iapetus Ocean crust (see Chapter 10).

9.5 Subsidence history of the Welsh Basin and Midland Platform

Stratigraphic successions such as those in Fig. 9.3 can be used to reconstruct the subsidence history of their sedimentary basement, after correcting for compaction and the effect of sediment loading (see Chapter 1). Subsidence curves in turn provide clues to the timing of crustal stretching events on the Gondwana margin. Representative curves for different parts of the Welsh Basin and its margin (Fig. 9.4) reveal spasmodic subsidence through Cambrian and Ordovician time, but important similarities from area to area. Steep segments of the curves reflect episodes of rapid subsidence and possible crustal stretching at four discrete times:

1 In the Early Cambrian, with an apparent time difference between the platform and its margin that may be an artefact of uncertainties in the time-scale.
2 In Late Cambrian to early Tremadoc time, with associated normal faulting confirmed by field and seismic observations of fault-bounded depositional troughs of this age (see also Fig. 9.7).
3 During late Arenig and early Llanvirn time, associated with early back-arc volcanism on the basin margin.
4 In the Caradoc, synchronous with the climax of back-arc volcanism in the Welsh Basin.

The last two phases postdate the late Tremadoc onset of arc volcanism and subduction of Iapetus lithosphere. They are conveniently explained as the back-arc rifting that separated Eastern Avalonia, the crustal fragment that included southern Britain, away from the main Gondwana continent (see Chapter 10). The earlier phases pre-date major volcanism, and their plate tectonic driving forces are less clear. However, they may represent a continuation of the regional transtensional tectonics that characterized the late stages of the Avalonian margin of Gondwana (see Chapter 8).

9.6 Early to Mid-Cambrian transgression and subsidence

A more detailed account of the 'presubduction' history of the Gondwana margin is conveniently divided into two parts: Early to Mid-Cambrian time (this section) during rising sea levels and early basin formation, and Late Cambrian to Tremadoc time (see

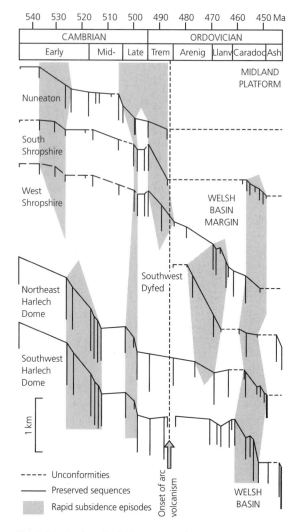

Fig. 9.4 Backstripped subsidence curves for representative areas of the Welsh Basin and the bordering Midland Platform. Localities are labelled in Fig. 9.3a. Vertical bars are errors in water-depth estimation (after Prigmore *et al.* 1997, with permission from the Geological Society of America 2000).

Section 9.7) during high or falling sea levels and active rift-related subsidence.

A palaeogeographic map for Early to Mid-Cambrian time shows the four distinct areas already discussed (see Section 9.4): the Midland Platform, the Welsh Basin, the Irish Sea Platform and the Leinster Basin (Fig. 9.5). The boundaries of the Midland Platform are not as strongly constrained by the available data as they are in Ordovician and Silurian time.

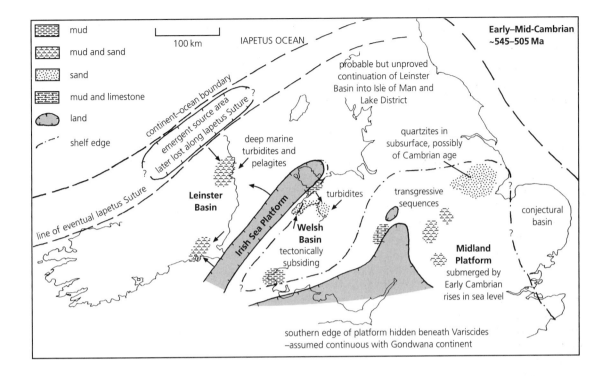

Fig. 9.5 Palaeogeographic map of southern Britain and Ireland for Early to Mid-Cambrian time (modified after Brasier *et al.* 1992, with permission of the Geological Society, London 1999).

During the Cambrian the platform may have extended north-east of its later boundaries. The Welsh and Leinster Basins are shown separated by an Irish Sea Platform that acted as a sediment source and was to persist through most of Early Palaeozoic time. A sediment source area to the north-west of the Leinster Basin is also implied by palaeocurrent data. Its nature is unknown, and it need not have survived even into Late Cambrian time. It must have been removed by later subduction beneath or strike-slip along the continental margin. Further north-west lay the presumed Iapetus Ocean crust.

During Early Cambrian time the formerly emergent Neoproterozoic crust south-east of the Irish Sea Platform was transgressed as global sea levels rose. Nearshore marine sandstones and thin limestones on the Midland Platform tend to pass upwards into offshore mudstones (Figs 9.3f,g, 9.5). Three deepening/shallowing cycles have been recognized, but it is uncertain whether they are controlled eustatically or tectonically. By contrast, early Cambrian deepening sequences in the Welsh Basin are thicker and, in North Wales (Fig. 9.3c,d), culminate in deep-marine turbidites derived southwards from the Irish Sea Platform. This platform also supplied turbidity currents taking sand and mud into the Leinster Basin, augmented by debris from the putative landmass outboard on the Gondwana margin.

Unconformities cut out parts or all of the Middle Cambrian succession over some of the Midland Platform and even in Dyfed and north-west Wales (Fig. 9.3c,e–g). Tilting of Lower Cambrian rocks below the Middle Cambrian in the Welsh Borderland shows that tectonic movements were partly responsible, along with any eustatic sea-level fall at the beginning and end of Mid-Cambrian time (Fig. 9.2). A switch in sediment supply direction in the northern Welsh basin also suggests some tectonic reorganization of basin geometry at this time. The turbidite systems in the Leinster Basin varied in the proportion of sand to mud that they deposited (Fig. 9.3b), but dating of these successions is too poor to correlate these trends with events inboard on the Gondwana margin.

9.7 Late Cambrian to earliest Ordovician rifting, thermal subsidence and uplift

A regression and transgression near the beginning of Late Cambrian time is marked by a widespread disconformity on the Midland Platform and the shallow parts of the Welsh Basin (Fig. 9.3c,f,g) and by a transition to shallow-water deposits in the deeper parts of the basin (Fig. 9.3d). However, abnormally large thicknesses of shallow marine shales and sandstones were deposited early in Late Cambrian time in several areas: up to 800 m in the Harlech Dome, 600 m in Dyfed and north-west Gwynedd, and 350 m even on the eastern Midland Platform at Nuneaton. These rates of sediment accumulation in persistently shallow water imply rapid subsidence of the underlying basement (Fig. 9.4), probably due to lithospheric exten-

sion. This extension therefore affected parts of both the Welsh Basin and the Midland Platform (Fig. 9.6). The relative abundance of fine clastic debris to keep the subsiding areas filled almost to sea level probably required complementary uplift of areas such as the Irish Sea Platform and, in early Late Cambrian time, a zone along the Welsh Borderland Fault System.

By the later part of Late Cambrian time, the rate of sediment supply failed to keep pace with the rate of basement subsidence, and shallow marine sequences were overlain by a blanket of carbonaceous shales. These were deposited in anoxic bottom waters, induced either by transgression over platform areas or more likely by restricted circulation of marine water in the small fault-controlled basins of this part of the Gondwana margin.

Mudstone deposition continued into Tremadoc time, but sediments are less carbonaceous and more bioturbated, implying that bottom waters had become better oxygenated. Widespread deposition of Tremadoc mudstones was promoted by a probable eustatic rise in sea level (Fig. 9.2). The sedimentary

Fig. 9.6 Palaeogeographic map of southern Britain and Ireland for Late Cambrian to early Tremadoc (Early Ordovician) time (modified from Brasier *et al.* 1992, with permission of the Geological Society, London 1999).

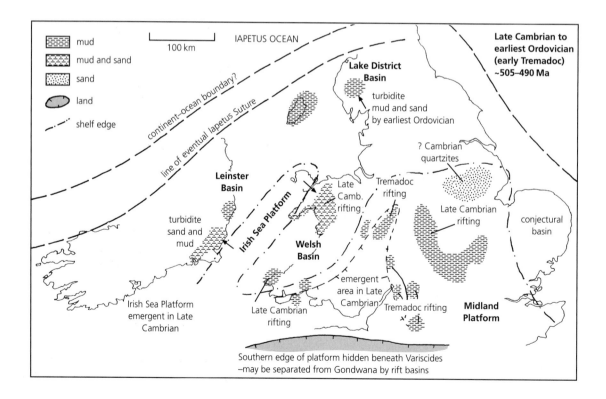

effects were augmented on the western part of the Midland Platform by a new pulse of extensional rifting. This rifting is confirmed by thickness variations of Tremadoc mudstones across contemporaneously active faults, reaching 2000 m in parts of Shropshire and Gloucestershire (Fig. 9.7). Thinner deposition persisted on the east of the platform and in the Harlech Dome, probably accommodated by thermal subsidence.

A shallowing trend during Tremadoc time is seen in relatively full successions south-east of the Irish Sea Landmass, for instance in the Harlech Dome and Shropshire (Fig. 9.3d,f). Other successions, such as in Dyfed and north-west Gwynedd, show a strong unconformity, with Tremadoc rocks missing (Fig. 9.3c,e). A late Tremadoc eustatic sea-level fall contributed to these relationships (Fig. 9.2). However, the extent of erosion or non-deposition seems implausibly large at a time when the whole area should have been undergoing thermal subsidence in the aftermath of Late Cambrian and early Tremadoc rifting. A possible explanation is regional thermal uplift associated with the impending volcanic activity of late Tremadoc time.

Turbidite sedimentation continued in the Leinster Basin through Late Cambrian and Early Ordovician time, with the succession apparently becoming finer upwards (Fig. 9.3b). Large thicknesses of sediments accumulated, requiring either contemporaneous crustal extension or deposition on already thin crust—perhaps the outboard edge of the Gondwana continent—or even oceanic crust of the Iapetus. Imprecise dating precludes accurate subsidence analysis to test between these possibilities. There is also the further possibility that the Leinster Basin does not contain a stratigraphically continuous succession, but rather a series of thinner successions separated by major thrusts. Early Ordovician volcanic rocks occur in the Ribband Group and may represent the earliest products of the Iapetus subduction zone, seen more clearly in the Welsh Basin.

Further north-west, the first evidence occurs of sedimentation in the Isle of Man and Lake District areas. Like the Ribband Group, these rocks of the Manx Group and the Skiddaw Group are mostly deep marine turbidites, and pass continuously upwards into Arenig rocks. They will be fully described in Chapter 10.

9.8 Summary: a non-volcanic continental margin of Gondwana

The segment of the Gondwana margin now comprising England, Wales and south-eastern Ireland was fashioned by tectonic assembly of fragments of vol-

Fig. 9.7 Seismic profile across the eastern edge of the Worcester Graben, on the south-west of the Midland Platform (after Smith & Rushton 1993, with permission from Cambridge University Press 2000).

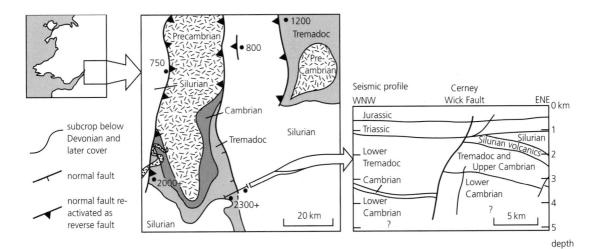

depth

canic arcs and continental crust in the late Neoproterozoic Cadomian Orogeny. By Early Cambrian time, this margin faced the wide Iapetus Ocean to its north. Inboard areas of the margin had subsided and eroded enough for sedimentation to resume over it, aided by a major eustatic rise in sea level (Fig. 9.8). Through Cambrian and earliest Ordovician time the Gondwana margin was passive, in the sense that it hosted little volcanic activity and that Iapetus Ocean crust was not subducting beneath it. But it had suffered sporadic localized crustal extension, maybe on a transform-faulted margin similar to that inherited from the Cadomian events. Subduction resumed beneath the margin by late Tremadoc time (Early Ordovician), heralding a new phase in the evolution of southern Britain (see Chapter 10).

Throughout Cambrian and Tremadoc time, the Irish Sea Platform separated a deep-marine outboard sector of the margin, preserved in the Leinster and Lake District Basins from a generally shallower inboard sector comprising the Welsh Basin and Midland Platform. No stratigraphic links between these sectors can be established and their contiguity cannot therefore be proved. The outboard sector was less responsive to the tectonic and eustatic events that influenced the inboard sector (Fig. 9.8). Inboard at least, three phases of localized rifting can be identified. However, sediment supply was sufficient to maintain shallow marine environments over most of the area, except in the Harlech Dome, where the Welsh Basin became deep enough to host thick turbidite deposits.

At no time in the margin's history is there evidence of large volumes of coarse sediment being supplied from the vast Gondwana continent assumed to be continuous with the south edge of the Midland Platform. Either there were other depositional troughs, like the Welsh Basin, trapping sediment further inboard on the margin, or the Gondwanan drainage systems were transporting eroded sediment to some other part of the continental margin. Indeed, intimate attachment of southern Britain to its parent continent cannot be assumed. The consensus view is that a

Fig. 9.8 Tentative chronology of events on the passive margin of Gondwana.

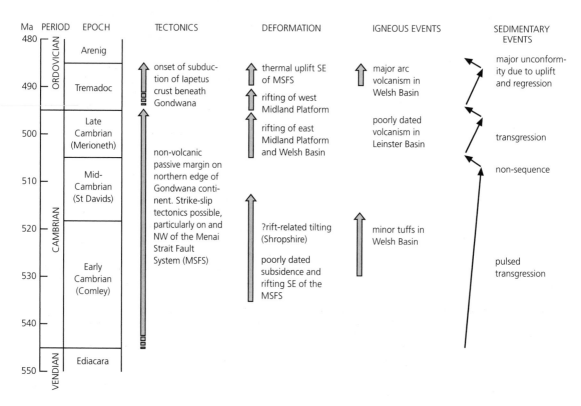

separate microcontinent of Eastern Avalonia detached itself by subduction-related rifting only during Arenig or Llanvirn time (see Section 10.1, p. 153). However, any of the prevolcanic rift events could have accompanied a modest separation from Gondwana sufficient to sever sedimentary links, although maintaining faunal links.

References

Brasier, M.D., Ingham, J.K. & Rushton, A.W.A. (1992) Cambrian. In: *Atlas of Palaeogeography and Lithofacies* (eds J. C. W. Cope, J. K. Ingham & P. F. Rawson), Memoir 13, pp. 13–18. Geological Society, London.

Conway Morris, S. & Rushton, A.W.A. (1988) Precambrian to Tremadoc biotas in the Caledonides. In: *The Caledonian Appalachian Orogen* (eds A. L. Harris & D. J. Fettes), Special Publication 38, pp. 93–109. Geological Society, London.

Fortey, R.A. (1984) Global earlier Ordovician transgressions and regressions and their biological implications. In: *Aspects of the Ordovician System* (ed. D. L. Bruton), pp. 37–50. Universitetsforlaget, Oslo.

Leggett, J.K., McKerrow, W.S., Cocks, L.R.M. & Rickards, R.B. (1981) Periodicity in the early Palaeozoic marine realm. *Journal of the Geological Society, London* **138**, 167–176.

Prigmore, J.K., Butler, A.J. & Woodcock, N.H. (1997) Rifting during separation of Eastern Avalonia from Gondwana: evidence from subsidence analysis. *Geology* **25**, 203–207.

Smith, N.J.P. & Rushton, A.W.A. (1993) Cambrian and Ordovician stratigraphy related to structure and seismic profiles in the western part of the English Midlands. *Geological Magazine* **130**, 665–671.

Torsvik, T.H., Smethurst, M.A., Meert, J.G. *et al.* (1996) Continental break-up and collision in the Neoproterozoic and Palaeozoic—a tale of Baltica and Laurentia. *Earth-Science Reviews* **40**, 229–258.

Tucker, R.D. & McKerrow, W.S. (1995) Early Paleozoic chronology: a review in light of new U–Pb zircon ages from Newfoundland and Britain. *Canadian Journal of Earth Sciences* **32**, 368–379.

Further reading

Bowring, S.A., Grotzinger, J.P., Isachsen, C.E., Knoll, A.H., Pelechaty, S.M. & Kolosov, P. (1993) Calibrating rates of Early Cambrian evolution. *Science* **261**, 1293–1298. [An interesting discussion of the problems of dating events and assessing geological rates in the early Cambrian.]

Brasier, M.D. (1992) Global ocean–atmosphere change across the Precambrian–Cambrian transition. *Geological Magazine* **129**, 161–168. [A stimulating review of environmental changes leading into the Early Cambrian.]

Brasier, M.D., Ingham, J.K. & Rushton, A.W.A. (1992) Cambrian. In: *Atlas of Palaeogeography and Lithofacies* (eds J. C. W. Cope, J. K. Ingham & P. F. Rawson), Memoir 13, pp. 13–18. Geological Society, London. [An authoritative summary of the Cambrian geological history of Britain and Ireland.]

Harland, W.B. (1989) Palaeoclimatology. In: *The Precambrian–Cambrian Boundary* (eds J. W. Cowie & M. D. Brasier), pp. 199–204. Clarendon Press, Oxford. [A review of climatic changes leading into the Early Cambrian.]

Max, M.D., Barber, A.J. & Martinez, J. (1990) Terrane assemblage of the Leinster Massif, SE Ireland, during the Lower Palaeozoic. *Journal of the Geological Society, London* **147**, 1035–1050. [A multi-terrane view of SE Ireland, to be contrasted with the views of Murphy *et al.* (1991).]

Murphy, F.C., Anderson, T.B., Daly, J.S. *et al.* (1991) An appraisal of Caledonian suspect terranes in Ireland. *Irish Journal of Earth Sciences* **11**, 11–41. [A view of the Early Palaeozoic of SE Ireland that minimizes the number of its constituent terranes, to be contrasted with the views of Max *et al.* (1990).]

Prigmore, J.K., Butler, A.J. & Woodcock, N.H. (1997) Rifting during separation of Eastern Avalonia from Gondwana: evidence from subsidence analysis. *Geology* **25**, 203–207. [Contains Cambrian–Ordovician subsidence curves and discusses their relationship to rifting of Eastern Avalonia from Gondwana.]

Rushton, A.W.A. (1974) The Cambrian of Wales and England. In: *Cambrian of the British Isles, Norden and Spitzbergen. Lower Palaeozoic Rocks of the World.* Vol. 2 (ed. C. H. Holland), pp. 43–121. John Wiley, London. [A formation-scale description and interpretation of Cambrian sequences in England and Wales.]

Smith, N.J.P. & Rushton, A.W.A. (1993) Cambrian and Ordovician stratigraphy related to structure and seismic profiles in the western part of the English Midlands. *Geological Magazine* **130**, 665–671. [Evidence for rift-related subsidence and sedimentation in the late Cambrian and Early Ordovician of the Midland Platform.]

Tietzsch-Tyler, D. & Phillips, E. (1989) Correlation of the Monian Supergroup in NW Anglesey with the Cahore Group in SE Ireland. *Journal of the Geological Society, London* **146**, 417–418. [A fascinating correlation of possible Cambrian lithostratigraphy from SE Ireland to N Wales.]

Torsvik, T.H., Smethurst, M.A., Meert, J.G. *et al.* (1996) Continental break-up and collision in the Neoproterozoic and Palaeozoic—a tale of Baltica and Laurentia. *Earth-Science Reviews* **40**, 229–258. [A valuable series of global palaeogeographic maps, with a discussion of the evidence behind them.]

Tucker, R.D. & McKerrow, W.S. (1995) Early Paleozoic chronology: a review in light of new U–Pb zircon ages from Newfoundland and Britain. *Canadian Journal of Earth Sciences* 32, 368–379. [An integrated Early Palaeozoic time-scale, based on high-precision age determinations.]

10 Ordovician volcanism and sedimentation on Eastern Avalonia

N. H. WOODCOCK

10.1 Palaeocontinental setting: northward drift of Eastern Avalonia

In Early Ordovician time the crust bearing the southern parts of Britain and Ireland broke free of its position against the margin of the major Gondwana continent. The resulting continental fragment—Eastern Avalonia—moved northward, opening the Rheic Ocean in its wake (Fig. 10.1). Ahead of it, the Iapetus Ocean closed progressively until, by earliest Silurian time, Eastern Avalonia began to impinge on the Laurentian continent to the north. The closure of this segment of Iapetus involved oceanic subduction at both its Laurentian and Avalonian margins. Abundant igneous rocks are the most characteristic geological signature of this subduction beneath Eastern Avalonia. This chapter describes the 50-Myr time-span of this subduction-related magmatism.

The Early Ordovician proximity of southern Britain and Ireland to the Gondwana continent is confirmed both by faunal links and by palaeomagnetic data, which place the margin at high southern latitudes (about 60°S, Fig. 10.1). By Llanvirn time (Mid-Ordovician) Eastern Avalonia had moved to about 50°S and by Caradoc time to 40°S or less, whereas the NE margin of Gondwana remained in higher latitudes. The rift from Gondwana probably happened during Arenig time, although faunal contrasts do not appear until late in the Llanvirn. Other pieces of the Gondwana margin also moved independently northward: Western Avalonia (fragments of eastern North America), Armorica and Iberia. Baltica had a drift path of its own, moving northward by about 20° latitude and rotating anticlockwise by over 90° (Fig. 10.1).

Faunal contrasts between Eastern Avalonia and Baltica broke down in Late Ordovician time, although the evidence for hard, forceful collision between the two continental fragments is sparse. The end of major volcanism on Eastern Avalonia might in part be related to this collision. Another possibility is that the microcontinent overran the Iapetus spreading ridge (see Section 10.9). For whatever reason, subduction beneath Eastern Avalonia seems to have ended before its collision with Laurentia, which began in early Silurian time (see Chapter 11).

10.2 The Ordovician time-scale

For most of the 20th century the Ordovician system has been subdivided into five series, all defined by successions on Eastern Avalonia: Arenig, Llanvirn, Llandeilo, Caradoc and Ashgill. However, the chronostratigraphic time-scale used here (e.g. Fig. 10.2) incorporates two more recent modifications:

1 The Tremadoc series (or epoch), formerly the uppermost part of the Cambrian, is now included as the lowest division of the Ordovician.

2 The Llandeilo series/epoch has been downgraded. Its lower part now makes up the upper division of the Llanvirn series, and its middle and upper parts form the lowest part of the Caradoc series.

The chronometric calibration of Ordovician time is more accurate than that of the Cambrian, aided by the abundance of easily dated igneous rocks. However, there are still uncertainties in parts of the time-scale, particularly in the relative lengths of finer subdivisions within each epoch.

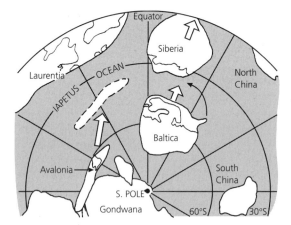

Fig. 10.1 Palaeocontinental reconstruction for Early Ordovician (late Tremadoc) time. The arrows on Baltica and Siberia show their directions of movement at this time. The broken outline of Avalonia shows its position in Caradoc time. Gondwana and Laurentia moved relatively little during the Ordovician (modified from Torsvik *et al.* 1996, with permission from Elsevier Science 2000).

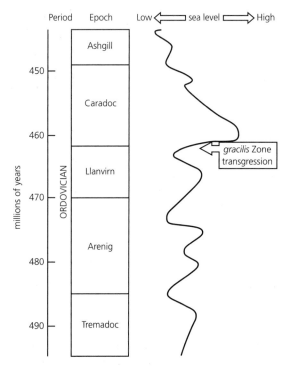

Fig. 10.2 A global sea-level curve for Ordovician time (modified from Fortey & Cocks 1986).

10.3 Changes in global sea level and climate

The Ordovician geological record on Eastern Avalonia is strongly influenced by the abundance of subduction-related igneous and tectonic activity. Localized crustal uplift and subsidence was driven by lithospheric stretching or by thermal effects close to igneous centres. However, these local controls on sediment accumulation or erosion were superimposed on a globally changing pattern of sea level. Attempts have been made to reconstruct this global curve by correlating major sedimentary transgressions and regressions on Eastern Avalonia with those recognized worldwide (Fig. 10.2). This curve shows:

1 Periods of low sea level at the end of the Tremadoc, Arenig and Llanvirn.

2 A major transgression in the Early Caradoc (the *gracilis* Zone trangression).

3 Intermittently falling sea level to lows at the end of Caradoc time and in the late Ashgill.

The close correspondence between the sea-level lowstands and the boundaries of Ordovician epochs is no coincidence. These subdivisions of Ordovician time were first placed by 19th-century geologists at the resulting unconformities or regressive facies in the southern British successions. Of the intervening transgressions, that in early Caradoc time is much the largest and best documented worldwide. It had obvious consequences on Eastern Avalonia, flooding the Midland Platform more extensively than before and promoting a major phase of carbonaceous shale deposition in the surrounding basins. From its highstand in Caradoc time, sea level probably fell intermittently towards its eventual low during the late Ashgill (end-Ordovician) ice age (see Chapter 11).

10.4 Large-scale stratigraphic signatures

Ordovician rocks of Arenig, Llanvirn or Caradoc age crop out particularly in Wales and the Welsh Border-land, the Lake District, the Isle of Man and south-east Ireland (Fig. 10.3a). They have also been penetrated by boreholes in eastern England. All these areas lie between the Variscan Front in the south and the Iapetus Suture to the north-west. However, rocks of this age may also have accumulated on the portion of

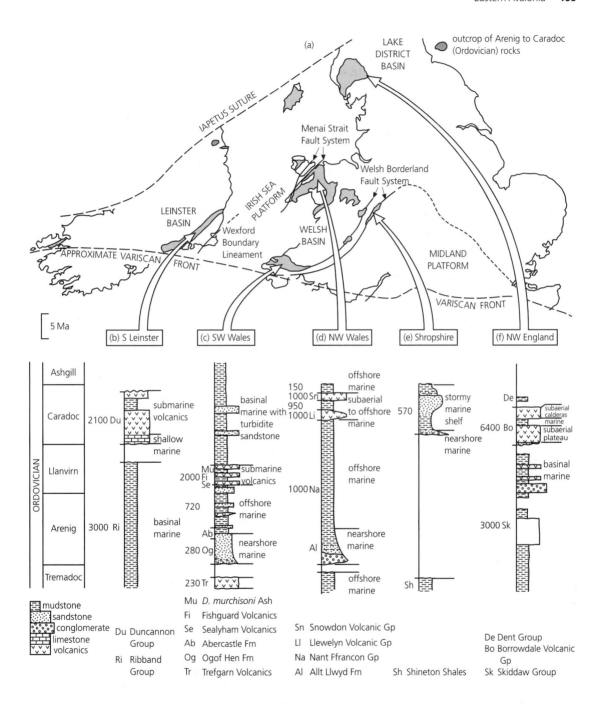

Fig. 10.3 Representative stratigraphic columns from southern Britain and Ireland for Ordovician (Arenig, Llanvirn and Caradoc) rocks.

Eastern Avalonia to the south of the Variscan Front, only to be overthrust by younger rocks in Carboniferous and Permian time. Blocks of late Llanvirn or early Caradoc quartzite contained in an upper Devonian breccia in south Cornwall are the only exposed remnant of such rocks.

The present geographical relations of Ordovician rocks may only partly conform with those during their accumulation. Deformation during the Acadian Orogeny (see Chapter 12) shortened the former sedimentary basins in Devonian time, and significant strike-slip displacements may have further reorganized the Avalonian crust then and during Ordovician time. Three of the possible zones of movement were the *Welsh Borderland Fault System*, the *Menai Strait Fault System* and the *Wexford Boundary Lineament*, which now separate four areas of contrasting Ordovician stratigraphy, depositional and igneous environments, and tectonic setting (Fig. 10.3a).

The *Midland Platform*, south-east of the Welsh Borderland Fault System, records local subsidence during Tremadoc (Early Ordovician) time (see Chapter 9), but its subsequent Ordovician history is that of stable continental crust. It preserves only thin patches of shallow marine sediment, and includes no volcanic centres. The best-known succession, from Shropshire (Fig. 10.3e) contains a large unconformity between Tremadoc mudstones and sandstones and mudstones deposited during the Caradoc highstand in sea level (Fig. 10.2).

The *Welsh Basin* lay between the Welsh Borderland and Menai Strait Fault Systems. It accumulated thicker, vertically more continuous successions than the Midland Platform and hosted numerous volcanic centres during late Tremadoc to Caradoc time. These features suggest an area of actively stretched and thinned continental lithosphere bordering the north-western edge of the stable platform. The stretching in the upper crust took place on large faults so that rates of subsidence and types of sedimentary facies varied markedly from place to place. The faults guided magma to the surface and localized volcanic piles, providing further scope for lateral thickness and facies variations. Stratigraphic successions from north-west and south-west Wales (Fig. 10.3c,d) both show an unconformity at the base of the Arenig that characterizes most Welsh Basin successions. The succeeding sedimentary and volcanic successions are some kilometres thick. Accumulation was mostly submarine except, as in the Caradoc of north-west Wales, when the volume of extrusive rocks was sufficient to maintain volcanoes above sea level. The area

in eastern England lying to the north-east of the Midland Platform includes subsurface volcanics and sediments resembling those in the Welsh Basin. The two areas may have been a continuous tract of Ordovician subsidence and igneous activity around the northern rim of the Midland Platform.

The *Irish Sea Platform* was bounded to the south-east by the Menai Strait Fault System and to the north-west by the Wexford Boundary Lineament. This relatively buoyant strip of crust separated the more rapidly subsiding Welsh and Leinster Basins. It apparently persisted through much of Cambrian to Silurian time. Sometimes, as in the Cambrian, it was gently uplifting and acting as a source of sediment. Sometimes, as in Ordovician time, parts of it subsided for long enough to preserve patches of shallow marine sediment. Localized sediment thicknesses of over a kilometre on Anglesey (north-west Wales) suggest fault control between, as well as on, the bounding faults to the platform. Volcanic centres border the Irish Sea Platform, but do not occur on it (Fig. 10.4).

The *Leinster and Lake District Basins* lie north-west of the Wexford Boundary Lineament, and its possible continuation to the north-west of Anglesey. Many Ordovician successions here are continuous from the Tremadoc to Llanvirn, in contrast to most of the area of Eastern Avalonia further south-east. Such continuity, and the pervasive basinal marine sedimentation, suggests an outboard setting on the Avalonian margin, which continued that of Cambrian time. The Iapetus Ocean crust lay further north-west, across an actively subducting trench. Until Llanvirn time, accumulation in Leinster, the Isle of Man and the Lake District was dominated by turbidites: the Ribband, Manx and Skiddaw Groups, respectively. However, in Leinster and the Lakes, a late Llanvirn unconformity is followed by a succession dominated by volcanics: the Duncannon and Borrowdale Groups. These successions record a volcanic arc built on the outer part of the Avalonian margin during Caradoc time.

10.5 Distribution of Ordovician igneous rocks

The igneous rocks that dominate the Ordovician geology of the old Avalonian microcontinent are highly variable in composition, age and palaeogeo-

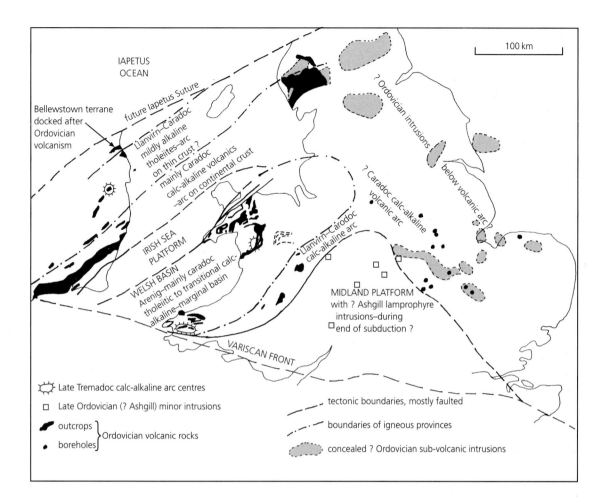

Fig. 10.4 Distribution, age and petrology of the main Ordovician volcanic provinces.

graphic position. The tectonic significance of this variation is only partly understood. However, the volcanic and plutonic rocks can be grouped into a number of igneous provinces (Fig. 10.4) each with some petrological coherence. Each province can tentatively be assigned to a distinctive position in a subduction system (Fig. 10.5). The provinces conform broadly to the palaeogeographic areas discussed in Section 10.4, but the warning issued there must be repeated: the spatial relationship of these areas may have been tectonically shuffled since Ordovician time. The provinces are described from the outboard areas of the margin in toward the Midland Platform.

The north-western parts of the *Leinster and Lake District Basins* contain basic and intermediate volcanics and minor intrusives, mostly of Llanvirn to Caradoc age. These rocks, such as the Eycott Group in the Lake District, have mildly alkaline tholeiitic compositions. The most likely site of this magmatism is in an immature volcanic arc built on the thin outer edge of a continent (Fig. 10.5c). The substrate to this segment of the arc comprises deep-water sediments of the Ribband, Manx and Skiddaw Groups, possibly already deformed during early subduction (Fig. 10.3; see also Fig. 9.3, p. 143). The Ribband Group, south-east Ireland, contains volcanics inferred to be of Early Ordovician (Tremadoc or Arenig) age, suggesting that parts of the arc were initiated here well before its Caradoc climax.

Further anomalies in this sector are the Llanvirn

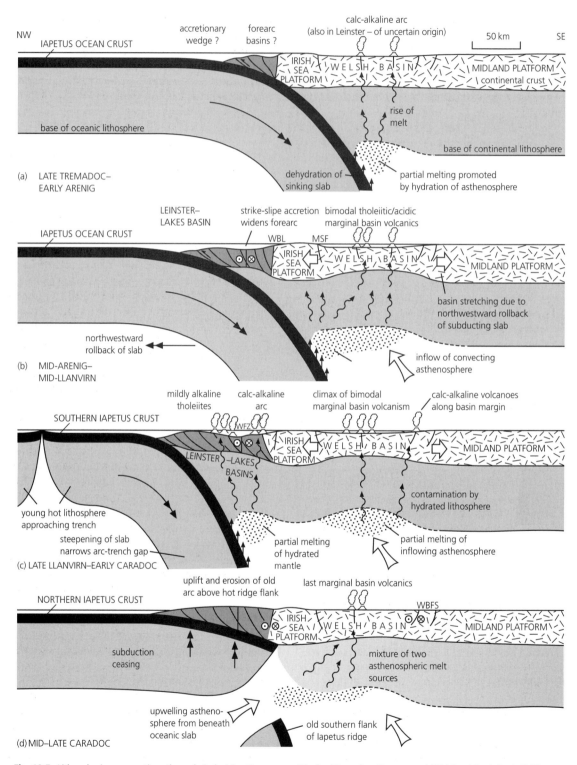

Fig. 10.5 Lithospheric cross-sections through Ordovician time, emphasizing different modes of magma generation. WBL, Wexford Boundary Lineament; MSF, Menai Strait Fault; WFZ, Wicklow Fault Zone; WBFS, Welsh Borderland Fault System.

volcanics at Grangegeeth and Bellewstown, which occur in two fault-bounded slivers close to the Irish sector of the Iapetus Suture. The ocean–island compositions of these rocks contrast with each other and with other volcanics in the north-western Leinster Basin. The slivers probably represent small exotic terranes originally positioned elsewhere along, or offshore from, the Avalonian margin, and accreted to it during the Acadian Orogeny.

The south-eastern parts of the *Leinster and Lake District Basins* contain large volumes of Llanvirn to mainly Caradoc volcanics, part of the Duncannon Group in Leinster and the Borrowdale Volcanic Group in the Lake District. In both areas, the proportion of acidic rocks increases upwards. For instance, the lower 2.5 km of the Borrowdale Group are dominated by andesite lavas and the upper 4 km by intermediate to acid pyroclastic rocks infilling major calderas. The dominant chemistry is calc-alkaline and characteristic of a mature volcanic arc developed on continental crust (Fig. 10.5c). This magmatism was broadly coeval with that of the immature arc to the north-west, but it is uncertain whether the two components of the arc were active on the same segment of the Ordovician margin. The north-western and south-eastern terranes are separated by fault zones with significant post-Ordovician and possibly syn-Ordovician displacement; the Causey Pike Fault in the Lake District and the Wicklow Fault Zone in Leinster.

Boreholes in *eastern England* prove an extensive suite of calc-alkaline volcanic rocks (Fig. 10.4) of Late Ordovician age (440–460 Ma). Geochemical signatures distinguish this probable Caradoc suite from older, late Precambrian, volcanics beneath the same area (see Chapter 8). The chemical character of the eastern England Caradoc rocks matches that of the Borrowdale Volcanic Group in north-west England. It is possible that the mature volcanic-arc terranes are continuous at depth between the two areas, as was the subducting ocean crust, near the confluence of the Iapetus Ocean and Tornquist Sea (Fig. 10.1).

A further link between the two segments of the mature Avalonian arc is provided by a chain of large intrusions, shown by geophysical evidence to underlie eastern and northern England. These intrusions probably mark the arc's magmatic core. In north-western England, they evolved from a Llanvirn basic complex with tholeiitic chemistry, into Late Ordovi-

cian to Devonian acidic bodies with calc-alkaline chemistry. This compositional trend parallels that in the overlying volcanic rocks, although it continues later in time. The plutonics in eastern England give Late Ordovician ages and have intermediate to acidic calc-alkaline compositions that are compatible with the volcanic rocks in the same area.

The *Welsh Basin*, like north-west Leinster, preserves evidence for Early Ordovician arc volcanism. The late Tremadoc Rhobell Volcanic Complex in north Wales and the late Tremadoc or early Arenig Trefgarn Volcanic Group in the south (Fig. 10.4) comprise basic, intermediate and acidic rocks with a calc-alkaline chemistry. They are presumed to record the onset of southward subduction beneath Eastern Avalonia and the construction of arc volcanoes on continental crust (Fig. 10.5a). However, this episode was relatively short lived. Subsequent Arenig, Llanvirn and Caradoc volcanic suites in Wales tend to have bimodal basalt and rhyolite compositions and to lack intermediate rocks. Their tholeiitic chemistry suggests eruption in a back-arc basin, with a component of the magmas derived from subcontinental mantle weakly contaminated by fluids and magmas from the subducting slab (Fig. 10.5b,c). The Arenig to Caradoc Welsh Basin is therefore seen as stretched continental lithosphere behind the Lake-District–Leinster volcanic arc.

An unexplained feature is the persistence of calc-alkaline components in Llanvirn to Caradoc volcanics along the south-eastern margin of the Welsh Basin (Fig. 10.4). Structural evidence makes it unlikely that this terrane was faulted against the Welsh Basin only in post-Caradoc time. More probably the steep faults of the Welsh Borderland Fault System were tapping sources of magma that were contaminated by lithosphere already hydrated above the earlier Ordovician subduction zone (Fig. 10.5c).

The *Midland Platform* lacks major Ordovician volcanic rocks (Fig. 10.4): Early Silurian volcanic centres in the southern part of the platform are discussed later (see Chapter 11). However, sills and dykes of latest Ordovician or earliest Silurian age (442 ± 3 Ma, U–Pb age) were intruded into the northern apex of the platform, just after the main volcanism on Eastern Avalonia had ended. These intrusions are lamprophyres, crystallized from wet alkalic basic magmas. Such magmas are not generated in the normal spectrum of igneous processes across a subduction system (Fig. 10.5a–c). However, they can be derived when subduction ends, the subducting slab detaches, and a window appears into the asthenosphere beneath the slab (Fig. 10.5d). The alkalic magmas are derived by mixing partial melts from this suboceanic asthenosphere with those from the subplatform mantle already chemically altered above the former subduction zone. The last Caradoc basaltic volcanics in North Wales have a chemistry compatible with a similar source through a slab window. These volcanics are mid-Caradoc in age, giving an estimate for the end of active subduction beneath Eastern Avalonia.

10.6 Volcanic-arc initiation and rifting (late Tremadoc to early Arenig)

A detailed description of the subduction-related history of Eastern Avalonia is divided into four parts:

Fig. 10.6 Palaeogeographic map of southern Britain and Ireland for late Tremadoc to Arenig time (modified after Bevins *et al.* 1992, with permission of the Geological Society, London 1999).

1 Late Tremadoc to Early Arenig time (this section) during the onset of arc volcanism.
2 Mid-Arenig to Mid-Llanvirn (see Section 10.7) when a marginal basin developed in Wales.
3 Late Llanvirn to Early Caradoc time (see Section 10.8) during the climax of arc volcanism on the north-western edge of Eastern Avalonia.
4 Mid-Caradoc time (see Section 10.9) when arc volcanism shut down, and subduction beneath Eastern Avalonia ceased.

A palaeogeographic map for Late Tremadoc to Early Arenig time (Fig. 10.6) shows the initiation of a subduction zone along the formerly passive north-western margin of Eastern Avalonia (compare with Fig. 9.6, p. 148, for Late Cambrian to Early Tremadoc). The trench is assumed to have lain close to the present trace of the Iapetus Suture, and to have allowed Iapetus Ocean lithosphere to subduct approximately southward beneath Eastern Avalonia (Fig. 10.5a).

The first arc volcanism above the subduction zone started in late Tremadoc time, at two centres in Wales and one in Ireland. The contrast in arc–trench distance between the Welsh and Irish centres suggests

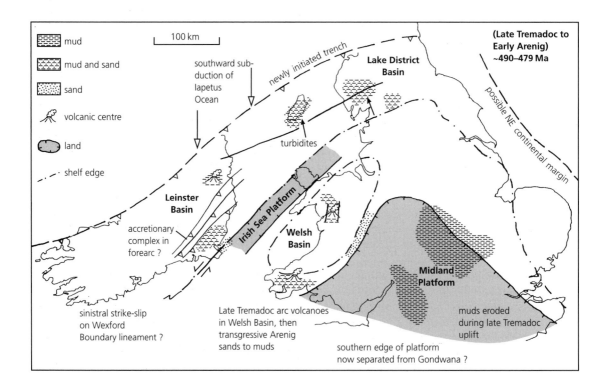

that they may not now be in their original position. In any case, subduction must have been active for several million years before the late Tremadoc, to allow the Iapetus oceanic slab to descend to the depths where its dehydration promoted melting of the mantle. The onset of subduction may be recorded over a wider area of Wales by late Tremadoc uplift, for instance on the Midland Platform and on the south-western edge of the Irish Sea Platform. However, the resulting late Tremadoc to early Arenig unconformity (Fig. 10.3) may also be due partly to a eustatic sea-level fall (Fig. 10.2).

The first phase of arc volcanism in Wales was short lived, maybe only 5–10 Myr in duration. As the volcanic centres expired and subsided, they were transgressed by Arenig sediments. These record deepening of a marine environment in their upward transition from nearshore sandstones to offshore mudstones (e.g. Allt Llwyd Formation and Ogof Hen Formation, Fig. 10.3). Arenig successions elsewhere in the Welsh Basin are also typically transgressive, probably promoted by a eustatic sea-level rise as well as by the thermal subsidence of the Avalonian margin. The Midland Platform does not preserve any Arenig rocks, and probably remained emergent, though low lying, for another 30 Myr into the Caradoc.

In the extreme south-east of Ireland, shallow marine transgressive sandstones of Arenig age lie directly on Precambrian rocks of the edge of the Irish Sea Platform (Fig. 10.6). However, most successions north-west of the Irish Sea Platform comprise deep marine mudstones and turbidite sandstones of the Skiddaw, Manx and Ribband groups of the Lake District, Isle of Man and Leinster, respectively (Fig. 10.3b,f). Although still inadequately dated, they apparently lack the internal unconformities and shallow marine facies of the Welsh Basin. The deep-water turbidites were supplied north-westward from the direction of the Irish Sea Platform, and probably record sedimentation in the fore-arc of the subduction system. They are cut by major faults, which may juxtapose successions that are dissimilar in detail, and may even relate to accretionary deformation in the fore-arc.

The palaeogeographic map shows the subduction vector as southward, and therefore sinistral-oblique to the north-west margin of Eastern Avalonia. This obliquity is suggested by field evidence of a component of sinistral strike-slip across the margin through-

out the history of Ordovician volcanism. In particular, the Rhobell volcanism and later extrusive activity in north Wales were localized along north–south striking fractures and graben, suggesting east–west extension oblique to the trench. Sinistral strike-slip displacement along the Wexford Boundary Lineament may be as late as Tremadoc or earliest Arenig and be a component of this oblique subduction. Potential displacements along this line are substantial, and it has been suggested that the terrane comprising the Leinster, Manx and Lake District successions may have been slid into place along such strike-slip faults during or after the first burst of arc volcanism. This terrane accretion would explain why the distance from the trench to the Welsh arc is now at least 200 km (Fig. 10.6), compared with as little as 100 km in present subduction systems.

10.7 Marginal basin formation (Mid-Arenig to Mid-Llanvirn)

A brief period of volcanic quiescence followed the early Arenig shutdown of the arc volcanoes in Wales. Mid-Arenig marine mudstones in the Welsh Basin contain only occasional tuffs and bentonite clays. However, by late Arenig time new volcanic centres were forming in north and south-west Wales (Fig. 10.7). These centres were not simply rejuvenated arc volcanoes, fed only by the same source of subduction zone magmas; their erupted products tend to have bimodal basic/acid compositions rather than the intermediate compositions of the arc rocks. They also show a more tholeiitic than calc-alkaline geochemistry, suggesting derivation from asthenospheric mantle more primitive than that below a subduction arc. The favoured tectonic setting is in a marginal basin formed by lithospheric extension within or behind a volcanic arc (Fig. 10.5b). During the Llanvirn these volcanoes were augmented by others in the Welsh Basin, and by new centres along its south-east margin (Fig. 10.7).

Marine sediments continued to accumulate while the marginal basin formed. Mudstones predominate, with laterally impersistent sandstones and conglomerates that comprise resedimented debris eroded from local volcanic edifices. Regressive intervals or unconformities in the late Arenig and late Llanvirn record relative sea-level falls (Fig. 10.2), but the intervening successions mostly contain deep-marine graptolite

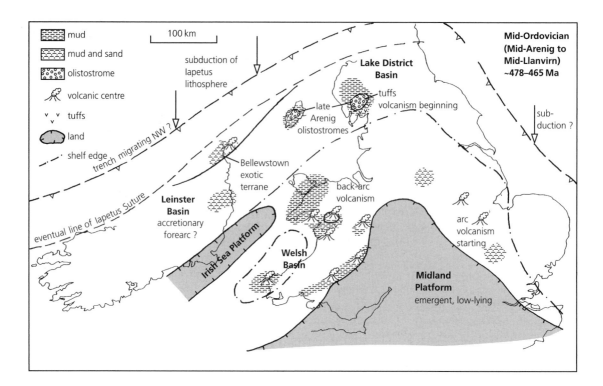

Fig. 10.7 Palaeogeographic map of southern Britain and Ireland for Llanvirn time (modified after Bevins *et al.* 1992, with permission of the Geological Society, London 1999).

and trilobite faunas. There is no evidence of substantial sediment supply from outside the Welsh Basin. By the late Arenig, Eastern Avalonia had probably rifted completely from Gondwana to the south, removing its connection with major continental source terranes. Much of the Midland Platform was probably emergent, but must have had a low relief and sediment-supply potential.

Boreholes show that marine shelf mudstones and thin sandstones were being deposited on the north and north-east edges of the Midland Platform during Llanvirn time. This depositional area in eastern England was probably continuous with the Welsh and Lake District Basins, around the northern edge of the Midland Platform. It is assumed to deepen northeastward towards the margin of Eastern Avalonia with the Tornquist Sea.

Deep marine sedimentation continued to the northwest of the Irish Sea Platform through late Arenig and early Llanvirn time (Figs 10.5b, 10.7). The turbidite

successions of the Skiddaw and Manx Groups are punctuated in late Arenig time by large submarine slump sheets derived from the south-east, indicating major instability on the Avalonian margin. By mid-Llanvirn time, this area was undergoing uplift and erosion. These events may have been precursors to the onset of substantial arc volcanism in late Llanvirn time. The Ribband Group of the Leinster Basin shows an analogous but slightly later sequence of events. Here, deep-marine sedimentation pre-dates a late Llanvirn unconformity before arc volcanism began again in the Caradoc.

10.8 Volcanic climax (Late Llanvirn to Early Caradoc)

About 20 million years after the episode of arc volcanism had ceased in the Welsh Basin, eruption of calc-alkaline rocks began again further outboard on the Avalonian margin, in much greater volumes than before (Fig. 10.8). The arc volcanoes restarted during late Llanvirn time in the Lake District Basin, represented by the Eycott and Borrowdale Volcanic Groups, and during early Caradoc time in the

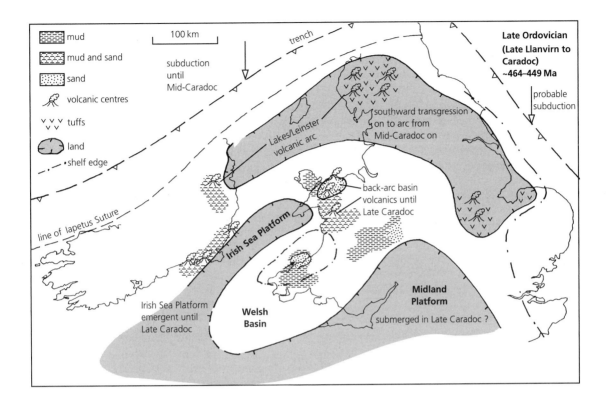

Fig. 10.8 Palaeogeographic map of southern Britain and Ireland for Caradoc time (modified after Bevins *et al.* 1992, with permission of the Geological Society, London 1999).

Leinster Basin, with the Duncannon Group (Fig. 10.3b,f).

In both basins, the volcanoes were constructed on the uplifted and partially eroded piles of Tremadoc to Llanvirn deep-marine sediments. Angular unconformities beneath the volcanic successions have been interpreted in various ways. In south-east Ireland a cleavage-forming deformation is thought to precede the volcanism. In the Lake District, the consensus favours only superficial soft-sediment deformation and tilting of fault blocks before igneous activity began. Probably the tectonic history of this fore-arc region varied both along- and across-strike. Certainly the variation in chemistry of the arc rocks suggests that the fore-arc was thinner in the north-west than in the south-east (Fig. 10.5c). This change occurs across the Causey Pike Fault in the Lakes and the Wexford

Boundary Lineament in Ireland, possible lines of major displacement both during and after arc volcanism.

The basal parts of the Lake District arc became submarine, and were interbedded with siltstones and tuffaceous sandstones. However, the volcanoes later maintained a topography above sea level and most of their deposits are subaerial. Andesite lavas and andesitic to rhyolitic ignimbrites are interbedded with laterally variable terrestrial, fluvial and lacustrine sediments: volcaniclastic sandstone, siltstone, conglomerate and breccia. A major change in eruptive style is represented within the Borrowdale Volcanic Group from an early plateau andesite field to later ignimbrites associated with caldera collapse. By contrast, the volcanoes in the Leinster Basin formed more subdued topography in an area of marine deposition. Here, the Duncannon Group transgressive sequences of conglomerate, limestone then shale were deposited above a late Llanvirn unconformity. This latest Llanvirn to early Caradoc '*gracilis* transgression' is recorded right across the Avalonian margin and may have a eustatic origin (Fig. 10.2).

What was the tectonic cause of the north-westward jump in arc volcanism during the Early Ordovician? Subduction was continuing during the arc jump, recorded by the persistent back-arc volcanism in the Welsh Basin. Some possibilities are that the subduction zone steepened between Tremadoc and Llanvirn time, or that the fore-arc widened by orthogonal or lateral accretion of slivers in front of the arc (Fig. 10.5b).

By Caradoc time the volcanic arc seems to have continued south-eastward from the Lake District, into eastern England, around the presumed northern edge of Eastern Avalonia (Fig. 10.8). The position of the old arc is marked by the Caledonian granitic intrusions inferred from gravity anomalies, and by boreholes in the east Midlands into calc-alkaline volcanic and plutonic rocks (Chapter 10e, Fig. 10.4). The swing in the arc presumably records a similar swing in the trench around the northern edge of the microcontinent.

Back-arc extension and volcanism continued in the Welsh Basin while the Lakes–Leinster arc was active. Large volumes of material were erupted here in Caradoc time, most notably in the Llewelyn and Snowdon Volcanic Groups in north Wales (Fig. 10.3d). The volcanoes formed islands in the otherwise marine basin, and were rimmed by a range of shallow marine sedimentary rocks predominantly composed of igneous debris. Deep-marine mudstones were deposited away from the volcanic centres, in the deeper parts of the Welsh Basin. Lower Caradoc mudstones tend to be high in organic carbon, due to deposition in conditions of bottom-water anoxia. This anoxia is probably in turn related to the *gracilis* transgression in early Caradoc time (Fig. 10.2).

The *gracilis* transgression flooded at least the north-western part of the Midland Platform, which had been emergent for much of Ordovician time. A succession of conglomerate and sandstone lies unconformably on Tremadoc and older rocks, and fined upwards in response to deepening marine shelf conditions. An analogous transgressive sequence was deposited on parts of the Irish Sea Platform.

10.9 Volcanic-arc shutdown (Mid–Late Caradoc)

Volcanism on the margin of Eastern Avalonia waned dramatically mid-way through Caradoc time, both in the volcanic arc of the Lake District and Leinster Basins and in the back-arc Welsh Basin. The volcanic edifices were no longer maintained by erupted material, and erosion and thermal subsidence combined to lower them below sea level in a short time. Each centre was transgressed by the fine-grained marine deposits that characterize Late Caradoc sedimentation. This transgression was accelerated by a late Caradoc pulse of probable eustatic sea-level rise (Fig. 10.2). The post-volcanic sedimentation is detailed in Chapter 11.

What caused the abrupt volcanic shutdown on Eastern Avalonia? The microcontinent was fast approaching Laurentia, but palaeomagnetic evidence suggests that the intervening Iapetus Ocean still spanned at least 20° of latitude (Fig. 10.2). There is no evidence of collisional deformation during the Caradoc, nor a supply of coarse sedimentary debris from any uplifting collision zone. A more plausible hypothesis is that, during the Mid-Caradoc, the Avalonian margin overran the spreading ridge in the Iapetus Ocean. The Californian margin of North America provides an active analogue of this process, from which views of Mid-Caradoc history can be reconstructed (Figs 10.5d, 10.9).

A cross-sectional view of the subduction system (Fig. 10.5d) shows how the slab of southern Iapetus lithosphere might have decoupled from the lithosphere north of the ridge. The southern plate continued to fall into the asthenosphere, pulled by the excess density of its older, cooler part. The leading edge of the northern Iapetus plate was too young and buoyant to subduct, and became lodged beneath the Avalonian lithosphere. An important consequence of this geometry is that the upwelling asthenosphere, which had fed the igneous activity at the ridge, now rose through a widening window behind the subducting slab. This asthenosphere was chemically distinct from the mantle above the old slab that had fed the subduction zone volcanoes. The new source of magma is recorded geochemically in the latest mid-Caradoc volcanics in the Welsh back-arc basin and perhaps in the suite of Late Ordovician minor intru-

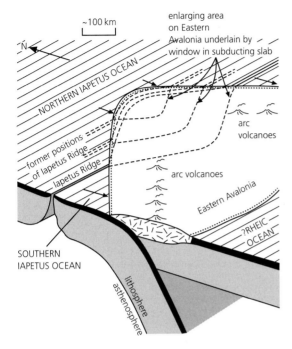

Fig. 10.9 Schematic representation of interaction of the Iapetus ridge and the Avalonian subducting margin to leave oblique-slip margins and a growing slab window.

sives into the northern part of the Midland Platform (Fig. 10.4).

A three-dimensional view (Fig. 10.9) demonstrates the complexity that would have arisen if the subducting margins of Eastern Avalonia were oblique to the Iapetus ridge that they overran. Triple junctions would have migrated along the Avalonian margins, so that the end to arc volcanism was diachronous. The apparently later end to arc volcanism in south-eastern Ireland compared with the Lake District suggests that this triple junction migrated south-westwards. A diamond-shaped slab window opened progressively southwards, through which were fed the more primitive magmas that mark the last burst of volcanism. The plate boundary formed in the wake of the triple junction, between the northern Iapetus plate and the continental margin, probably had a strike-slip component, analogous to present-day California. This component may be that recorded in a late Caradoc to early Llandovery phase of strike-slip

faulting along the Welsh Borderland Fault System (see Section 11.5, p. 172).

Palaeomagnetic data suggest that, by Caradoc time, the continent of Baltica was close to Eastern Avalonia. Faunal contrasts had been eradicated, but there is no indication on the north-eastern edge of Eastern Avalonia of a collisional orogeny. There is no strong metamorphism, deformation or uplift, nor even a pulse of clastic sediment that might indicate a collision further north-east. The encounter with Baltica must have been gentle enough, or sufficiently dominated by strike-slip tectonics, not to leave a strong mark on Eastern Avalonia.

10.10 Summary: rift and drift of the Eastern Avalonia microcontinent

For much of Ordovician time, the southern British Isles was part of the discrete microcontinent of Eastern Avalonia. Its rifting from Gondwana and its northward drift towards Laurentia coincided with, and were partly driven by, the subduction of Iapetus oceanic lithosphere beneath its northern margin.

Subduction had begun by late Tremadoc time (Fig. 10.10), initiating calc-alkaline arc volcanoes in the Welsh and Leinster Basins. This episode was short lived and for 20 million years during Arenig and most of Llanvirn time there was no arc volcanism. When it resumed, in late Llanvirn and Caradoc time, the arc was further outboard on the margin, in the area of north-western England and south-eastern Ireland.

Voluminous marginal basin volcanism, typically with a bimodal acid/basic composition, began in the Welsh Basin during the hiatus in arc activity. The volcanic centres were often controlled by north–south faults resulting from east–west basin extension, a continuing effect of sinistral-oblique subduction. The marginal basin volcanism continued while the arc rejuvenated further north-west. During the Caradoc the Welsh Basin was a true back-arc basin. By this time, the arc rimmed not only the north-western edge of Eastern Avalonia, but probably its north-eastern edge too.

Volcanic activity ceased during mid-Caradoc time, probably as Eastern Avalonia overran the spreading ridge in the Iapetus Ocean. This interaction was responsible for a phase of strike-slip tectonics on the

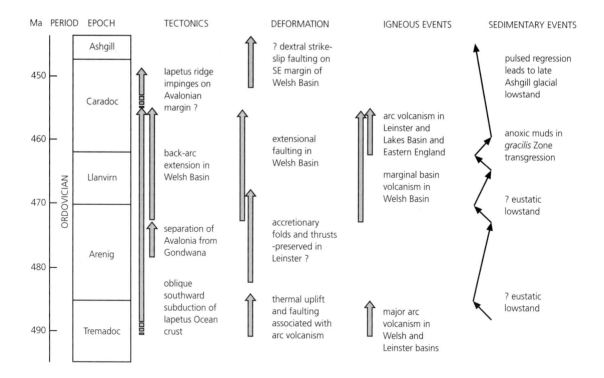

Fig. 10.10 Tentative chronology of events on the active margin of Eastern Avalonia.

margin, and suites of chemically distinctive volcanics and minor intrusions.

References

Bevins, R.E., Bluck, B.J., Brenchley, P.J. *et al.* (1992) Ordovician. In: *Atlas of Palaeogeography and Lithofacies* (eds J. C. W. Cope, J. K. Ingham & P. F. Rawson), Memoir 13, pp. 19–36. Geological Society, London.

Fortey, R.A. & Cocks, L.R.M. (1986) Marginal faunal belts and their structural implications, with examples from the Lower Palaeozoic. *Journal of the Geological Society, London* **143**, 151–160.

Torsvik, T.H., Smethurst, M.A., Meert, J.G. *et al.* (1996) Continental break-up and collision in the Neoproterozoic and Palaeozoic—a tale of Baltica and Laurentia. *Earth-Science Reviews* **40**, 229–258.

Further reading

Bevins, R.E., Bluck, B.J., Brenchley, P.J. *et al.* (1992)

Ordovician. In: *Atlas of Palaeogeography and Lithofacies* (eds J. C. W. Cope, J. K. Ingham & P. F. Rawson), Memoir 13, pp. 19–36. Geological Society, London. [Detailed palaeogeographic maps and accompanying commentary.]

Branney, M.J. & Soper, N.J. (1988) Ordovician volcano-tectonics in the English Lake District. *Journal of the Geological Society, London* **145**, 367–376. [Reinterprets the angular unconformities bounding the Borrowdale Volcanic Group as due to volcano-tectonic uplift, faulting and tilting rather than to regional shortening.]

Cooper, A.H., Millward, D., Johnson, E.W. & Soper, N.J. (1993) The early Palaeozoic evolution of northwest England. *Geological Magazine* **130**, 711–724. [An excellent concise review of the geological history of the Lake District Basin.]

Cooper, A.H., Rushton, A.W.A., Molyneux, S.G., Hughes, R.A., Moore, R.M. & Webb, B.C. (1995) The stratigraphy, correlation, provenance and palaeogeography of the Skiddaw Group (Ordovician) in the English Lake District. *Geological Magazine* **132**, 185–211. [A definitive revision of the depositional history of the early Ordovician rocks of the Lake District.]

Fortey, R.A. & Cocks, L.R.M. (1986) Marginal faunal belts and their structural implications, with examples from the Lower Palaeozoic. *Journal of the Geological Society,*

London **143**, 151–160. [An attempt to position continents and to define global sea-level changes from fossils.]

Fortey, R.A., Harper, D.A.T., Ingham, J.K., Owen, A.W. & Rushton, A.W.A. (1995) A revision of Ordovician series and stages from the historical type area. *Geological Magazine* **132**, 15–30. [The redefinition of the traditional subdivisions of the British and Irish Ordovician.]

Kokelaar, B.P. (1988) Tectonic controls of Ordovician arc and marginal basin volcanism in Wales. *Journal of the Geological Society, London* **145**, 759–776. [A stimulating interpretative review of Welsh Basin volcanism, concentrating on north Wales.]

Kokelaar, B.P., Howells, M.F., Bevins, R.E., Roach, R.A. & Dunkley, P.N. (1984) The Ordovician marginal basin of Wales. In: *Marginal Basin Geology* (eds B. P. Kokelaar & M. F. Howells), Special Publication 16, pp. 245–269. Geological Society, London. [Persuasively makes the case for the Mid–Late Ordovician Welsh Basin being a back-arc basin on thinned continental crust.]

Leat, P.T. & Thorpe, R.S. (1989) Snowdon basalts and cessation of Caledonian subduction by the Longvillian. *Journal of the Geological Society, London* **146**, 965–970. [Attributes the distinctive geochemistry of the last north Welsh volcanics to new magma sources caused by the end of subduction.]

Max, M.D., Barber, A.J. & Martinez, J. (1990) Terrane assemblage of the Leinster Massif, SE Ireland, during the Lower Palaeozoic. *Journal of the Geological Society, London* **147**, 1035–1050. [Detailed review of SE Ireland geology. Includes an accretionary fore-arc interpretation for the Early–Mid Ordovician Leinster Basin.]

Murphy, F.C., Anderson, T.B., Daly, J.S. *et al.* (1991) An appraisal of Caledonian suspect terranes in Ireland. *Irish Journal of Earth Sciences* **11**, 11–41. [A concise but informative review of Irish stratigraphy, used to define major Early Palaeozoic terranes.]

Pharaoh, T.C., Brewer, T.S. & Webb, P.C. (1993) Subduction-related magmatism of late Ordovician age in eastern England. *Geological Magazine* **130**, 647–656. [Interprets the Ordovician igneous rocks in the subsurface of eastern England as a remnant of a subduction-related volcanic arc.]

Torsvik, T.H., Smethurst, M.A., Meert, J.G. *et al.* (1996) Continental break-up and collision in the Neoproterozoic and Palaeozoic—a tale of Baltica and Laurentia. *Earth-Science Reviews* **40**, 229–258. [A valuable series of global palaeogeographic maps, with a discussion of the evidence behind them.]

Woodcock, N.H. (1990) Sequence stratigraphy of the Welsh Basin. *Journal of the Geological Society, London* **147**, 537–547. [A review of Welsh Basin stratigraphy, highlighting the subdivision into pre-, syn-, and post-volcanic sequences that correlate broadly across Eastern Avalonian basins.]

11 Late Ordovician to Silurian evolution of Eastern Avalonia during convergence with Laurentia

N. H. WOODCOCK

11.1 Palaeocontinental setting: Eastern Avalonia approaches Laurentia

For much of Ordovician time, the southern parts of Britain and Ireland lay on the discrete continental fragment of Eastern Avalonia (Fig. 11.1). After its Early Ordovician rifting from Gondwana (see Chapter 10), this fragment had moved rapidly northward towards Laurentia, as the intervening Iapetus Ocean was narrowed by subduction at both its northern and southern margins. Subduction-related volcanism on Eastern Avalonia waned in Late Ordovician (Mid-Caradoc) time, perhaps because of collision with Baltica or because the Iapetus Ridge was subducted beneath Avalonia (see Chapter 10). Whatever its tectonic cause, this end to widespread volcanism is a convenient bookmark in the pages of Eastern Avalonian history.

From Caradoc time onwards, Eastern Avalonia was increasingly influenced by its larger neighbours: Laurentia to the north-west and Baltica to the north-east. Firstly, Eastern Avalonia was colonized by Laurentian and Baltican organisms, next flooded by their sedimentary debris, then compressed against their margins and finally welded to them along the seam of the Caledonian mountain belt (Fig. 11.1). This chapter describes the stratigraphic, sedimentological and volcanic record of this progressive continental collision through latest Ordovician and Silurian time.

By late in the Silurian, the former deep marine basins of Eastern Avalonia had shallowed dramatically. The predominantly non-marine sedimentation that followed in Devonian time is described in Chapter 13, whereas the late Caledonian (Acadian) deformation, metamorphism and associated plutonic activity that dominated Late Silurian and Devonian history is described in Chapter 12. By the end of this orogenic event, in Mid-Devonian time, the southern parts of Britain and Ireland lay close to their northern counterparts, all components of the new continental entity of Laurussia (Fig. 11.1).

11.2 Changes in global sea level and climate

The interactions of continents provide one major control on the Late Ordovician to Silurian geological history of southern Britain and Ireland. The next most potent influences are global changes in sea level and climate. The Silurian was a time of relatively high sea levels on the first-order Phanerozoic curve (see Chapter 2), probably due to high rates of sea-floor spreading. However, within this highstand, shorter-term variations affected the proportion of continental platforms that were flooded. The stratigraphic record preserves the consequent fluctuations in the amount of land subjected to erosion and in the volume of shallow organically productive sea water.

Global sea-level fluctuations must be distinguished from more local relative changes by matching shallowing and deepening events across a number of different palaeocontinents. A curve derived using this approach (Fig. 11.2) shows the following important features.

1 A pulsed shallowing during later Caradoc and Ashgill time.

2 A rapid shallowing during latest Ashgill (Hirnantian) time. This coincides with the maximum extent of a southern-hemisphere ice cap, then centred on the North African region of Gondwana.

3 A rapid post-glacial deepening in Early Llandovery

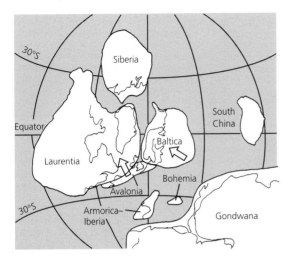

Fig. 11.1 Palaeocontinental reconstruction for mid-Silurian (Wenlock) time. The open arrows on Baltica and Eastern Avalonia show their directions of movement up to this time (modified from Torsvik *et al.* 1996, with permission from Elsevier Science 2000).

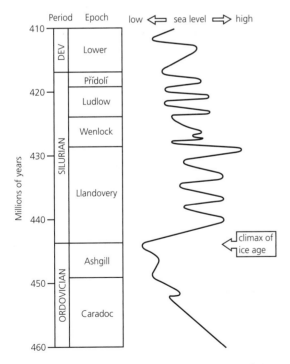

Fig. 11.2 A global sea-level curve for Late Ordovician to Silurian time (modified from Johnson 1996).

time followed by fluctuating sea levels through the Silurian. These variations may have been a response to changes in the volumes of ice on Gondwana at least into Wenlock time.

4 A pulsed shallowing during Late Silurian time and into the Devonian.

The stratigraphic effects of these sea-level changes are visible both on a broad time-scale (see Section 11.3) and in more detailed depositional events (see Sections 11.4, 11.7, 11.8, 11.11, 11.12).

11.3 Large-scale stratigraphic signatures

Before charting the detailed stratigraphic record of Avalonia's collision with Laurentia, it is helpful to view the broad perspective through Late Ordovician to Early Devonian time. The areas where rocks of this age were deposited group into three main palaeogeographic settings (Fig. 11.3):

The Midland Platform was the relatively stable continental core of Eastern Avalonia. This had been delineated after the Early Ordovician rifting off Gondwana, but had since accumulated only thin patches of post-Tremadoc sediment. During Silurian time it was flooded more persistently by shallow seas. The Midland Platform remained largely undeformed by the late Caledonian Orogeny, although it received large volumes of erosional debris from its resultant mountains.

Basinal areas deformed by the late Caledonian (Acadian) Orogeny comprise the Leinster, Lake District and Welsh Basins to the north-west of the Midland Platform and the Anglian Basin to the north-east. These areas all subsided more rapidly than the Midland Platform during Late Ordovician and Silurian time, accumulated larger thicknesses of sediment, and hosted deeper-water environments. However, the basins also contained the cooling remnants of the Ordovician volcanic centres, capable of supporting shallow marine or non-marine environments for some millions of years after their demise. A small elongate Irish Sea Platform separated the Leinster and Welsh Basins. The basinal areas were all strongly folded and cleaved during the Acadian Orogeny.

Basinal areas south of the Variscan Front may have existed during Ordovician or Silurian time. However, their deposits are now hidden beneath the Devonian and Carboniferous sediments of the Cornubian Basin

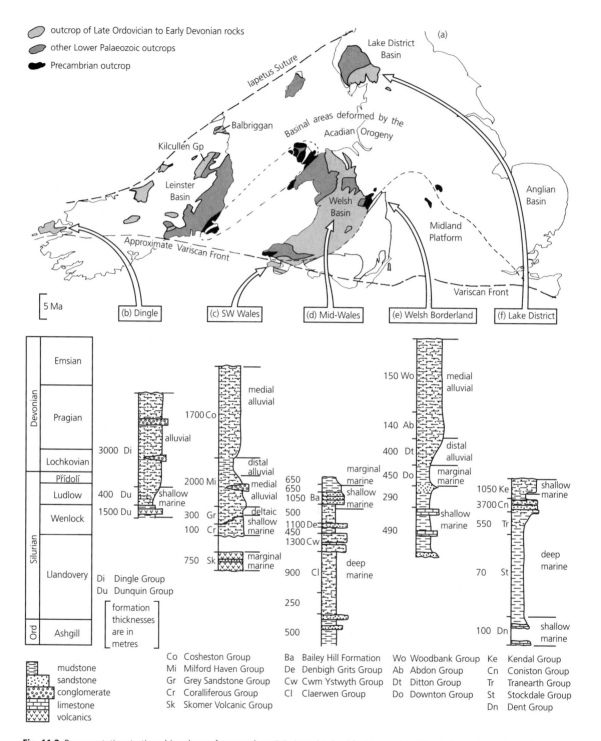

Fig. 11.3 Representative stratigraphic columns from southern Britain and Ireland for Upper Ordovician to Lower Devonian rocks.

in south-west England and its possible continuation beneath southernmost Ireland.

Detailed stratigraphies clearly vary from area to area, particularly between basins and platforms. However, some important generalizations can be made about the upper Ordovician (Ashgill) and Silurian successions on the Midland Platform and the basins to the north (Fig. 11.3).

1 The successions are predominantly sedimentary, lacking the thick volcanic accumulations that characterize most of Ordovician time. This absence reflects the shutdown of the subduction-related volcanic centres in the Lake District, Wales and Leinster.

2 Successions preserved during Late Ordovician and Early Silurian (Llandovery) time tend to be marine but to host unconformities. These breaks can be due to the residual relief of cooling Ordovician volcanoes, to gentle tectonic tilting and uplift, or to the effects of glacio-eustatic falls in sea level.

3 Later Silurian successions (Wenlock, Ludlow and Přídolí) tend to be more conformable, perhaps due to more stable sea levels and to lack of contractional tectonics.

4 All successions show an essentially conformable upward transition from marine sediments into paralic or non-marine sediments, at latest by Přídolí time. This transition heralds the demise of the basinal areas during the Acadian Orogeny. These basins, even if still subsiding, were overfilled by large volumes of sediment newly supplied from already uplifting areas to the north.

Early Devonian intervals are included in Fig. 11.3, to emphasize the predominantly conformable upward transition from Silurian marine deposits. The Acadian climax is marked by the unconformity at the top of these successions, not by the marine to non-marine facies change within them—the base of the 'Old Red Sandstone'. It is not that the Late Ordovician and Silurian were periods of tectonic inactivity on Eastern Avalonia (see Sections 11.5, 11.10); they merely lacked episodes of crustal compression on an orogenic scale.

The Late Ordovician to Silurian geological history of Eastern Avalonia will now be charted in more detail below (see Sections 11.4–11.14). Devonian sedimentation is fully described in Chapter 13.

11.4 Post-volcanic transgressions and mud blankets (late Caradoc to Ashgill)

For much of Ordovician time, Eastern Avalonia had hosted the persistently emergent Midland Platform, the emergent or shallow marine Irish Sea Platform, and a series of volcanic centres that intermittently gave shallow marine or emergent conditions in otherwise basinal settings (see also Figs 10.6–10.8, pp. 160–163). In late Caradoc time, many of these land areas were submerged and marine sediment was deposited on the platforms and the extinct volcanic piles as well as in the basins (Fig. 11.4). The main reason for this submergence was the cooling and thermal subsidence of the extinct volcanic areas. However, a regional, and possibly global, late Caradoc sea-level rise flooded the platforms and enhanced the water depth over the old volcanoes (Fig. 11.2).

The environment of late Caradoc and Ashgill deposition was strongly influenced by its palaeogeographic setting, and particularly by its relationship to the extinct Ordovician volcanoes.

Shallow marine mudstones and sandstones were deposited on the Midland Platform, founded on relatively stable continental crust without local volcanic centres. In the Welsh Borderland, the Early Caradoc sea-level rise had initiated deposition that lasted through Caradoc time. Sediment eroded from gently emergent areas of the platform was reworked by shallow marine waves and tides and redeposited in gently subsiding areas. Mud deposition predominated over sand during late Caradoc time, as the platform source areas became progressively submerged. The Ashgill part of this succession is not preserved, probably due to uplift and erosion during the lowered sea levels of Ashgill to Early Llandovery time (see Section 11.5).

Shallow marine mudstones and limestones were deposited on old volcanic piles that were subsiding only gently, such as those in Leinster and the Lake District. The base of the Windermere Supergroup (Lake District) is diachronous, and is followed by patchy, lenticular sandstones and conglomerates, recording the transgression over an irregular surface of the Borrowdale Volcanic Group. Submergence of the volcanic rocks cut the supply of coarse clastic debris. Mudstone deposits then dominated, with interbedded limestones containing organisms living

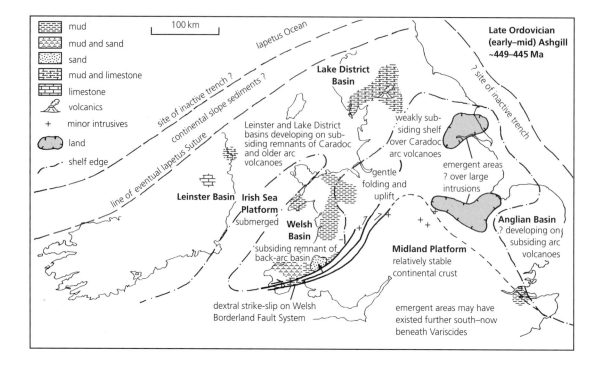

Fig. 11.4 Palaeogeographic map of southern Britain and Ireland for late Caradoc to early Ashgill time (modified after Bassett *et al.* 1992, with permission of Geological Society, London 1999).

in the warm, shallow, well-oxygenated waters of the offshore marine shoals. Intermittent regression during the Ashgill gave disconformities above the more gently subsiding parts of the old volcano.

Deep marine mudstones and sandstones blanketed rapidly subsiding volcanoes or areas in the already deep part of the Ordovician basin. Late Caradoc mudstones tend to be carbonaceous and well laminated, reflecting anoxic bottom waters. This anoxia was due to rapid fallout of the increased volume of planktonic organisms promoted by progressive flooding of the adjacent shelves. During the regressions of the Ashgill, burrowed mudstones record oxygenated bottom waters. The deeper parts of the Welsh Basin accumulated modest volumes of sandstone and conglomerate, supplied by turbidity flows from the basin margins. Some of these coarse clastic units probably correlate with regressions on the shelf, but others may have been sourced from tectonic uplifts newly created by intra-Ashgill deformation (see Section 11.5).

11.5 Late Ordovician tectonics

The sedimentary pattern that resulted from Caradoc and Ashgill sea-level changes is also affected by two types of postvolcanic tectonic activity.

Caradoc volcano-tectonic faulting and tilting is responsible for an angular unconformity between some of the volcanic centres and their uppermost Ordovician or Silurian cover. The clearest example is between the Borrowdale Volcanic Group of the Lake District and the overstepping Windermere Supergroup (Fig. 11.5). Structures in the volcanic rocks, originally assigned to crustal shortening, have been reinterpreted as faults and tilted blocks due to caldera collapse and foundering of the subaerial volcanic pile. Similar structures have been identified in the Caradoc centres of North Wales, although without such marked unconformity.

Ashgill (or ?early Llandovery) strike-slip faulting and gentle folding is recorded along the Welsh Borderland Fault System (Fig. 11.5). Structures are preserved in the Builth and Shelve inliers of Ordovician rocks, where they are unconformably overstepped by Middle or Upper Llandovery rocks. The predominance of strike-slip movement indicators on the faults

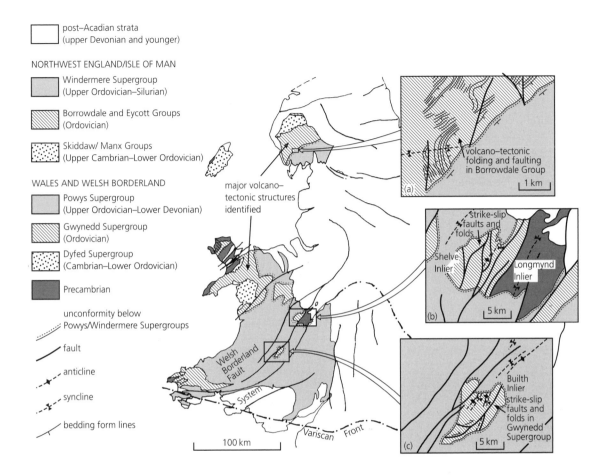

Fig. 11.5 Maps showing unconformities below the Powys and Windermere Supergroups as evidence for Late Ordovician deformation. ((a) from Branney & Soper 1988, with permission from the Geological Society, London (1999); (b) and (c) from Woodcock 1984, with permission from the Geological Society, London 1999).

precludes an origin by simple volcanic foundering, but some component of Acadian displacement cannot be ruled out. A persistent disconformity or angular unconformity around the Welsh Basin in the Ashgill, together with contemporaneous turbidite fans shed into the basin, may coincide with this strike-slip deformation and its localized uplifts. The deformation has been termed the Shelveian event.

The foundering and subsidence of the Ordovician volcanic centres was a direct isostatic consequence of their shut-down and cooling, mostly in Caradoc time. Southward subduction ended below Eastern

Avalonia (see Section 10.9, pp. 164–165), perhaps because of collision of Eastern Avalonia with Baltica or of the Iapetus ridge with the Eastern Avalonian margin. By analogy with present-day California, strike-slip faulting would also be a predictable consequence of ridge/margin collision (see also Fig. 10.9, p. 165).

Despite the local stratigraphic and sedimentological effects of the Late Ordovician tectonic phases, they produced neither folds nor cleavage comparable in intensity or extent with the overprinting Acadian deformation of Devonian time.

11.6 Late Ordovician volcanism

To describe the post mid-Caradoc history of Eastern Avalonia as 'postvolcanic' is not quite accurate. Although most volcanic centres shut down at this time, sporadic volcanic rocks and related minor intrusives provide evidence for both local and remote igneous activity throughout the uppermost Ordovician to lowest Devonian record.

Late Ordovician volcanism occurred in north-west England, an area that had already hosted Caradoc and earlier centres (Fig. 11.6). By contast, Silurian volcanism was initiated in new areas (Fig. 11.6). The thin rhyolites and rhyolitic tuffs in north-west England occur spatially above the Borrowdale Volcanic Group, although they are stratigraphically separated from them within upper Caradoc or upper Ashgill shallow marine mudstones and limestones. The simplest interpretation of this latest Ordovician volcanism is as a late-stage eruption of magmas above the old southward-dipping Iapetus subduction zone. This model implies a time lag of up to 8 Myr after the initial detachment of the subducting slab in mid-Caradoc time (see Chapter 8).

11.7 The Late Ordovician ice age

Towards the end of Ashgill (Late Ordovician) time there was a brief but strong lowering of global sea level (Fig. 11.2) that had a marked effect on the palaeogeography and stratigraphic record on Eastern Avalonia (Fig. 11.7). The cause was the glacial event centred on the Gondwana continent, which had straddled the Earth's south pole for much of the Ordovician.

Late Ordovician continental glacial deposits are known from a number of localities on the old Gondwana continent: Saharan Africa, South Africa, South America and Arabia. Striated pavements show that land ice reached from the south pole up to between 70° and 50° latitude and water-lain tilloids suggest that floating ice reached to about 40°S. Tilloids are preserved in Iberia and France, then in mid-latitude, but Eastern Avalonia was already in a low latitude, beyond the reach of ice. Here the glacial signature was a 100-m fall in sea level, forced by large volumes of water being locked up in the continental ice sheet. This sea-level fall can be dated precisely to within the Hirnantian Stage of the Ashgill (latest Ordovician), probably lasting no more than a million years.

The geological effects of the lowering of sea level varied with palaeogeographical setting.

The Midland Platform became emergent. Much of the sediment deposited on it earlier in Ordovician

Fig. 11.6 Location and age of Late Ordovician (late Caradoc and Ashgill) and Early Silurian volcanics in relation to mid-Caradoc and earlier volcanic areas (bentonite distribution data from Fortey *et al.* 1996).

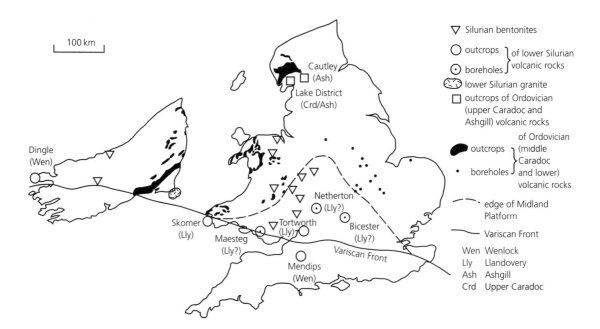

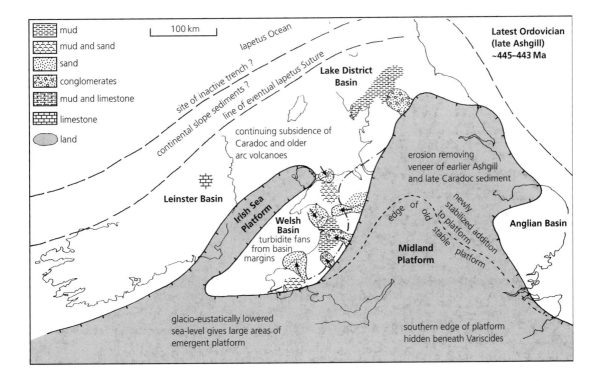

mud

mud and sand

sand

conglomerates

mud and limestone

limestone

land

100 km

Iapetus Ocean

site of inactive trench ?

continental slope sediments ?

line of eventual Iapetus Suture

Lake District Basin

Latest Ordovician (late Ashgill) ~445–443 Ma

continuing subsidence of Caradoc and older arc volcanoes

erosion removing veneer of earlier Ashgill and late Caradoc sediment

Leinster Basin

Irish Sea Platform

Welsh Basin

turbidite fans from basin margins

edge of old stable platform

newly stabilized addition to platform

Anglian Basin

Midland Platform

glacio-eustatically lowered sea-level gives large areas of emergent platform

southern edge of platform hidden beneath Variscides

Fig. 11.7 Palaeogeographic map of southern Britain and Ireland for late Ashgill (Hirnantian) time (modified after Bassett *et al.* 1992, with permission from the Geological Society, London 1999).

time was now eroded and transported towards the platform edge. The emergence may have been enhanced on parts of the platform that had suffered deformation during Ashgill time (see Section 11.5). The stratigraphic result is an unconformity, spanning at least the Ashgill, in all successions from the platform (Fig. 11.3).

Basin margins and deep basin floors received the sedimentary debris eroded from the platforms and derived by slumping of the upper slopes of the basin (Fig. 11.7). This sediment was redeposited by mass-flow processes as fans of sand and conglomerate. Most examples were shed north-westwards from the edge of the Midland Platform, but fans in North Wales may have been supplied south-eastward from a newly emergent Irish Sea Platform.

Some shallow basinal or outer platform areas were by-passed by coarse sediment and continued to accumulate mud-dominated successions. However, lime-

stones became less common (Fig. 11.3). This was partly due to reduced organic productivity in the colder waters of the glacial climax, but also to the lowered biological diversity during the glacially related extinctions in groups such as trilobites, brachiopods, echinoderms and corals.

11.8 Post-glacial transgressions (Llandovery)

The sedimentary effects of the Late Ordovician ice age were terminated by a sea-level rise near the end of Hirnantian time. This rise triggered a pulsed transgression over the platforms (Fig. 11.8) that was to last through the Llandovery and into the Wenlock epoch of the Silurian period. Sea-level changes were probably caused by the fluctuating volume of a persistent ice sheet in the southern hemisphere. Early Silurian marine successions are therefore initiated or permeated by characteristic signs of marine transgression.

On the Midland Platform the unconformity produced by the glacial regression is overlain by successions of shallow marine mudstones and limestones,

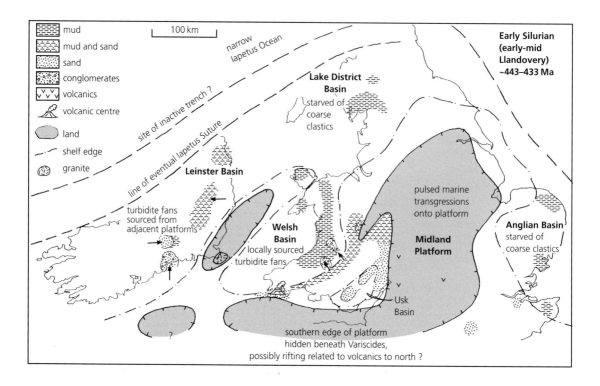

Fig. 11.8 Palaeogeographic map of southern Britain and Ireland for Early Silurian (early to mid-Llandovery) time (modified after Bassett *et al.* 1992, with permission from the Geological Society, London 1999).

often with transgressive sandstones at their base. The pulsed nature of the transgression is reflected in the diachroneity of the sediments overlying the unconformity, varying up to latest Llandovery time. The transgression across the platform can also be tracked by charting the spread of depth-related assemblages of brachiopods and associated benthonic organisms across the flooded part of the platform. Near-shore areas were characterized by *Lingula*, more offshore areas successively by the *Eocoelia*, *Pentamerus*, *Stricklandia* and *Clorinda* assemblages, and basinal areas by the dominance of graptolites. The biological gradients revealed across the platform (Fig. 11.9) may have been controlled by water turbulence, the type of substrate, or nutrient availability rather than strictly by depth. However, they demonstrate how an abrupt shelf to basin transition at the beginning of the

transgression was transformed into a shelf with a width of 75 km by the beginning of late Llandovery time and of 125 km by the end of the Llandovery.

In the basinal areas the most common record of post-glacial transgression was a blanket of carbonaceous anoxic shales. These organic shales were caused by fallout of the increased volume of marine phytoplankton produced as the area of warm shallow shelves increased, a situation identical to that during the Caradoc transgression (see Section 11.4). Once again, rising sea levels progressively drowned the sources of coarse clastic debris, so that only isolated turbidite fans were now supplied from the Midland Platform or its possible continuation south of Ireland.

11.9 Silurian volcanism

Silurian volcanic and high-level intrusive rocks have a distribution different from that of the Caradoc and older volcanic centres (Fig. 11.6).

Early Silurian volcanic centres occurred in an

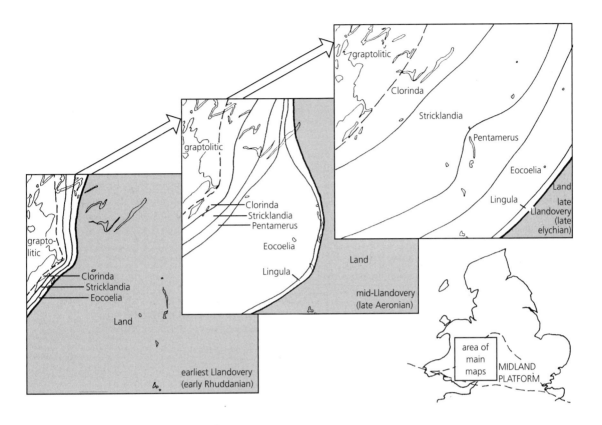

Fig. 11.9 The Llandovery transgression of the Welsh Borderland charted by brachiopod communities (after Ziegler *et al.* 1968, with permission from the Palaeontological Association 1999).

east–west zone through southern Eastern Avalonia, close to or just within the later Variscan Belt. The best-known products are the Skomer Volcanic Group of south-west Wales, a mainly basaltic succession interstratified with terrestrial or paralic sedimentary rocks and thin rhyolitic tuffs. The upper part of the succession at least is of middle Llandovery age. Petrographically similar basic and intermediate volcanics outcrop further east, within the upper Llandovery at Tortworth and in the middle Wenlock of the Mendip Hills. The Tortworth volcanics may be part of a more extensive igneous succession of Llandovery age, mainly intermediate tuffs, that has been imaged on seismic profiles and encountered in boreholes (Fig. 11.6). On the basis of the Skomer Group, the Silurian volcanics have a geochemistry that contrasts with their Ordovician counterparts, suggesting a domi-

nant source from within-plate rifting and only minor contributions from a relict or active subduction zone. An episode of extension or transtension through this southern belt of Eastern Avalonia is suggested. Wenlock volcanics on the Dingle Peninsula of southwest Ireland may represent a westward continuation of this belt, though with a stronger subduction zone chemical signature. The Carnsore and Saltees Granites of south-east Ireland are Early Silurian age, but of uncertain affinities.

Thin tuffs from remote volcanic centres occur throughout the latest Ordovician and Silurian successions of Eastern Avalonia (Fig. 11.6). Most are very fine grained and have altered to bentonitic clays. Some are widespread enough to be used for correlation on their physical or geochemical characteristics, but few can be traced back to a specific volcanic source. They demonstrate the persistence of volcanic activity capable of providing volcanic dust to Eastern Avalonia, but not necessarily centred on the microcontinent. The source volcanoes could well have been those on the southern rim of Laurentia.

11.10 Silurian tectonic controls

The continuing volcanism on Eastern Avalonia is one indication that its crust was not just responding passively as it approached Laurentia to the north. This section describes other evidence that Eastern Avalonia was tectonically active after subduction ended beneath it but before the climax of collisional Acadian deformation in Devonian time. None of this pre-Acadian deformation is strong when compared with the Acadian event itself, but it has important effects on regional stratigraphic relationships and sedimentary patterns.

Llandovery crustal extension is recorded, particularly in the Welsh Basin, by episodes of sediment accumulation that begin rapidly and decrease with time. These episodes represent lithospheric stretching events followed by thermal subsidence of the stretched lithosphere. The crustal stretching was accommodated on mappable normal faults that delimited intervening tilted blocks. Early and Mid-Llandovery stretching has been recognized on the west side of the Midland Platform (Fig. 11.8). However, more important is late Llandovery movement on Welsh Basin faults that exerted a major control on the thickness and depositional sites of major turbidite systems (see Section 11.11). Most of these tilt blocks were submarine and were continually draped with sediment. However, along the margin of the Welsh Basin, some uplifted fault footwalls were eroded and unconformably overlain by post-tilting sediment, typically of upper Llandovery age. The late Llandovery to Wenlock stretching event does not seem to have affected the Lake District Basin, and remains unproven in Ireland.

Ludlow and later crustal flexure is suggested by rapidly increasing rates of sediment accumulation through this time period, particularly in the Lake District and North Wales basins. The subsidence pattern is consistent with the progressive depression of crust by an advancing load, probably the overriding Laurentian continent, or at least the accretionary sediment prism on its leading edge. The southward advance of this load gives a corresponding southward diachroneity in the depression of the Eastern Avalonian crust: from Early Ludlow in the Lake District, to Mid-Ludlow in North Wales and Pridoli in South Wales. This diachroneity is also reflected in the pattern of sediment supply and deformation in the Eastern Avalonian basins (see Section 11.11).

The first substantial collision of Eastern Avalonia with Laurentia probably happened during Llandovery time (Fig. 11.10). This collision reactivated old fault systems within Eastern Avalonia, causing compression and uplift of new sediment sources in some areas and extension and subsidence in others such as the Welsh Basin. Through the rest of Silurian time, the microcontinent was drawn towards and into the Laurentian trench. The accretionary prism on the leading edge of Laurentia was progressively overthrust onto Eastern Avalonia, causing flexural subsidence and increasing sediment accumulation rates. By early Devonian time the major compression and uplift associated with the Acadian Orogeny was beginning (see Section 11.13 and Chapter 12).

11.11 Mid-Silurian platforms and turbidite basins

The changed palaeogeography (Fig. 11.11) of mid-Silurian (Late Llandovery to Early Ludlow) time resulted from three main factors:
1 the culmination of the eustatic sea-level rise that had flooded much of the previously emergent Midland Platform;

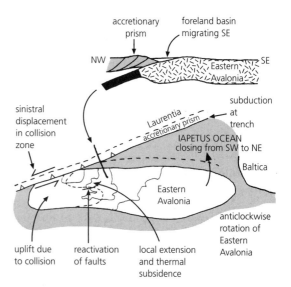

Fig. 11.10 Possible tectonic controls on Silurian geology depicted in about Wenlock time.

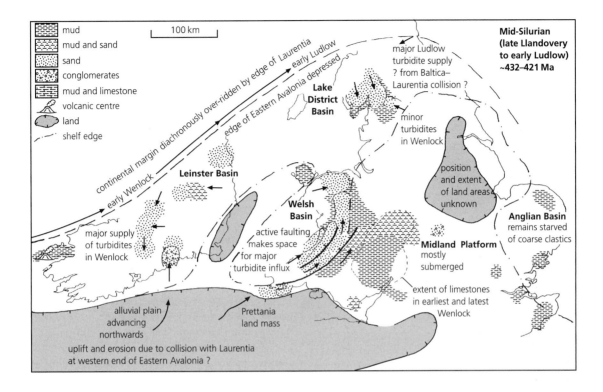

Fig. 11.11 Palaeogeographic map of southern Britain and Ireland for mid-Silurian (late Llandovery to early Ludlow) time (modified after Bassett *et al.* 1992, with permission from the Geological Society, London 1999).

2 a phase of localized crustal extension and subsequent thermal subsidence, which created more submarine space to accommodate sediment, at least in the Welsh Basin and on the Midland Platform;
3 the impingement of Eastern Avalonia with Laurentia, which progressively restricted connections between the marine basins and the open ocean system.

The Midland Platform accumulated a continuous succession of marine sediments through late Llandovery to Ludlow time. Mudstones, often calcareous, predominate and preserve a rich fossil assemblage of shelfal marine organisms—brachiopods, trilobites, bivalves, crinoids, ostracods, stromatoporoids, corals and bryozoans (Fig. 11.3). However, for two short periods in earliest and latest Wenlock time, conditions favoured high carbonate productivity and the local establishment of patch reefs built by stromatoporoids, corals, algae and bryozoans. These times

correlate with sea-level falls on the eustatic curve (Fig. 11.2), suggesting that at other times the shelf water was too deep for reef growth. Followed south-westwards, the platform deposits become less calcareous and their clastic component coarser, implying a persistent source area, termed Prettania, that survived the earlier Silurian transgression.

The Lake District Basin and, from the limited available evidence, *the Anglian Basin* accumulated mostly mudstones and siltstones through late Llandovery and Wenlock time. The lack of sand-grade debris is consistent with rising sea levels having submerged most emergent source areas on the platform. The type of basinal mudstone is also closely related to sea-level changes. Laminated anoxic mudstones are abundant, particularly during Wenlock time, but are interrupted by bioturbated oxic mudstones in the earliest and latest Wenlock, coincident with the short regressions that favoured limestones on the platform. Sandstone turbidites were deposited in the southeastern part of the Lake District Basin during Wenlock time, but the main influx of coarse clastic material came in the Ludlow from north-westerly and north-easterly directions (Fig. 11.11). This was the time when the Laurentian plate was beginning to load the edge of the downgoing Avalonian plate opposite the Lake District (Fig. 11.10). The sediment for the Ludlow turbidite fans may have been derived from the Laurentia–Baltica collision zone further northeast, but its accumulations effectively represent the toe of the Laurentian accretionary prism encroaching onto Avalonian crust.

The Leinster Basin probably had a broadly similar history to that of the Lake District Basin, except that the extensive influx of northerly derived turbidite sandstones occurred earlier—in Early Wenlock time. This time difference may reflect the earlier impingement of Avalonia with Laurentia further south-west down the old plate boundary. A diachronous closure of this sector of the Iapetus Ocean is implied, with the trench being progressively transformed into a foreland basin on depressed Eastern Avalonian crust. The required anticlockwise rotation of Eastern Avalonia

with respect to Laurentia is also indicated by palaeomagnetic data (Fig. 11.10).

In *the Welsh Basin* and the southern Leinster Basin the supply of turbidite sandstones began even earlier, in late Llandovery time. This volume of clastic debris could not have been sourced from the mainly flooded Midland Platform. Palaeoflow structures show that it was being derived from the west (Fig. 11.11).The sediment may have been sourced from a strongly uplifted 'Pretannia' or a new land area further west. In either case, the tectonic driving force for uplift may have been the initial impingement of the western end of Eastern Avalonia with the Laurentian margin (Fig. 11.10).

In the Lake District and Leinster Basins the space for thick Silurian sediment was probably made by crustal flexure. In the Welsh Basin there was a clear phase of crustal extension, coincident with the Late Llandovery influx of turbidite sands. A cross-section of the basin restored to its Silurian geometry (Fig. 11.12) shows the major normal faults that underlay the basin, and the thickness changes in upper Llandovery and Wenlock units that record the syn-

Fig. 11.12 Restored cross-section across Mid Wales (after Woodcock *et al.* 1995, with permission from the Geological Society, London 1999).

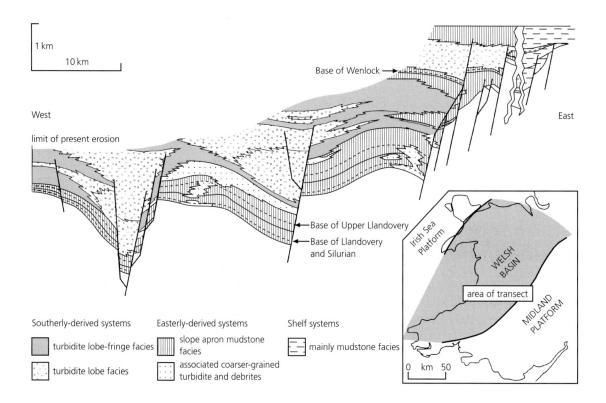

depositional displacement on these faults. The fault scarps were draped by sediment and only rarely cut the sediment surface. However, as turbidity flows were supplied parallel to the faults, the sea-floor relief was sufficient to constrain them to the downthrow sides of the faults.

11.12 Late Silurian basin shallowing

The predominantly marine deposition on Eastern Avalonia came to an end during the Late Silurian and earliest Devonian. Basins and shelves alike were progressively transformed into areas of non-marine deposition. There were three main causes of this environmental change:
1 the fall in global sea level during late Ludlow and Přídolí time (Fig. 11.2);
2 the increasing rate of sediment supply from uplifted areas within the developing collision zone between Eastern Avalonia and Laurentia;
3 the decreasing submarine accommodation space for sediments as subsidence in the basins first slowed, then gave way to shortening and uplift.

A palaeogeographic map for Přídolí (latest Silurian) time shows how the marine shoreline had retreated from its mid-Silurian position (Fig. 11.13). By the end of the Silurian, non-marine deposition was affecting all the basins around the northern rim of the Midland Platform, including the remnants of the Iapetus Ocean itself. These non-marine sediments are commonly referred to as the 'Old Red Sandstone', an informal term embracing a wide range of clastic sediments from mudstones to conglomerates. A shallow sea still encroached onto the south-east of the Midland Platform, and perhaps onto the south of the old Anglian Basin. This sea probably deepened southwards into a marine basin along the southern edge of Eastern Avalonia, the first firm evidence for which occurs in Lower Devonian successions (see Chapter 13).

The transition to non-marine conditions differs from area to area in its succession of facies and in its timing. Marginal marine facies occur in south-west

Fig. 11.13 Palaeogeographic map of southern Britain and Ireland for latest Silurian (Přídolí) time (modified after Bassett *et al.* 1992, with permission from the Geological Society, London 1999).

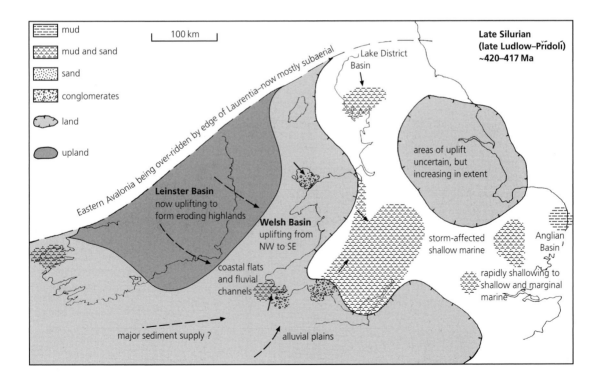

Wales by early Ludlow time, in south-west Ireland by
the Late Ludlow, in north and mid-Wales in the Early
Přídolí, in north-west England in late Pridoli time, but
in eastern England not until the Early Devonian. This
crude eastward diachroneity might correlate with the
distance of the basins from the initial impact zone of
Avalonia with Laurentia. Another sedimentological
marker of the final closure of the Iapetus Ocean is the
influx onto Eastern Avalonia of northerly derived
sediment from the uplifting collision zone. This influx
is also diachronous, with basins near the Iapetus
Suture being affected before those further to the south
(Fig. 11.14). This southward diachroneity continues
a similar trend in the Laurentian accretionary prism
of the Southern Uplands before the impingement of
Avalonia (Fig. 11.14).

A range of shallow marine to marginal marine
sedimentary facies is developed during basin shallowing. For instance, the Lake District basin shallows
from basinal marine mudstones with thin turbiditic
sandstones, through inner slope and shelf sediments

dominated by storm-deposited sandstones, up to
interlaminated siltstones and mudstones deposited
under fairweather wave influence on the inner shelf.
Here, later parts of the succession were eroded before
the Acadian Orogeny, but in South Wales a comparable succession continues into the mudstones and
sandstones of a tidal estuary or lagoon before passing
up into Lower Devonian alluvial deposits. Analogous
transitions are seen in the Dingle Group of western
Ireland and in the subsurface remnants of the Anglian
Basin.

Biostratigraphical dating of these transitional successions is increasingly problematical as the graptolites, brachiopods and trilobites of the fully marine
rocks become less common and reliance has to be
placed first on ostracods and vertebrates and then on
spores. Lithological correlation is aided by extensive
calcretes—horizons of calcite cementation during
subaerial evaporation—and occasional volcanic
tuffs.

11.13 Summary: basin formation and inversion in the Avalonia–Laurentia collision zone

Late Ordovician and Silurian time saw the closure of

Fig. 11.14 Timing of key events on Laurentia and Eastern
Avalonia during the closure of the Iapetus Ocean (from King
1994, with permission from the Geological Society, London
1999).

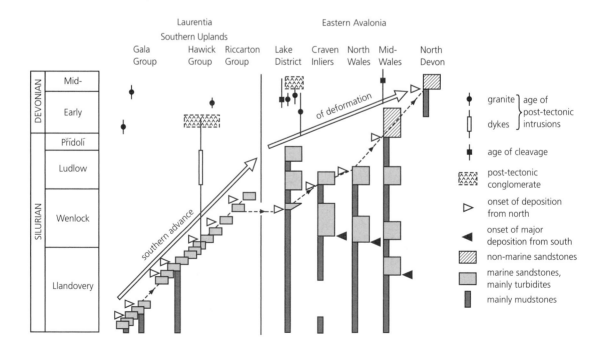

the Iapetus Ocean. In Late Ordovician time, before any collision with Laurentia, Eastern Avalonia probably overrode the mid-Iapetus ridge, ending subduction-related volcanism and inducing a phase of strike-slip fault tectonics. A weak collision with Baltica also occurred at this time. In Silurian time, Eastern Avalonia and Baltica approached and collided with the larger Laurentian continent to the north. Sedimentary basins on the northern rim of Eastern Avalonia were deepened as the Avalonian crust flexed beneath the over-riding edge of Laurentia. Some were subjected to local crustal extension, but all were eventually compressed and uplifted as a wave of collisional deformation spread southward from the Iapetus Suture zone.

A chart of these events and their geological record (Fig. 11.15) is made more complex because of the protracted and diachronous nature of continental collision. The leading edge of Eastern Avalonia approached the Laurentian margin obliquely, so that

impingement effects are diachronous along the Iapetus Suture as well as normal to it.

Preserved upper Ordovician and Silurian sediments are mostly marine, reflecting generally high global sea levels. Only in latest Silurian and Early Devonian time were large thicknesses of non-marine sediments preserved, reflecting falling sea levels and high sedimentation rates ahead of the advancing Acadian mountain front.

Volcanoes on Eastern Avalonia mostly shut down during Caradoc time with the end of southward subduction of the Iapetus ocean crust, although minor activity persisted later in Ordovician time. New Early Silurian volcanic centres seem to reflect crustal extension through the southern flank of Eastern Avalonia. This may relate to a major outstanding problem of Avalonian geology: the history of the southern edge of the microcontinent. Was it a passive margin formed by rifting from Gondwana in Early Ordovician time, or was it formed by later rifting? Did it face a major 'Rheic' Ocean to the south, or a narrower mainly intracontinental basin? Answers to these questions lie within the deep crust of the later Variscan Orogenic Belt (see Chapter 15).

Fig. 11.15 Tentative chronology of events in the Avalonia–Laurentia collision zone.

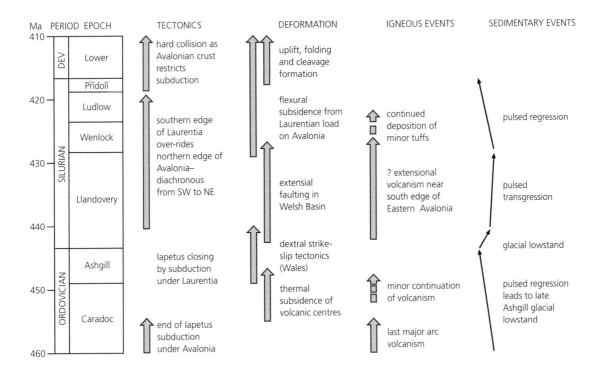

References

Bassett, M.G., Bluck, B.J., Cave, R., Holland, C.H. & Lawson, J.D. (1992) Silurian. In: *Atlas of Palaeogeography and Lithofacies* (eds J. C. W. Cope, J. K. Ingham & P. F. Rawson), Memoir 13, pp. 37–56. Geological Society, London.

Branney, M.J. & Soper, N.J. (1988) Ordovician volcano-tectonics in the English Lake District. *Journal of the Geological Society, London* **145**, 367–376.

Fortey, N.J., Merriman, R.J. & Huff, W.D. (1996) Silurian and late Ordovician K-bentonites as a record of late Caledonian volcanism in the British Isles. *Transactions of the Royal Society of Edinburgh* **86**, 167–180.

King, L.M. (1994) Subsistence analysis of Eastern Avalonian sequences: implications for Iapetus closure. *Journal of the Geological Society, London* **151**, 647–657.

Johnson, M.E. (1996) Stable cratonic sequences and a standard for Silurian eustasy. *Geological Society of America, Special Paper* **306**, 203–211.

Torsvik, T.H., Smethurst, M.A., Meert, J.G. *et al.* (1996) Continental break-up and collision in the Neoproterozoic and Palaeozoic—a tale of Baltica and Laurentia. *Earth-Science Reviews* **40**, 229–258.

Woodcock, N.H. (1984) The Pontesford Lineament, Welsh Borderland. *Journal of the Geological Society, London* **141**, 1001–1014.

Woodcock, N.H., Butler, A.J., Davies, J.R. & Waters, R.A. (1995) Sequence stratigraphical analysis of late Ordovician and early Silurian depositional systems in the Welsh Basin: a critical assessment. *Special Publication of the Geological Society, London* **103**, 197–208.

Ziegler, A.M., Cocks, L.R.M. & McKerrow, W.S. (1968) The Llandovery transgression of the Welsh Borderland. *Palaeontology* **11**, 736–782.

Further reading

Bassett, M.G., Bluck, B.J., Cave, R., Holland, C.H. & Lawson, J.D. (1992) Silurian. In: *Atlas of Palaeogeography and Lithofacies* (eds J. C. W. Cope, J. K. Ingham & P. F. Rawson), Memoir 13, pp. 37–56. Geological Society, London. [Detailed palaeogeographic maps and accompanying commentary.]

Brenchley, P.J. & Newall, G. (1984) Late Ordovician environmental changes and their effect on faunas. In: *Aspects of the Ordovician System* (ed. D. L. Bruton), pp. 65–79. Universitetsforlaget, Oslo. [A survey of the sedimentological and faunal impacts of the Late Ordovician ice age.]

Fortey, N.J., Merriman, R.J. & Huff, W.D. (1996) Silurian and late Ordovician K-bentonites as a record of late Caledonian volcanism in the British Isles. *Transactions of the Royal Society of Edinburgh* **86**, 167–180. [A comprehensive discussion of the origin of Silurian volcanism on both sides of the Iapetus Suture.]

Hambrey, M.J. (1985) The Late Ordovician–Early Silurian glacial period. *Palaeogeography, Palaeoclimatology, Palaeoecology* **51**, 273–289. [A global review of the Late Ordovician ice age.]

Johnson, M.E. (1996) Stable cratonic sequences and a standard for Silurian eustasy. *Geological Society of America, Special Paper* **306**, 203–211. [Derives a eustatic sea-level curve by comparing Silurian successions on the North American platform with similar successions worldwide.]

Kemp, A.E.S. (1991) Mid Silurian pelagic and hemipelagic sedimentation and paleoceanography. *Special Papers in Palaeontolology, London* **44**, 261–299. [An important review of the effect of sea-level change on Silurian basinal facies.]

King, L.M. (1994a) Subsidence analysis of Eastern Avalonian sequences: implications for Iapetus closure. *Journal of the Geological Society, London* **151**, 647–657. [Uses quantitative subsidence curves to distinguish extensional and flexural subsidence of Eastern Avalonian sedimentary successions.]

King, L.M. (1994b) Turbidite to storm transition in a migrating foreland basin: the Kendal Group (Upper Silurian), northwest England. *Geological Magazine* **131**, 255–267. [Sedimentological study of the end of Late Silurian shallowing of the Lake District basin.]

Kneller, B.C. (1991) A foreland basin on the southern margin of Iapetus. *Journal of the Geological Society, London* **148**, 207–210. [The proposal that Silurian sedimentation along the northern edge of Avalonia was controlled by flexural subsidence.]

Soper, N.J. & Woodcock, N.H. (1990) Silurian collision and sediment dispersal patterns in southern Britain. *Geological Magazine* **127**, 527–542. [An instructive attempt to deduce the collision geometry of Avalonia and Laurentia from sediment supply patterns.]

Thorpe, R.S., Leat, P.T., Bevins, R.E. & Hughes, D.J. (1989) Late-orogenic alkaline/subalkaline Silurian volcanism of the Skomer Volcanic Group in the Caledonides of south Wales. *Journal of the Geological Society, London* **146**, 125–132. [A detailed study of the most informative of Silurian volcanic centres.]

Ziegler, A.M., Cocks, L.R.M. & McKerrow, W.S. (1968) The Llandovery transgression of the Welsh Borderland. *Palaeontology* **11**, 736–782. [A classic study of the palaeogeographic implications of apparent depth control of Silurian faunal communities.]

Part 4
The End of the Iapetus Ocean

Illustration overleaf: The Caledonian Moine Thrust Zone, Sutherland. Upper picture: view northeast across Loch Glencoul. The hummocky ground on the far left is underlain by Late Archaean orthogneisses of the Lewisian Complex. These are overlain unconformably by Cambrian quartzites of the Eriboll Sandstone Formation which form the steep faces; bedding can be seen to dip gently to the right. The Caledonian Glencoul thrust is located in the poorly exposed ground immediately above the quartzites; the hangingwall comprises Lewisian gneisses. Lower picture: a sharp brittle thrust within the Cambrian An t-Sron Formation emplaces dolomitic sandstones of the Fucoid Beds over quartz arenites assigned to the Salterella Grit (location near Loch Eriboll).

12 The Caledonian Orogeny: a multiple plate collision

N. H. WOODCOCK AND R. A. STRACHAN

12.1 Extent and timing

The Caledonian Orogen straddles Britain and Ireland, from north-west Scotland to the Welsh Borders (Fig. 12.1). The north-western boundary of the orogen is usually taken along the Moine Thrust Zone, where metamorphic rocks of the Scottish Highlands were thrust westward over the Hebridean Platform. Its opposing boundary lies along the Welsh Borderland Fault System, where folded and cleaved rocks of the Welsh Basin are upfaulted against the weakly deformed Midland Platform. This deformation front has been traced in the subsurface around a northern promontory of the platform and south-eastwards beneath eastern England. The southern Caledonian deformation front is obscured at both its ends beneath the northern edge of the Variscan Belt (Fig. 12.1). The extent and history of the Caledonian orogen is uncertain south of the Variscan Front.

The Caledonian Orogeny was responsible for consolidating most of the British and Irish crust into the present pattern of fault-bounded blocks or terranes, each with a distinctive geological history (see Chapter 2). The internal geometry of the orogen is therefore dominated by the major block-bounding faults, mostly having the north-east to south-west 'Caledonoid' strike (Fig. 12.1). The trend of folds and cleavage in the intervening terranes is also broadly Caledonoid (Fig. 12.1), but with some important swings, particularly through East Anglia.

The 'Caledonian' orogenic belt had a protracted constructional history through early Palaeozoic time. Early ideas of a simple 'Caledonian Cycle' of sedimentation followed by deformation, metamorphism and igneous activity have proved too simplistic. The existence of a wide Ordovician ocean between the

north-western and south-eastern parts of the orogen implies that they had independent tectonic histories at least until Silurian time. The chapters in Parts 2 and 3 of this book have detailed complex sequences of crustal extension and contraction on both margins of the Iapetus Ocean. Some early contractional events are intense or widespread enough to have their own names. The Grampian Orogeny on the Laurentian margin (see Chapter 6) and the Cadomian Orogeny on the Avalonian margin (see Chapter 8) are the main examples. The term 'Caledonian Orogeny' has therefore acquired a more specific meaning, restricted to the Silurian and Devonian events that resulted from the collision of Laurentia with Eastern Avalonia and other former fragments of Gondwana.

The Caledonian Orogeny, even in its restricted sense, still has a complex history. Several component phases of deformation or magmatism can be distinguished in space and time. One component, the Early Devonian deformation south of the Iapetus Suture, has been separately named as the Acadian deformation. The various components, outlined below (see Section 12.3), may have had different driving mechanisms within the orogen.

12.2 Palaeocontinental framework

This chapter will focus on the Iapetus Ocean in trying to identify the plate tectonic causes of the Caledonian Orogeny. However, it is important to recall that the Iapetus-bordering continents of Eastern Avalonia and Laurentia were only two of a collage of fragments that were converging in mid-Palaeozoic time (Fig. 12.2). Both Baltica and Western Avalonia were also advancing on Laurentia during the Silurian. Further south, Armorica probably collided with the assembled

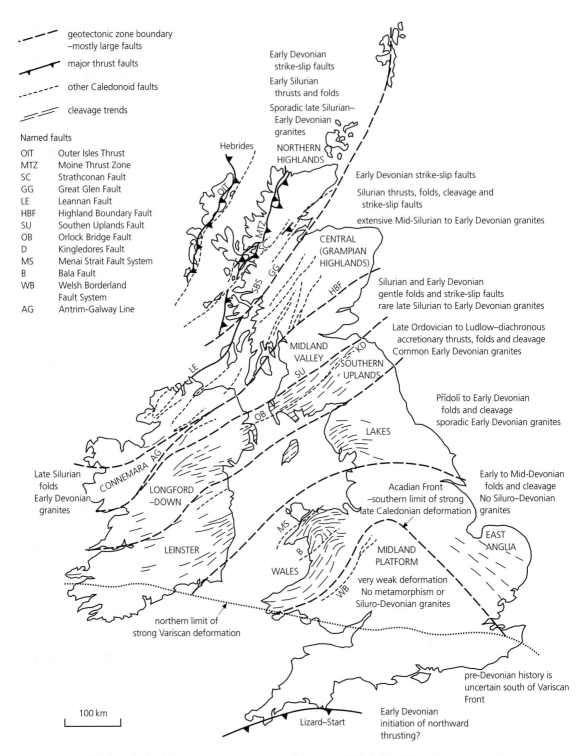

Fig. 12.1 Map of Britain and Ireland showing the extent and general character of Caledonian deformation, metamorphism and igneous activity (faults and cleavage traces from Soper 1986.)

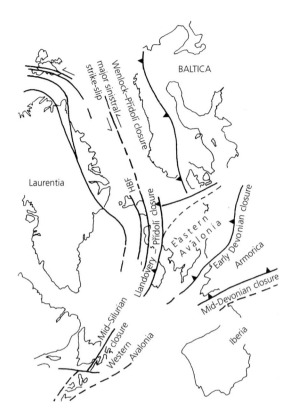

Fig. 12.2 Palaeocontinental positions during Early Devonian time (after Soper *et al.* 1992, with permission from the Geological Society, London 1999). HBF, Highland Boundary Fault.

northern continents during Early Devonian time and Iberia with Armorica in the Mid-Devonian.

A crucial component of the palaeo-continental collage (Fig. 12.2) is the Highland Boundary Fault and associated structures such as the Great Glen Fault. The large Caledonian sinistral strike-slip displacements postulated on these faults imply that the Scottish Highlands were located further north-east during early Silurian time, probably opposite Scandinavia. This palaeogeography makes it likely that Caledonian displacements on the Moine Thrust Zone resulted from the collision of Laurentia and Baltica, rather than from contemporaneous Caledonian processes south of the Highland Boundary Fault. Only during later Silurian and Devonian time did sinistral strike-slip displacements juxtapose these two contrasting parts of the Caledonides.

Further south, the Iapetus Ocean was essentially closed by the end of Silurian time (see Sections 11.1,

11.11). Although the Acadian phase of the Caledonian Orogeny is generally ascribed to this closure, it is not clear why the shortening and uplift of this event should occur as late as Early Devonian time. Forceful northward collision of Eastern Avalonia may have continued for this further 20 million years after the last remnant of Iapetus crust was subducted (see also Fig. 12.12b). However, Acadian deformation might have been enhanced by collision of Armorica with Eastern Avalonia in Early Devonian time.

Mid-Devonian extension between Armorica and Avalonia (see Chapter 13) accompanied the collision of Iberia with Armorica, closely followed by parts of the main Gondwana continent. These collisions drove the earlier part of the Variscan Orogeny (see Chapter 15). On a regional scale, therefore, there is no major time gap between the Caledonian and Variscan episodes, although much of Britain and Ireland escaped strong Variscan deformation until later in Carboniferous time. Chapters 13 and 14 will show how Late Devonian and Carboniferous sequences accumulated in the tectonically unstable environments that intervened between the two orogenic events.

12.3 Regional timing and character

This section outlines the general character of Caledonian events in the component zones of the orogen, beginning in the south and working northwards. The pattern of events is complex, and some components of this complexity are analysed in more detail in Sections 12.4–12.10, before a synthesis is attempted in Section 12.11.

South of the Midland Platform, pre-Devonian rocks affected by any Caledonian events have been obscured by later sedimentation and Variscan overthrusting. Accumulation of lower Devonian sedimentary and volcanic rocks continued, in the Cornubian basins of south-west England, throughout the climax of Caledonian deformation further north (see Chapter 13). Late Devonian northward thrusting of the Lizard–Start complexes (Fig. 12.1) is assigned to an early stage of the Variscan Orogeny (see Chapter 15).

The Midland Platform itself is only weakly affected by Caledonian deformation.

Between the Midland Platform and the Iapetus Suture, the Caledonian Orogeny is clearly recognizable by a major intra-Devonian unconformity (Fig. 12.3). This unconformity records the uplift and erosion associated with crustal thickening in the orogen. Typically, upper Devonian or Carboniferous strata overlie lower Devonian or earlier rocks that have been folded, cleaved, weakly metamorphosed, and intruded by dykes or granitic plutons. This stratigraphic relationship is similar to that in the original westward continuation of the orogen into the Canadian Appalachians. There, the Devonian orogeny is termed the *Acadian* event, and the same name is now applied in the former Eastern Avalonian part of Britain and Ireland. Textural evidence around the *c.* 400-Ma granites in the Lake District and south-east Ireland suggests that their intrusion overlapped the

end of cleavage formation in the country rocks, thereby dated as late Early Devonian (Emsian). However, at the same time, non-marine lower Devonian rocks were still being deposited in the remnants of the Welsh Basin and probably the Anglian Basin. A synchronous Acadian deformation implies that this sedimentation was continuing at the surface while folding and cleavage formation proceeded at depth. Although this is likely, the evidence favours a slightly earlier climax to the Acadian deformation in the Lakes–Leinster Terrane than in Wales (Fig. 12.3). Such diachroneity becomes even more marked in the Laurentian terranes to the north-west.

In the Southern Uplands–Longford–Down zone, the hypothesis of a synchronous Acadian deformation becomes impossible to sustain (Figs 12.1, 12.3). For instance, post-cleavage strata were being laid down in the Southern Uplands of Scotland before the end of the Early Devonian, while predeformation

Fig. 12.3 Chronostratigraphic chart showing constraints on the timing of Caledonian events.

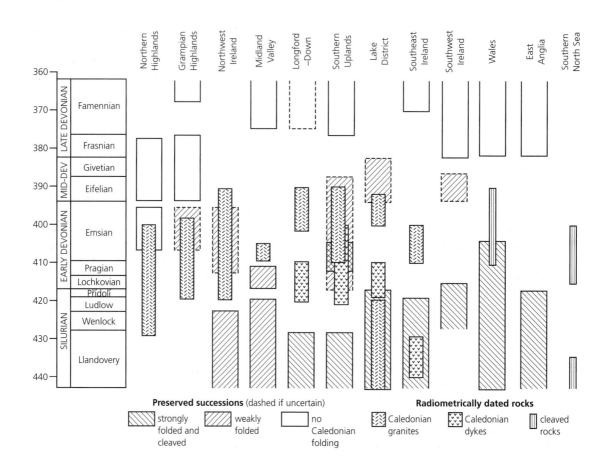

sediments were still accumulating in Wales. In any case, the accretionary prism model for the Southern Uplands (see Chapter 7) implies that this area was sequentially thrust, folded and cleaved throughout Late Ordovician and Silurian time (see also Fig. 11.14, p. 182). Here, the Devonian event is probably represented by north-west-verging folds and related thrusts that cross-cut the older structures formed within the accretionary prism.

In the Midland Valley of Scotland, local unconformity between lowest Devonian and Silurian sequences marks the beginning of weak Caledonian folding. This deformation lasted through Early Devonian time, while sedimentation of non-marine sequences continued. The boundaries of the Midland Valley are major steep faults that probably had an important sinistral strike-slip component during the Early Devonian. These faults may have been the main control both on the formation of the Early Devonian basins and on their concurrent deformation (see Chapter 13).

The Scottish Highlands, and their along-strike equivalents in Ireland, are also cut by Early Devonian sinistral strike-slip faults. Displacement on these faults was coeval with the intrusion of the later group of granitic plutons and dykes in this zone. In the Connemara–Mayo region of western Ireland, folding postdated Mid-Silurian sequences (Fig. 12.1). However, there is little evidence for Caledonian folding in the Scottish Highlands. The most conspicuous Caledonian event is the major phase of thrusting in the north-west Highlands, constrained to early or mid-Silurian time by radiometric dating of pre- and post-tectonic intrusions and of metamorphic minerals. The Moine Thrust Zone is the main example of this structural phase and is commonly taken as the north-western limit of the orogen. However, Silurian faulting also affected the Outer Isles Thrust, and possibly further thrusts on the continental margin to the north-west. All these thrust movements are much later than the main Mid-Ordovician (Grampian) metamorphic peak in the Highlands and clearly represent a separate event.

Amidst the complexity of timing and character of Caledonian deformation, one unifying factor is the generally low grade of associated metamorphism. In the Scottish Highlands, Caledonian pressure and temperature conditions never reached those achieved during the Grampian event. Most rock sequences south of the Highland Boundary Fault experienced pressures and temperatures below those of the mid-greenschist facies.

In summary, the evidence points to at least three contrasting components of Caledonian deformation:

1 an Early Devonian 'Acadian' folding and cleavage formation south of the Iapetus Suture;

2 a more prolonged and diachronous Silurian to Early Devonian folding and cleavage formation between the Iapetus Suture and the Highland Boundary Fault;

3 a Mid-Silurian thrusting and strike-slip faulting north of the Highland Boundary Fault.

12.4 Pre-Silurian inheritance

Major faults have been responsible for assembling at least the Laurentian segments of the Caledonides in approximately their present geometry. It is not surprising therefore to find that Caledonian events were strongly influenced by the geology inherited from each crustal block (Fig. 12.4). Particularly important factors of this legacy are as follows.

1 The age of the last high-grade metamorphism and deformation associated with crustal thickening. Such events first heat and soften the lithosphere, then allow it to strengthen over a period of about 50 Myr.

2 The age of any rapid post-metamorphic sediment accumulation. Such events, due to subduction accretion, lithospheric stretching or flexure, produce a thick volume of easily deformed sediments in the upper crust, sometimes overlying a thin or absent continental basement.

Significant correlations can now be seen between Caledonian events and pre-Devonian geology (compare Figs 12.3 and 12.4):

1 *Cratonic areas*, which resisted the Caledonian deformation and contain no synchronous intrusions, are those that suffered a Proterozoic tectonothermal event, but did not subsequently undergo major crustal stretching, shortening or heating. These crustal areas were relatively cold and strong by Silurian time. They reacted to crustal shortening only by local reactivation of pre-existing faults. The Midland Platform acted in this way on Eastern Avalonia, with the Irish Sea Platform suffering more pervasive deformation consistent with its smaller extent. The Hebridean Platform acted as the opposing foreland to

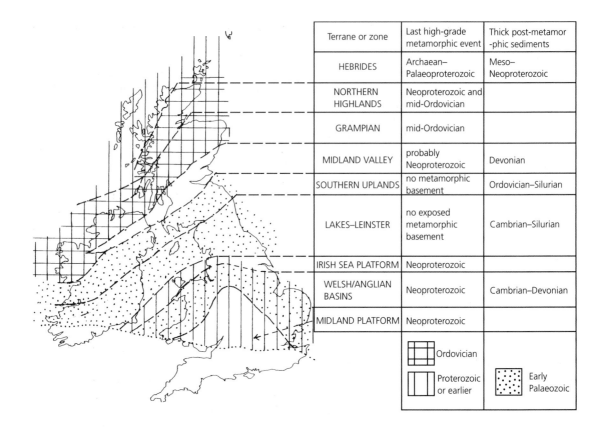

Terrane or zone	Last high-grade metamorphic event	Thick post-metamor -phic sediments
HEBRIDES	Archaean–Palaeoproterozoic	Meso–Neoproterozoic
NORTHERN HIGHLANDS	Neoproterozoic and mid-Ordovician	
GRAMPIAN	mid-Ordovician	
MIDLAND VALLEY	probably Neoproterozoic	Devonian
SOUTHERN UPLANDS	no metamorphic basement	Ordovician–Silurian
LAKES–LEINSTER	no exposed metamorphic basement	Cambrian–Silurian
IRISH SEA PLATFORM	Neoproterozoic	
WELSH/ANGLIAN BASINS	Neoproterozoic	Cambrian–Devonian
MIDLAND PLATFORM	Neoproterozoic	

Ordovician

Proterozoic or earlier

Early Palaeozoic

Fig. 12.4 Map of Britain and Ireland showing major pre-Devonian crustal features that influenced Caledonian deformation and igneous activity.

the Caledonian belt, and records only localized reactivation along the Outer Isles Thrust Zone.

2 *Strongly folded and cleaved zones* of Caledonian age are restricted to those segments of the orogen that inherited thick sequences of early Palaeozoic (Cambrian to Silurian) sediments. These 'slate belts' were developed on the sites of earlier deep marine basins formed dominantly by lithospheric stretching (Welsh and Anglian Basins), by subduction accretion of sediments deposited on oceanic crust (Southern Uplands and ?Leinster Basins), or by a combination of stretching and lithospheric flexure (Lake District Basin).

3 *Caledonian granitic plutons* occur most commonly in crust that had undergone a major Ordovician thermal event, that is, in the Central Highlands. The

origin of these granites is still controversial (see Section 12.10), and it is unclear whether the residually hot lithosphere promoted the generation of granitic magmas, or their subsequent uprise and emplacement or both. Certainly granites become less common in the colder zones of the orogen, and are absent in the cratonic areas of the Hebridean and Midland Platforms.

12.5 Silurian thrusting at the north-west Caledonian Front

The Moine Thrust Zone is generally taken as the north-western boundary to the Caledonides. Its northernmost exposed part, at Loch Eriboll (Fig. 12.1), is classic ground in Highland geology and was the subject of major controversy over 100 years ago. Were the meta-sedimentary rocks of northern Scotland (now known as the Moine Supergroup) in stratigraphic sequence with the underlying Cambro-Ordovician strata of the west coast, and thus of Silurian age (as proposed by Murchison)? Alternatively, were they in tectonic contact and thus perhaps of Precambrian age (as proposed by Lapworth)? Following the demonstration by Nicol of north-west-directed overthrusting at Eriboll the Geological Survey, under the leadership of Peach and Horne, began a mapping programme that established the Moine Thrust Zone as a classic example of a marginal thrust belt.

The Moine Thrust Zone has an on-land length of about 200 km (Fig. 12.1), but its total length may exceed 500 km. The thrust zone at present dips at about 8–12° to the east as a result of late Palaeozoic or Mesozoic tilting. The structures are invariably complex, but the thrust zone at Loch Eriboll can generally be subdivided into two parts (Fig. 12.5).

The structurally highest part of the thrust zone comprises a thick belt (hundreds of metres) of mylonites formed from the intense ductile shear and recrystallization of Moine meta-sediments and associated slices of Lewisian basement. The mylonites are characterized by the widespread recrystallization of quartz and the growth of new muscovite and chlorite within the greenschist facies. Formation of the mylonites records early west-north-west-directed thrusting onto the Hebridean Platform at mid-crustal levels. The thrust sheets were partially eroded as they formed, so that deformation occurred at progressively higher crustal levels and therefore at lower pressures and temperatures.

The structurally lower part of the thrust zone is a complex stack of thrust sheets that incorporates

Fig. 12.5 Cross-sections through the Moine Thrust Zone east of Loch Eriboll (after Coward 1984, with permission from the Geological Society, London 1999).

Lewisian basement and Cambro–Ordovician sedimentary rocks (Fig. 12.5). The thrusts are sharp, brittle structures, lacking much mylonite. This part of the thrust zone clearly developed at higher crustal levels than the mylonite belt, and the Cambro–Ordovician rocks record peak temperatures of only about 275°C, in the upper anchizone.

Thrusts generally developed in a foreland-propagating order, becoming sequentially younger towards the west. Older thrusts may therefore be folded and cut through by structurally lower younger thrusts (Fig. 12.5). This pattern is, however, complicated in some areas where later low-dip 'out-of-sequence' faults cut through previously thrust and folded strata. Such faults may either be late thrusts or extensional faults due to gravitational instability of the evolving pile of thrust nappes. Restored cross-sections suggest that the Cambro–Ordovician shelf sequence formerly extended for about 50–60 km east of the present outcrop of the Moine Thrust Zone. To this minimum displacement must be added the displacement recorded in the intense ductile shearing of the mylonite belt, generally assumed to be at least 100 km. A total displacement on the Moine Thrust Zone of at least 150 km therefore seems likely.

The timing of movements on the Moine Thrust Zone has been the subject of long debate. Isotopic dating of recrystallized micas within the mylonite belt suggests that thrusting of the Moine rocks onto the foreland probably occurred at 435–430 Ma. Displacement was then transferred onto lower thrusts within the foreland stratigraphy. A series of syenite plutons in the Assynt area, assumed to have been intruded during thrusting, have yielded U–Pb zircon ages ranging from 439 ± 4 Ma to 430 ± 4 Ma. However, the syenites pre-date some of the brittle

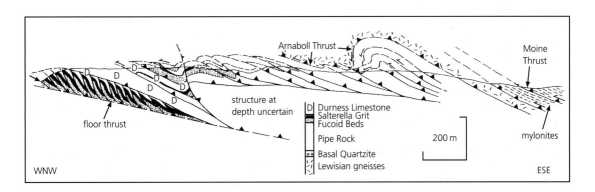

thrusts, showing that thrusting continued after intrusion. The isotopic dates span much of Silurian time (early Llandovery to Wenlock, Fig. 12.3). They show that the Moine Thrust Zone was active long after the main Grampian Orogeny in the Scottish Highlands (see Chapter 6) and demand another tectonic cause. The collision of Baltica with Laurentia is a probable driving mechanism for the thrusting.

The Moine Thrust Zone has traditionally been viewed as the north-west limit of Caledonian deformation in the Scottish Highlands. Nevertheless, dating of fault rocks in the Outer Hebrides, west of the Moine Thrust, has shown that the Outer Isles Fault Zone also records important displacements at around 430 Ma. However, this zone was probably initiated in the Proterozoic (see Chapter 4). Its parallelism to Caledonian structures made it prone to reactivation, even though it is located well within the foreland, beyond the main deformation front.

12.6 Siluro–Devonian strike-slip faulting on Laurentia

The former Laurentian margin is pervaded by a series of major steep faults that strike approximately north-east–south-west and cut most other Caledonian structures (Fig. 12.1). Their strikes swing to nearly east–west in western Ireland and to north–south in Shetland. There is also a cross-strike variation, the north-east strikes of the Highland Boundary Fault and Southern Uplands Fault contrasting with more north-north-east strikes in the Grampian and Northern Highlands, parallel to the Great Glen Fault. Geophysical evidence indicates that the Great Glen Fault and the Highland Boundary Fault both reach a depth of at least 40 km.

Many of the faults have a long history. The Highland Boundary Fault, for example, may originally have defined the south-east limit of rifted Laurentian crust during the Neoproterozoic development of Iapetus (see Chapter 5). It was later transformed into a collisional suture by accretion of the Midland Valley basement and arc during the Grampian Orogeny. Similarly, the Orlock Bridge Fault in the Southern Uplands may have initiated as a low-dip thrust developed within the accretionary prism. It was progressively rotated into a steep orientation, favouring reactivation as a strike-slip fault during the Caledonian Orogeny. Many of these faults experi-

enced further reactivation during the late Palaeozoic and Mesozoic.

Four principal lines of evidence have been used to show that most of these steep faults had sinistral strike-slip displacements during the Caledonian Orogeny:

1 correlation across faults of older geological features, for instance fold axial traces formed during the Grampian Orogeny;
2 field studies of the kinematic indicators developed within fault rocks;
3 structural analysis of the intrusion histories of igneous plutons emplaced along major faults (see Section 12.10);
4 palaeomagnetic studies.

Proven sinistral displacements demonstrated by the matching of older geological features range from about 10 km or less, on the Loch Tay and Strathconan faults in Scotland, to about 40 km, on the Leannan Fault in north-west Ireland. Displacements along the Highland Boundary and Great Glen faults are likely to be much greater, because in neither case can geological markers be correlated across the fault. The consensus is that both faults experienced at least several hundred kilometres of Caledonian sinistral strike-slip, followed by a smaller post-Devonian dextral movement.

It is not easy to constrain the timing of strike-slip faulting by the isotopic dating of fault rocks. The mylonites and cataclasites developed along the faults are characterized by low-grade metamorphic assemblages, which are generally not amenable to isotopic dating, and in any case are highly altered as a result of the passage of hydrous fluids during faulting. One indirect way of establishing the ages of fault movements is by the isotopic dating of igneous intrusions thought to have been emplaced during sinistral displacement. Such studies indicate that faulting occurred as a series of discrete events between the Mid-Silurian and Early Devonian. Two examples of this evidence are shown in Fig. 12.6:

1 The Donegal Batholith in north-west Ireland was intruded between about 405 Ma and 388 Ma, within the Early Devonian. Whereas most of the eight component plutons of the batholith are relatively undeformed, the Main Donegal granite has a strong foliation formed during recrystallization and cooling. A late schistosity within the Dalradian country rocks has a sigmoidal trend, which indicates that the pluton

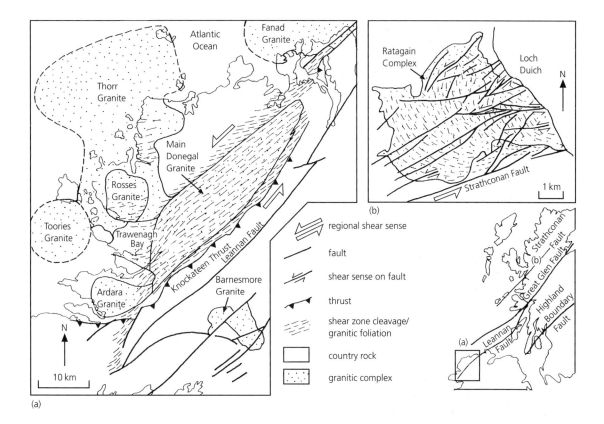

Fig. 12.6 Relationships of Caledonian granites to strike-slip faults or shear zones: (a) The Main Donegal Granite, north-west Ireland (after Hutton 1982); (b,c) the Ratagain Complex, northern Scottish Highlands (after Hutton & McErlean 1991).

was emplaced into an active strike-slip shear zone with a sinistral sense (Fig. 12.6a). Such shear zones would have passed upwards and maybe laterally into more brittle strike-slip fault belts in cooler parts of the Early Devonian crust.

2 The Ratagain granite complex was intruded next to the Strathconan Fault in the Northern Highlands. It also contains a steep foliation with a pattern suggesting sinistral strike-slip along its south-eastern margin during emplacement at 425 ± 3 Ma (mid-Silurian) (Fig. 12.6b). This date is similar to that of the Clunes tonalite, which was intruded during sinistral shear along the Great Glen Fault at 428 ± 2 Ma. Displacement along the two faults was therefore essentially contemporaneous. Additionally, the Ratagain granite is cut by brittle sinistral strike-slip faults

that were active during intrusion of a later dyke swarm, between 410 Ma and 395 Ma (Early Devonian). Lack of intervening deformation suggests that there were two discrete strike-slip episodes rather than one progressive event.

Analysis of the former Eastern Avalonian margin (see Section 12.8) will show that the pattern of sinistral strike-slip is recognizable there too, suggesting that a major orogen-parallel shear operated throughout the Iapetus collision zone.

12.7 The Iapetus Suture

The suture between the former Laurentian and Eastern Avalonian continents can justifiably be regarded as the most important Caledonian structural feature of Britain and Ireland. It is boldly drawn on most Palaeozoic tectonic maps, running from south-west Ireland, north-westward to Ireland's east coast, then north-west of the Isle of Man and into the Solway Firth along the England–Scotland border

(Fig. 12.1). Yet the Iapetus Suture is the least well exposed of the region's Palaeozoic terrane boundaries. Along most of its length it is obscured by post-Caledonian strata, and only in eastern Ireland does it cross exposed Ordovician and Silurian rocks. Consequently, there is still debate about its exact surface trace.

Along the Scottish border the suture is constrained by faunal contrasts in Ordovician rocks to lie between the Lake District and the northern belt of

the Southern Uplands (Fig. 12.7a). The accretionary model for the Silurian rocks of the Southern Uplands implies that they were scraped onto the Laurentian margin before collision, and that the suture lies to the the south-east of them. Deep seismic surveys in the Irish Sea and North Sea are an additional guide to the position of the suture in this region. These surveys show a prominent reflector separating southern crust containing reflecting surfaces from less reflective crust to the north (Fig. 12.7b). Contours on this reflector project up to the surface at about the conventional 'Solway Line' (Fig. 12.7a). The seismic data show that the Iapetus Suture is a northward-dipping plane, and that the crust of Eastern Avalonia has underthrust the Laurentian margin. Reflectors in the Avalonian crust may be northward-dipping reverse faults developed during this underthrusting. A minimum age for final displacement on the

Fig. 12.7 (a) Map of the course of the Iapetus Suture showing key areas referred to in the text, and contours on the seismic reflector that may represent the suture at depth. (b) Interpreted line drawing of the main reflectors on the NEC seismic profile, offshore south-east Scotland (profile and contours in (a) from Soper *et al.* 1992, with permission from the Geological Society, London 1999).

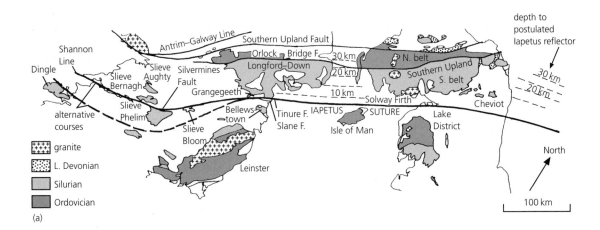

(a)

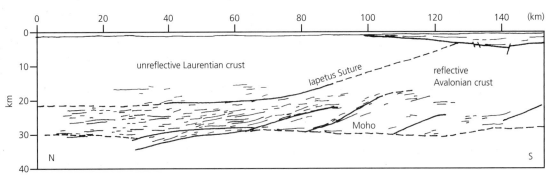

(b)

Iapetus Suture is given by the overstepping Devonian Cheviot volcanics and associated granite (396 ± 3 Ma, Rb–Sr).

In north-east Ireland the Southern Uplands accretionary prism continues in the Longford–Down massif. Minor intrusions here contain xenoliths of schistose and mylonitic volcanic rocks, some of which match the Borrowdale Volcanic Group of the Lake District. This evidence is further confirmation that Avalonian crust has been overthrust by Laurentian rocks at the Iapetus Suture.

In the north-west Isle of Man Ordovician rocks of the Manx Group, deposited on the Avalonian margin, are overthrust by mid-Silurian (Wenlock) units that may match those of similar age in the Southern Uplands. In this case, the thrust contact could be one of the fault strands in the suture zone, confirming the geophysical evidence that the Iapetus Suture lies close to the north-west coast of the island.

In eastern Ireland the suture is drawn between the Ordovician inliers of Grangegeeth and Bellewstown (Fig. 12.7a), which show, respectively, Laurentian and Avalonian affinities in their brachiopod and trilobite faunas. However, the Grangegeeth fauna shows some contrasts with that of the Longford–Down Ordovician rocks to the north, as does the Bellewstown fauna with that of the Balbriggan and Leinster areas to the south. The Grangegeeth and Bellewstown terranes are therefore regarded as transported slivers caught along the obliquely colliding suture zone. Such complications are to be expected along a major palaeocontinental suture, which is better regarded as a wide zone than as a discrete plane. This point is reinforced by attempts to locate the suture by contrasts in structural style, particularly the vergence direction of the folds. The most obvious contrast is along the Tinure Fault, within the Grangegeeth area, rather than along the Slane Fault at its southern margin.

In central Ireland the suture is usually taken along the Silvermines Fault (Fig. 12.7a). Certainly it must lie south of the Slieve Aughty and Slieve Bernagh inliers, which contain Ordovician rocks with Laurentian faunas. Silurian rocks here are similar to those of the Laurentian Longford–Down region, as are those of Slieve Phelim and Slieve Bloom to the south of the Silvermines Fault. This similarity may be due to the diminishing faunal and sedimentological contrasts across the fault as the remnant Iapetus Ocean was finally subducted. Alternatively, it can be used to argue a more southerly course for the suture (Fig. 12.7a). Such a course is supported by the uniform southerly vergence of folds throughout the Silurian inliers.

In south-western Ireland the suture is conventionally drawn along the line of the Shannon Estuary (Fig. 12.7a). However, offshore seismic profiles to the west have imaged a dipping reflector just north of the Dingle Peninsula, rather than at the Shannon Line, consistent with the southerly course for the suture. Structural vergence in the Silurian rocks in the Dingle area suggests only that this area lies to the south of the suture.

12.8 Acadian fold and cleavage patterns on Eastern Avalonia

To the north of the Iapetus Suture, the Caledonian structural grain is predominantly north-east–south-west, the so-called 'Caledonoid' trend. To the south of the suture, on the old Eastern Avalonian continent, the grain, expressed mainly as the trend of folds and the strike of cleavage, shows more large-scale variation (Fig. 12.1). This fabric is east–west in south-west Ireland and south-west Wales. Followed north-eastward it then swings anticlockwise to become north-east–south-west through much of eastern Ireland, the Isle of Man and the western Lake District, but reaches north–south in Central Wales. It then swings back clockwise, reaching east–west in North Wales and the eastern Lake District. Geophysical and borehole data suggest that it swings further to a north-west–south-east trend through the buried Silurian rocks of East Anglia. Variable trend patterns are also apparent on a small scale, particularly in North Wales (Fig. 12.1).

The cleavage responsible for this structural grain is thought to be predominantly Late Silurian to Mid-Devonian in age, that is, the product of the Acadian phase of the Caledonian Orogeny. There has long been a controversy about the existence of an earlier cleavage that deformed the Skiddaw Group (Lake District), before accumulation of the Borrowdale Volcanic Group. However, the folds of this age lack an associated cleavage, and are now thought to have formed within downslope slides in soft sediment. A stronger case for an Ordovician cleavage event can be made for the Ribband Group of south-east Ireland (see Section 10.8, pp. 162–164) possibly deformed in a subduction accretion complex on the Avalonian margin. However, even here the main regional cleavage seems to be Acadian.

The Leinster Granite, dated at 404±24 Ma (Rb–Sr), demonstrates both the Acadian age of the main cleavage and one cause of the sigmoidal and arcuate cleavage patterns. The batholith was intruded during cleavage formation and both retains a fabric itself and deforms the fabric in the country rock (Fig. 12.8a). It comprises five individual plutons, apparently arranged *en echelon*. This geometry suggests intrusion into a steep north-north-east-striking shear zone with a sinistral displacement sense. The broadly north-east-striking cleavage is deflected towards parallelism with the shear zone within and bordering the granite. This pattern of cleavage is reminiscent of that associated with the Main Donegal Granite (see Section 12.5). It implies that the major sinistral shear that pervaded the old Laurentian margin during the Early Devonian also affected the Avalonian side of the collision zone.

A more pervasive indication of sinistral shear across the old Avalonian basins is the common tendency for the Acadian cleavage to cut gently across the associated folds in a clockwise-deflected sense. These transected folds occur throughout much of southern Ireland, Wales, the Lake District and indeed the Southern Uplands. They arise from the different degrees of rotation experienced by folds and cleavage during non-coaxial strain, and are a valuable indicator that the collision direction of Avalonia with Laurentia was not orthogonal to their former continental margins.

As in Leinster, the cleavage arcs in North Wales are probably also related to granites, but here to already

Fig. 12.8 Maps of arcuate cleavage and fold geometries in: (a) Leinster (after Cooper & Bruck 1983); (b) north-west England (after Soper *et al.* 1987); (c) Wales (after Woodcock *et al.* 1988).

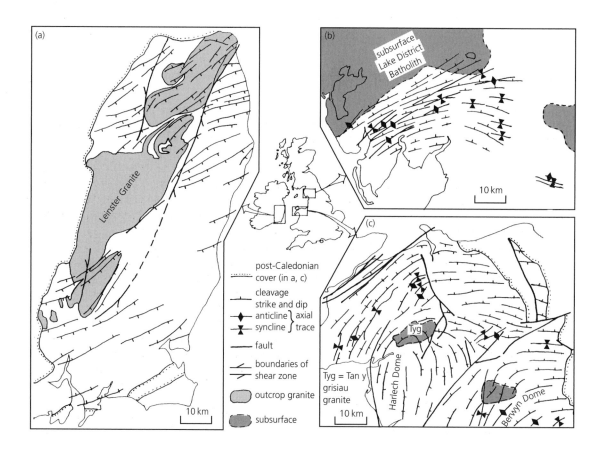

consolidated Ordovician plutons (Fig. 12.8c). The arc around the north end of the Harlech Dome is centred on the Tan y Grisiau Granite, and that around the Berwyn Dome on a postulated buried granite suggested by a gravity low. In both areas the soft Late Ordovician and Silurian cover has apparently been moulded around and over the resistant highs in the basement. The cleavage flattens over the top of each pluton and is deflected around its lateral margins. Analogous controls by pre-existing basement faults were probably responsible for the major sigmoidal swing in Welsh structure (Fig. 12.1). In Mid Wales the folds are clockwise transected by their cleavage, indicating sinistral shear. In North Wales, on the opposite limb of the sigmoid, the folds are weakly anticlockwise transected, indicating dextral shear. This pattern is consistent with shortening of the Welsh Basin in a NNW–SSE direction.

A closely analogous situation developed in the Lake District, with moulding of the Acadian structures against buried Ordovician granites (Fig. 12.8b). The resulting arcuate folds are also transected by cleavage, clockwise in the west and weakly anticlockwise in the east. The inferred shortening direction is just east of north. The Lake District arc may be part of the major swing in structural trends around the northern edge of Eastern Avalonia and southeast into East Anglia (Fig. 12.3). On this scale, the whole pattern of Acadian deformation is seen to be strongly controlled by the triangular shape of the old rigid continental block underlying the Midland Platform. The original arcuate trend of basins around the northern corner of this Midland Microcraton was enhanced as the block indented northward towards Laurentia. This indentation probably also caused wholesale rotation of crustal blocks as they wrapped against the platform. These rotations have been deduced from the swings in declination of the palaeomagnetic field in Ordovician and Silurian rocks.

12.9 Caledonian metamorphism

The Caledonides of Britain and Ireland are still sometimes divided into a 'metamorphic' zone to the north of the Highland Boundary Fault and a 'nonmetamorphic' zone to the south. This subdivision is based on the predominance of greenschist and amphibolite facies metamorphic rocks in the Scottish Highlands and north-west Ireland compared with their relative rarity further south. However, radiometric dating has shown that the time of peak metamorphism is different in the two areas. The high-grade metamorphism of the Highlands dates from the Grampian Orogeny (see Chapter 6) or earlier events. Caledonian metamorphism is predominantly low grade, producing retrogression of mineral assemblages in the Highlands.

With improved techniques for detecting and calibrating low-grade metamorphism, it is clear that the early Palaeozoic rocks south of the Highland Boundary Fault have also suffered a Caledonian metamorphism at temperatures above those acquired during diagenesis. Moreover, the pattern of grade variation preserves information on burial depths and crustal temperature gradients that cannot be otherwise obtained (Fig. 12.9). Metamorphism in low-grade terrains is most commonly quantified using the spacing of silicate sheets in white micas, supplemented by the colour index of conodont fossils, the reflectance of graptolites or the index minerals in basic rocks. On this basis, diagenetic conditions are succeeded by the *lower anchizone* grade at about 200°C, then by the *upper anchizone* grade at about 250°C, and by the *epizone* grade at about 300°C (Fig. 12.9).

A metamorphic map, excluding Ireland, where grade patterns are not yet well determined, shows the following major features.
1 An extensive diagenetic zone in central England and the Welsh Borderland corresponding closely to the Midland Platform. This zone records the area where Lower Palaeozoic sequences were thin, and which avoided strong deformation and increased heat flow during Caledonian events.
2 Predominantly anchizone or epizone grades in the areas previously occupied by Early Palaeozoic marine basins. These higher-temperature zones reflect the thicker overburden and probably raised thermal gradients achieved in the basin areas. The peak temperatures were reached during *syn-deformation metamorphism* because, at least in the Welsh Basin, the metamorphic isograds were superimposed across strata that had already been moderately folded.

More detailed variations are superimposed on this broad pattern.
1 Older rocks tend to preserve higher grades, for instance in the former Welsh Basin. This correlation

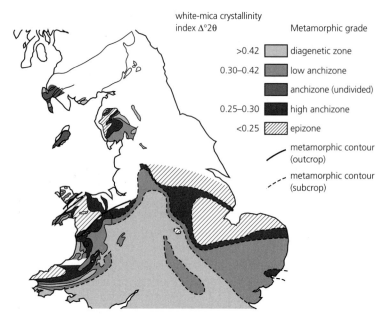

white-mica crystallinity
index Δ°2θ

Metamorphic grade

>0.42 ◻ diagenetic zone

0.30–0.42 ◻ low anchizone

◻ anchizone (undivided)

0.25–0.30 ◻ high anchizone

<0.25 ◨ epizone

— metamorphic contour (outcrop)

-- metamorphic contour (subcrop)

Fig. 12.9 Metamorphic grade in Devonian and earlier rocks.

reflects the higher burial potential of older rocks. It has given rise to the alternative hypothesis of *diastathermal metamorphism*, in which peak temperatures were reached during basin stretching and filling rather than during basin shortening and deformation.

2 Highly deformed rocks tend to preserve higher grades, not necessarily in older rocks. Such examples suggest that high strain rates may promote recrystallization or generate additional heat, both during basin deformation.

3 A marked correlation occurs in the Southern Uplands between metamorphic grade and position in the complex of imbricate thrust sheets. Higher grades occur in the north, where the tectonic thickening of the thrust stack was greatest, although there are some contrasts in grade across thrusts. These relationships suggest that metamorphism peaked during deformation.

The weight of evidence suggests that the preserved metamorphic grade south of the Midland Valley was acquired mainly during Caledonian deformation, or at least during the progressive accretion in the Southern Uplands that led up to it. This conclusion is supported by a suite of Rb–Sr whole-rock dates at around 399 ± 9 Ma (Early Devonian) on cleaved Ordovician volcanic and intrusive rocks in North Wales.

The detailed pattern of Caledonian metamorphism north of the Midland Valley of Scotland is difficult to discriminate. It is characterized by retrograde, hydrated and lower-temperature, mineral assemblages developed patchily in higher-grade rocks formed during the Grampian event (see Chapter 6).

12.10 The Caledonian 'Newer' granites

An extensive suite of granitic plutons and associated dykes intruded the Caledonian Orogen during Silurian and Devonian time (Fig. 12.10a). These intrusions are known collectively and informally as the Newer granites. Many intrusions occurred at a late stage in folding and cleavage formation in surrounding country rocks (e.g. Figs 12.6, 12.8a). Radiometric age determinations show different granite age patterns in different zones of the orogen (Fig. 12.10b). In the Central Highlands, Midland Valley and Southern Uplands zones, the mean age is around 400 Ma, and few granites range back before 410 Ma. In the flanking zones of the Northern Highlands and Leinster Lakes, there is a more protracted record of intrusion back from 400 Ma to about 440 Ma.

There are marked geochemical differences across the Iapetus Suture. Granites immediately to the north of the suture contain a lower proportion of radiogenic strontium (^{87}Sr) than those to the south, and a greater proportion of low ^{87}Rb/^{86}Sr values. These differences reflect contrasts in the magma chemistry

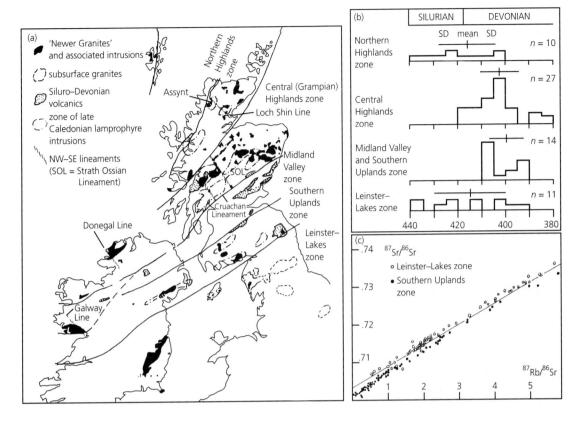

Fig. 12.10 (a) Location of the Siluro-Devonian 'Newer' granites. (b) Age distributions (with means and standard deviations) of granites from different tectonic zones (data from Harris 1985). (c) Rubidium–strontium isochron diagram for granites bordering the Iapetus Suture (from Todd *et al.* 1991, with permission from the Geological Society, London 1999).

across the Suture, probably a greater contribution from mantle rocks beneath the Southern Uplands of Scotland.

The precise mechanisms by which granitic magmas are generated and subsequently ascend through the crust are geotectonic problems that have preoccupied geologists for many years. Detailed field studies of granite plutons in the Caledonides of the British Isles have contributed greatly to a clearer understanding of the interrelationship between tectonics and magmatism. It is now clear that the deep-seated Caledonian strike-slip faults exerted a fundamental control on the siting and ascent of magmas derived from the upper mantle and lower crust (see Section 12.6). These magmas probably ascended the fault zones as narrow dykes that coalesced at mid to high crustal levels to

form the elongate to subrounded, steep-sided plutons characteristic of the Newer granite suite.

Contemporaneous lamprophyre and appinite intrusions associated with many of the Newer granites are thought to derive from partial melts of subcontinental mantle. To preserve their primitive chemistry, these melts must have risen rapidly through the crust. Rapid ascent along crustal-scale strike-slip faults is the most likely mechanism.

Examples of plutonic complexes emplaced during strike-slip faulting include the Strontian and main Donegal granites (along, respectively, the Great Glen Fault and associated splays in north-west Ireland), the Ratagain granite (Strathconan Fault) and the Ox Mountains granodiorite (Fair Head–Clew Bay Line). Many of the Newer granites in the Grampian Highlands are either elongate or form linear arrays, and their emplacement is thought to have been influenced by north-west–south-east-trending lineaments such as the Cruachan and Strath Ossian Lineaments (Fig. 12.10a). A similarly orientated lineament north-west of the Great Glen, the Loch Shin Line, may have controlled emplacement of the Rogart granite and associated appinites. In Ireland, the appinites associated with the Ardara pluton lie along a NNW–SSE-trending Donegal Line. These lineaments may represent older pre-Caledonian structures in the lower crust which, although having little influence on structural development of the cover, played an important part in determining sites of Caledonian magmatism.

Despite the important role of strike-slip faults in Newer granite emplacement, other structures evidently controlled the final emplacement geometries of some plutons. The syenite plutons of the Assynt area (Fig. 12.10a) intrude the Caledonian foreland, and were apparently emplaced as a series of low-angle sills during WNW-directed overthrusting and development of the Moine Thrust Zone. By contrast, the Loch Loyal Syenite Complex of north Sutherland appears to have been intruded during northwest–south-east late Caledonian extension of the Moine country rocks, perhaps at the same time as gravitational spreading of the Moine Thrust Zone onto the foreland.

Volcanic sequences of Siluro–Devonian age occur in the Cheviots, the Midland Valley and the Grampian Highlands (see Chapter 13). They are genetically related to the granites, and similar sequences probably overlay many of the now-exhumed granitic centres. The large number of granites exposed in the Grampian Highlands is partly related to the present erosion level, with greater Devonian and later uplift having occurred there compared with the Midland Valley.

The Siluro–Devonian igneous suite has a generally calc-alkaline geochemistry, most readily explained if it was generated in one or more subduction-related volcanic arcs (Fig. 12.11a). One such arc, in the Midland Valley and Grampians, was related to northward subduction of Iapetus Ocean crust beneath Laurentia. However, this model does not easily explain the plutons close to the suture, in the Southern Uplands, and fails to account for those south of the suture, in Leinster, the Isle of Man and the Lake District. One suggestion is that these southern granites were intruded above a second subduction zone, dipping northward under Eastern Avalonia from the Rheic Ocean (Fig. 12.11a). However, three other processes of magma generation, not necessarily related to a subduction arc, have been invoked to

explain the lower-crustal and uppermost-mantle melting required to source the Newer granites.

1 Melting by raised temperatures during tectonic thickening of the crust during the Caledonian Orogeny (Fig. 12.11b). However, the evidence is now that the main thickening in the Scottish Highlands occurred 50 Myr years earlier, during the Grampian event.

2 Melting by lowered pressures during rapid uplift and erosion (Fig. 12.11c). This hypothesis involves

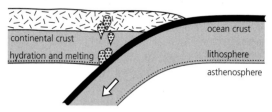

(a) Subduction

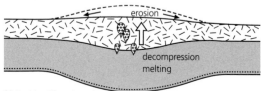

(b) Crustal shortening and thickening

(c) Rapid uplift and erosion

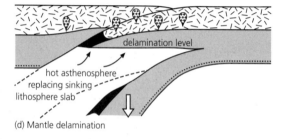

(d) Mantle delamination

Fig. 12.11 Alternative origins for the Caledonian granites. See text for discussion.

far more uplift than has occurred anywhere south of the Highland Boundary Fault, where low-grade metamorphic sequences are still exposed at the surface rather than high-grade rocks exhumed from the mid-crust.

3 Melting by delamination of the lithosphere, replacing old cold lithosphere by hotter rocks that could promote partial melting (Fig. 12.11d). Such reorganization of the deep lithosphere is known to occur

during the death of subduction zones, below both sides of the site of the former trench.

The relative contribution of these processes to the formation of the Newer granites has yet to be assessed.

12.11 Summary: welding of Gondwanan continental fragments to Laurentia

The effects of the Caledonian Orogeny are complex, and any attempt to arrange them in a simple Silurian and Devonian geological history must be tentative. Spatial relationships across the orogen are summa-

Fig. 12.12 Schematic tectonic cross-sections across the Caledonian Orogen during Mid-Silurian (a) and Early Devonian (b) times.

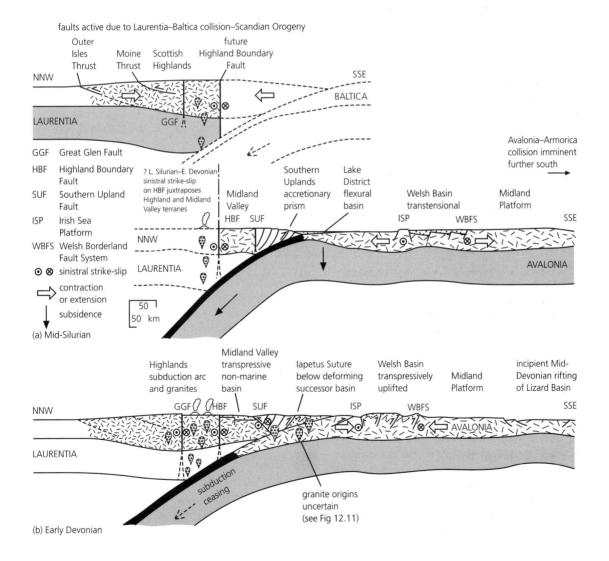

rized in two time-frames in Fig. 12.12, and time-relations are summarized in Fig. 12.13:

1 In *earliest Llandovery* time, major deformation was probably restricted to the Southern Uplands of Scotland, where sediments deposited on Iapetus Ocean crust were being scraped back northward against the Laurentian margin.

2 By *Mid-Llandovery* time, the western end of Eastern Avalonia was colliding with some part of the Laurentian continental margin south-west of the present Southern Uplands (see also Fig. 11.15, p. 183), and beginning the diachronous closure of the Iapetus Ocean. The Scottish Highlands began to overthrust their foreland to the north-west, as a result of the 'Scandian' collision of Laurentia with Baltica (Fig. 12.12a).

Fig. 12.13 Tentative chronology of Caledonian events.

3 By *Late Llandovery time,* major north-westward thrusting was well developed on the Laurentian margin. Thrusting along the Outer Hebrides Fault Zone and late-stage gravitational movements along the Moine Thrust Zone were accompanied by the onset of major sinistral strike-slip displacements.

4 In *Wenlock and Ludlow* time, large-scale sinistral strike-slip faults were the dominant control on emplacement of the Newer Granite suite of plutons. Further south, the Southern Uplands began to override the northern edge of the thick Eastern Avalonian crust (Fig. 12.12a). Accretionary deformation effectively propagated southwards as a wave of folding and cleavage formation.

5 By the *Přídolí and earliest Devonian (Lochkovian),* this deformation wave was affecting the Lake District and Wales, and depressing the crust ahead of it to form a flexural foreland basin.

6 Later in the *Early Devonian (Pragian, Emsian),* the

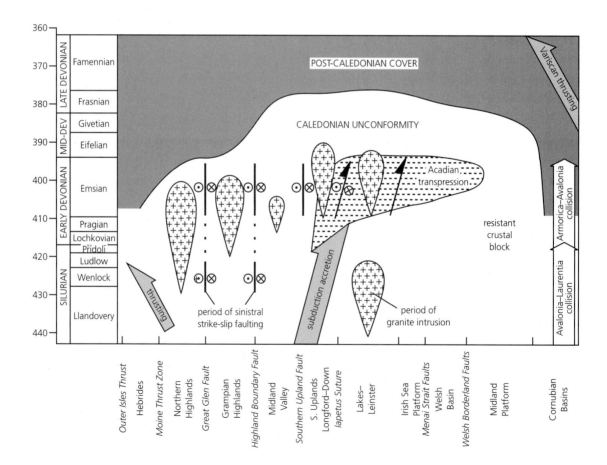

Armorican continent may have collided with the southern edge of Eastern Avalonia. Whether due to this collision or not, the whole orogenic belt underwent sinistral strike-slip shear. This shear was expressed to the north of the Southern Uplands as major displacement on the existing strike-slip faults, juxtaposing the Highlands against the southern terranes across the Highland Boundary Fault (Fig. 12.12b). In the slate belts to the south, sinistral transpression formed oblique folds and cleavage. Granites were now being intruded throughout the Caledonides, as far south as the Lake District.

7 Crustal shortening was waning by *Mid-Devonian* time. However, isostatic uplift, encouraged by recent granite intrusion, was affecting most of the resulting orogen to produce the major Caledonian unconformity (Fig. 12.13). Possible collision of Iberia with Armorica was too distant to cause shortening in Britain and Ireland, except in the extreme south.

References

Cooper, M.A. & Bruck, P.M. (1983) Tectonic relationships of the Leinster granite, Ireland. *Geological Journal* **18**, 351–360.

Coward, M.P. (1984) A geometrical study of the Arnaboll and Heilam thrust sheets, NW of Ben Arnaboll, Sutherland. *Scottish Journal of Geology* **20**, 87–106.

Harris, A.L. (1985) The nature and timing of orogenic activity in the Caledonian rocks of the British Isles. *Memoirs of the Geological Society, London* **9**, 1–53.

Hutton, D.H.W. (1982) A tectonic model for the emplacement of the main Donegal granite, N.W. Ireland. *Journal of the Geological Society, London* **139**, 615–632.

Hutton, D.H.W. & McErlean, M. (1991) Silurian and Early Devonian sinistral deformation of the Ratagain granite, Scotland: constraints on the age of Caledonian movements on the Great Glen fault system. *Journal of the Geological Society, London* **148**, 1–4.

Soper, N.J. (1986) The Newer Granite problem: a geotectonic view. *Geological Magazine* **123**, 227–236.

Soper, N.J., Webb, B.C. & Woodcock, N.H. (1987) Late Caledonian (Acadian) transpression in North West England: timing, geometry and geotectonic significance. *Proceedings of the Yorkshire Geological Society* **46**, 175–192.

Soper, N.J., England, R.W., Snyder, D.B. & Ryan, P.D. (1992) Iapetus suture zone in England, Scotland and eastern Ireland: a reconciliation of geological and deep seismic data. *Journal of the Geological Society, London* **149**, 697–700.

Todd, S.P., Murphy, F.C. & Kennan, P.S. (1991) On the trace of the Iapetus suture in Ireland and Britain. *Journal of the Geological Society, London* **148**, 869–880.

Woodcock, N.H., Awan, M.A., Johnson, T.E., Mackie, A.H. & Smith, R.D.A. (1988) Acadian tectonics in Wales during Avalonia/Laurentia convergence. *Tectonics* **7**, 483–495.

Further reading

British Geological Survey (1996) *Tectonic Map of Britain, Ireland and Surrounding Areas*, 1 : 500 000 map. British Geological Survey, Keyworth. [A reliable and thought-provoking compilation of the tectonic features of Britain and Ireland.]

Halliday, A.N., Aftalion, M., Parsons, I., Dickin, A.P. & Johnson, M.R.W. (1987) Syn-orogenic magmatism and its relationship to the Moine Thrust Zone and the thermal state of the lithosphere in NW Scotland. *Journal of the Geological Society, London* **144**, 611–617. [Provides reliable dating for the movements on the Moine Thrust.]

Harris, A.L. (1985) The nature and timing of orogenic activity in the Caledonian rocks of the British Isles. *Memoir of the Geological Society, London* **9**, 1–53. [A major compilation of data, including maps of deformation style and timing, metamorphic grade and timing, and igneous rocks.]

McKerrow, W.S. (1988) Wenlock to Givetian deformation in the British Isles and the Canadian Appalachians. In: *The Caledonian Appalachian Orogen* (eds A. L. Harris & D. J. Fettes), Special Publication 38, pp. 437–448. Geological Society, London. [A useful review of stratigraphic evidence for the timing of Caledonian events. Argues for a synchronous late Emsian (latest Early Devonian) deformation climax.]

Piper, J.D.A. (1997) Tectonic rotation within the British paratectonic Caledonides and Early Palaeozoic location of the orogen. *Journal of the Geological Society, London* **154**, 9–13. [A summary of palaeomagnetic evidence for rotation of crustal blocks during the Caledonian deformation.]

Roberts, B., Merriman, R.J., Hirons, S.R., Fletcher, C.J.N. & Wilson, D. (1996) Synchronous very low-grade metamorphism, contraction and inversion in the central part of the Welsh Lower Palaeozoic Basin. *Journal of the Geological Society, London* **153**, 277–285. [Demonstrates that the low-grade metamorphism in Wales occurred during deformation rather than deposition.]

Soper, N.J., Webb, B.C. & Woodcock, N.H. (1987) Late Caledonian (Acadian) transpression in North West England: timing, geometry and geotectonic significance. *Proceedings of the Yorkshire Geological Society* **46**, 175–192. [Interprets the arcuate cleavage trend around the Lake District.]

Soper, N.J., England, R.W., Snyder, D.B. & Ryan, P.D. (1992) Iapetus suture zone in England Scotland and eastern Ireland: a reconciliation of geological and deep seismic data. *Journal of the Geological Society, London* **149**, 697–700. [A reassessment of the geophysical evidence for the location of the Iapetus Suture.]

Soper, N.J., Strachan, R.A., Holdsworth, R.E., Gayer, R.A. & Greiling, R.O. (1992) Sinistral transpression and the Silurian closure of Iapetus. *Journal of the Geological Society, London* **149**, 871–880. [An integrated tectonic model for the accretion of continental fragments onto Laurentia in the mid-Palaeozoic.]

Todd, S.P., Murphy, F.C. & Kennan, P.S. (1991) On the trace of the Iapetus suture in Ireland and Britain. *Journal of the Geological Society, London* **148**, 869–880. [Argues for a southerly course for the suture in Ireland, and presents geochemical contrasts across the suture.]

Vaughan, A.P.M. (1996) A tectonomagmatic model for the genesis and emplacement of Caledonian calc-alkaline lamprophyres. *Journal of the Geological Society, London* **153**, 613–623. [A review and interpretation of Caledonian lamprophyre intrusions.]

Watson, J.V. (1984) The ending of the Caledonian Orogeny in Scotland. *Journal of the Geological Society, London* **141**, 193–214. [A perceptive review of the main Caledonian events on the former Laurentian margin.]

13 Devonian sedimentation and volcanism of the Old Red Sandstone Continent

N. H. WOODCOCK

13.1 Regional and global palaeogeography of Devonian events

The progressive closure of the Iapetus Ocean has been charted through Silurian time from the preserved record on both its northern and southern margins (see Chs 7, 11). The resulting Caledonian Orogeny was well under way by Early Devonian time (see Chapter 12). The basement rocks of both the old Laurentian and Avalonian continents were shortened and their existing faults reactivated. The overlying Palaeozoic cover rocks were deformed and weakly metamorphosed. The marine basins of the former Silurian collision zone were uplifted above Devonian sea level, and were either accumulating non-marine sediments or were being actively eroded. At the climax of Caledonian uplift, during Mid-Devonian time, most parts of Britain and Ireland were shedding rather than receiving sediment. Subsiding basins persisted only in parts of northern Scotland and particularly along the southern margin of Eastern Avalonia, now represented by south-west England.

The continent formed by the amalgamation of Laurentia with Baltica and Avalonia (Fig. 13.1) has been called Laurussia or, more informally, the 'Old Red Sandstone Continent'. This second name arises from the widespread non-marine sedimentary facies preserved over much of the eastern half of the continent (Fig. 13.1). The British and Irish parts of the continent lay at about 15–20°S latitude. Armorica may have been sutured to the south edge of Eastern Avalonia by Devonian time, and possibly Iberia to the south again. However, the critical Siluro-Devonian palaeomagnetic record from these fragments has been mostly overprinted by later Variscan events. Yet further south, the large Gondwana continent had its southern edge in south polar latitudes, but its northern parts could have reached similar low latitudes to Laurussia. Consequently, there was probably no wide ocean separating Laurussia and Gondwana during Devonian time.

13.2 Changes in global sea level and climate

Global changes in sea level were a less important influence on Devonian deposition than they were on Silurian events in Britain and Ireland. Most of the Devonian basins were non-marine, and the drainage through many of these basins probably adjusted to a local base level rather than to contemporary sea level. However, the marine Devonian record in south-west England is affected by global sea-level change, as are the marginal marine and alluvial deposits immediately to the north, in England and southern Ireland.

There is agreement about the gross features of the Devonian eustatic curve (Fig. 13.2). From a lowstand near the Silurian–Devonian boundary, sea level generally rose through Early and Mid-Devonian time to a highstand within the Late Devonian (late Frasnian). A subsequent fall in the later part of Late Devonian time was reversed before the Carboniferous began. This second-order curve is smoothed from a more complex record of third-order variation (Fig. 13.2). This more detailed curve shows a number of rapid transgressions, some of which are visible in the British and Irish Devonian record. Examples occur in the latest Emsian, Mid-Givetian, latest Frasnian and Late Famennian.

The causes of the third-order variations in the Devonian sea-level curve are uncertain. Continental

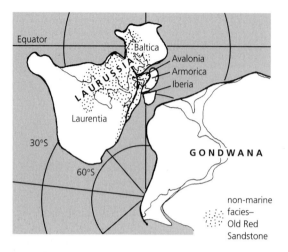

Fig. 13.1 Palaeocontinental reconstruction for Late Silurian to Early Devonian time, showing the approximate extent of the non-marine Old Red Sandstone facies (reconstruction modified from Channell *et al.* 1992).

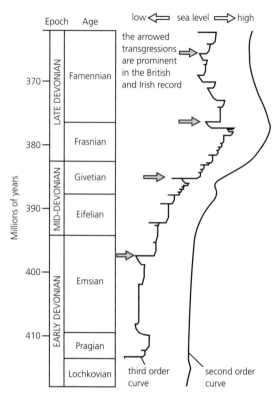

Fig. 13.2 Detailed (third-order) and smoothed (second-order) estimates of eustatic sea-level variation, based primarily on facies changes in the western United States (data from Johnson *et al.* 1985, time-scale from Tucker *et al.* 1998). See Chapter 2 for a first-order curve.

ice sheets are directly evidenced in the Famennian, and changes in their volume are a possible cause of the marked fluctuations then. However, the sea-level change in the rest of the Devonian has to be attributed to other mechanisms, of which the variation in rate of spreading at mid-ocean ridges is most commonly invoked.

13.3 Correlation problems

The Devonian System was originally defined by Adam Sedgwick and Roderick Murchison in the county of Devon. The structural complexity of these rocks has made them unsuitable as an international type section. Nevertheless, the marine rocks of south-west England can be readily correlated with the European reference sections, mainly by using ammonoids and conodonts. These fossils are, of course, absent from the non-marine successions further north, and so correlation of the Old Red Sandstone with the marine Devonian is problematical. Miospores are increasingly valuable, but are not preserved in all facies.

Because of these correlation difficulties, the informal terms Lower, Middle and Upper Old Red Sandstone are still widely used, equating only very crudely with the chronostratigraphic series. Many Lower Old Red Sandstone successions begin in or at the base of the Přídolí (uppermost Silurian), where the predominant facies change from marine to marginal marine rocks occurs (see Section 11.12, pp. 181–182). In some areas, such as south-west Wales, south-west Ireland and eastern Scotland, this facies change may occur even earlier, within or before the Ludlow. Similarly the non-marine rocks assigned to the Upper Old Red Sandstone typically include some units of Early Carboniferous age below the first marine sediments.

13.4 Tectonic setting of Devonian basins

Devonian rocks crop out over substantial areas of onshore Britain and Ireland and have also been extensively recorded from well and seismic data in offshore areas to the north and north-east of Scotland (Fig. 13.3a). Stratigraphic similarities between out-

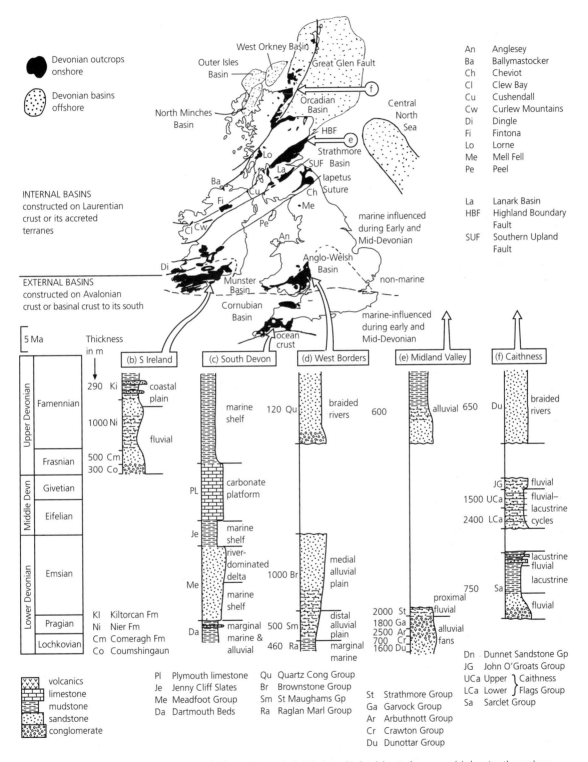

Fig. 13.3 Representative stratigraphic logs (b–f) of Devonian rocks in Britain and Ireland, located on a map (a) showing the onshore Devonian outcrop and the known offshore extent of Devonian basins.

crops within some areas prove that they formed part of a coherent sedimentary basin, with an interconnected drainage system and a shared geological history. The Orcadian Basin of north-east Scotland is an example of such a basin. However, geological contrasts over Britain and Ireland as a whole show that, on this scale, Devonian rocks were deposited in a number of separate basins (Fig. 13.3a). Different basins had distinct subsidence and uplift histories and were separated from each other by continental drainage divides or, in the extreme south, by submarine horsts.

The preserved Devonian basins can be divided into two main groups (Fig. 13.3a), with respect to the Caledonian mountain belt. The *internal* or *intramontane basins* formed within the main mountain chain and did not have a direct drainage connection to the sea for most of their history. By contrast, the *external* or *extramontane basins* formed in more subdued topography, south of the main mountain front, and comprise both non-marine basins and depositionally contiguous marine basins to the south.

The Iapetus Suture is a convenient dividing line between the internal and external basins. North-west of the Suture lay the late Proterozoic crust of Laurentia and its accreted terranes. During Early Devonian time this Laurentian crust was being shortened, added to by large volumes of granites and related volcanic rocks (see Section 12.10, pp. 200–204), and displaced along pre-existing north-east–south-west faults, predominantly by strike-slip movements (see Section 12.6, pp. 194–195). The regional tendency of the shortening and magmatism to produce crustal uplift was locally counteracted by subsidence of fault-bounded sedimentary basins such as the Strathmore Basin. The elevated Caledonian topography may also have induced lateral gravitational spreading of the mountain belt, and promoted upper-crustal extension within it. The internal basins were probably formed by some combination of strike-slip faulting and gravitational collapse.

The Avalonian crust to the south-east of the Iapetus Suture was distinguished by widespread basins with thick Lower Palaeozoic marine fill. Even though these basins were folded and cleaved during the Caledonian Orogeny, their less common Devonian intrusive activity and their more readily eroded lithologies meant that the Avalonian crust never supported mountain belts as high as those further north. The

Caledonian foothills between the Iapetus Suture and the Caledonian deformation front (Fig. 13.3a) probably hosted some intramontane basins, later eroded. However, the preserved deposits of the Anglo-Welsh and Munster Basins seem to have accumulated on alluvial plains between the foothills and the seaway to the south. The marine basins were the product of fault-controlled rifting, possibly with a strong strike-slip component, within the Cornubian basins.

13.5 Stratigraphic fill pattern of the Devonian basins

Stratigraphic logs through representative Devonian basins in Britain and Ireland reveal a variety of fill patterns through time (Fig. 13.3b–f). The most continuous successions are understandably preserved in the predominantly marine Cornubian basins of Devon and Cornwall (Fig. 13.3c). Here, basin subsidence approximately kept pace with sediment supply throughout Devonian time.

When the marine successions are traced northwards, a major unconformity develops to separate lower Devonian from upper Devonian non-marine intervals. This unconformity is best displayed in the Anglo-Welsh Basin (Fig. 13.3d) and the Strathmore Basin (Fig. 13.3e), but the stratigraphic pattern is probably similar in all the Old Red Sandstone basins as far north as the Highland Boundary Fault. On this hypothesis, the upper Devonian Munster Basin (Fig. 13.3b) would be underlain by a non-marine lower Devonian succession. The Mid-Devonian unconformity presumably records the time of maximum uplift in the external and southern internal parts of the Caledonian mountain belt (see Section 12.11, pp. 204–205). Any mid-Devonian deposits must have been only locally deposited and rapidly removed by subsequent erosion. This erosion also stripped off much or all of the pre-existing lower Devonian succession, especially over the strongly uplifted areas of the former marine basins; the Welsh Basin and Lake District Basin for instance. Only when uplift rates waned in Late Devonian time did deposition resume prior to the marked marine transgression of the early Carboniferous.

The main Devonian basins north of the Highland Boundary Fault differ from the non-marine basins further south in preserving a thick mid-Devonian succession. This succession is directly seen in the Orcadian Basin (Fig. 13.3f) and inferred in the smaller offshore basins to the north-west (Fig. 13.3c). These basins overlap some of the most strongly uplifted and deeply dissected crust in the Caledonian Belt, and the preservation of a thick, let alone fairly continuous, succession through the time of maximum uplift is paradoxical. Possible explanations are discussed in the following sections.

The strong stratigraphic differences between the Devonian basins, themselves often geographically isolated from each other, makes it difficult to describe an integrated Devonian geological history for the whole of Britain and Ireland. Such a history will be attempted in Section 13.11. Firstly, however, the local histories will be detailed of each separate basin or group of basins, beginning in the north and working southwards. The generalized palaeogeographic maps of Figs 13.9–13.11 will be referred to, along with more detailed maps of each region.

Fig. 13.4 Stratigraphic logs (a) from onshore outcrops in the Orcadian Basin, located on a base map (b) in which the syn- and post-Mid-Devonian fault displacements have been removed (modified from Marshall *et al.* 1996).

13.6 Orcadian, West Orkney, Outer Isles and North Minches basins

The Orcadian Basin is represented by onshore outcrops around the Moray Firth, in Caithness, Orkney and as far north as Shetland, and by well records defining the wide offshore extension of the basin (Fig. 13.3a). A Devonian fill to the offshore West Orkney, Outer Isles and North Minches basins has not been proved directly from well records, but is inferred from seismic interpretation and correlation with onshore outcrops. The Orcadian Basin is the best guide to the facies that might occur in these more westerly basins.

During much of Early Devonian time, the Scottish Highlands were dominated by uplift and erosion rather than by sedimentation. The intrusion of Caledonian granites, culminating around 400 Ma (see Fig. 12.10, p. 201), was mainly responsible for this uplift, together with renewed Early Devonian strike-slip activity on the major north-north-east-trending faults. Only local deposits of Lower Old Red Sandstone are preserved above the unconformity with the Highland metamorphic rocks (Fig. 13.4). The predominant sediments are conglomerates

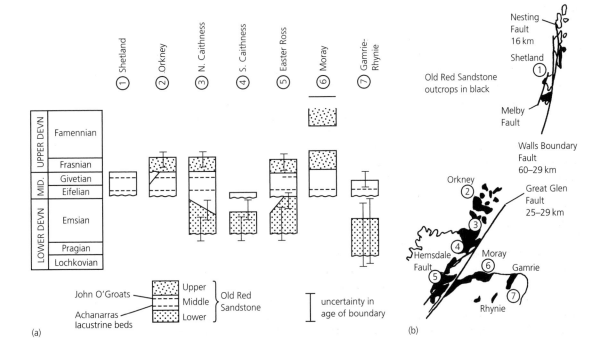

and sandstones deposited on alluvial fans that typically bordered active faults. Some of these faults may still have been dominated by sinistral strike-slip. However, the seismic evidence from the offshore basins suggests instead that, by latest Early Devonian time, normal dip-slip displacements were controlling the sites of sedimentary basins. Most of the active faults were east-south-east dipping, and possibly controlled by old Caledonoid structures, such as the Outer Isles and Moine Thrusts, in the underlying basement. However, direct reactivation is not supported by onland observations. In any case, the normal faulting may simply record a switch from the Caledonian shortening to regional crustal extension. However, another possibility is that extension at a high level in the crust may have been driven or enhanced by lateral gravitational spreading, orogenic collapse, of the Caledonian mountain belt.

The main fill of the Orcadian Basin is of Mid-Devonian age (Fig. 13.4) and reaches 4 km thick in Caithness and at least twice this in Shetland. The fill is conformable on Lower Old Red Sandstone rocks in the basin centre but in angular unconformity at the basin margin, where the older rocks had been rotated above curved normal faults. Alluvial fans were still active in the south but most of the Orcadian Basin was occupied by shallow lakes and fringing alluvial plains (Fig. 13.4, see also Fig. 13.10). Variations in the level of the lakes produced deposits that cycle between lacustrine and alluvial facies over thicknesses of 5–10 m. A few widespread lake transgressions are valuable for correlation between outcrop areas (Fig. 13.4). There was localized volcanism in Orkney and Shetland.

The first evidence of marine influence in the Orcadian Basin occurs high in the Middle Old Red Sandstone of Orkney, probably in Mid- to Late Givetian time. Siltstones containing pseudomorphs of halite crystals suggest brief incursions of sea water. This water probably came from the south-east, where marine limestones and mudstones in wells in the Argyll Field may range down into the Mid-Devonian (see also Fig. 13.10). This marine connection became more important in Late Devonian time, when further marine incursions have been recorded in wells east of Orkney. The marine intervals probably correspond to prominent transgressions on the global sea-level curve (Fig. 13.2) in Mid-Givetian and Frasnian time.

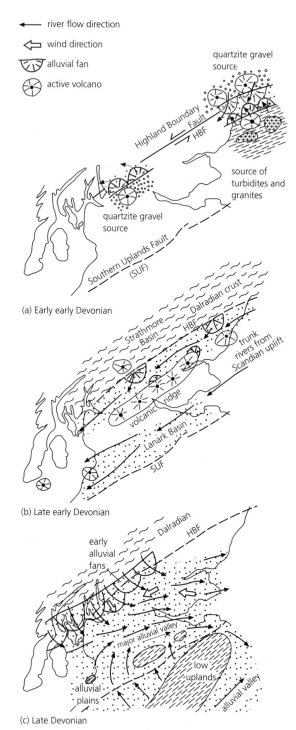

Fig. 13.5 Palaeogeographic maps of the Midland Valley, to emphasize the shifting location, extent and fill direction of successive Old Red Sandstone basins ((a, b) after Haughton & Bluck 1988; (c) after Mykura 1991).

The evidence for less active alluvial fans and more widespread lakes during Mid-Devonian time suggests that crustal extension on normal faults was declining, to be replaced by more regional thermal subsidence. Extension may have continued later in the West Orkney Basin and contemporaneous basins to the west. Localized tilting of the Middle Old Red Sandstone in the Orcadian Basin, before deposition of the Upper Old Red Sandstone, has been interpreted as a phase of basin inversion by reverse movements on the former normal faults. However, over most of the basin, the transition to the Upper Old Red Sandstone is conformable (Fig. 13.4) and merely records a change in depositional environment. The lakes were replaced in the basin centre by sabkha plains, flooded by water from the hinterland and, during marine highstands, by sea water from the south-east. Near the southern basin margin, lakes were overwhelmed by the deposits of braided and meandering rivers.

Tens of kilometres of latest Devonian or post-Devonian dextral strike-slip displacement can be proved on the Walls Boundary Fault, the Great Glen Fault and related faults (Fig. 13.4b). Strong sinistral strike-slip displacements along the same fault complex are likely in the Early Devonian. However, the deposition of the Old Red Sandstone of the Orcadian Basin was apparently dominated only by extensional faulting. In this respect it contrasts strongly with the Lower Old Red Sandstone Basins of southern Scotland, a contrast reflected in major facies differences.

13.7 South Scottish and north Irish basins

Middle Old Red Sandstone rocks are conspicuously absent south of the Scottish Highlands. Here a thick Lower Old Red Sandstone succession, ranging in places down into the Silurian, is overlain by a thinner Upper Old Red Sandstone sequence, much of which is lowest Carboniferous in age (Fig. 13.3e). The most extensive outcrops of Old Red Sandstone in this region occur in the Midland Valley; that is, between the Highland Boundary and Southern Uplands faults. However, there has been considerable debate as to whether these faults were active in bounding depositional basins during Devonian sedimentation or whether they merely define post-Devonian limits to originally more extensive basins. In any case, the

Midland Valley can be used as a reference area for understanding depositionally comparable, though less extensive, outcrops in surrounding areas: the west Highlands, north Ireland and the Scottish Southern Uplands (Fig. 13.3a).

The Midland Valley is most simply visualized in terms of a pair of Early Devonian basins — the Strathmore and Lanark basins — filled predominantly from the north-east (Fig. 13.5b), overlain by a single Late Devonian basin — the Midland Valley Basin — mostly filled from the south-west (Fig. 13.5c). However, there are complications to this pattern, particularly early on in the regional history. At the base of the Old Red Sandstone succession, the Stonehaven Group has been shown to be of Mid-Silurian (Wenlock–Ludlow) age, and to record a separate episode of basin fill. The overlying conglomeratic Dunottar and Crawton groups (Fig. 13.3e) are remnants of a sub-basin fed laterally by alluvial fans encroaching from both the south-east and north-west (Fig. 13.5a). Active volcanoes provided a local source for some of the clastic debris. In this respect the Crawton Basin is similar to other early Lower Old Red Sandstone deposits, further south-east along the Highland border, beneath the Lorne Plateau Lavas, and in the isolated basin remnants in the northern part of Ireland. Caledonian sinistral strike-slip activity on the Highland Boundary and adjacent faults has been suggested as an important control on the subsidence of these early basins.

The basaltic and andesitic volcanism that characterizes the Lower Old Red Sandstone of the Midland Valley reached a climax during deposition of the Arbuthnott Group (Fig. 13.3e). A line of volcanoes formed a topographic barrier down the Midland Valley and induced a more axial (north-east to south-west) flow of the Devonian rivers. The overlying Garvock and Strathmore groups of this Strathmore Basin record a cessation of volcanism, the waning of locally supplied alluvial fans and their replacement by a major trunk river system supplied from the north-east (Fig. 13.5b). This river carried sand, probably derived from the Scandian collision of Baltica with Laurentia.

The volcanic hills through the Midland Valley separated the Strathmore Basin from the more poorly dated Lanark Basin. The Lower Old Red Sandstone here shows the same associated volcanic rocks and the important south-west-directed axial flow,

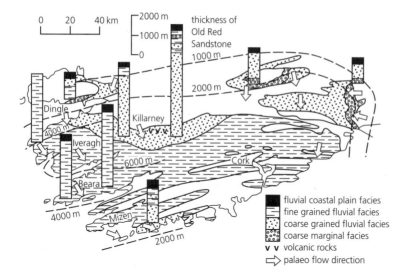

Fig. 13.6 Sedimentary facies and thickness of the Upper Old Red Sandstone of the Munster Basin (after Graham 1983).

although lower units are more variable. Lower Old Red Sandstone successions of conglomerate, sandstone and lava also outcrop south of the Southern Uplands Fault as far as the Cheviot. The angular unconformity of these probable Lower Devonian rocks on deformed Ordovician and Silurian units provides an important constraint on the age of uplift and planation of the Southern Uplands thrust stack.

A long period of Devonian time, spanning the Mid-Devonian, is not recorded by deposits in the Midland Valley. When deposition resumed during the Late Devonian, a northern zone of alluvial fans bordered an extensive area of finer-grained deposits containing important calcrete horizons (Fig. 13.5c). These sediments were deposited by meandering and braided rivers on an alluvial plain. Accumulation was slow enough for evaporation to form carbonate cements within the arid alluvium. The Late Devonian rivers tended to flow eastwards or north-eastwards, towards the seaway identified from offshore well data. This flow direction is opposite to that in the Early Devonian, emphasizing the major topographic adjustments that occurred during the time of Mid-Devonian uplift.

13.8 Lower Devonian remnants near the Iapetus Suture

The lavas of the Cheviot (Fig. 13.3a) record events just north of the Iapetus Suture, during its final activity in Early Devonian time. Across the Suture to the south, the next extensive outcrops of Lower Devonian rocks are over 250 km across-strike in the Welsh Borderland. However, the metamorphic grade of intervening Silurian rocks, for instance in the Lake District and North Wales, suggests that a substantial cover of Lower Old Red Sandstone was deposited in this zone to the south of the suture, only to be eroded during the Mid-Devonian uplift event.

A few isolated outcrops of this missing basin fill remain. These include conglomeratic alluvial fan successions in the Lake District (Mell Fell, Fig. 13.3a), and sandstone-dominated fluvial sequences on the Isle of Man (Peel) and Anglesey. Further south-west, in south-west Ireland (Dingle, Fig. 13.3a), the Dingle Group is transitional above the marine Silurian (Ludlow) rocks of the Dunquin Group, and probably ranges up into Early Devonian time. It records an early phase of lakes fringed by debris fans, and a later fluvial phase.

The Early Devonian deposits to the south of the Iapetus Suture were probably formed in a foreland basin, produced by the southward-advancing load of Laurentia depressing the Avalonian lithosphere. This enhanced subsidence is evidenced in the marine Silurian basins (see also Fig. 11.10, p. 178), but probably continued into Early Devonian time. As thrust displacement ceased on the Suture, however, subsidence gave way to shortening, uplift and erosion of the foreland basin fill.

13.9 Munster Basin

The resumption of sedimentation across the now inactive Iapetus Suture Zone is recorded by thin Upper Old Red Sandstone deposits in central Ireland (Fig. 13.3a). This fluvial sequence is continuous upwards into marine Carboniferous deposits. Indeed, much of the non-marine facies may itself be of Carboniferous rather than Devonian age, and is therefore analogous to the non-marine clastic intervals that typically herald Carboniferous deposition in England and Wales (see Chapter 14). This alluviation of the old Caledonian land-surface was a product of rising sea level late in Late Devonian (Famennian) time (Fig. 13.2).

Followed southwards, the thin Upper Old Red Sandstone deposits of central Ireland expand into a thicker and longer-ranging succession deposited on the rapidly subsiding crust of the Munster Basin (Figs 13.3a, 13.6). Stratigraphic thicknesses reach over 6 km in the centre of this basin, compared with a maximum of 0.5 km on its northern flanks. The fill to the Munster Basin shows a broadly fining-up trend, from coarse-grained through fine-grained fluvial facies into fluvial coastal plain facies (Figs 13.3b, 13.6). However, there is also a strong lateral facies variation, with fine-grained fluvial facies dominating the basin centre and coarse facies the northern and eastern basin margin (Fig. 13.6). Conglomeratic alluvial fan facies also occur low in the basin margin successions, and suggest that these margins were controlled by active faults. Sediment transport was broadly southwards, away from the relict Caledonian Mountains and towards a seaway to the south. A brief marine incursion from this seaway occurred in Frasnian time.

The oldest strata proved in the main Munster Basin lie close to the base of the Upper Devonian. However, subsurface preservation of older units is likely (see also Fig. 13.10), probably represented at the northern basin margin by poorly dated units such as the Caherbla Group on the Dingle Peninsula. In any case, the evidence is clear that Late Devonian sedimentation in southern Ireland did not simply record a transgression over an inactive post-Caledonian land surface. Rather, a major phase of fault-controlled basin rifting seems to have occurred in Late Devonian Munster (see also Fig. 13.11). This rift activity matches more closely the Late Devonian tectonic setting of the marine Cornubian basins (Fig. 13.3c) than the non-marine Anglo-Welsh Basin.

13.10 Anglo-Welsh Basin

The Devonian deposits of the Anglo-Welsh Basin now crop out in South Wales and the Welsh Borderland, and in the subsurface beneath central and eastern England (Fig. 13.3a). These rocks comprise a 2–4-km-thick Lower Old Red Sandstone succession separated by the Mid-Devonian unconformity from a much thinner (100–300 m) Upper Old Red Sandstone interval (Fig. 13.3d). However, the remnants of Devonian successions between here and the Iapetus Suture (see Section 13.8) make clear that the non-marine Early Devonian basin was originally much more extensive, stretching from the Suture southwards to a shoreline with the marine Cornubian basins. The northern limit of preservation of Anglo-Welsh Basin rocks is approximately the final position of the Caledonian mountain front in Mid-Devonian time.

Non-marine environments had already replaced marine conditions over most of the basin by the beginning of Silurian time, as the vigorous supply of sediment from the growing Caledonian uplands outpaced the subsidence of the basins to the south (see Section 11.12, pp. 181–182). This subsidence was now being caused predominantly by the advancing load of the Laurentian continent, depressing it in a zone that spread southward from the Iapetus collision zone. The composition of the latest Silurian (Přídolí) sediments in the Anglo-Welsh Basin suggests that they were derived from eroding metamorphic rocks such as those in the Scottish Highlands, but cross-strike transport routes for such sediment are not obvious (see also Fig. 13.9).

By Mid-Lochkovian (earliest Devonian) time, the remote supply of metamorphic debris was waning. An interval of relatively slow sedimentation on distal alluvial plains allowed extensive calcretes to form by evaporation in the arid Devonian climate. The rest of the Lower Devonian succession records the deposition of river-borne debris that was probably eroded from the uplifting lower Palaeozoic rocks of northern Wales and England. South Wales and the Welsh Borderland preserve distal alluvial facies, dominated by floodplain mudstones, overlain by medial facies dominated by sandstones. This upward coarsening probably records the southward advance of the Caledonian mountain front and its attendant alluvial systems (Figs 13.3d, 13.7). The sediments bear debris derived from progressively lower in the uplifting stratigraphy.

By the end of the Early Devonian, the Caledonian uplift was affecting most of the old Avalonian continent, and deposition only continued in the Cornubian basins to the south (Fig. 13.7, see also Fig. 13.10). The succeeding Mid-Devonian hiatus records the time of maximum uplift and erosion during the thermal and contractional climax of the Caledonian Orogeny (see Fig. 12.13). An emergent landmass along the Bristol Channel fed alluvial fans locally into south-west Wales. However, deposition of alluvial and marginal marine facies only began again across most of the Anglo-Welsh Basin during Late Devonian time. An important factor in this alluvation was the final decay of any Caledonian contractional and thermal activity, and the transition to the active extensional subsidence of the Early Carboniferous (see Chapter 14). However, the tectonic influence on Upper Old Red Sandstone deposition is weak enough for the more

important pulses of global sea-level rise to be evident in the preserved stratigraphy (Figs 13.2, 13.7). These pulses drove marine transgressions onto the edge of the Anglo-Welsh shelf, and consequent accumulation of alluvial deposits as far north as the relict uplands of Wales and East Anglia (see also Fig. 13.11). The global sea-level rise eventually flooded much of this alluvial plain, as it did in the Munster Basin (see Section 13.9), and the Old Red Sandstone deposits pass conformably up into marine Carboniferous sediments (see Chapter 14).

13.11 Cornubian basins

The northward marine transgression into the Munster and Anglo-Welsh basins is a tangible link with the Devonian seaway that lay along the southern edge of Eastern Avalonia. These marine deposits, together with the distal equivalents of the Munster and Anglo-Welsh alluvial systems, are preserved in a series of basins now forming much of Devon and Cornwall—the Cornubian basins (Fig. 13.3). However, reconstructing these basins has proved problematical. The Variscan deformation, which affected the southern part of the Anglo-Welsh Basin and much of the Munster Basin, was more intense towards the south (see Chapter 15). The deposits of the Cornubian basins were repeated by faults and upthrust over their own margins. More confusing

Fig. 13.7 Stratigraphic summary of preserved Devonian rocks on a south to north transect across south-west England and Wales (after Allen 1979, with permission from the Palaeontological Association 1999). Arrows show prominent global sea-level rises marked in Fig. 13.2.

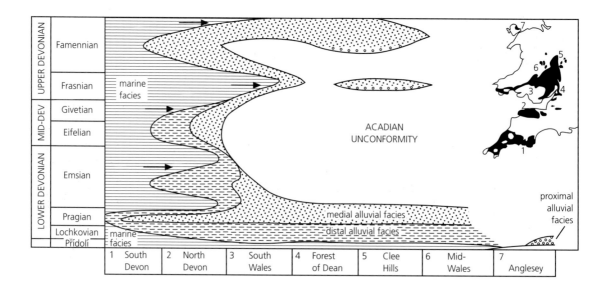

still, strike-slip faults—striking north-west and east-west—segmented the basins and obscured the original connections between them. In particular, the presence of several east–west fault zones with major dextral displacement makes it probable that, during Devonian time, the basins lay considerably further east, and not opposite Wales at all.

Fig. 13.8 Schematic map (a) and cross-section (b) showing facies relationships in south-west England during Devonian and into Carboniferous time. Deformation and fault displacements during deposition and the subsequent Variscan Orogeny make the Devonian relationship speculative (map based on Bluck *et al.* 1992; section based on Bluck *et al.* 1988).

Because of these uncertainties, it is possible to deduce only a generalized palaeogeography of the Cornubian basins (Fig. 13.8a). They all lay to the south of a zone of shallow, occasionally emergent, Precambrian basement along the Bristol Channel. To the south lay the Exmoor Basin, containing marine and marginal marine deposits interfingering with tongues of northerly derived fluvial sediments. These facies probably passed southwards into marine shelf deposits, but this transition is now obscured by the overlying Culm Basin of Carboniferous age. When Devonian rocks re-emerge along the southern edge of the Culm Basin, they contain slumps, debris flows and volcanics supposed to mark the northern fault-

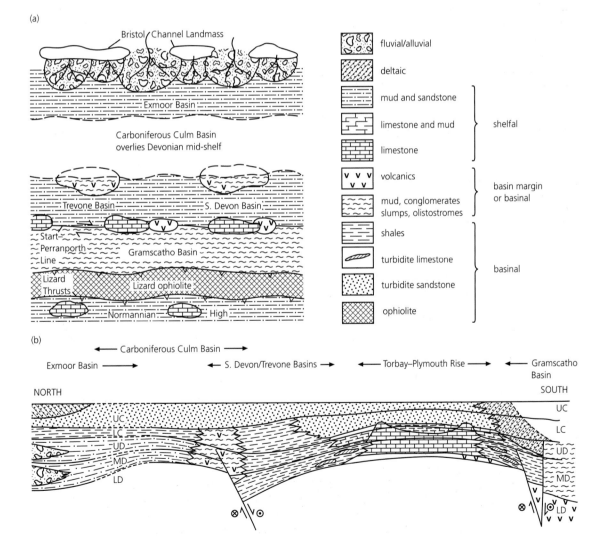

controlled margin of the deeper-marine Trevone and South Devon basins. These basins must have shallowed gently again to the south, where they are rimmed by substantial carbonate reefs and their associated facies. Another major deformation zone then bounds the Gramscatho Basin, filled by a deep-marine succession of mudstones, turbidites and olistostromes. Some of these sediments were supplied from the south. Volcanic rocks in the basin have geochemical affinities with oceanic crust, as does the large slice of the Lizard ophiolite that is thrust northwards over the Gramscatho Basin. Further south still is the Normannian High, part of the Armorican continent, postulated as a source area for the carbonate and siliceous debris in the Gramscatho Group.

This palaeogeographic view of the Cornubian basins does not fully display their evolution through Devonian time. A schematic cross-sectional view (Fig. 13.8b) shows that fluvial sequences, the presumed southern correlatives of the Old Red Sandstone, covered much of Cornubia throughout the Early Devonian. The implication is that the Trevone and South Devon basins had not developed at this time. The tectonic distinction between basins and bordering rise areas developed during the Mid-Devonian, a time when the Gramscatho Basin was also active. Extension or transtension is proposed on the bordering faults, which acted as conduits for basic magmas, and provided the sea-floor relief that triggered slumps and debris flows. Active basin extension may have

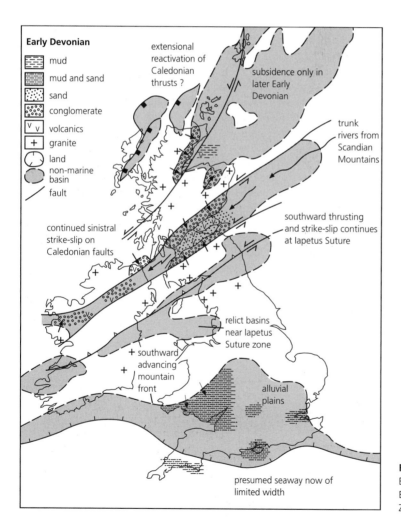

Fig. 13.9 Palaeogeographic map of Britain, Ireland and surrounding areas for Early Devonian time (modified from Ziegler 1990).

continued until early Carboniferous time, when a renewed cycle of basinal clastic sedimentation was promoted by flexural subsidence ahead of the northward-advancing Variscan thrust stack (see Chapters 14, 15).

The Cornubian basins record events near the southern margin of the Eastern Avalonian continent. For the later part of Ordovician time and for most of the Silurian, this margin is presumed to have bordered a sizeable Rheic Ocean, separating Gondwana in the south from its former fragments of Avalonia, Iberia and Armorica. However, the evidence from the Devonian successions is that this seaway was essentially closed by then. Although the Lizard ophiolite may be part of an oceanic crustal basement to the Gramscatho Basin, this basin was, by Mid-Devonian time, being supplied mainly from the south rather than from Avalonia to the north. This source suggests that Armorica had already docked with the 'Old Red Sandstone Continent', and that the Lizard crust was formed in an intervening zone of transtension rather than as part of a major ocean. By Carboniferous time, even this small basin was being deformed and thrust northwards in the early phases of Variscan shortening (see Chapter 15).

13.12 Summary: sedimentation during a collisional orogeny and its aftermath

Any summary of the Devonian depositional history

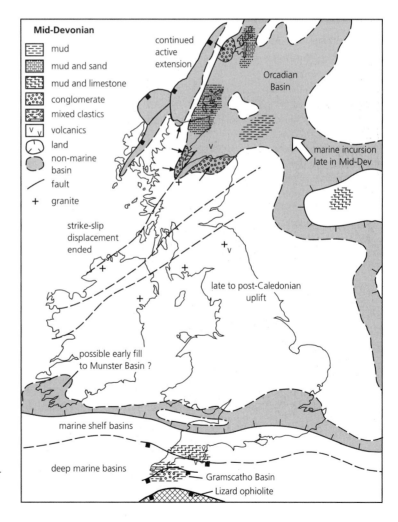

Fig. 13.10 Palaeogeographic map of Britain, Ireland and surrounding areas for Mid-Devonian time (modified from Ziegler 1990).

of Britain and Ireland risks being an oversimplification of a complex story. The transition from the culminating early Devonian events of the Caledonian Orogeny to the mainly extensional basin-forming later Devonian events ensured complexity through time. Geographical complexity arose because of the varying tectonic response to Caledonian deformation of different sectors of the British and Irish crust, and because the southernmost part of the region was, by later Devonian time, already being deformed by early 'Variscan' movements.

Palaeogeographic maps summarize the spatial complexity of sedimentation during the Early, Middle and Late Devonian (Figs 13.9–13.11) and a chrono-

logical chart sets depositional history against contemporary tectonic and igneous events (Fig. 13.12).

In *Early Devonian* time, any remnants of the former marine basins bordering the Iapetus Ocean were rapidly over-filled by sediment, except possibly in the southernmost Cornubian basins (Fig. 13.9). However, the south-advancing load of the Laurentian continent down-flexed the Avalonian lithosphere and allowed continued accumulation of northerly derived alluvial sediments in the external basins. As the Caledonian deformation front moved south, the alluvial fill became coarser, then was uplifted, eroded and recycled further south. The Caledonian deformation is generally taken to result from the collision of

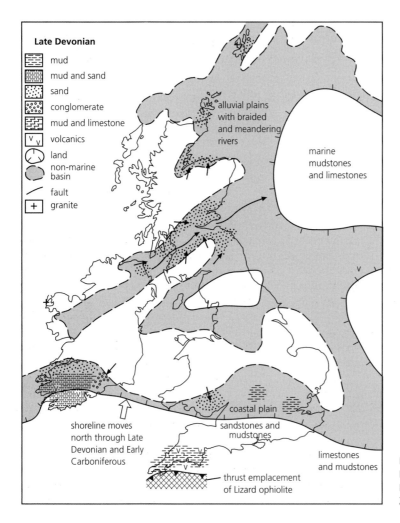

Fig. 13.11 Palaeogeographic map of Britain, Ireland and surrounding areas for Late Devonian time (modified from Ziegler 1990).

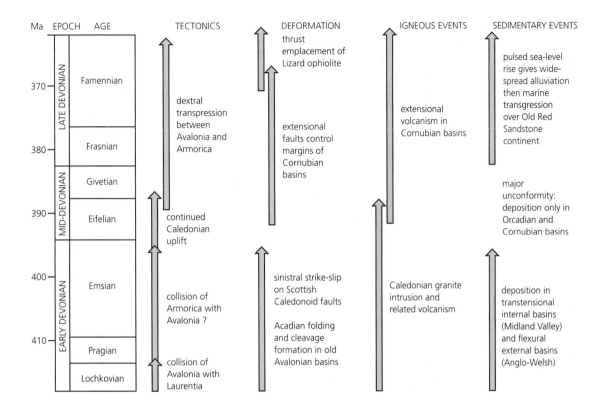

Fig. 13.12 Generalized chronology of Devonian events in Britain and Ireland (time-scale from Tucker *et al.* 1998).

Avalonia with Laurentia. However, a component of north–south compression due to collision of Armorica with Avalonia cannot be discounted, provided that it was not so strong as to prevent the continuous sedimentation in the Cornubian basins.

The collisional tectonics caused sinistral strike-slip along the major Laurentian faults. This displacement created rapidly subsiding basins in the Scottish Midland Valley, infilled at first by locally supplied alluvial fans. Later in Early Devonian time, a major axial drainage system in the Midland Valley was fed from the more remote source of the Scandian Orogen, between Baltica and Laurentia. Large volumes of magma intruded the Scottish crust throughout the Early Devonian, leaving a record of granite plutons, originally overlain by extensive volcanic complexes. This magmatism ensured that the Scottish Highlands

were actively uplifting and eroding, deposition beginning in the Orcadian Basin only late in Early Devonian time.

By *Mid-Devonian* time, the Caledonian crustal shortening and transcurrent deformation seem finally to have ended. However, the isostatic consequences of the orogeny lasted longer, with continued exhumation and erosion over most of Britain and Ireland (Fig. 13.10). The stratigraphic result is the widespread unconformity between Lower and Upper Old Red Sandstone sequences. Only in the Orcadian Basin, and adjacent basins to the west, are Mid-Devonian non-marine sediments preserved. These basins formed by crustal extension, perhaps due to gravitational spreading of this part of the orogenic belt. Late in the Mid-Devonian, an intermittent marine connection developed between the Orcadian Basin and a seaway in the central North Sea. This seaway may itself have connected southwards with the Cornwall–Rhenish sea (Fig. 13.10). The marine and marginal marine deposits of the northern edge of this small ocean are preserved in the actively rifting Cornubian basins. Despite preserving a remnant slice of ophiolitic crust, this ocean was probably only a narrow transtensional rift between Avalonia and Armorica.

By the Late Devonian, the topography of the Caledonian mountain belt had been substantially peneplained. Pulsatory rises in the global sea level flooded the lower-lying parts of this eroded topography, for instance in the Orcadian and Anglo-Welsh basins, and allowed alluvial sequences to be preserved (Fig. 13.11). The Midland Valley Basin was fed by alluvial fans from bordering uplands, as well as by an axial river system. Not all Late Devonian accumulation was due to passive post-Caledonian transgression. Active normal faulting of the Munster Basin created space for a large thickness of Upper Devonian non-marine sediment. This extension is paralleled in the northern Cornubian basins. However, the Mid- to Late Devonian thrust stacking of the southern Cornubian basins is an important precursor to the north-directed Variscan deformation that was to dominate events in southern Britain and Ireland through Carboniferous time.

References

Allen, J.R.L. (1979) Old Red Sandstone facies in external basins, with particular reference to southern Britain. *Special Papers in Palaeontology* **23**, 65–80.

Bluck, B.J., Haughton, P.D.W., House, M.R., Selwood, E.B. & Tunbridge, I.P. (1988) Devonian of England, Wales and Scotland. In: *Devonian of the World* (eds N. J. McMillan, A. F. Embry & D. J. Glass), Memoir of the Canadian Society of Petroleum Geologists 14 (1), pp. 305–324.

Bluck, B.J., Cope, J.C.W. & Scrutton, C.T. (1992) Devonian. In: *Atlas of Palaeogeography and Lithofacies* (eds J. C. W. Cope, J. K. Ingham & P. F. Rawson), Memoir 13, pp. 57–66. Geological Society, London.

Channell, J.E.T., McCabe, C. & Woodcock, N.H. (1992) An Early Devonian (pre-Acadian) magnetization component recorded in the Lower Old Red Sandstone of South Wales (UK). *Geophysical Journal International* **108**, 883–894.

Graham, J.R. (1983) Analysis of the Upper Devonian Munster Basin, an example of a fluvial distributary system. In: *Modern and Ancient Fluvial Systems* (eds J. D. Collinson & J. Lewin), Special Publication of the International Association of Sedimentologists 6, pp. 473–483. Blackwell Scientific Publications, Oxford.

Haughton, P.D.W. & Bluck, B.J. (1988) Diverse alluvial sequences from the Lower Old Red Sandstone of the Strathmore region, Scotland—implications for the relationship between late Caledonian tectonics and sedimentation. In: *Devonian of the World* (eds N. J. McMillan, A. F. Embry & D. J. Glass), Memoir of the Canadian Society of Petroleum Geologists 14(1), pp. 269–293.

Johnson, J.G., Klapper, G. & Sandberg, C.A. (1985) Devonian eustatic fluctuations in Euramerica. *Bulletin of the Geological Society of America* **96**, 567–587.

Marshall, J.E.A., Rogers, D.A. & Whiley, M.J. (1996) Devonian marine incursions into the Orcadian Basin, Scotland. *Journal of the Geological Society of London.* **153**, 451–466.

Mykura, W. (1991) Old Red Sandstone. In: *The Geology of Scotland* (ed. G. Y. Craig), pp. 297–346. Geological Society, London.

Tucker, R.D., Bradley, D.C., Ver Straeten, C.A., Harris, A.G., Ebert, J.R. & McCutcheon, S.R. (1998) New U–Pb zircon ages and the duration and division of Devonian time. *Earth and Planetary Science Letters* **158**, 175–186.

Ziegler, P.A. (1990) *Geological Atlas of Western and Central Europe*, 2nd edn. Shell International Petroleum Maatschappij B.V., The Hague.

Further reading

Allen, J.R.L. (1979) Old Red Sandstone facies in external basins, with particular reference to southern Britain. *Special Papers in Palaeontology* **23**, 65–80. [An accomplished synthesis of the sequence stratigraphy of the external ORS basins in relation to the marine Cornubian basins.]

Bluck, B.J., Haughton, P.D.W., House, M.R., Selwood, E.B. & Tunbridge, I.P. (1988) Devonian of England, Wales and Scotland. In: *Devonian of the World* (eds N. J. McMillan, A. F. Embry & D. J. Glass), Memoir of the Canadian Society of Petroleum Geologists 14 (1), pp. 305–324. [A useful overview, particularly strong on Scotland and Cornubia.]

Bluck, B.J., Cope, J.C.W. & Scrutton, C.T. (1992) Devonian. In: *Atlas of Palaeogeography and Lithofacies* (eds J. C. W. Cope, J. K. Ingham & P. F. Rawson), Memoir 13, pp. 57–66. Geological Society, London. [Detailed palaeogeographic maps and accompanying commentary.]

Coward, M.P., Enfield, M.A. & Fischer, M.W. (1989) Devonian basins of Northern Scotland: extension and inversion related to Late Caledonian–Variscan tectonics. In: *Inversion Tectonics* (eds M. A. Cooper & G. D. Williams), Special Publication 44, pp. 275–308. Geological Society, London. [A stimulating analysis of offshore seismic data suggesting a Devonian fill to basins off North Scotland.]

Graham, J.R. & Clayton, G. (1988) Devonian rocks in

Ireland and their relation to adjacent regions. In: *Devonian of the World* (eds N. J. McMillan, A. F. Embry & D. J. Glass), Memoir of the Canadian Society of Petroleum Geologists 14 (1), pp. 325–340. [A readable yet comprehensive review of the Devonian basins of Ireland.]

Haughton, P.D.W. & Bluck, B.J. (1988) Diverse alluvial sequences from the Lower Old Red Sandstone of the Strathmore region, Scotland—implications for the relationship between late Caledonian tectonics and sedimentation. In: *Devonian of the World* (eds N. J. McMillan, A. F. Embry & D. J. Glass), Memoir of the Canadian Society of Petroleum Geologists 14 (1), pp. 269–293. [A detailed study of one of the sub-basins of the Midland Valley, illustrating the methodology for interpreting Old Red Sandstone facies.]

Mykura, W. (1991) Old Red Sandstone. In: *The Geology of Scotland* (ed. G. Y. Craig), pp. 297–346. Geological Society, London. [A comprehensive review of the Devonian sedimentary rocks of Scotland.]

Rogers, D.A., Marshall, J.E.A. & Astin, T.R. (1989) Devonian and later movements on the Great Glen fault system, Scotland. *Journal of the Geological Society, London* **146**, 369–372.

Selwood, E.B. (1990) A review of basin development in central south-west England. *Proceedings of the Ussher Society* **7**, 199–205. [A rigorous yet accessible review of the complex Cornubian basins.]

Todd, S.P., Boyd, J.D. & Dodd, C.D. (1988) Old Red Sandstone sedimentation and basin development in the Dingle Peninsula, southwest Ireland. In: *Devonian of the World* (eds N. J. McMillan, A. F. Embry & D. J. Glass), Memoir of the Canadian Society of Petroleum Geologists 14 (1), pp. 251–268. [A study that unravels the complicated stratigraphy of this important terrane next to the Iapetus Suture.]

Part 5
The Variscan Cycle: Consolidation of Pangaea

Illustration overleaf: Recumbent chevron folds within the Carboniferous Crackington Formation at Millook Haven, Cornwall. Folding occurred during the Variscan Orogeny in the late Carboniferous.

14 Carboniferous sedimentation and volcanism on the Laurussian margin

P. D. GUION, P. GUTTERIDGE AND S. J. DAVIES

14.1 Introduction

Major palaeogeographic changes accompanied the Variscan Orogeny, which resulted from the collision of the African portion of Gondwana with European Laurussia during the Carboniferous (see Chapter 15). The collision created a Himalayan-scale mountain belt extending from Russia, through western Europe and into North America, welding Gondwana and Laurussia into a supercontinent. Britain and Ireland lay north of the collision zone for much of the Carboniferous, but the advancing orogen fundamentally controlled the tectonic evolution of the region.

The wealth of economic resources in Carboniferous rocks, particularly coal, promoted the dramatic economic expansion during the 18th and 19th centuries, and much of our understanding of the Carboniferous is founded on studies to exploit these resources. The basic, regional stratigraphy was established by the early 20th century, although it is still being refined. In the late 20th century, the discovery of oil and gas in Carboniferous sandstones, particularly in the Southern North Sea, rekindled interest in rocks of this period.

The Carboniferous of the British Isles is divided into two major successions: the Lower Carboniferous or Dinantian, and the Upper Carboniferous or Silesian (Fig. 14.1). The Dinantian series are the Tournaisian and Viséan, and the Namurian, Westphalian and Stephanian series comprise the Silesian. There are proposals to redefine the Lower and Upper Carboniferous, based on a boundary within the Chokierian stage of the Namurian (Fig. 14.1), but the existing nomenclature is retained here.

14.2 Tectonic setting

A northward-dipping subduction zone had been established through central France and southern Germany by the Dinantian. To the north of this subduction zone, a back-arc seaway floored by oceanic crust extended across south-west England into northern France and Germany (Rheno-Hercynian Zone; see also Fig. 15.1, p. 272). Important, long-lived palaeogeographical barriers, the Wales–London–Brabant High and the Southern Uplands High, created three separate elongate areas of Carboniferous sedimentation in Britain and Ireland. Interconnections existed between the main areas at various times, particularly during the later Carboniferous. Each area had a different tectonic setting. The British Isles is subdivided into five separate 'provinces', separated by important tectonic or palaeogeographical features (Fig. 14.2).

1 The Pennine province of central and northern England and its lateral continuation into the Southern North Sea, bounded to the north by the Southern Uplands High and to the south by the Wales–London–Brabant High.
2 The Scottish province situated north of the Southern Uplands High.
3 The Irish province in the west.
4 The Southern province, south of the Wales–London–Brabant High, including South Wales.
5 The 'Culm Basin' of Devon and Cornwall, separated from the rest of the British Isles by the Bristol Channel Fault Zone.

The Pennine province includes the most studied and most complete sequences, and hence is described first. The other provinces are described in the order shown in each of the stratigraphical sections that

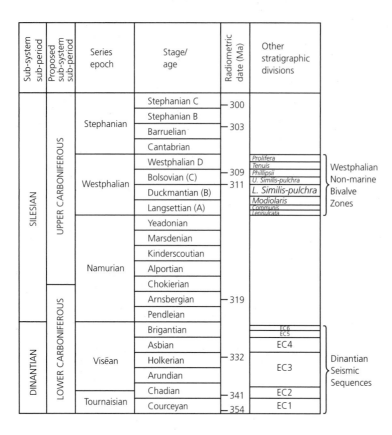

Sub-system sub-period	Proposed sub-system sub-period	Series epoch	Stage/ age	Radiometric date (Ma)	Other stratigraphic divisions		
SILESIAN	UPPER CARBONIFEROUS	Stephanian	Stephanian C	300			
			Stephanian B	303			
			Barruelian				
			Cantabrian				
		Westphalian	Westphalian D	309	*Prolifera* / *Tenuis* / *Phillipsii* / *U. Similis-pulchra*	Westphalian Non-marine Bivalve Zones	
			Bolsovian (C)	311	*L. Similis-pulchra*		
			Duckmantian (B)		*Modiolaris*		
			Langsettian (A)		*Communis* / *Lenisulcata*		
		Namurian	Yeadonian				
			Marsdenian				
			Kinderscoutian				
			Alportian				
			Chokierian				
	LOWER CARBONIFEROUS		Arnsbergian	319			
			Pendleian				
DINANTIAN		Viséan	Brigantian		EC6 / EC5	Dinantian Seismic Sequences	
			Asbian		EC4		
			Holkerian	332	EC3		
			Arundian				
		Tournaisian	Chadian	341	EC2		
			Courceyan	354	EC1		

Fig. 14.1 Major Carboniferous stratigraphical divisions (synthesized from a variety of sources, including Ramsbottom *et al.* 1978; Leeder 1988a; Riley *et al.* 1993; Besly 1998; radiometric data from Hess & Lippolt 1986 (Westphalian and Namurian) and Claoué-Long *et al.* 1995 (Dinantian; Seismic Sequences from Fraser *et al.*, 1990).

follow (Dinantian, Namurian, Westphalian). The Wales–London–Brabant High extended from southeast Ireland across central England into Belgium with the Midlands Microcraton (Fig. 14.2) forming its core. The Midlands Microcraton consisted of a small, rigid, triangular-shaped terrane of Neoproterozoic rocks overlain by relatively undeformed earlier Palaeozoic platform deposits (see Chapters 11, 13), with a northernmost apex situated in the west Midlands. The Southern Uplands High, which extended west into Ireland as the Longford–Down Massif, and east into the North Sea as the Mid-North Sea High (Fig. 14.2), originated as a Lower Palaeozoic accretionary prism bordering the Iapetus Suture (see Chapter 7).

In Britain and Ireland, back-arc extension north of the Variscan subduction zone during the Late Devonian to Early Carboniferous produced a series of fault-bounded half-grabens in the Pennine province (Fig. 14.2), which had a major control on Dinantian subsidence, bathymetry and sedimentary facies distribu-

tion. The position and orientation of these grabens and their bounding faults was strongly controlled by inherited Caledonian features (Fig. 14.2). To the west of the apex of the Midlands Microcraton, the Welsh Caledonides and later Variscan structures are dominated by north-east–south-west ('Caledonoid') trends, whereas to the east, north-west–south-east ('Charnoid') trends of the eastern Caledonides are important. In the central part of the microcraton, a north–south ('Malvernoid') trend possibly originated as a Neoproterozoic suture zone. Several of the blocks between the fault-bounded sub-basins, such as the Askrigg and Alston blocks, are cored by late Caledonian granites, whose buoyancy caused these areas to be relatively elevated and covered only by thin Carboniferous platform sequences.

Active rifting north of the Wales–London–Brabant High became less important by early Silesian times, and was gradually replaced by thermal subsidence. Recent radiometric dates suggest that the Namurian was far shorter than earlier estimates, implying that

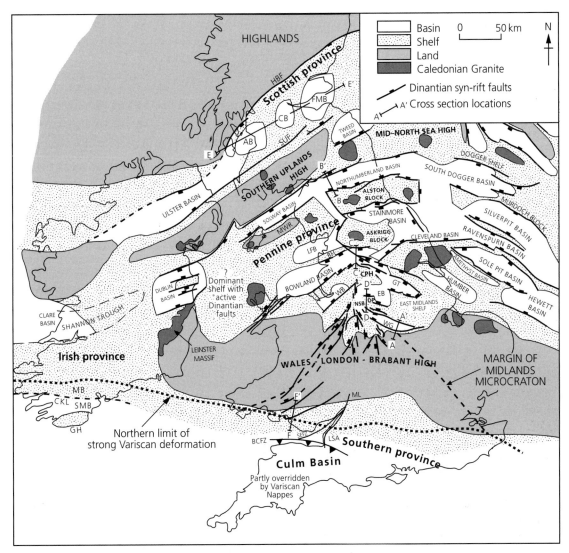

Pennine province

BH	Bowland High
CPH	Central Pennine High
DP	Derbyshire Platform
EB	Edale Basin
GT	Gainsborough Trough
LFB	Lancaster Fells Basin
MWR	Manx Whitehaven Ridge
NSB	North Staffordshire Basin
WG	Widmerpool Gulf
WB	Widnes Basin

Irish province

CKL	Cork Kenmare Line
GH	Glandore High
MB	Munster Basin
SMB	South Munster Basin

Southern province

BCFZ	Bristol Channel Fault Zone
LSA	Lower Severn Axis
ML	Malvern Line
SEFZ	Severn Estuary Fault Zone

Scottish province

AB	Ayrshire Basin	HBF	Highland Boundary Fault
CB	Central Basin	SUF	Southern Uplands Fault
FMB	Fife Midlothian Basin		

Fig. 14.2 Dinantian palaeogeography showing main 'provinces' and fault-bounded extensional basins and platforms that influenced sedimentation through much of the Carboniferous. Locations are shown of the seismic line A–A' (Fig. 14.4); sections B–B', the Northumberland Basin (Fig. 14.5); C–C', the Bowland Basin (Fig. 14.6); D–D', Derbyshire carbonate platform (Fig. 14.7); E–E', volcanics of the Scottish province (Fig. 14.8) and F–F', South Wales (Fig. 14.9) (based mainly on the synthesis of Corfield *et al.* 1996, with additions).

high rift-related subsidence rates may have continued into the Namurian from the Dinantian. The continuing movement of basin-bounding faults, combined with relict Dinantian bathymetry and the continuing compaction of mud-dominated sequences within the basins, meant that Namurian sedimentation patterns follow the Dinantian template. The influence of the blocks and basins was more subtle by the Westphalian, and the much more uniform distribution of sedimentary facies in Central and Northern England and the Southern North Sea shows that active rifting was much less important.

Tectonic control on Carboniferous sedimentation and magmatism was also important in the Scottish Midland Valley. Although there is a consensus that the overall setting was extensional during the early Carboniferous, a number of differing proposals have been put forward for the dominant stress orientations, including east–west extension, north–south extension, NNW–ESE extension, sinistral transtension and dextral transtension. By the Late Carboniferous, the overall regime was probably one of dextral transpression, with inversion of some earlier structures.

The Culm and South Wales basins, which formed important depositional areas south of the Wales–London–Brabant High, were strongly influenced by the northerly propagating Variscan Orogen, and eventually developed into flexural foreland basins. The South Wales and Culm basins are now separated by the Bristol Channel Fault Zone (Fig. 14.2; see also Fig. 15.3, p. 276). Some workers have argued for dextral strike-slip along this zone during and after the Silesian, and that the present positions of the South Wales and Culm basins are the result of lateral tectonic juxtaposition. However, the zone forms a southward-dipping seismic reflector, which has been interpreted as a Variscan thrust.

The Variscan deformation front progressively migrated northwards through Cornubia (Devon and Cornwall) from the Late Devonian onwards, such that thrust sheets reached central south-west England by the late Namurian. Although Namurian sediments occur within the nappe complex in the south, the majority of the Silesian rocks are preserved north of the complex in the 'Culm Trough' and in north Devon. Some workers have claimed that these occur in two distinct sequences with different thermal histories: the Crackington Formation passes up into the Bude Formation in the more southerly 'Culm Trough', whereas in north Devon, the Westward Ho! Formation is overlain by the Bideford Group.

14.3 Stratigraphy and correlation schemes

The range of Carboniferous depositional environments, from continental to deep marine, has caused considerable problems in correlation throughout Britain and Ireland because many index fossils are facies controlled. The challenges of hydrocarbon and coal exploration require additional correlation techniques. Successful correlation requires a multi-tool, interdisciplinary approach combining biostratigraphy (fossils), radiometric dating, volcanic horizons, facies analysis, well-log interpretation, seismic stratigraphy and geochemistry (C–S ratios, Sr isotopic ratios of marine bands, chemostratigraphy, etc.).

The Lower Carboniferous was originally subdivided using corals and brachiopods for platform carbonates, and goniatites and bivalves for basinal successions. However, these fossils are strongly facies controlled and type sections often contain gaps. In the 1970s, Ramsbottom used the cyclic nature of Dinantian and Namurian sedimentation as a basis of correlation. He recognized six major cycles (mesothems) of transgression and regression in the Dinantian and 11 in the Namurian, each containing smaller-scale cycles (cyclothems). Although Ramsbottom's work was more a model of Carboniferous sedimentation than a correlation scheme, it stimulated a great deal of sedimentological and biostratigraphical research. Current Dinantian stages, based on foraminifera, conodonts and macrofauna linked to lithological changes, derive directly from Ramsbottom's work (Fig. 14.1).

The majority of Silesian (Upper Carboniferous) rocks in the British Isles are either Namurian or Westphalian in age. Stephanian rocks are poorly represented onshore, but occur in the Southern North Sea. The Silesian divisions are being reviewed and it is not certain whether they will ultimately be applied on a world-wide basis (Fig. 14.1). The Carboniferous radiometric time-scale is poorly constrained, as the abundant volcanic rocks are rarely suitable for dating because of severe alteration.

In the Silesian, highly fossiliferous 'marine bands' represent periods of condensed sedimentation and marine flooding. These contain goniatites, which were nektonic cephalopods that evolved rapidly and were widely distributed, providing an excellent basis for correlation. Microfossils, including conodonts, ostracods, foraminifera and palynomorphs (products of terrestrial vegetation) are also used to date and correlate successions.

Some 60 goniatite-bearing marine bands occur within the Namurian, but there are far fewer within the Westphalian when marine transgressions were much rarer. Marine bands are generally more radioactive than adjacent beds, some because of uranium or thorium concentrations, and others because of potassium-bearings clays, and are therefore commonly recognizable on gamma-ray well logs. In marginal areas of the basins, marine bands may be represented by brackish water fossils, such as *Lingula*, or trace fossils.

Other schemes are used for Westphalian correlation where marine bands are uncommon. Non-marine bivalve assemblages and plants (macroflora) require large samples and may be facies controlled, but are important when other schemes are inadequate or non-diagnostic. Palynology, the study of fossil spores, is invaluable in Upper Carboniferous correlation, particularly in subsurface exploration where macrofossil recovery is rare.

Thin, altered volcaniclastic horizons (bentonites in the Namurian and tonsteins in the Westphalian) sometimes provide correlation markers and some contain zircons that can be used for radiometric dating. Tonsteins usually occur within coal seams and may be useful for seam identification and correlation. In coal exploration, seam properties can be used for correlation despite the lateral variability of individual seams. The petrographic and chemical character of seams, for instance any distinctive and laterally continuous dull (durain-rich) bands, clastic partings or cannel horizons (sapropelic lacustrine coals) and sulphur distribution can 'fingerprint' individual seams.

The uppermost Silesian of Britain presents the most difficult correlation challenge. These 'barren' red-beds contain sparse, environmentally controlled fauna and flora. Palynomorphs are rare and have variable lateral and vertical distributions. Traditional tools such as biostratigraphy, gross lithostratigraphy and palaeocurrent analysis are complemented by newer techniques such as chemostratigraphy and magnetostratigraphy to resolve correlation problems.

14.4 Cycles and controls on sedimentation

Carboniferous siliciclastic and carbonate successions show pronounced cyclicity. The origin and controls of this cyclicity have been much debated since its initial description in the 1930s. Carboniferous sedimentation was undoubtedly influenced by the interplay of autocyclic (intrinsic) processes such as delta switching and river avulsion, and allocyclic (external) controls including tectonics, climate, eustasy and sediment supply. Marine bands have been interpreted since the 1960s as records of sea-level rise, but the record of Carboniferous sea-level falls is disputed. In the 1970s, Ramsbottom proposed that major breaks (hiatuses) in sedimentation were associated with sea-level lowstands. He termed successions between the hiatuses 'mesothems', each comprising smaller cycles (cyclothems).

Glacial sediments in South America, South Africa and Australia indicate that the Carboniferous was a period of glacio-eustasy. A major ice sheet, initiated over the southern-hemisphere Gondwana landmass during Devonian to early Carboniferous time, grew progressively until it reached an estimated area of 21 $\times 10^6$ km^2 during the Carboniferous and Permian—similar to that of the Pleistocene ice sheet (23.5 $\times$ 10^6 km^2). Many Late Dinantian shelf carbonate successions comprise repeated shallowing-upward cycles deposited under the influence of eustatic sea-level variations estimated to be of the order of tens of metres. This high-frequency, low-amplitude eustasy reflects the repeated growth and melting of small continental ice caps during the early stages of accretion of the major southern-hemisphere ice sheet. Climate modelling suggests that the eustatic effects of glaciation were most important during the Namurian and early Westphalian. These Mid- to Late Carboniferous sea-level fluctuations have been estimated at 60 $\pm$ 15 m, approximately half the 120 m sea-level rise that has taken place since the Pleistocene. It is believed that ice sheets built up slowly, but melted rapidly over hundreds to thousands of years, producing an asymmetrical sea-level curve. These 'icehouse' conditions persisted into the Permo–Triassic.

Cyclothems within the Namurian have generally been interpreted to represent progradation of depositional systems (e.g. deltas or shallow marine environments). The transition from pro-delta deposits through delta front and into fluvial distributaries implies that all the subenvironments are linked and related. However, erosion surfaces at the base of some multistorey fluvial systems can be demonstrated to cut downwards into pro-delta deposits and may also remove underlying marine bands. In these examples, the fluvial systems may not be an integral component of the underlying depositional system. Periods of falling sea level may thus have given rise to regional fluvial erosion, creating valleys incised into the pre-existing marine shelves as the shoreline retreated beyond the shelf edge.

Such models have implications for the distribution of sandstones (potential hydrocarbon reservoirs) and mudrocks. Turbidite sandstones may have been deposited predominantly during periods of sea-level fall, when sediment bypassed the shelf and was transported into the basins through the incised valleys. Fluvial reservoirs would form as sea level began to rise and as the valleys filled with sediment, and marine mudrocks (source rocks) would be restricted to periods of rising sea level.

Ramsbottom's ideas prompted attempts to construct a global Carboniferous correlation framework based on sea-level changes. These early attempts were not initially widely accepted until work by Exxon geologists in the 1980s introduced the concept of unconformity-bound sequences that led to the application of sequence stratigraphy to Carboniferous sedimentation. The influence of glacio-eustatic fluctuations in sea level on facies architecture in Carboniferous successions is now largely accepted.

14.5 Palaeoclimate

The Carboniferous was a period of progressive climate change influenced by the collision of Gondwana and Laurussia, creating mountain belts that caused major changes in atmospheric circulation. The development of the southern-hemisphere ice sheet and the gradual northward drift of continents across the equatorial belt also affected the climate.

Throughout the Dinantian the climate was seasonal, possibly monsoonal. Clay minerals in pre-Arundian palaeosols suggest that the climate was dry with seasonal wetting, whereas after the Arundian the climate became humid with seasonal drying. Abundant meteoric cements in Late Dinantian shelf carbonates record humid conditions during emergence, whereas leached evaporites in Brigantian peri-tidal carbonates in Derbyshire imply a semiarid climate.

The Namurian and Westphalian climate was tropical ever-wet to seasonally wet, punctuated by short-term fluctuations that were inextricably linked with the southern-hemisphere glaciation. The effect of high-latitude glaciation on the tropical climate is controversial, but seasonally drier climates are most likely during periods of intense glaciation and lowered sea level. Melting of ice and consequent rising sea level promoted wetter equatorial climates with peat formation, giving rise to the abundant coal deposits in the British Isles. The prevailing conditions in the source regions for the major fluvial systems extending across Britain and Ireland were of fundamental importance for the erosion and transport of sediment to the Carboniferous basins, but little is known of these hinterland climates.

Arid conditions in the late Westphalian and Stephanian are indicated by the calcretes and red-beds of the English Midlands and the Southern North Sea. The increasing aridity at the end of the Carboniferous was caused by the development of a rain shadow situated north-west of the growing Variscan mountains, which gave shelter from the dominant trade winds blowing from the east.

14.6 Dinantian syn-rift carbonates, clastics and volcanics

The Dinantian was characterized by carbonates that were produced biogenically on carbonate platforms. Carbonate ramps (a carbonate platform passing from emergent to basinal without a marked break of slope) were a common, if not predominant, platform type throughout the early Carboniferous. Ramp development is favoured by constant carbonate production with depth, a common feature of carbonate systems following mass extinctions. The Early Carboniferous marine ecosystem was recovering from a series of extinctions related to global anoxic episodes during the Late Devonian. This stimulated the diversification of microbial carbonate mud mounds that filled niches previously occupied by reef communities. Dinantian frame-built reefs are rare.

14.6.1 Pennine province

Dinantian basin evolution between the Wales–

London–Brabant Landmass and the Southern Uplands was controlled by the reactivation of late Caledonian (Acadian) basement structures in response to episodic north–south extension (Fig. 14.2). East–west trending structures underwent extensional reactivation and often show major changes in sediment thickness and facies across them. Rifting began in the Late Devonian–earliest Dinantian and continued into the Namurian. Three episodes of active rifting (Late Devonian–early Dinantian, Chadian–Arundian and Asbian–Brigantian) have been recognized throughout northern England and Ireland (Fig. 14.3), followed by inversion resulting in local uplift during the late Brigantian.

Rift events alternating with regional subsidence produced contrasting sedimentary sequences that can be recognized on seismic sections (Fig. 14.4). The approximate ages of these seismic sequences are shown by Figs 14.1 and 4.3. During rift events, tilting of intrabasin slopes produced down-slope thickening and an increase of siliciclastic sediment into the basins. Carbonate production on structural highs was reduced as a consequence of subaerial exposure and some structural highs were eroded. Seismic sequences deposited during rift episodes typically thin over structural highs and thicken into basinal sections.

Between rift events, sedimentation was influenced by glacio-eustasy and thermal subsidence. Carbonate platform and fluvio-deltaic systems show a well-developed cyclicity reflecting a combination of glacio-eustatic and autocyclic processes. This cyclicity, however, cannot be resolved on seismic sections. The reduced siliciclastic input allowed carbonate platforms to become established and to grow. The increased carbonate production over highs and reduced basinal sedimentation produced seismic sequences that are typically thick at the basin margins and thin towards the basin centre.

The evolution of two contrasting basins and a carbonate platform from northern England will be examined in more detail.

The Northumberland and Solway basins

These basins lie between the Southern Uplands and the Alston Block (Fig. 14.2). The evolution of the Northumberland Basin is shown in Fig. 14.5. Early syn-rift sedimentation was dominated by mature fluvial sediment derived from the Southern Uplands. Fluvial deposits are intercalated with basaltic lavas and widespread dolomitic limestone 'cementstones' of fluvio-lacustrine origin. During the Courceyan to Chadian, deltas periodically advanced from the north-east. Seismic reflection profiles across the southern basin margin indicate differential subsidence of the hangingwall of the Stublick–Ninety Fathom Fault system at this time.

During the Arundian to Holkerian a braided fluvial system, channelled by active intrabasinal faults, advanced from the north-east building south-westward into a marine environment. Siliciclastic sediment was also shed from the fault-controlled northern margin of the Solway Basin. Soft sediment deformation features show that these faults were active during sedimentation.

Asbian and Brigantian sedimentation was dominated by Yoredale cycles that represent episodic delta progradation from the north-east along the basin axis. Equivalent facies to the north-east are much thinner and largely composed of delta-top or fluvial facies with many thick coals. Yoredale cycles are recognized in the Asbian to early Namurian of northern England, the Midland Valley of Scotland and in Ireland. Individual cycles are between 5 and 80 m thick and comprise a thin marine limestone overlain by fluvio-deltaic or shoreline siliciclastic sediments. Limestones accumulated in areas starved of siliciclastic sediment by delta lobe switching. The origin of this cyclicity is complex and attributed to the interplay of tectonic, eustatic and autocyclic processes.

The Bowland Basin

The Bowland Basin, in contrast to the Northumberland Basin, was a marine basin in which distal siliciclastics and resedimented carbonates accumulated. The Bowland Basin is a complex half-graben with the northern margin controlled by the North and Middle Craven Faults (Figs 14.2, 14.6). The Pendle Monocline is the surface expression of a major basement fault (now inverted) that controlled the southern basin margin. The Bowland Basin was surrounded by carbonate platforms over the Askrigg Block, the Bowland High and the Central Pennine High; the latter two concealed beneath upper Carboniferous rocks.

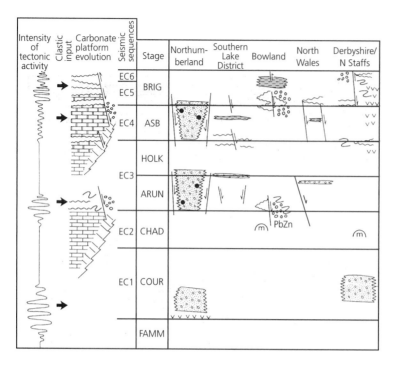

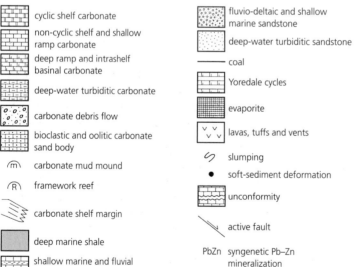

cyclic shelf carbonate

non-cyclic shelf and shallow ramp carbonate

deep ramp and intrashelf basinal carbonate

deep-water turbiditic carbonate

carbonate debris flow

bioclastic and oolitic carbonate sand body

⌒m⌒ carbonate mud mound

⌒R⌒ framework reef

carbonate shelf margin

deep marine shale

shallow marine and fluvial mudstone with calcrete

fluvio-deltaic and shallow marine sandstone

deep-water turbiditic sandstone

—— coal

Yoredale cycles

evaporite

v v / v lavas, tuffs and vents

⌐ slumping

● soft-sediment deformation

unconformity

active fault

PbZn syngenetic Pb–Zn mineralization

Fig. 14.3 Summary of tectono-sedimentary events in Late Devonian–Dinantian basin evolution in northern England and North Wales, and key to Figs 14.5, 14.6 and 14.7 (from Gawthorpe *et al.* 1989).

The earliest sediments are Courceyan and Chadian limestones and carbonate mud mounds deposited on a deep carbonate ramp, with local facies and thickness variations indicating active intrabasinal structures. Extension during the late Chadian–Arundian resulted in the reactivation of intrabasinal faults, producing local unconformities over highs. Sedimenta-tion became dominated by terrigenous mud with local influxes of siliciclastic turbidites and sedimentary slides. Syn-depositional lead–zinc mineralization was also associated with this phase of extension. During the Late Arundian to Early Asbian, carbonate platforms grew on surrounding highs, whereas deep marine shales and carbonate and siliciclastic tur-

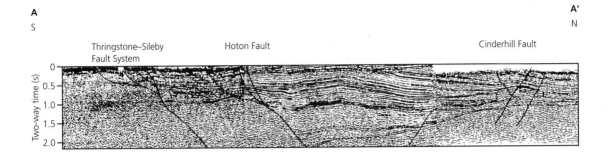

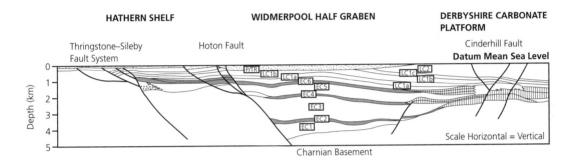

Fig. 14.4 Composite seismic section (A–A') across the Widmerpool Gulf showing the underlying half-graben basement structure and the development of seismic sequences in the Dinantian. Location is given in Fig. 14.2 (from Fraser *et al.* 1990, with permission from BP Amoco Exploration).

bidites were deposited in the basin. Further extension during the Mid–Late Asbian to Brigantian resulted in the deposition of basin-wide units of limestone conglomerates derived from basin margin and intrabasinal sediments.

The Derbyshire Carbonate Platform

The Derbyshire Carbonate Platform formed around a basement high composed of Precambrian and lower Palaeozoic rocks divided into two fault blocks by the Bonsall Fault (Figs 14.2, 14.7). The northern margin of the carbonate platform is fault controlled, whereas the southern and western margins are located over the gently dipping basement flanking the Widmerpool Gulf and North Staffordshire Basin, respectively (Figs 14.2, 14.4).

During the Late Devonian to early Dinantian, terrestrial red-beds were deposited around the flanks of

the Derbyshire High, followed by carbonates and marginal marine evaporites during the Tournaisian marine transgression. Tournaisian to Holkerian shallow marine carbonates onlapped the basement high, which became buried by a continuous carbonate platform by the Arundian or Holkerian. Carbonate sedimentation continued until Mid- to Late Brigantian times, when carbonate production was progressively limited by the increasing supply of suspended sediment, terrestrial nutrients and freshening of the water as delta systems prograded from the north. After the cessation of carbonate deposition, the platform was onlapped by Late Brigantian–Early Namurian organic-rich marine mudstones.

The Derbyshire Carbonate Platform illustrates the contrasting tectonic control on platform margin facies types. The northern fault-controlled margin grew by vertical accretion; marginal slopes of 30–40° contain marine cements that reflect slow sedimentation on the steep slope. By contrast, the southern margin prograded some 1–2 km into the Widmerpool Gulf during the Asbian; marginal slopes of 5–10° are characterized by fine-grained carbonates that contain slump deposits formed due to faster sedimentation on the less steep slope. Extensive carbonate sand bodies

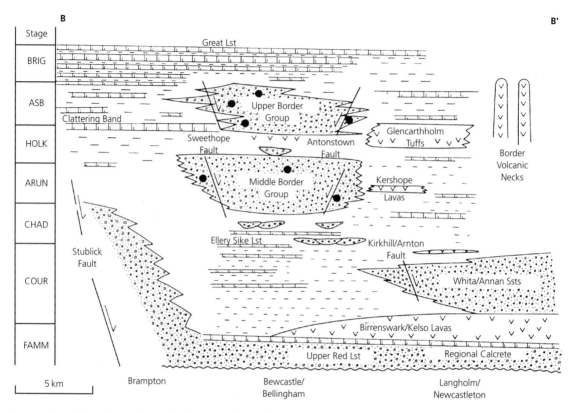

Fig. 14.5 Section B–B', summarizing the Dinantian evolution of the Northumberland Basin. Location of section is given by Fig. 14.2; key is given in Fig. 14.3 (from Gawthorpe *et al.* 1989).

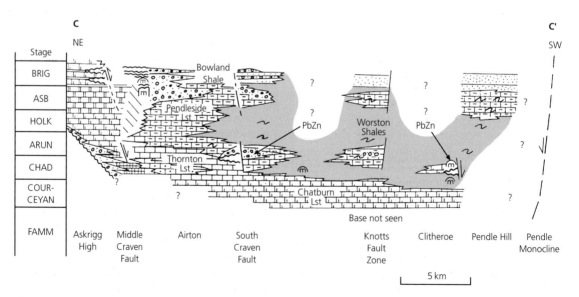

Fig. 14.6 Section C–C', summarizing the Dinantian evolution of the Bowland Basin. Location of section is given by Fig. 14.2; key is given in Fig. 14.3 (from Gawthorpe *et al.* 1989).

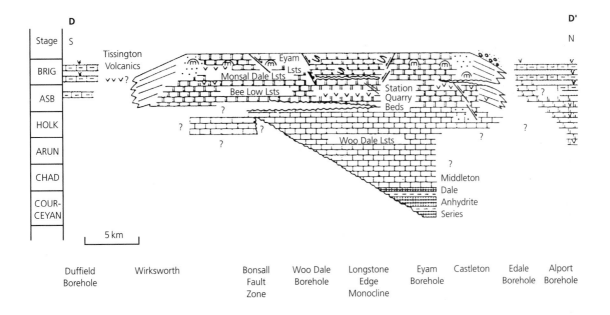

Fig. 14.7 Section D–D', summarizing the Dinantian evolution of the Derbyshire carbonate platform. Location of section is given by Fig. 14.2; key is given in Fig. 14.3 (from Gawthorpe *et al.* 1989).

deposited at the shelf edge are associated with both platform-margin types.

An episode of rifting at the Asbian–Brigantian boundary caused differential subsidence of the carbonate platform, forming an intrashelf basin. During tectonically quiescent periods, sedimentation was controlled by glacio-eustasy; however, the expression of this cyclicity differs in shelf interior, shelf margin and intrashelf basinal settings.

1 Shelf carbonates comprise repeated shallowing-upward cycles capped by emergent surfaces.

2 Carbonate sand bodies at the carbonate platform margins grew during transgressions and prograded over the shelf margin, shedding bioclastic carbonates into the surrounding basins during sea-level high-stands.

3 In the intrashelf basin, fine-grained carbonates and bioclastic turbidites were deposited during high-stands. During minor sea-level lowstands the margin of the intrashelf basin prograded with subsequent transgressions being marked by basin-wide coral beds. The basin was almost completely drained during major sea-level lowstands and thin dolomitized intertidal carbonates were deposited.

14.6.2 Scottish province

Sedimentation in the Midland Valley of Scotland took place in a series of linked sub-basins separated from basins in northern England by the Southern Uplands High. Deposition was influenced by volcanism, eustasy, extension and strike-slip movements along the Highland Boundary Fault and Southern Uplands Fault (Fig. 14.2). The earliest Dinantian successions are dominated by calcareous mudstones, sandstones and argillaceous dolomitic limestones ('cement-stones'). The cementstones contain calcified nodular evaporites, halite and gypsum pseudomorphs and desiccation features indicating deposition in hypersaline conditions.

Basic volcanism began in the Tournaisian and continued through the Carboniferous. Widespread outpouring of lavas and tuffs, represented by the Clyde Plateau Lavas, took place during the Viséan (Fig. 14.8). The resulting volcanic topography created separate depocentres in the east and west of the Midland Valley (Fig. 14.2). These restricted basins contained stratified water columns resulting in the preservation of organic matter and the accumulation of oil shales. Thin carbonate beds were deposited in

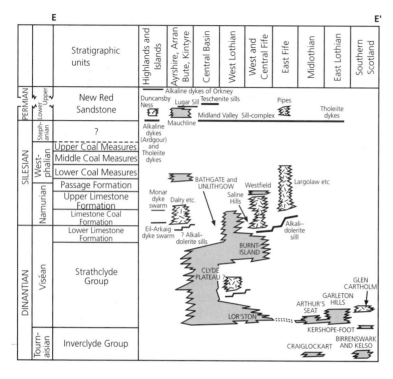

Fig. 14.8 Section E–E', showing the stratigraphical and geographical distribution of Carboniferous volcanics in Scotland. Location of section is given by Fig. 14.2 (from Francis 1991b).

marginal marine and lacustrine settings in widely varying conditions of salinity and oxygenation. These contain a unique biota, including vertebrates, spirorbids, ostracods, shrimps, coprolites, stromatolites and a conodont animal. These basins received occasional input of fluvio-deltaic sediments deposited by south-west-flowing rivers.

During the Late Dinantian, the Lower Limestone Formation was deposited in three basins: the Fife–Midlothian Basin, the Central Basin and the Ayrshire Basin, separated by topographic highs formed by the underlying Clyde Plateau Lavas (Fig. 14.2). Sedimentation was dominated by episodic progradation and abandonment of delta lobes, which were supplied from the north-east and west. Facies include marine shales, pro-delta siltstones and distributary and fluvial sandstones. Marine carbonates were deposited in interdistributary bays, and carbonate mud mounds formed over depositional highs from which siliciclastic sediment was excluded.

14.6.3 Irish province

A major transgression during the Tournaisian flooded the Old Red Sandstone Continent that included much of Ireland (see Chapter 13). This transgression spread northwards, reaching northern Ireland by the Chadian. Ireland was divided into two subprovinces: a shelf area over much of central and north Ireland; and the South Munster Basin, which formed an area of higher subsidence in southern Ireland (Fig. 14.2). The northern margin of the South Munster Basin was a persistent palaeogeographic feature called the Cork–Kenmare Line, which influenced sedimentation throughout the Dinantian and was probably controlled by basement structure.

Carbonate sedimentation was established over south and central Ireland during the Late Tournaisian with the deposition of Waulsortian limestones and associated deep marine shales and resedimented limestones. The term 'Waulsortian' refers to a specific type of carbonate mud mound first described from Waulsort in Belgium. Mud mounds of this type form by the *in situ* production of carbonate mud by an algal bacterial community, probably as a mat that bound the mound surface. A complex system of internal cavities contains multiple generations of marine cement and geopetal sediment. Waulsortian carbonate mud mounds cover much of central and southern

Ireland, where they occur in platform interior, margin and basinal settings.

Waulsortian limestones and their associated facies also form the host rocks of lead–zinc sulphide ore bodies such as those at Navan and Tynagh, which are associated with NE–SW to ENE–WSW-trending basement faults. These faults influenced Dinantian basin evolution and had a long history of movement. A syn-sedimentary model linking exhalative mineralization with high heat flow during rifting and the reactivation of basement faults has been proposed; however, recent work favours mineralization during early burial.

The onset of basin development in central Ireland took place in the Late Tournaisian to Early Chadian with the formation of the Shannon Trough and Dublin Basin. Although these basins are closely associated, they show contrasting sedimentary and subsidence histories. Deposition of Waulsortian limestones in the Shannon Trough was followed by major volcanism. After this, a westward-dipping carbonate ramp was established with an oolitic barrier that sheltered shallow-water lagoonal carbonates. Resedimented bioclastic limestones and carbonate mud mounds were deposited over the deeper part of the ramp, which persisted until the Asbian–Brigantian when carbonate deposition was terminated by a further phase of volcanism.

The Dublin Basin shows a similar sedimentary history during the Courceyan–Chadian with carbonate ramp sedimentation taking place over subdued intrabasinal topography. A rift event during the Late Chadian, coinciding with the first phase of volcanism in the Shannon Trough, produced fault-controlled topography in the Dublin Basin. Carbonate platforms nucleated over structural highs that were surrounded by deeper basins. This sedimentary style continued until the Brigantian, when carbonate production ceased following a basinwide influx of fine-grained siliciclastic sediment.

14.6.4 Southern province

In the South Wales–Bristol area, the Dinantian succession was deposited on the southern flanks of the Wales–London–Brabant High. The Old Red Sandstone Continent (see Chapter 13) was flooded by the Tournaisian transgression (Fig. 14.9) with the deposition of siliciclastic mudstones and thin bioclastic limestones in an offshore storm-dominated setting. The pulsed nature of the transgression is indicated by several progradational episodes of near-shore oolitic and bioclastic carbonate barriers. Restricted lagoons with local emergent and hypersaline conditions formed behind these barriers.

A southward-dipping carbonate ramp was established during the Courceyan to Late Arundian (Fig. 14.9). The north crop of South Wales represents the shallowest part of the ramp on which oolites were deposited in near-shore settings. The latter sheltered

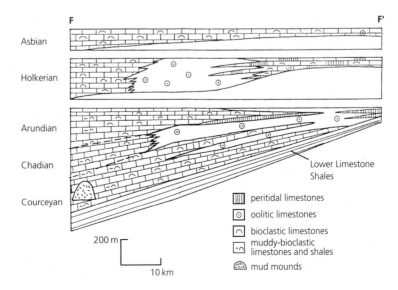

Fig. 14.9 Section F–F′, summarizing the Dinantian evolution of South Wales. Location of section is given by Fig. 14.2 (from Wright 1987).

areas of intertidal carbonate deposition with calcretes and minor fluvial sediments; numerous unconformities and non-sequences reflect the proximal ramp setting. Mid-ramp facies are represented by hummocky cross-stratified bioclastic and oolitic limestones deposited in a storm-dominated subtidal environment. The Cannington Park borehole in Somerset contains equivalent argillaceous bioclastic mudstones and a Waulsortian carbonate mud mound deposited in several hundreds of metres of water. Local tectonic control of thickness variations and facies distribution resulted from strike-slip reactivation of basement faults along the Severn Estuary Fault Zone and the Usk Axis (Fig. 14.2).

During the Holkerian, a major oolitic carbonate sand body prograded southwards over the mid-part of the ramp, on which storm-dominated subtidal bioclastic limestones were deposited. The carbonate ramp evolved into a carbonate shelf by the Asbian and Brigantian, with the deposition of carbonate cycles similar to shelf carbonates in northern England and North Wales. Shelf carbonates pass shoreward into oolitic shoreface and peritidal carbonates with minor terrigenous sands.

14.6.5 Culm Basin

In south-west England, Dinantian sediments are present in Variscan thrust sheets; transport directions of these thrust sheets indicate that the sediments were originally deposited in basins some distance to the south. These successions, continuous with those of Upper Devonian age (see Chapter 13), include black shales, carbonate turbidites, radiolarian cherts, pillow lavas and olistolithic facies deposited in deep marine slope, rise and pelagic settings. Shallow marine and deltaic siliciclastics are present in some nappes. In north Devon, Dinantian deposits include offshore shelf mudstones with minor sandstones and carbonates.

14.7 Namurian deltas and rivers

14.7.1 Pennine province

The diachronous transition from mixed carbonate–clastic to predominantly clastic deposition occurs close to the base of the Namurian in many areas of Britain and Ireland. Namurian deposits fall into three broad environmental types: fluvial, deltaic and deep-water clastics. The pronounced bathymetry in northern England and the Southern North Sea, which followed Dinantian rifting, gradually became more subdued during the Namurian (Fig. 14.10). Early Namurian water depths of up to a few hundred metres in many of the sub-basins were reduced to less than 100 m by the later Namurian as they were infilled by southerly prograding fluvio-deltaic systems. Extension was progressively replaced by post-rift thermal subsidence during the Namurian and Westphalian, although successions up to 2 km thick imply high subsidence rates and continuing tectonism.

Shallow marine sedimentation was restricted to the Askrigg and Alston blocks and intervening Stainmore and Northumberland basins of northern England (Fig. 14.2). Paralic, predominantly siliciclastic, sedimentation persisted from the end Brigantian into the early Namurian. The basal Namurian, Main (Great) limestone, representing the final prolonged period of carbonate deposition, was succeeded by Yoredale-type coastal plain and shallow marine deposition cycles. The shallow marine environments are represented initially by thin, regionally persistent limestones. Above each limestone a coarsening-upward siliciclastic succession records shoreline progradation into a shallow marine bay. Single-storey fluvial channels on low-lying coastal plains fed the shoreline systems. Larger-scale, eastward-flowing braided fluvial channel deposits up to 12–15 m thick and 0.85 km wide are also present. Wave action was limited within the bays of the coastal plain, but tidal drapes and lateral accretion surfaces occur as small-scale features within the fluvial systems. In the sub-basins further to the south (e.g. Derbyshire) condensed black mudstones accumulated in deeper-water conditions with reduced sediment supply (Figs 14.11, 14.12). A number of laterally persistent shallow marine limestones within the northern England successions are broadly correlative with goniatite-bearing marine bands in the southern basins.

In the Late Pendleian to Early Arnsbergian, coarse-grained southerly flowing rivers entered the north of the Pennine province (Fig. 14.11). These major rivers cut valleys across the 'high' areas (e.g. the Askrigg Block) abruptly terminating shallow marine deposition. Sand fed through these valleys was probably

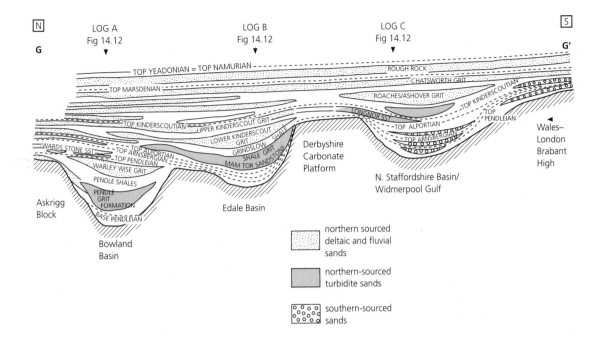

Fig. 14.10 Generalized distribution of major sedimentary units in Namurian sub-basins of the Pennine province. The topography created during the Dinantian was gradually infilled. Fluvio-deltaic sedimentation was predominant by Mardenian times. Location of section G–G′ is given in Fig. 14.11 (based on Collinson 1988; Brandon *et al.* 1995).

transported by turbidity currents into deep-water basins south of the Askrigg Block. Alternatively, rivers flowing eastwards across the Lake District block may have sourced turbidity flows that turned south-west into the basin. These early deep-water systems (Pendle Grit Formation) are represented by 250–450-m-thick successions comprising both thin turbidite sandstone beds and thick composite (amalgamated) sandstone units interbedded with mudstones (Fig. 14.12a). Sand-rich successions pass upward into siltstones and fine sandstones, with more sporadic coarse-grained units representing feeder channels. Abundant marine trace fossils, including *Rhizocorallium* and *Curvolithus* on the sandstone beds and *Sanguinolites* bivalve horizons within intervening mudstones are characteristic of deposition on the lower slope and basin floor.

The sediment-starved basins to the south (Edale, North Staffordshire sub-basins, the Widmerpool

Gulf and the Gainsborough Trough, Fig. 14.2) have some of the most complete thick-shelled goniatite-bearing marine bands (Fig. 14.12b). These southern areas remained starved of sediment throughout much of the Namurian, and coarse-grained turbidite deposition did not begin until the upper Kinderscoutian (Fig. 14.12b) with thinly bedded, distal turbidites (e.g. Mam Tor sandstones). These may be the lateral or down-dip equivalents of sand-rich, probably more proximal, deep-water systems (e.g. Shale Grit).

The bypass of sediment through the major river systems on the Askrigg Block into the basin may have occurred during periods of lowered sea level. During periods of higher sea level, the inherited bathymetry, including some deep sub-basins, promoted the development of unstable shelf-edge deltas fed by the river systems. The delta systems prograded from the north during the Namurian, moving the focus of turbidite deposition southwards. During the Kinderscoutian, the sediment supply began to exceed the accommodation space, and the basins were progressively infilled. As a result, sheet-like deltaic systems developed in the shallow northern Pennine Basin, whereas turbidite-fronted deltas characterized the underfilled southern Pennine sub-basins (Fig. 14.10).

Top Kinderscoutian and Marsdenian basin-floor

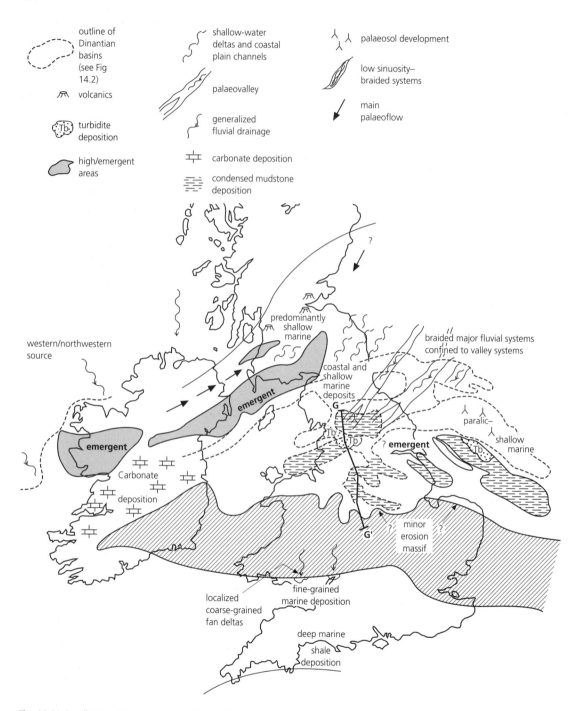

Fig. 14.11 Pendleian palaeogeography, and key to Figs 14.13 and 14.14 (based on data from a number of sources, including Collinson *et al.* 1991; Martinsen 1993; Brandon *et al.* 1995; Martinsen *et al.* 1995).

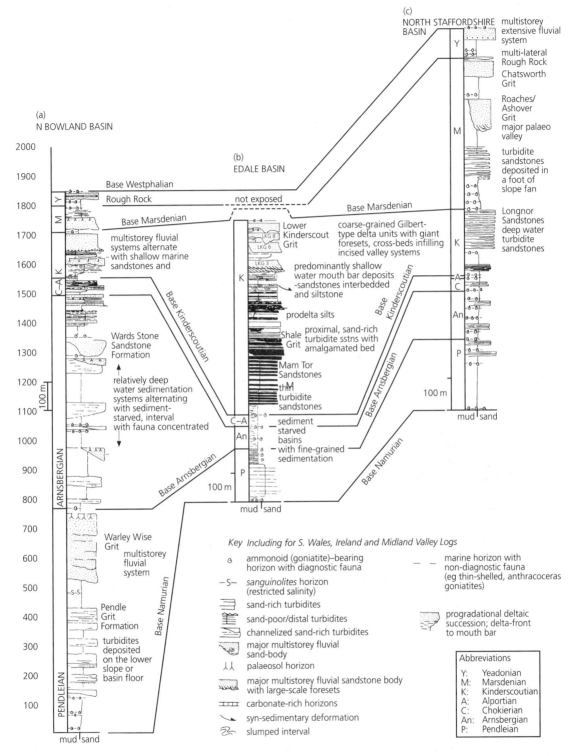

Fig. 14.12 Representative Namurian stratigraphical sections from the: (a) North Bowland Basin; (b) Edale Basin; (c) South Staffordshire Basin. Location of stratigraphical sections is given by Fig. 14.13 (based on unpublished data and Arthurton et al. 1988; Davies & McLean 1996; Hampson 1997; Jones & Chisholm 1997).

fan systems in the southern Pennine region (North Staffordshire sub-basin) represent the youngest deep-water clastics in central and northern England (Fig. 14.12c). These fans are located close to the basinward limit of delta progradation and probably formed a detached or 'foot-of-slope' system (e.g. the Ashover Grit equivalent). The major sand-rich turbidite packages are separated by fine-grained intervals. However, goniatite-bearing horizons are much rarer than in the basinal condensed successions and prodeltaic deposits. *Sanguinolites* bivalves are the predominant fauna in many of these fine-grained intervals (Fig. 14.12c). Correlation both within these turbidite packages and with equivalent up-depositional dip successions remains problematical.

Pendleian and Arnsbergian multistorey fluvial sandstone bodies are restricted to the northern platform areas and northern Pennine Basin (e.g. Warley Wise Grit). Major river systems, sourced predominantly from the north-east, were active throughout most of central and northern England and the Southern North Sea from the Kinderscoutian to the Yeadonian (Figs 14.13, 14.14).

Many Namurian multistorey fluvial sandstone complexes are a few tens of metres thick, comprising vertically and laterally stacked channel-fill units. Large, in-channel barforms are common but lateral accretion surfaces are rare, suggesting deposition in low-sinuosity channels, which were braided at low-flow stage. Individual channel fills may occasionally reach 15 m in thickness and contain 'giant cross-beds' interpreted as deep fluvial scour-and-fill structures. Much larger foresets (up to 35 m high) with Gilbert-style geometries occur at down-dip terminations of the fluvial complexes, and have been interpreted as coarse-grained deltas located at valley mouths (e.g. the Lower Kinderscout Grit and the Roaches Grit, Fig. 14.12). The multistorey complexes are sand dominated and rarely show a fining-upward trend. Evidence for a transition from fluvial to estuarine deposition is relatively uncommon, and tidally influenced sediments are not widely recognized in the majority of Namurian fluvial sandstone bodies in the basins of northern England and South Wales.

Some multistorey fluvial sandstone complexes have been interpreted as incised valleys created during periods of sea-level fall. These sandstone bodies are characterized by basal erosion surfaces, which extend laterally for up to 60 km, and may erosionally remove underlying marine horizons. The sandstone body thicknesses generally ranges between 20 and 40 m, although 80 m has been documented (e.g. the Marsdenian Roaches Grit, Fig. 14.12c). The concept of downcutting into pre-existing strata and then subsequent back-filling of a valley contrasts with the idea of gradual progradation and aggradation of the fluvial system. Sandstones in the Pennine province that have been interpreted as valley fills include the Marsdenian-aged Roaches and Chatsworth Grits (Fig. 14.12). The Yeadonian Rough Rock of the Central Pennine Basin is extremely widespread and appears to have a sheet-like, as opposed to valley-like, geometry (Fig. 14.12c); however, the coeval Rough Rock on the East Midlands platform has a valley geometry.

Progradation directions of the main delta systems were usually coincident with palaeoflow indicators in the overlying major fluvial systems; for example, the south-east-prograding delta represented by the Rough Rock Flags in the central Pennines. An exception is the Yeadonian elongate delta system represented by the Haslingden Flags. Palaeoflow and provenance indicators indicate progradation to the east in contrast to the overlying south-flowing fluvial system.

14.7.2 Scottish province

The east-north-east-trending Midland Valley and its lateral continuations extend for approximately 300 km along strike (Fig. 14.2). Each of its component sub-basins may contain up to 5 km of Carboniferous sediments and volcanics. Fluvio-deltaic systems flowing along the axis of the basin towards the south-west were predominant throughout the Namurian. From the latest Dinantian to earliest Namurian, the sub-basins in the Midland Valley reflect differences in rates of subsidence and differential fault block movements, although the main depocentres of the Limestone Coal, Upper Limestone and Passage Formations (Fig. 14.15) are broadly coincident with those that existed during the Dinantian.

The Pendleian Limestone Coal Formation, which occurs above the Top Hosie limestone, is characterized by cyclic alternations of fluvio-deltaic sandstones and mudstones (Fig. 14.15). This was one of the two main coal-forming periods in the Midland Valley. Non-marine bivalves and *Lingula* bands are common

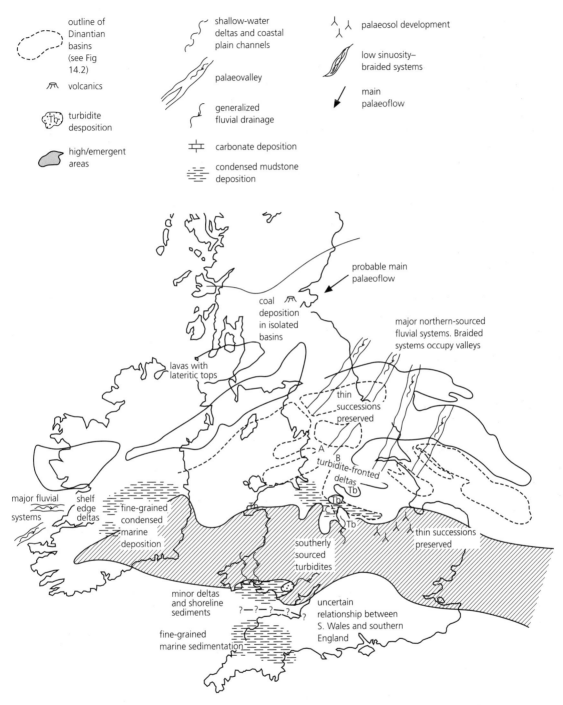

outline of Dinantian basins (see Fig 14.2)

volcanics

turbidite desposition

high/emergent areas

shallow-water deltas and coastal plain channels

palaeovalley

generalized fluvial drainage

carbonate deposition

condensed mudstone deposition

palaeosol development

low sinuosity–braided systems

main palaeoflow

probable main palaeoflow

coal deposition in isolated basins

lavas with lateritic tops

major northern-sourced fluvial systems. Braided systems occupy valleys

thin successions preserved

A

B

turbidite-fronted deltas

Tb

major fluvial systems

shelf edge deltas

fine-grained condensed marine deposition

Tb

C

Tb

southerly sourced turbidites

thin successions preserved

minor deltas and shoreline sediments

?—?—?—?—?

uncertain relationship between S. Wales and southern England

fine-grained marine sedimentation

Fig. 14.13 Kinderscoutian palaeogeography, showing locations of stratigraphical parts (a–c) of Fig. 14.12. (based on data from a number of sources, including: Bristow 1988; Leeder 1988b; Li 1990; Collinson *et al.* 1991; Aitkenhead & Riley 1996; Jones & Chisholm 1997).

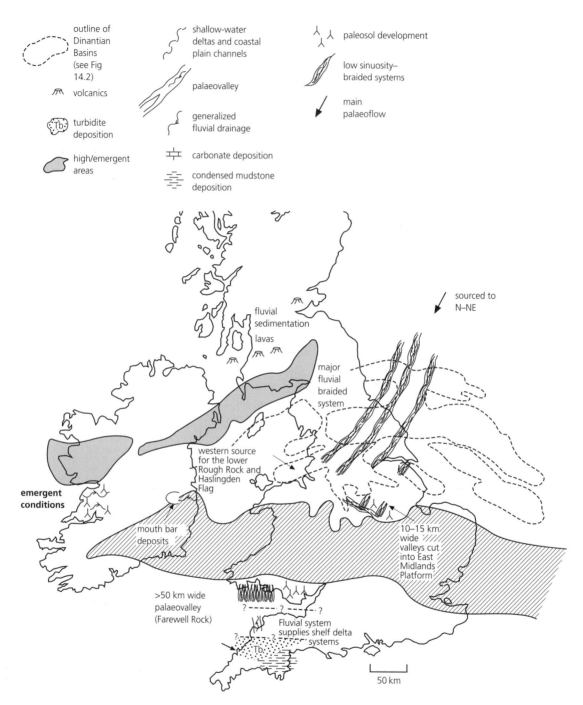

Legend:

- outline of Dinantian Basins (see Fig 14.2)
- volcanics
- Tb: turbidite deposition
- high/emergent areas
- shallow-water deltas and coastal plain channels
- palaeovalley
- generalized fluvial drainage
- carbonate deposition
- condensed mudstone deposition
- paleosol development
- low sinuosity–braided systems
- main palaeoflow

Map labels:

- sourced to N–NE
- fluvial sedimentation
- lavas
- major fluvial braided system
- western source for the lower Rough Rock and Haslingden Flag
- emergent conditions
- mouth bar deposits
- 10–15 km wide valleys cut into East Midlands Platform
- >50 km wide palaeovalley (Farewell Rock)
- Fluvial system supplies shelf delta systems
- Tb
- 50 km

Fig. 14.14 Upper Yeadonian palaeogeography. (Based on data from a number of sources, including: Collinson *et al.* 1991; Maynard 1992; Church & Gawthorpe 1994; Hampson *et al.* 1996.)

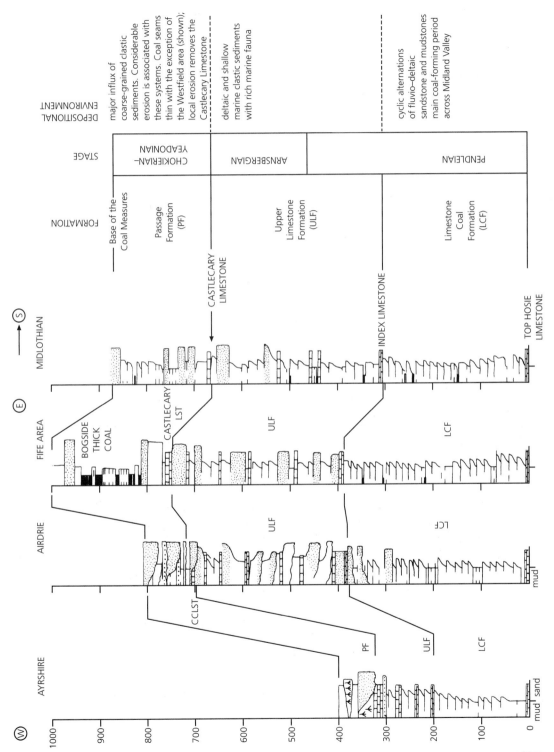

Fig. 14.15 Comparative Namurian stratigraphic sections in the Scottish Midland Valley (after Browne *et al.* 1985).

but few fully marine horizons are recognized. The Ayrshire and Glasgow regions in the west were the most marine influenced.

The Upper Limestone Formation of Arnsbergian age is dominated by deltaic and shallow marine clastic sediments, with six major marine flooding surfaces and additional thinner marine horizons. Rich marine faunas include brachiopods, bivalves and bryozoa. Both south-westerly and north-easterly flowing rivers fed prograding delta systems. The uppermost marine horizon is the Castlecary Limestone, although this bed has been locally removed by erosion by later fluvial systems (Fig. 14.15).

A sudden influx of predominantly coarse clastic sediments is represented by the Passage Formation (Arnsbergian to Langsettian, Fig. 14.15), which includes both braided and meandering fluvial deposits. Coals are generally thin except locally in the Westfield sub-basin in the east. Several marine bands and one or two thin limestones indicate short-lived marine transgressions, but diagnostic faunas that would allow correlation with successions in northern England are lacking. Kinderscoutian, Marsdenian and Yeadonian stages are recognized from miospore occurrence, but there is no faunal or floral evidence for the Chokierian or Alportian.

Namurian volcanism was concentrated in West Lothian and north Central Fife (Fig. 14.8). Most eruptions were localized phreato-magmatic ash rings and contrast with the thick accumulations of basaltic lava extruded during the Dinantian. Extrusive lavas and basaltic plugs occur locally throughout the Limestone Coal and Upper Limestone Formations, with interbedded extrusive rocks providing evidence for intermittent but continuing igneous activity. Alkaline dolerite sills intruded into the Limestone Coal and Upper Limestone Formations are only slightly younger than their host sediments. There are no dykes identified, and the magma must have reached the intrusion levels through volcanic pipes. Many of the sills in Fife show signs of interaction with the host sediments, which may be plastically deformed. Intrusion into moist peat was common and tuffisite horizons, with basalt clasts and coal fragments, are abundant. Numerous intrusions occur along SW–NE and WSW–ENE trending major tectonic lineaments (e.g. the Ardross Fault in Fife).

14.7.3 Irish province

In Western Ireland, the transition from limestone to black shale deposition occurred during the early Namurian (Fig. 14.16a) Continuous deposition in the central basin contrasts with condensed and incomplete black shale successions unconformably overlying carbonates on the marginal shelves. Turbidite sedimentation commenced in the central trough during the Alportian (Ross Sandstone) and continued through into the Kinderscoutian (Fig. 14.16b). These turbidites were sourced from the west and may correlate with early Namurian fluvio-deltaic deposits in the offshore Porcupine Basin. The turbidite succession is sand dominated, with both sheet-like and channelized turbidites vertically separated by goniatite-bearing shales, representing condensed flooding surfaces. An upward transition into a mudstone-dominated package with thin turbidites, containing much evidence for instability, such as major slumps and slides, indicates slope progradation into the basin (Gull Island Formation) (Fig. 14.16c). Further east in the Leinster area, the Namurian was dominated by condensed black shale deposition, until delta systems were established in the late Kinderscoutian to Marsdenian.

During the Kinderscoutian and Marsdenian, major unstable shelf-edge deltas and more minor shelf deltas prograded into the Clare Basin. Predominantly fine-grained, river-dominated delta systems were characterized by a high degree of instability indicated by growth faults, slumps and slides (Fig. 14.16d). The thinner shelf delta systems are characterized by more stable 9–10-m-thick progradational sequences but diapirs are present. Three major fluvial systems incise into the underlying shelf-edge deltas and may remove underlying marine bands. These incised river systems, up to 35 m thick, are equivalent in scale to those in central and northern England.

14.7.4 Southern province

During the early Namurian, the deep-water Rheno-Hercynian basin extended from Cornwall to Poland and the southern British part became isolated only during the last stages of continental collision. Subsequent tectonic shortening due to northward migration of the Variscan Orogen obscures the exact relationships between the predominantly shallow

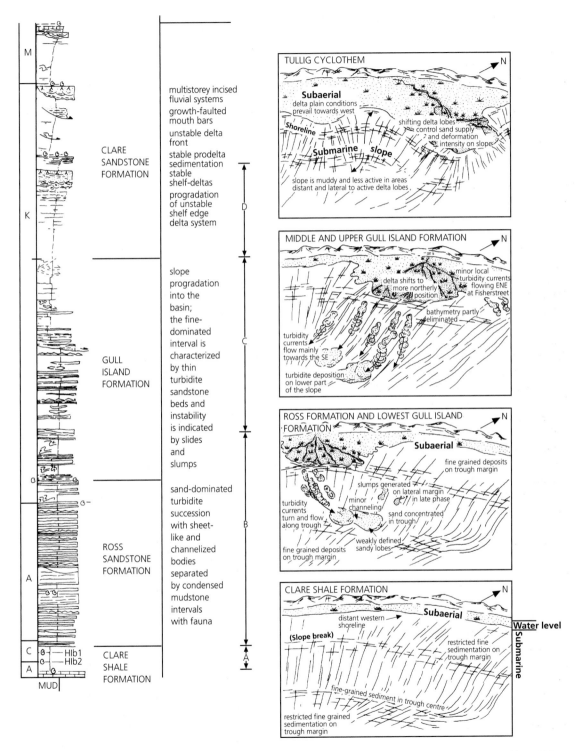

Fig. 14.16 Summary section for the Clare Basin in the west of Ireland, viewed from the east, with schematic illustrations of the inferred development on the basin-filling succession (based on several sources, including Gill 1979; Martinsen & Bakken 1990; palaeogeography from Collinson *et al.* 1991).

marine, deltaic and fluvial sediments of South Wales and those of Cornwall and Devon (see Section 14.2).

A pronounced unconformity exists across most of the South Wales basin between the Dinantian and Namurian, although Pendleian sediments directly overlie Dinantian carbonates in the Gower area. In parts of western Pembrokeshire, the Namurian rests directly on Ordovician formations. The majority of the Namurian consists of the Pendleian to Marsdenian 'Basal Grit Group' and the Late Marsdenian to top Yeadonian 'Middle Shale Group' (Fig. 14.17). The main Namurian depocentre was located in the southern part of the South Wales basin, where both groups reach their maximum thickness. The Basal Grit Group thins onto the northern margins of the basin and is not present in Pembrokeshire to the west or near the Bristol Channel in the east. Southward-flowing river systems supplied sediment to beach and near-shore systems on the basin margins. These shallow marine deposits comprise lenticular quartzitic sandstone bodies with abundant shelly material, separated by both marine goniatite-bearing mudstones and fine-grained coastal plain sediments, including carbonaceous siltstones and thin coals. Brachiopods, crinoids and bivalves characterize shallow water deposition in the northern and eastern basin. Shallow marine depositional systems pass southwards into quiet offshore marine environments with goniatite-bearing open marine mudstones. By the Early Marsdenian, major deltaic systems developed in the north-east whereas shallow marine sedimentation persisted in the west.

A major marine flooding event re-established the beach and shoreline systems along the margins of the basin. Fine-grained deposition, with a brackish fauna, continued basin-wide until the Early Yeadonian. By the Mid-Yeadonian, the southward progradation of large delta systems formed a number of coarsening-upward successions. Mouth bars deposits at the top of each of these sequences are succeeded by mudstone horizons containing *Anthracoceras*, a thin-shelled goniatite, indicating marine flooding across the delta. These deltaic sediments are cut by a multistorey, multilateral fluvial sand body, the 'Farewell Rock' (Fig. 14.17), which is the age equivalent of the Rough Rock of the Pennine province. This has been interpreted as an incised valley fill, characterized by

an upward decrease in grain size and cross-bed coset thickness, indicating a decrease in stream power accompanying a decrease in channel slope as sea level rose and the incised valley became infilled. An upper fine-grained fill is interpreted as a siltstone-dominated channel plug and bay-head delta. The valley is capped by a rooted horizon overlain by a marine flooding surface.

The Namurian of Devon and Cornwall is described later, along with the Westphalian fill of the Culm Basin (see Section 14.9).

14.8 Westphalian (Coal Measures) rivers, lakes and mires

14.8.1 Pennine province

The Pennine province includes the major coalfield areas of Yorkshire and the East Midlands; Northumberland and Durham; and Warwickshire, North and South Staffordshire in the West Midlands (Figs 14.18, 14.19). In addition, a number of small, isolated coalfields occur in north-west England and southern Scotland. These probably all originally formed in a continuous sedimentary basin, later separated by post-Westphalian tectonics. More than 2800 m of Westphalian strata are preserved in the basin depocentre that was situated in the North Staffordshire–Lancashire region. Within the Pennine coalfields, the majority of the productive seams are within the uppermost part of the Langsettian and the Duckmantian, but coals outside this range are occasionally worked.

The base of the Westphalian is marked by the widespread Subcrenatum Marine Band. The lowest Langsettian rocks above, up to about the top of the Lenisulcata Zone or the horizon of the Kilburn coal (Fig. 14.20), were formed on a marine-influenced delta plain. Repeated marine flooding events deposited thin, carbonaceous mudstones similar to the Namurian marine bands, with different faunal phases indicating a gradual increase in salinity to an acme, followed by a retreat. Although several of the marine bands are not recognizable throughout the British coalfields, some may be traced over wide areas of Britain and western Europe, and indicate important eustatic rises across an extensive area of subdued relief.

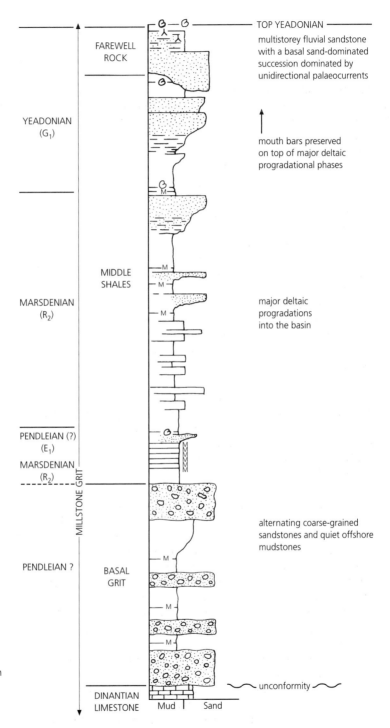

TOP YEADONIAN

multistorey fluvial sandstone
with a basal sand-dominated
succession dominated by
unidirectional palaeocurrents

mouth bars preserved
on top of major deltaic
progradational phases

major deltaic
progradations
into the basin

alternating coarse-grained
sandstones and quiet offshore
mudstones

unconformity

FAREWELL
ROCK

YEADONIAN
(G₁)

MIDDLE
SHALES

MARSDENIAN
(R₂)

PENDLEIAN (?)
(E₁)

MARSDENIAN
(R₂)

MILLSTONE GRIT

PENDLEIAN ?

BASAL
GRIT

DINANTIAN
LIMESTONE

Mud   Sand

Fig. 14.17 Namurian stratigraphy of South
Wales, no vertical scale implied (based on
Ramsbottom *et al.* 1978; George & Kelling
1982; Hampson 1998).

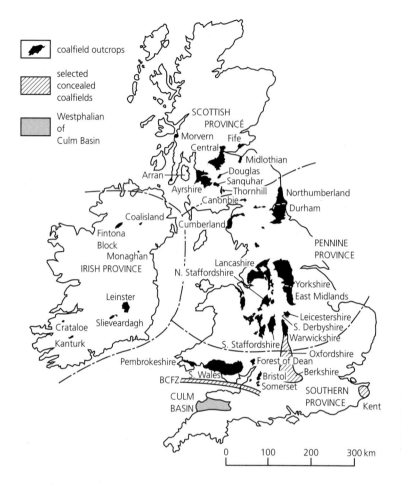

Fig. 14.18 British and Irish Westphalian coalfields and outcrop areas showing Carboniferous depositional provinces (note: not all concealed coalfields shown for sake of clarity) (modified from Calver 1969). BCFZ, Bristol Channel Fault Zone.

The marine transgressions punctuated episodes of progradation of shallow-water deltas fed by fluvial systems. Sandstones of the lower part of this interval were of northerly provenance, and are often coarse and feldspathic. The fluvial Crawshaw Sandstone, the lowest Westphalian sandstone, shows many similarities with the underlying Namurian Sandstones, and has been interpreted as an incised valley fill. Some coal seams formed in the lowermost Langsettian, but these are generally thin and of low quality, often being rich in sulfur from sulfate-rich seawater introduced onto the delta plain during marine flooding.

Marine transgressions ceased simultaneously across the whole of the Pennine Basin, such that from mid-Langsettian to upper Duckmantian marine influence was minimal, with only one major marine band, the Vanderbeckei Marine Band, which marks the Langsettian–Duckmantian (Westphalian A–B) boundary. The cessation of marine transgressions occurred at approximately the same time as a switch from northerly derived feldspathic sediments to finer-grained westerly derived sediments. The ensuing environment consisted of an extensive, low-lying water-logged plain with shallow lakes whose conditions were ideal for the repeated development of widespread peat mires, which ultimately formed economic coal deposits.

Clastic sediment and water were transported into the low-lying plain by a hierarchy of channels:

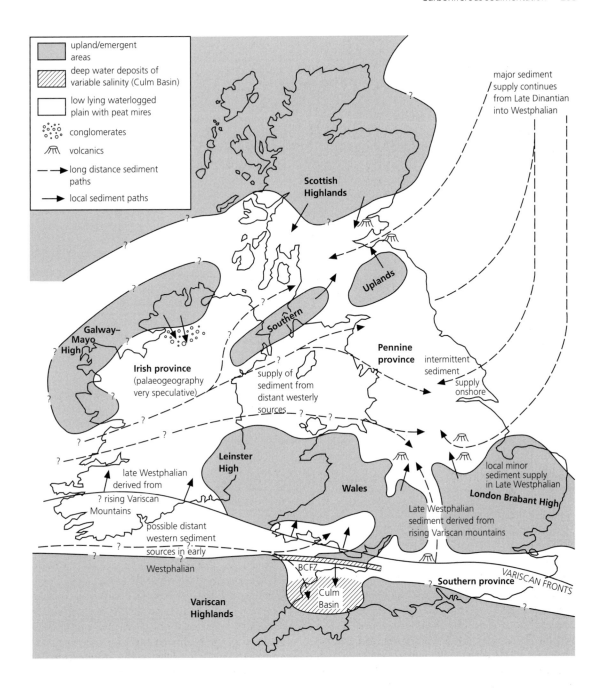

Fig. 14.19 Generalized early Westphalian palaeogeography in the BCFZ-Bristol Channel Fault Zone (compiled from a variety of sources, including: Cope *et al.* 1992; Rippon 1996).

major channels fed a distributive network of minor channels, ultimately depositing sediment in shallow lakes via crevasse splays, lacustrine deltas and overbank floods. Breaching of adjacent channel banks led to the introduction of crevasse splays or lacustrine deltas into many of the lakes, depositing upwards-

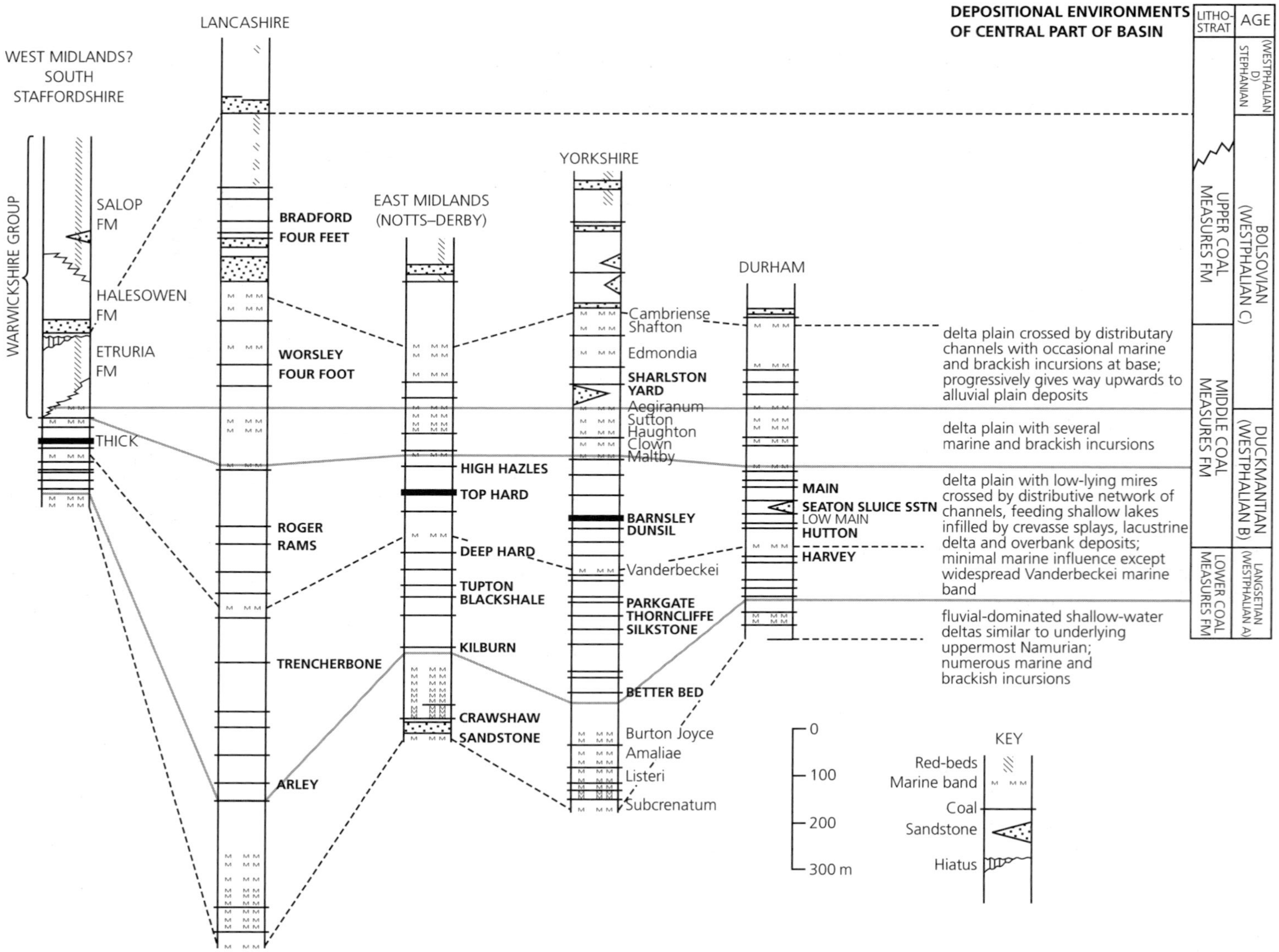

Fig. 14.20 Representative stratigraphical sections of the major coalfields of the Pennine province, showing main coal seams and marine bands, and selected sandstones referred to in the text. For location of the coalfields, see Fig. 14.19 (based on Ramsbottom et al. 1978; Powell et al. 2000).

coarsening sequences, typically 5–10 m thick (Fig. 14.21). The shallow lakes formed as the result of locally enhanced subsidence rates, induced in part by compaction of underlying peat and muds, superimposed on overall tectonic subsidence. Many of the lakes were initially starved of coarse clastic input, allowing organic-rich muds to accumulate in anoxic or suboxic conditions. Non-marine bivalves such as *Carbonicola*, *Anthracosia* or *Naiadites*, ostracods such as *Geisina* and *Carbonita*, and fish usually colonized the lakes. Early diagenetic siderite (iron carbonate) often formed in the reducing conditions on the floors of the lakes. The ensuing 'ironstone' beds are thin with low iron concentrations, but have been formerly worked commercially. The filling of lakes and abandonment of channel systems allowed the formation of palaeosols (fossil soils) and peat mires, leading to coal formation.

The coals initially formed in waterlogged reducing conditions, which favoured preservation of organic matter. The coal-forming floras were dominated by lycopsids such as *Lepidodendron* and *Sigillaria*, which form the bright coal (vitrain and clarain) within most seams. Other plants included arthrophytes such as *Calamites*, and pteridosperms such as *Neuropteris* and *Alethopteris*. Characteristic vertical sequences of spores and coal types indicate that, in some seams, peat built up above the surroundings, forming raised bogs. In these conditions, the peat mire was fed by rainwater, and organic decay

was enhanced such that mainly resistant, thick-walled spores (crassispores) were preserved. This formed the dull, hard (durain) bands characteristic of the middle to upper part of some seams. The elevation of raised mires above the delta plain inhibited clastic input, and thus the inorganic content of these dull bands is low. As a consequence, the seams containing dull bands of high-quality coal, such as the Top Hard of the East Midlands and the Barnsley Seam of Yorkshire (Fig. 14.20), were prized. The tops of seams are commonly bright coal with a high clastic content and lycopsid spores indicating progressive drowning and inundation of the peat mires. The dark mudstones that overlie many seams thus represent the lakes that formed after the peat deposits had become submerged. Although the majority of coals formed within mires, organic-rich sapropel also accumulated within shallow lakes. This environment often supported an abundant algal flora, which formed subaqueous cannel or boghead sapropelic coals.

When traced laterally, seams show variation in thickness and quality, as well as complex patterns of splitting. Thickness variation is often the result of peat accumulating above an irregular topography produced by different underlying sediment types: for instance, seams show elongate belts of thickened coal called 'swilleys' where they overlie abandoned channels.

Palaeosols beneath seams, referred to as

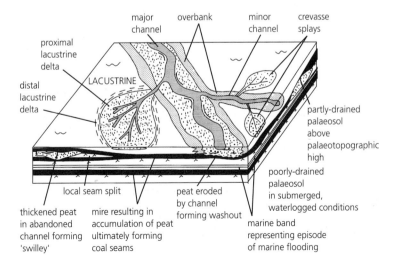

Fig. 14.21 Schematic facies model of the coal-bearing Westphalian (mid-Langsettian to late Duckmantian) of the Pennine province, showing relationships of the major facies (after Guion *et al.* 1995).

'seatearths', generally contain lycopsid roots such as *Stigmaria* and rootlets. The seatearths typically have an upper leached layer, with a lower layer of siderite nodules. Where leaching is extreme, the seatearths may be highly kaolinitic, and have been exploited commercially for refractory and pottery clays. The majority of seatearths accumulated in waterlogged conditions and are grey, resembling gley soils. Brown, red, cream or mottled seatearths represent better drained, more oxidizing conditions on the basin margins or overlying abandoned channels.

Major channels are represented by sand bodies typically 2–5 km wide and 10–20 m thick. The thick channel sandstones generally originated by gradual aggradation rather than occupying erosional palaeovalleys. Major channel sandstones form important hydrocarbon reservoirs in the onshore oilfields of Nottinghamshire and Lincolnshire, and similar sandstones are gas exploration targets in the Southern North Sea.

Minor channels range from about 10 m to 1 km in width and up to about 10 m thick, and some have been demonstrated to be offshoots of major channels that die out down-palaeocurrent, terminating as a lacustrine delta or crevasse splay. Basal erosion surfaces may be overlain by breccias and conglomerates, which include mudstone and ironstone clasts, masses of coal and plant remains. The lithologies within minor channels depend on their mechanisms of channel filling and evolutionary history, and can be very complex and variable. Minor channels cause considerable problems during coal mining, including the erosional removal of coal seams, resulting in 'washouts', leading to coal faces being abandoned. The variable nature of some fills may cause roof support problems in underground workings, and slope stability difficulties in opencast coal mines.

Overbank deposits are often present on channel margins, and generally consist of massive siltstones, or siltstones with thin interbeds of sandstones. This facies forms narrow, wedge-shaped elongate belts that fine, thin and dip gently away from associated channels. Drifted stems and fronds of plants such as *Neuropteris* are common, and other plants sometimes occur rooted in growth position. Where overbank deposits are immediately above a coal seam, *in situ* vertical trunks of lycopsids, such as *Lepidoden-*

dron, may be preserved infilled with siltstone, suggesting that high rates of sedimentation were required to entomb the trees.

The uppermost part of the Duckmantian and lowermost Bolsovian is characterized by a number of marine bands (Fig. 14.20). The Aegiranum Marine Band, which marks the Duckmantian–Bolsovian boundary, is widespread throughout western Europe and has been correlated with a similar horizon in the United States. It is particularly thick, locally attaining 10 m in the East Midlands, and unlike other Westphalian marine bands, it contains a diverse benthic fauna including brachiopods, crinoids, bivalves and ostracods, as well as nekton such as goniatites. The Cambriense Marine Band is the last Westphalian marine horizon known in Britain, but is absent in western Europe. Nevertheless, faunas indicate fully marine conditions.

The Bolsovian (Westphalian C) of the majority of the Pennine coalfields contains only thin coals. The paucity of well-developed coals in the late Westphalian is attributed to a gradual change to better-drained conditions and the evolution from deltaic to fluvial environments that were less favourable for thick peat development. However, in North Staffordshire and Lancashire, several economic coals occur in a thick Bolsovian sequence. This area is the Westphalian depocentre of the Pennine province, where higher subsidence rates and less emergent conditions allowed coal-forming conditions to continue later. The Bolsovian of the North Staffordshire Coalfield contains several beds of early diagenetic sideritic 'blackband' ironstone deposits, comparable with Recent bog iron ores.

Changes that took place in the Bolsovian (Westphalian C) relate to the onset of Variscan tectonism in the Rheno-Hercynian Zone (see Chapter 15), accompanied by gradual climatic change. The effects are manifested best on the extreme southern margins of the Pennine province, where Westphalian C to Stephanian strata are present.

The Etruria Formation, of mainly Bolsovian age (Fig. 14.20), is an important red-bed sequence in the West Midlands, which contains sandstones rich in volcanic clasts. It interdigitates and passes diachronously towards the north into grey coal-bearing strata. It has been interpreted as originating during localized intra-Westphalian extensional faulting and uplift of the Wales–London–Brabant High, which shed sediment to the north. The Etruria Formation is locally unconformable on folded late Duckmantian strata, forming the 'Symon unconformity', which is believed to represent the onset of local Variscan basin inversion. The simultaneous occurrence of locally extensional and compressive movements has led some workers to maintain that dextral strike-slip was superimposed on late Carboniferous compression.

The overlying Westphalian D Halesowen Formation rests unconformably on the Etruria Formation in the extreme south of the basin. The Halesowen Formation is interpreted as fluvial, with immature calcretes suggesting a moderately dry climate. The formation is the equivalent of a thick coal-bearing sequence in Oxfordshire, and the middle and upper portions of the southerly derived Pennant sandstones of South Wales. The southern derivation of the Halesowen Formation suggests that the Wales–London–Brabant Massif had been reduced to a subdued relief by the late Westphalian, allowing connections between the Southern and Pennine provinces. The mainly Westphalian D Salop Formation contains abundant calcretes and relatively mature sandstones containing clasts of Carboniferous limestone and reworked calcrete. The sandstones are also probably derived from the south, partly from reworking of Lower Carboniferous and Old Red Sandstone rocks.

Bolsovian to Westphalian D red-beds are also present in the north of the Pennine province. In Cumbria, the red 'Whitehaven Sandstone Series' overlies grey Coal Measures with the majority of the reddening being ascribed to pre-Permian oxidative weathering. In the Canonbie Coalfield in southern Scotland (Fig. 14.18), red-beds rest conformably on grey coal-bearing strata, except in the north-west of the coalfield, where an unconformity implies late Westphalian erosion and uplift on the northern basin margin.

Late Carboniferous alkaline and tholeiitic basalts in the southern part of the Pennine Basin include both extrusive lavas and volcaniclastics, and intrusive sills and dykes. The extrusive igneous activity that occurred intermittently in the Namurian continued until the late Langsettian. In the Vale of Belvoir area, south-east of Nottingham, a volcanic pile about 150 m thick developed, containing lavas, tuffs and agglomerates. After eruption ceased, the lava pile formed a volcanic landmass that was progressively onlapped by later Duckmantian deposits. A late Langsettian tuffaceous siltstone can be traced over a wide area of the East Midlands and contains numerous thin-graded beds and includes basaltic glass shards. It thickens to the east, where accretionary lapilli are present, and represents the ejecta from several vents situated east of Nottingham. Extrusive igneous rocks are absent above the Vanderbeckei Marine Band in the East Midlands, although there are extensive alkaline dolerite sills within the Langsettian and Duckmantian. These may post-date the extrusive volcanics, and were probably emplaced in the late Westphalian or Stephanian.

Westphalian igneous rocks of the West Midlands are entirely alkaline and of Bolsovian age, and include thin basic sills intruded into wet sediments and laccolithic masses. A major volcanic centre near Dudley erupted lapilli tuffs that entombed conifer stems *in situ*.

In the Pennine province, tonsteins (kaolinized volcanic ash bands) appear to be concentrated in the Langsettian and Bolsovian. The majority of these bands are basic and commonly show grading, suggesting an air-fall origin from nearby contemporaneous volcanic centres. This composition contrasts with the abundant acidic Westphalian tonsteins in mainland Europe, derived from Plinian eruptions associated with the Variscan destructive plate margin situated south of Britain. The Bolsovian supra-Wyrley–Yard tonstein of the West Midlands is a rare example of such an acid tonstein in Britain.

The dominant Westphalian alkaline basic magmatism may have been a consequence of mantle partial melting, with local crustal-thickness faults forming conduits for the magmas. This magmatism terminated in the late Westphalian, when regional north–south tension and transtension ceased. A single phase of tholeiitic basaltic magmatism dated at 301 Ma, near the Permian–Carboniferous boundary, resulted in the emplacement of the Whin Sill complex of northern England. The complex consists of a series of sills of varying thickness of up to about 100 m, together with dykes which trend ENE–WSW.

14.8.2 Scottish province

In Scotland, Westphalian rocks are concentrated in a number of isolated outcrops (Fig. 14.18), mainly within the Midland Valley. A low-relief emergent area in the present position of the Southern Uplands separated the Scottish and Pennine provinces. However, attenuated Westphalian sequences in the Sanquhar and Thornhill inliers south of the Southern Uplands Boundary Fault indicate that sedimentation also took place beyond the Midland Valley, and that connections existed between the Scottish and Pennine areas. The northern limit of the depositional areas is difficult to define, as Carboniferous deposits may have been eroded from the area of the Scottish Highlands. The isolated outcrop at Morvern on the north-west coast suggests that the depositional basin may have extended further north than is at present preserved (Figs 14.18, 14.19).

The Scottish Westphalian shows many similarities to that of the Pennine province, except that it is apparently less marine influenced. Several of the marine bands of the Pennine area have not been recognized, or are represented only by *Lingula*-bearing brackish horizons. Marine bands are particularly rare in the more proximal north-eastern part of the basin in the East Fife and Midlothian Coalfields.

The Westphalian A and B are traditionally referred to as the 'Productive Measures'. These are generally grey with many formerly economic seams, and separated by the Aegiranum (Skipsey's) Marine Band from the 'Barren Red Measures'. The Subcrenatum Marine Band, which marks the base of the Westphalian in other provinces, has not been formally recognized in Scotland, and hence the established practice was to place the base of the Coal Measures at an arbitrary horizon near the lowest worked seam. The subsequent recognition of *Lingula* or the use of other fossils has allowed the base of the Coal Measures to be redefined to correspond closely to the base of the Westphalian. This often extends down into the upper part of the Passage Formation, which was formerly assigned to the Namurian (Fig. 14.22). Coal-bearing strata generally follow conformably upwards from the Passage Formation, but in the south of the Sanquhar Coalfield, Westphalian strata rest upon Ordovician, and in Arran, successively higher horizons of the Coal Measures overstep onto the Passage Formation. In north Ayrshire, a thick Namurian to early Westphalian lava pile formed a topographical high, which was progressively onlapped by Langsettian sediments. It has also been suggested that the thick Dinantian Clyde Plateau Lavas may have formed upland areas situated between the Central and Ayrshire Coalfields during the Westphalian A and B.

Coals in the lowest part of the Langsettian are generally thin or of variable thickness, and some are high in sulfur if close to marine horizons. Consequently, few are economic and the majority of the workable seams are in the middle to upper part of the Langsettian. High-alumina fireclays have been worked in the northern part of the Central Coalfield. Other economic deposits include oil shales and ironstones formed in shallow lakes. Cannel coals, locally referred to as 'parrot', were also formerly distilled to produce oil. Sandstones tend to be thickest, coarsest and most common in the more proximal coalfields in the east, particularly in the lowermost Langsettian, with the sequences becoming progressively less sandy westwards. However, sandstones in the west of the province may include some of westerly derivation, similar to those of the Pennine province.

Duckmantian strata are similar to the underlying Langsettian, with horizons of marine or brackish faunas, especially *Lingula* bands that occur towards the top of the stage in the western and central coalfields. Strata are often partially reddened, and coals are altered ('lime-burnt') to black limestones.

The distribution of sandstones and marine bands suggests that dominant sediment transport was from north-east to south-west from a distant northerly source area during the Westphalian A and B, although the existence of westerly source areas is also a possibility. Old Red Sandstone, Dalradian and Moine rocks in the Scottish Highlands may have also contributed, and the Southern Uplands and Cheviots were intermittent source areas.

The Westphalian of Arran, believed to be mainly Langsettian, is partially reddened, lacks coal seams and marine bands, and is dominated by channel sandstones. It is interpreted as the deposits of an alluvial plain, close to the basin margin, draining a nearby source area to the north or north-west.

The 'Barren Red Measures' may extend locally up to the Stephanian. In the Douglas Coalfield, grey beds above the Aegiranum (Skipsey's) Marine Band may represent a continuation of conditions similar to

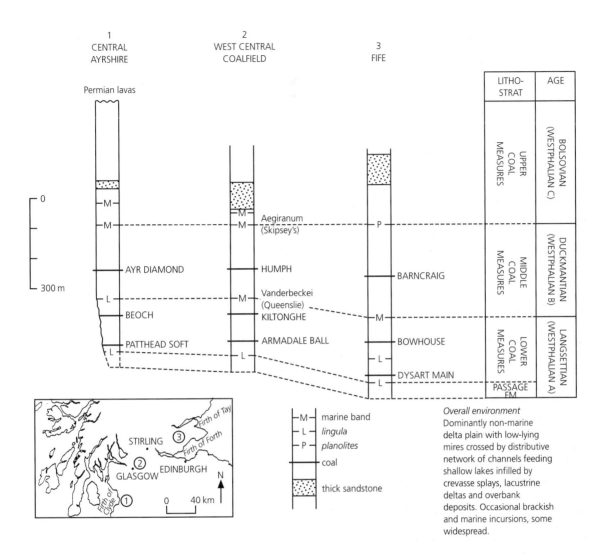

Fig. 14.22 Generalized stratigraphical sections of the Coal Measures of the Scottish Midland Valley, showing key coals and marine bands (based on information in Browne *et al.* 1996).

those in the Westphalian A and B. Elsewhere, channel fill sandstones with intraformational breccias, calcretes and desiccation cracks suggest deposition in a well-drained fluvial setting with a semiarid climate. Although these strata may have been in part subject to primary alteration, some of the reddening has been ascribed to secondary weathering beneath the pre-Permian unconformity.

Westphalian vulcanicity in Scotland was rather less than during the earlier Carboniferous, and was mainly restricted to Fife and the Firth of Forth, where volcanic activity continued from the Namurian, forming a thick volcanic pile. (Fig. 14.8). Tholeiitic dykes and a sill complex contemporaneous with the Whin Sill dated at 303 Ma were intruded during the Carbo–Permian period. The dykes, up to 300 km long and 100 m wide, were emplaced in east–west fractures, and acted as feeders to the quartz–dolerite Midland Valley Sill, which extends over 1600 km².

The Westphalian was a period of relative tectonic quiescence, compared with the underlying Passage Formation. The Westphalian A and B sequences in the south-western part of the Midland Valley indicate minor depocentres in the Douglas Coalfield and south Ayrshire, where faults continued to be active. Elsewhere, subsidence rates were more uniform, In mid-Ayrshire, an inversion occurred in the Langsettian–Duckmantian, when the Mauchline Basin formed over the site of an earlier high, and in north Ayrshire, the thick Passage Formation volcanic pile was not covered by sediments until the late Langsettian. The north–south-trending folds which affect some of the Midland Valley may have formed by east–west compression, probably as a consequence of dextral strike-slip during the late Carboniferous, with inversion of many earlier normal faults.

14.8.3 Irish province

Westphalian strata are restricted to a limited number of outcrop areas (Fig. 14.18), as a consequence of post-Carboniferous erosion. Langsettian strata occur in all coalfields, but younger strata are poorly represented. The base of the Westphalian has been identified in all the coalfields except Kanturk. In the northern areas, *G. subcrenatum* is present, but in the south, faunas are dominated by the brachiopods *Martinia* and *Lingula*. This faunal distribution may reflect the position of the Leinster High, a westerly continuation of the Wales–London–Brabant High (Fig. 14.19). Westphalian sediments have been recognized in the fault-bounded Fintona Block in Northern Ireland, consisting of red-mottled mudstone palaeosols and a pedogenic limestone of Langsettian age, overlain by about 1000 m of Duckmantian conglomerates containing volcaniclastic, granitic and metamorphic clasts. The thick conglomerate was deposited as a consequence of rapid uplift on the northern margin of the fault-bounded basin, which is contemporaneous with similar early Westphalian events in eastern Canada, where localized rapid subsidence occurred in narrow basins adjacent to dextral strike-slip faults.

The remainder of the Irish Westphalian is similar to that elsewhere in the British Isles, but marine bands are poorly developed. Coals have been formerly worked, including the Jarrow Coal Seam of the Leinster Coalfield (Fig. 14.18), whose roof shales have yielded a rich amphibian and fish fauna. A thick feldspathic sandstone is present above the Listeri Marine Band in the Leinster, Slieveardagh and Crataloe Coalfields, referred to as the Clay Gall Sandstone on account of containing numerous intraformational mud clasts. In the south, the coals are of high rank, including anthracite, and deformation increases southwards towards the Variscan Front. The Kanturk Coalfield is highly deformed, with tight folding and thrusting, and seams show major thickness variation of structural origin, rendering them uneconomic.

Westphalian D strata have been encountered in a mineral exploration borehole in Wexford in southeast Ireland. These contain recycled spores of Devonian to Westphalian D age, derived from the erosion of rising Variscan structures to the south. Tholeiitic dykes in north-west Ireland are currently dated at 305 Ma, suggesting that they are nearly contemporaneous with similar intrusions in northern England and the Scottish Midland Valley.

The Kish Bank Basin, a few kilometres offshore of Dublin in the Irish Sea, contains coal-bearing Westphalian B to D strata, suggesting that deposition originally occurred in a continuous belt extending from England and Wales to Ireland. Feasibility studies have been undertaken of the possibility of offshore mining or *in situ* gasification of these coals. Westphalian sediments have also been identified in offshore boreholes to the west of Ireland, where upper Carboniferous strata are preserved in Mesozoic rift basins. Strata of Late Westphalian or Stephanian age have been reported to contain bioclasts with marine fossils including brachiopods, trilobites and foraminifera. This suggests the possibility of late Carboniferous rifting to the west of Ireland, allowing the development of a seaway.

14.8.4 Southern province

The Southern province includes areas of Westphalian sedimentation south of the Wales–London–Brabant High, which is separated from the Culm Basin of Cornubia by the Bristol Channel Fault Zone (Figs 14.18, 14.19). Rocks in this province are more tectonically deformed than equivalent strata further north and the Variscan deformation front runs across the south of the South Wales Coalfield. This province not only includes the exposed coalfields of South Wales, Bristol–Somerset and the Forest of Dean, but concealed coalfields of Kent and Oxfordshire–Berkshire (Fig. 14. 18).

The Westphalian A–D succession of the South Wales Coalfield contains a thick, unbroken sequence exceeding 1800 m in the west (Fig. 14.23). The Westphalian A–B shows many similarities to the Pennine coalfields. Marine bands are particularly well developed, with the Subcrenatum Marine Band at the base attaining 25 m in thickness, and the Aegiranum (Cefn Coed) Marine Band containing the richest fauna in the British Isles, including corals and crinoids. The Westphalian thins towards the Usk Axis, which lies just beyond the south-eastern boundary of the present basin (Fig. 14.24). Here, a local unconformity within the Westphalian D suggests that this high sepa-

rated the South Wales basin from that of the Forest of Dean during much of the Westphalian.

The lowest Langsettian, in common with the Pennine province, is characterized by several marine bands and is deficient in economic coals, and was deposited within a marine-influenced delta plain environment. A centripetal pattern of palaeoflow into the basin has been claimed for this interval. An important sandstone above the Subcrenatum Marine Band has been termed the 'Farewell Rock' (Fig. 14.23), but this occupies a higher stratigraphical position than the uppermost Namurian 'Farewell Rock' of Pembrokeshire (see Section 14.7).

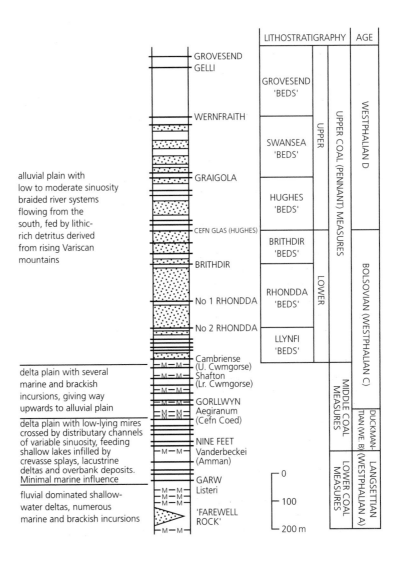

Fig. 14.23 Generalized stratigraphy of the South Wales Coalfield, showing key coals and marine bands (based on information in Ramsbottom *et al.* 1978).

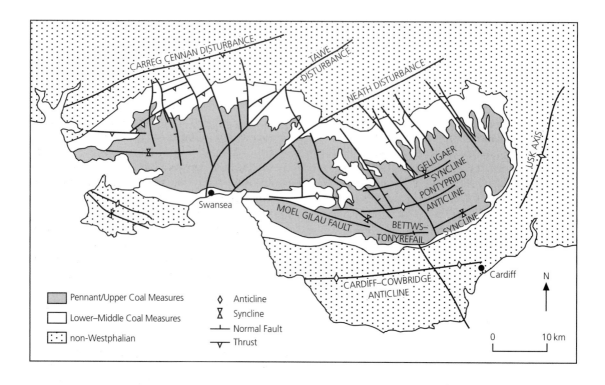

Fig. 14.24 Geological map of the South Wales Coalfield, showing main tectonic elements, some of which controlled sedimentation (simplified from Hartley 1993a).

Most of the mined seams are within the middle and upper Langsettian and lower and middle Duckmantian ('Productive Measures'). Deposition occurred on a low-lying plain with negligible marine influence. Typical sequences include coals overlain by organic-rich mudstones containing non-marine bivalves, followed by coarsening-upwards sequences representing lake fills. Major channels are on a smaller scale than in the Pennine coalfields. Mining data on seam splits and channel orientations suggest a dominant palaeo-flow from south-west to north-east (Fig. 14.19). Many of the sandstones in the South Wales productive measures have a distinctive dark grey colour, and may have a separate source area from those of other coalfields.

Patterns of seam splitting, thickness variation and channel orientation indicate tectonic control on sedimentation by a number of active structures. The north-east–south-west-trending Tawe (Swansea

Valley) and Neath 'disturbances' (Fig. 14.24) are considered to be active fault zones, founded on earlier Caledonian basement structures that were active over a long period. The Neath disturbance may have acted as a positive area, such that Westphalian sediments thin over it. Thinning occurs onto east–west trending anticlines such as the Pontypridd Anticline, and thickening into the Bettws–Tonyrefail Syncline, and these growth folds have been ascribed to the reactivation of east–west basement structures during early Variscan compression.

Thinner Westphalian A and B sequences are also present in the Bristol–Somerset, Oxfordshire–Berkshire, and Kent coalfields. In Oxfordshire and Berkshire, subaerial alkaline basic volcanics are interbedded with Langsettian sediments in the subsurface. Intrusive dolerite sills post-date the volcanics and are likely to be of late Westphalian age.

Several marine bands and coals occur in the lowermost Bolsovian in the South Wales, Bristol–Somerset and Kent coalfields, with the Cambriense (Upper Cwmgorse) Marine Band marking the last marine incursion. Soon after this event, a major influx of lithic sandstones took place from the south, and spread progressively north-eastwards. The resultant 'Pennant Measures' forms a thick Westphalian C and D sequence characterized by sandstones with subordinate mudstones, palaeosols and coals. Similar strata are present in the Kent, Forest of Dean, Oxfordshire–Berkshire and Bristol–Somerset coalfields, although in the two latter, the upper part is less arenaceous and includes primary red-beds.

The Pennant Measures were deposited by rivers of low to moderate sinuosity that flowed north into the actively subsiding basins. The detritus was derived from rising mountain chains being formed by nappe emplacement as the Variscan Orogen progressively migrated northwards. South Wales and other coalfields of the Southern province display the characteristics of a flexural foreland basin by late Westphalian times, with nappes to the south not only causing flexural loading, but supplying sediment.

The Southern province coalfields show evidence of considerable post-Westphalian Variscan deformation, with increasing intensity towards the south (see Chapter 15). In South Wales, the degree of deformation and coal rank also increase towards the west. In South Wales, the relatively incompetent Westphalian A and B Productive Measures are the most intensely deformed. The sand-dominated competent Westphalian C and D Pennant Measures are much less deformed, and acted as a passive roof to the thrust system. The deformation causes thickness variation, changes of orientation and repetitions or 'wants' in coal seams, and has contributed significantly to mining problems. Mining activity is now limited, but evaluation of coal-bed methane potential has indicated that South Wales may be prospective, but economic exploitation has yet to occur.

14.9 Namurian and Westphalian of the Culm Basin

The Bristol Channel Fault Zone (Figs 14.2, 14.19) separates the Culm Basin of Cornwall and Devon from the South Wales Basin. The original relationships between these depositional areas are unclear, and there are major differences between the Upper Carboniferous of the Culm Basin, where coals are rare, and that of the coal-bearing basins to the north. Two distinct Upper Carboniferous sequences are recognized in the Culm Basin, but the relationships between them are not unequivocally resolved.

In the south and central parts of the basin (Fig. 14.25), the mainly Namurian Crackington Formation extends upwards into the lowermost Westphalian. It is overlain by the Bude Formation, whose base is within the lowermost Langsettian, and whose highest recorded faunal horizon, the Warren Gutter Shale, is of Bolsovian age. The Westward Ho! and Bideford Formations, which outcrop in a narrow east–west belt on the northern side of the Culm Basin, are partly coeval with the Crackington and Bude Formations. Organic matter from the Bideford and Westward Ho! Formations shows a much lower vitrinite reflectance than that from the Bude and Crackington Formations, suggesting that they did not undergo such deep burial. For this reason, some workers believe that they lie within a separate tectonic unit, brought into its present position after deposition.

The lowest Namurian part of the Crackington Formation consists of thin turbidites with north-west- and north-eastward directed palaeoflows and contains scattered goniatites throughout the interbedded fine-grained intervals. In the south, shales contain Alportian and Kinderscoutian goniatites. In the northern Devon area, fauna is restricted to marine bands in the later Namurian. Diagnostic goniatites are generally absent from these horizons, which contain marine fauna such as *Canyella*. Only the base Westphalian marine band contains the marker goniatite *G. subcrenatum*.

Much of the Crackington Formation comprises fine-grained, thin-bedded mineralogically mature turbidites, which were transported eastwards along the basin. The depositional environment of the Crackington Formation is the subject of some controversy, although their sedimentation from gravity flows is not in dispute. The setting has been interpreted both as a high-gradient shelf margin or as the base of a slope. There is no evidence for emergence within the 1300-m-thick Crackington Formation and deposition in deep water, well below wave base, is envisaged.

The overlying Westphalian Bude Formation consists of about 1300 m of mudstones and laterally extensive sandstones, which show many similarities to the Crackington Formation. However, sandstone beds tend to be thicker, attaining 5 m, and may amalgamate to form sandstone bodies up to 20 m thick. Some sandstones are channelized with basal intrafor-

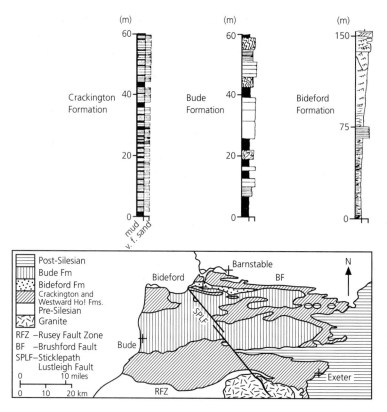

Fig. 14.25 Schematic representative logs of the Crackington, Bude and Bideford Formations. Note differences in scale. Inset map shows the distribution of main Silesian strata in the Culm Basin (after Hartley 1993b).

mational conglomerates. Flute casts and tool marks show palaeoflows from all quadrants except the south. Rare mudstones interbedded with the sandstones ('Index Shales') contain marine fossils, which are otherwise absent: thus fresh or brackish conditions are envisaged, with only occasional marine incursions. Interpretation of the depositional environment of the Bude Formation has proved controversial, with several hypotheses being put forward, including:

1 a shallow turbiditic setting, fed by a direct fluvio-deltaic supply with a southerly dipping palaeoslope;

2 a shallow lake margin subject to storms;

3 relatively deep-water subaqueous fans in a mainly low-salinity isolated basin.

The Westward Ho! Formation out crops on the northern side of the Culm Basin, and is partially equivalent in age to the Crackington Formation, and consists of coarsening-upward cycles indicating progressive shallowing. The lowest parts of the cycles contain thin-bedded turbidites and syn-depositional

deformation features. Wave ripples within the sequences suggest a shallow-water low-gradient offshore slope, comparable to that of the Mississippi Delta, subject to various slope failure phenomena.

The overlying Bideford Formation consists of nine large-scale (50–100 m) coarsening-upwards cycles, with dark shales at the base passing gradually upwards into feldspathic sandstones. Recent faunal evidence indicates that the lowermost Bideford Formation is Namurian, with the oldest dated horizon being Kinderscoutian. Immediately below the Westphalian–Namurian boundary, the Raleigh sandstone has a basal erosion surface which may cut out the underlying Cambriense Marine Band. This is the stratigraphic equivalent of similar erosively based uppermost Namurian sand bodies elsewhere in the British Isles that represent incised valley fills. The upper part of the Bideford Formation is entirely Langsettian, with the 'Culm Bed' consisting of impure coal being the highest horizon. Some workers have considered that another unit, the Greencliff Formation, overlies the Bideford Formation. However, vitrinite reflectance data suggest that it underwent much deeper burial, and that it is a Namurian deposit, faulted into its present position.

The coarsening-upwards cycles of the Bideford Formation have been interpreted as deposits of fluvial-dominated deltas that prograded southwards into moderately deep water, with the cross-bedded sandstones at the top of each cycle representing distributary channels.

It is difficult to make meaningful palaeogeographical reconstructions of the Upper Carboniferous of the Culm Basin, on account of post-depositional tectonic shortening and the unknown relative positions of the Bude–Crackington and Bideford–Westward Ho! sequences. However, most workers consider that the Bideford–Westward Ho! succession was deposited on the shallow northern margin of the deep-water Culm Basin, in which the Bude and Crackington Formations were deposited. Various tectonic models have been proposed to explain the relationships both within the Culm Basin and relative to South Wales involving either thin-skinned or thick-skinned tectonics, or strike-slip, but there is no consensus. One model envisages that the Culm Basin developed into a thrust-sheet-top basin by the early Westphalian (Fig. 14.26). By the late Westphalian, continuing Variscan compression meant that the area of Cornubia had become deformed and uplifted, and supplied the sediment for the Pennant fluvial systems that flowed northwards into South Wales.

14.10 Summary

The break-up of the stabilized and peneplained Old Red Sandstone Continent ('Laurussia') during the late Devonian to early Carboniferous was accompanied by marine transgression and the formation of a series of linked sedimentary basins. North-directed subduction south of Britain and Ireland promoted extension north of the Wales–London–Brabant High, and had a significant influence on basin development and sedimentation, summarized in Fig. 14.27. In the

Fig. 14.26 Schematic sections across Cornubia and South Wales, based on a thin-skinned tectonic model (after Hartley 1993b).(a) Namurian, (b) Westphalian A, (c) Westphalian C–D.

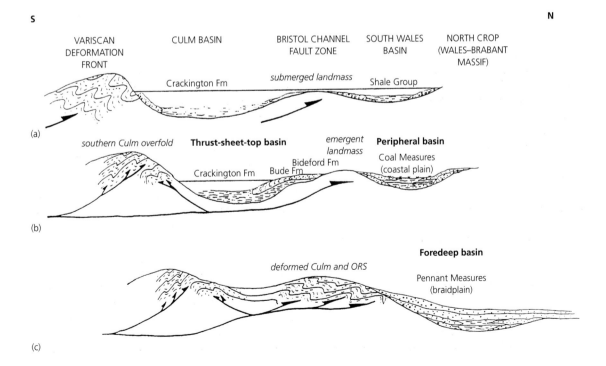

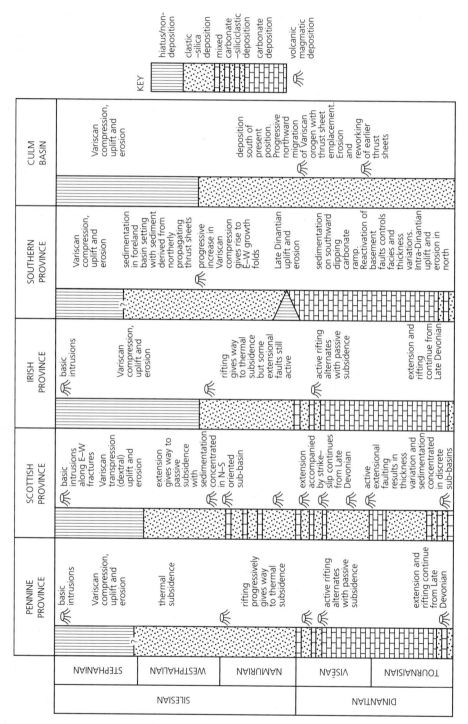

Fig. 14.27 Generalized summary of depositional, magmatic and tectonic events in the Carboniferous provinces of Britain and Ireland.

south, foreland basins of the Rheno-Hercynian zone included South Wales, Devon and Cornwall. Sedimentation was also strongly influenced by glacio-eustasy during much of the Carboniferous. Both intrusive and extrusive magmatism occurred throughout, with the magmas exploiting extensional fractures.

During the Dinantian, a major complex of extensional basins was present north of the Wales–London–Brabant High, covering northern England, North Wales and Ireland. In the north, deposition was dominated by siliciclastic sediments, whereas carbonate deposition was more important in the south. Sedimentation was influenced by episodic extension that caused reactivation of siliciclastic sources and uplift and erosion of carbonate platforms. Intervening periods of thermal subsidence were characterized by the nucleation and growth of carbonate platforms around basement highs and shedding of carbonates into basins. In the Scottish province, sedimentation took place in small linked basins influenced by volcanism and strike-slip tectonism. Deposition was dominantly in restricted lacustrine, marginal marine and fluvio-deltaic conditions. A south-dipping carbonate ramp was present south of the Wales–London–Brabant High throughout the Dinantian. Siliciclastic and carbonate sediments deposited in shallow and marginal marine, deep marine and pelagic settings in the Culm Basin were eventually displaced towards the north during Variscan thrusting.

Carbonate deposition continued into the lowermost Namurian in the Pennines and Ireland but the Silesian was dominated by siliciclastic sedimentation. The sub-basins established in the Dinantian continued to influence Namurian facies distribution. Gradual southward progradation of major fluvio-deltaic systems in Northern England deposited major sand bodies by the Early Namurian. Further south, sub-basins of the Pennine province were initially starved of coarse sediment and thick mudstone successions were deposited. Major deltas feeding turbidite systems reached the southern part of this area by the Late Namurian. Northern source areas were important throughout much of the Namurian and Westphalian, but source areas in the west and the Wales–London–Brabant High also contributed sediment.

The Westphalian was dominated by non-marine

fluvial and lacustrine sedimentation, although deeper water conditions persisted in the Culm Basin. Several marine incursions occurred in the early Westphalian but they gradually became less frequent. Extensive peat deposits accumulated in a humid climate on a low-lying plain that extended across much of Western Europe in the Westphalian A and B. By the Westphalian C and D thrust sheets had propagated northwards, forming a mountain chain across Cornwall and Devon. These rising Variscan mountains shed immature lithic sediment northwards into South Wales and southern England. The Wales–London–Brabant High was breached and fluvial systems transported this material into the southern part of the Pennine province. The climate became increasingly arid and, by the late Westphalian, peat development had been largely replaced by calcretes and other well-drained palaeosols in an alluvial setting. This aridity would be characteristic of the succeeding Permian period.

The Stephanian is poorly preserved in Britain and Ireland as uplift and erosion accompanied Variscan compression in the latest Westphalian and Stephanian. Some significant strike-slip movement occurred during this late phase, including transpressional movements in the Scottish Midland Valley. Late Carboniferous magmatism resulted in the intrusion of tholeiitic dykes and sills in Scotland, Ireland and Northern England. Controversy continues over Carboniferous plate configurations and the relative importance of tectonics, climate, sediment supply and glacio-eustasy in controlling depositional systems. However, the abundant economic resources contained in the Carboniferous will ensure that these rocks will continue to be a fertile testing ground for geological ideas.

References

Aitkenhead, N.R. & Riley, N.J. (1996) Kinderscoutian and Marsdenian successions in the Bradup and Hag Farm boreholes, near Ilkley, west Yorkshire. *Proceedings of the Yorkshire Geological Society* 51, 115–125.

Arthurton, R.S., Johnson, E.W. & Mundy, D.J.C. (1988) *Geology of the Country around Settle: memoir for 1 : 50 000 geological sheet 60 (England and Wales).* HMSO, London.

Besly, B.M. (1998) Carboniferous. In: *Petroleum Geology of the North Sea: basic concepts and recent advances*, 4th edn (ed. K. W. Glennie), pp. 104–136. Blackwell, Oxford.

Brandon, A., Riley, N.J., Wilson, A.A. & Ellison, R.A. (1995) Three new early Namurian (E_{1c}–E_{2a}) marine bands in central and northern England, UK, and their bearing on correlations with the Askrigg Block. *Proceedings of the Yorkshire Geological Society* 50, 333–355.

Bristow, C.S. (1988) Controls on the sedimentation of the Rough Rock Group (Namurian) from the Pennine Basin of Northern England. In: *Sedimentation in a Synorogenic Basin Complex: the Upper Carboniferous of Northwest Europe* (eds B. M. Besly & G. Kelling), pp. 114–131. Blackie, Glasgow.

Browne, M.A.E., Hargreaves, R.L. & Smith, I.F. (1985) *The Upper Palaeozoic Basins of the Midland Valley of Scotland: investigation of the geothermal potential of the UK*. British Geological Survey, Keyworth.

Browne, M.A.E., Dean, T.D., Hall, I.H.S., McAdam, A.D., Monro, S.K. & Chisholm, J.I. (1996) *A Lithostratigraphical Framework for the Carboniferous Rocks of the Midland Valley of Scotland*. British Geological Survey Technical Report WA/96/29. British Geological Survey, Keyworth.

Calver, M.A. (1969) Westphalian of Britain. *Compte Rendu Sixième Congrès International de Stratigraphie et de Géologie du Carbonifère, Sheffield 1967* 1, 233–254.

Church, K.D. & Gawthorpe, R.L. (1994) High resolution sequence stratigraphy of the late Namurian in the Widmerpool Gulf (East Midlands, UK). *Marine and Petroleum Geology* 11, 528–544.

Claoué-Long, J.C., Compston, W., Roberts, J. & Fanning, C.M. (1995) Two Carboniferous ages: a comparison of SHRIMP zircon dating with conventional zircon ages and $^{40}Ar/^{39}Ar$ analysis. In: *Geochronology, Time Scales and Global Stratigraphic Correlation* (eds W. A. Berggren, D. V. Kent, M. P. Aubry & J. Hardenbol), Society of Economic Paleontologists and Mineralogists. Special Publication no. 54, pp. 3–21, Tuba.

Collinson, J.D. (1988) Controls on Namurian sedimentation in the Central Province of northern England. In: *Sedimentation in a Synorogenic Basin Complex: the Upper Carboniferous of Northwest Europe* (eds B. M. Besly & G. Kelling), pp. 85–101. Blackie, Glasgow.

Collinson, J.D., Martinsen, O., Bakken, B. & Kloster, A. (1991) Early fill of the Western Irish Namurian Basin: a complex relationship between turbidites and deltas. *Basin Research* 3, 223–242.

Cope, J.C.W., Guion, P.D., Sevastopulo, G.D. & Swan, A.R.H. (1992) Carboniferous. In: *Atlas of Palaeogeography and Lithofacies* (eds J. C. W. Cope, J. Ingham & P. F. Rawson), Memoir 13, pp. 67–86. Geological Society, London.

Corfield, S.M., Gawthorpe, R.L., Gage, M., Fraser A.L. & Besly, B.M. (1996) Inversion tectonics of the Variscan foreland of the British Isles. *Journal of the Geological Society, London* 153, 17–32.

Davies, S.J. & McLean, D. (1996) Spectral gamma ray and palyological characterisation of Kinderscoutian marine bands in the Namurian of the Pennine Basin. *Proceedings of the Yorkshire Geological Society* 51, 103–114.

Francis, E.H. (1991) Carboniferous–Permian igneous rocks. In: *Geology of Scotland*, 3rd edn (ed. G. Y. Craig), pp. 393–420. Geological Society, London.

Fraser, A.J., Nash, D.F., Steele, R.P. & Ebdon, C.C. (1990) A regional assessment of the intra-Carboniferous play of Northern England. In: *Classic Petroleum Provinces* (ed. J. Brooks), Special Publication 50, pp. 417–440. Geological Society, London.

Gawthorpe, R.L., Gutteridge, P. & Leeder, M.R. (1989) Late Devonian and Dinantian basin evolution in northern England and North Wales. In: *The Role of Tectonics in Devonian and Carboniferous Sedimentation in the British Isles* (eds R. S. Arthurton, P. Gutteridge & S. C. Nolan), Yorkshire Geological Society Occasional Publication 6, pp. 1–23. A Wigley & Sons Ltd., Bradford.

George, G.T. & Kelling, G. (1982) Stratigraphy and sedimentology of Upper Carboniferous sequences in the coalfields of southeast Dyfed. In: *Geological Excursions in Dyfed, Southwest Wales* (ed. M. G. Bassett), pp. 175–201. National Museum of Wales, Cardiff.

Gill, W.D. (1979) *Syndepositional Sliding and Slumping in the West Clare Namurian Basin, Ireland*. Geological Survey of Ireland Special Paper 4.

Guion, P.D., Fulton, I.M. & Jones, N.S. (1995) Sedimentary facies of the coal-bearing Westphalian A and B north of the Wales–Brabant High. In: *European Coal Geology* (eds M. K. G. Whateley & D. A. Spears), Special Publication 82, pp. 45–78. Geological Society, London.

Hampson, G.J. (1997) A sequence stratigraphic model for deposition of the Lower Kinderscout Delta, an Upper Carboniferous turbidite-fronted delta. *Proceedings of the Yorkshire Geological Society* 51, 273–296.

Hampson, G.J. (1998) Evidence for relative sea-level falls during deposition of the Upper Carboniferous Millstone Grit, South Wales. *Geological Journal* 33, 243–266.

Hampson, G.J., Elliott, T. & Flint, S.S. (1996) Critical application of high resolution sequence stratigraphic concepts to the Rough Rock Group (Upper Carboniferous) of northern England. In: *High Resolution Sequence Stratigraphy: innovations and applications* (eds J. A. Howell & J. F. Aitken), Special Publication 104, pp. 221–246. Geological Society, London.

Hartley, A.J. (1993a) A depositional model for the Mid-Westphalian A to late Westphalian B Coal Measures of South Wales. *Journal of the Geological Society, London* 150, 1121–1136.

Hartley, A. J. (1993b) Silesian sedimentation in south-west Britain: sedimentary responses to the developing Variscan orogeny. In: *Rhenohercynian and Subvariscan Fold Belts* (eds R. Gayer, R. O. Greiling & A. K. Vogel), pp. 160–196. Vieweg, Weisbaden/Braunschweig.

Hess, J.C. & Lippolt, H.J. (1986) ^{40}Ar/^{39}Ar ages of tonstein and tuff sanidines: new calibration points for the improvement of the Upper Carboniferous time scale. *Chemical Geology (Isotope Geoscience Section)* **59**, 143–154.

Jones, C.M. & Chisholm, J.I. (1997) The Roaches and Ashover Grits: sequence stratigraphic interpretation of a 'turbidite-fronted' delta. *Geological Journal* **32**, 45–68.

Leeder, M.R. (1988a) Recent developments in Carboniferous geology: a critical review with implications for the British Isles and NW Europe. *Proceedings of the Geological Association* **99**, 73–100.

Leeder, M.R. (1988b) Devono-Carboniferous river systems and sediment dispersal from the orogenic belts and cratons of NW Europe. In: *The Caledonian–Appalachian Orogen* (eds A. L. Harris & D. J. Fettes), Special Publication 38, pp. 549–558. Geological Society, London.

Li, X. (1990) Changes in deltaic sedimentation in the Upper Carboniferous Westward Ho! Formation and Bideford Group of SW England. *Proceedings of the Ussher Society* **7**, 232–236.

Martinsen, O.J. (1993) Namurian (late Carboniferous) depositional systems of the Craven–Askrigg area, northern England: implications for sequence-stratigraphic models. In: *Sequence Stratigraphy and Facies Associations* (eds H. W. Posamentier, C. P. Summerhayes, B. U. Haq & G. P. Allen), Special Publication, International Association of Sedimentologists 18, pp. 247–281. Blackwell Science Ltd., Oxford.

Martinsen, O.J. & Bakken, B. (1990) Extensional and compressional zones in slumps and slides in the Namurian of County Clare, Ireland. *Journal of the Geological Society, London* **147**, 153–164.

Martinsen, O.J., Collinson, J.D. & Holdsworth, B.K. (1995) Millstone Grit cyclicity revisited, II: sequence stratigraphy and sedimentary responses to changes in relative sea-level. In: *Sedimentary Facies Analysis: a tribute to the research and teaching of Harold D Reading* (ed. A. G. Plint), Special Publication, International Association of Sedimentologists 22, pp. 305–331. Blackwell Science Ltd., Oxford.

Maynard, J.R. (1992) Sequence stratigraphy of Upper Yeadonian of Northern England. *Marine and Petroleum Geology* **9**, 197–207.

Powell, J.H., Chisholm, J.I., Bridge, D. McC., Rees, J.G., Glover, B.W. & Besly, B. M. (2000) *Stratigraphical Framework for Westphalian to Early Permian Red-Bed Successions of the Pennine Basin*. British Geological Survey Research Report RR001. British Geological Survey, Keyworth.

Ramsbottom, W.H.C., Calver, M.A., Eagar, R.M.C. *et al.* (1978) *A Correlation of Silesian Rocks in the British Isles*. Special Report 10, pp. 1–81. Geological Society, London.

Riley, N.J., Claoué-Long, J.C., Higgins, A.C. *et al.* (1993) Geochronometry and geochemistry of the European Mid-Carboniferous boundary global stratotype proposal, Stonehead Beck, North Yorkshire, UK. *Annales de la Société Géologique de Belgique* **116**, 275–289.

Rippon, J.H. (1996) Sand body orientation, palaeoslope analysis and basin fill implications in the Westphalian A–C of Great Britain. *Journal of the Geological Society, London* **153**, 881–900.

Wright, V.P. (1987) The evolution of the early Carboniferous Limestone of South Wales. *Geological Magazine* **124**, 477–480.

Further reading

Besly, B.M. (1998) Carboniferous. In: *Petroleum Geology of the North Sea: Basic Concepts and Recent Advances*, 4th edn (ed. K. W. Glennie), pp. 104–136. Blackwell, Oxford. [Provides a comprehensive account of Carboniferous geology, not only offshore but onshore.]

Besly, B.M. & Kelling, G. (eds) (1988) *Sedimentation in a Synorogenic Basin Complex: the Upper Carboniferous of Northwest Europe*. Blackie, Glasgow. [A volume containing important, but somewhat dated, reviews of Upper Carboniferous sedimentation and its controls. See particularly Chs 1, 4, 6, 9, 13, 15, 16.]

Bridges, P.H., Gutteridge, P. & Pickard, N.A.H. (1995) The environmental setting of early Carboniferous mud-mounds. In: *Carbonate Mud-Mounds, Their Origin and Evolution* (eds C. L. V. Monty, D. W. J. Bosence, P. H. Bridges & B. R. Pratt), Special Publication, International Association of Sedimentologists 23, pp. 171–190. Blackwell Science Ltd., Oxford. [Discusses the oceanographic conditions in which Lower Carboniferous sedimentation took place and reviews the nature of 'reef-building' communities present during the Dinantian.]

Cope, J.C.W., Guion, P.D., Sevastopulo, G.D. & Swan, A.R.H. (1992) Carboniferous. In: *Atlas of Palaeogeography and Lithofacies* (eds J. C. W. Cope, J. Ingham & P. F. Rawson), Memoir 13, pp. 67–86. Geological Society, London. [Summary of palaeogeographies of parts of the Carboniferous, in need of some revision in the light of new data.]

Francis, E.H. (1991) Carboniferous. In: *Geology of Scotland*, 3rd edn (ed. G. Y. Craig), pp. 347–392. Geological Society, London. [A recent review of the Carboniferous of Scotland.]

Fraser, A.J., Nash, D.F., Steele, R.P. & Ebdon, C.C. (1990) A regional assessment of the intra-Carboniferous play of Northern England. In: *Classic Petroleum Provinces* (ed. J. Brooks), Special Publication 50, pp. 417–440. Geological Society, London.

Gawthorpe, R.L., Gutteridge, P. & Leeder, M.R. (1989) Late Devonian and Dinantian basin evolution in northern England and North Wales. In: *The Role of Tectonics in Devonian and Carboniferous Sedimentation in the British Isles* (eds R. S. Arthurton, P. Gutteridge & S. C. Nolan), Yorkshire Geological Society Occasional Publication 6, pp. 1–23. A. Wigley & Sons Ltd., Bradford. [A review of Dinantian extensional basin evolution, summarizing sedimentary history and identifying the main tectonic events and their effects on sedimentation.]

Guion, P.D., Fulton, I.M. & Jones, N.S. (1995) Sedimentary facies of the coal-bearing Westphalian A and B north of the Wales–Brabant High. In: *European Coal Geology* (eds M. K. G. Whateley & D. A. Spears), Special Publication 82, pp. 45–78. Geological Society, London. [A summary of the main facies, their organization and their controls in the coalfields of the Pennine province.]

Hampson, G.J., Elliott, T. & Davies, S.J. (1997) The application of sequence stratigraphy to Upper Carboniferous fluvio-deltaic strata of the onshore UK and Ireland: implications for the southern North Sea. *Journal of the Geological Society, London* 154, 719–733. [A review of the application of sequence stratigraphic concepts to the Namurian from basins in Ireland, Wales and England.]

Hartley, A.J. (1993) A depositional model for the Mid-Westphalian A to late Westphalian B Coal Measures of South Wales. *Journal of the Geological Society, London* 150, 1121–1136. [Description of facies, discussion of controls and comparison with other coalfields for the main coal-bearing sequence in South Wales.]

Higgs, R., Reading, H.G. & Burne, R.V. (1998) Return of 'The fan that never was'; Westphalian turbidite systems in the Variscan Culm Basin; Bude Formation (south-west England); discussion and reply. *Sedimentology* 45, 961–975. [One of the many contributions to the continuing debate about the origin of the Bude Formation.]

Kelling, G. & Collinson, J.D. (1992) Silesian. In: *Geology of England and Wales* (eds P. McL. D. Duff & A. J. Smith), pp. 239–273. Geological Society, London. [Although not completely up to date, provides a detailed, comprehensively referenced account of the Upper Carboniferous.]

Leeder, M.R. (1988) Recent developments in Carboniferous geology: a critical review with implications for the British Isles and NW Europe. *Proceedings of the Geological Association* 99, 73–100. [Although a little dated, gives a comprehensive overview of Carboniferous geology in a wider context.]

Rippon, J.H. (1996) Sand body orientation, palaeoslope analysis and basin fill implications in the Westphalian A–C of Great Britain. *Journal of the Geological Society, London* 153, 881–900. [Although this paper is mainly restricted to the Westphalian, it has interesting implications for Carboniferous palaeogeography and provenance.]

Sevastopulo, G.D. (1981) Upper Carboniferous. In: *A Geology of Ireland* (ed. C. H. Holland), pp. 173–187. Scottish Academic Press, Edinburgh. [One of the few reviews of the Silesian of Ireland.]

Strogen, P., Somerville, I.D. & Jones, G. Ll. (eds) (1996) *Recent Advances in Lower Carboniferous Geology.* Special Publication 106. Geological Society, London. [Contains a useful collection of papers on the Dinantian of Ireland; see especially those by Johnson J.D. *et al.*, Strogen P. *et al.* and Naylor D. *et al.*]

15   The Variscan Orogeny: the welding of Pangaea

L. N. WARR

15.1 Definition

The Variscan (or Hercynian) Orogeny is defined as a Late Palaeozoic collisional episode in Europe which commenced in the Devonian, reached its climax during the Late Carboniferous, and ended in the Early Permian. In southern Britain and Ireland, it produced a fold-and-thrust belt that developed to the south of the main region of Caledonian deformation (see Chapter 12). The high-grade metamorphic core of the orogen runs through central and north-west Spain, France, Germany and the Bohemian area of the Czech Republic (Fig. 15.1a). The Variscan Belt forms part of a globally extensive east–west-trending orogenic zone extending from the Gulf of Mexico to eastern Europe. It includes the Ouachita Belt of Texas, the Alleghenian Belt of the Appalachians and the Mauritanian Belt of north-west Africa (Fig. 15.1b).

This episode of orogenic activity records the convergence of the southern continent of Gondwana, with the northern 'Old Red Sandstone' continent of Laurussia (Laurentia, Baltica and Avalonia), to form the supercontinent of Pangaea (see Chapter 2). The collision was complicated by the existence of a number of intermediate Gondwana-derived microplates and terranes (with intervening oceans) which, like Avalonia, migrated northwards to be successively accreted onto the Laurussian continental margin by the end of the Palaeozoic. The Variscan Orogeny is considered to represent a classical obduction–collision zone characterized by a particularly long period of intracontinental deformation that spanned over 100 Myr.

15.2 Geotectonic zones

The Variscan Belt is notably broad, with an across-strike width of about 1000 km, has a complex curvilinear shape, and contains relatively small volumes of dismembered oceanic crust or mantle. Despite its breadth, the reconstructed geometry of the orogen can be divided into a number of distinct geotectonic zones that show a general continuity around the belt (Fig. 15.1a). Many of these zones are separated by steeply dipping faults or shear zones, which record an oblique collision across the orogen. A number of these boundaries are also considered to represent suture zones, the sites of Palaeozoic oceans that separated fragments of continental crust in the form of microplates or smaller terranes. Variscan suture zones are marked by the occurrence of ophiolite assemblages and calc-alkaline continental arc volcanics of Devonian or Carboniferous age. Older ophiolite assemblages of Cambrian–Silurian age are preserved along some lines, and relate to the initial stages of continental fragmentation that occurred along the northern margin of Gondwana during the Early Palaeozoic (see Chapter 2). Although it is unclear exactly how many continental plates were involved in the Variscan mosaic, Armorica, Iberia and perhaps an Alpine fragment are commonly regarded as discrete microplates. Each microplate is also composed of a number of tectono-stratigraphic units that are best referred to as terrane assemblages.

Four principal phases of deformation and exhumation that led to the amalgamation of Palaeozoic continental fragments can be recognized across continental Europe (Fig. 15.2). Each phase lasted between 20 and 30 Myr, was restricted in geographical extent, and probably resulted from the successive docking of

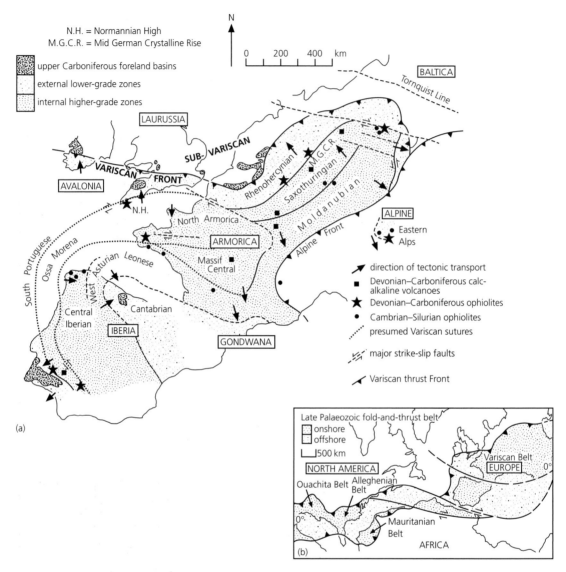

N.H. = Normannian High
M.G.C.R. = Mid German Crystalline Rise

upper Carboniferous foreland basins
external lower-grade zones
internal higher-grade zones

direction of tectonic transport
Devonian–Carboniferous calc-alkaline volcanoes
Devonian–Carboniferous ophiolites
Cambrian–Silurian ophiolites
presumed Variscan sutures
major strike-slip faults
Variscan thrust Front

Late Palaeozoic fold-and-thrust belt
onshore
offshore
500 km

Fig. 15.1 (a) Geotectonic zones of the Variscan orogenic belt of Europe on a pre-Mesozoic reconstruction. The names of Palaeozoic microplates are set in boxes. The Iberian Peninsula has been rotated 30° clockwise to close the Bay of Biscay and the post-Variscan cover has been stripped away (modified after Franke 1989, with permission from the Geological Society of America (1999)). (b) A pre-Mesozoic reconstruction of late Palaeozoic fold-and-thrust belts of the North Atlantic area (after Ziegler 1989, with permission from Kluwer Academic Publishers (1999)).

lithospheric plates along the southern margin of Laurussia. The Ligerian phase (Late Silurian to Early Devonian), well developed in the South Armorican Massif of France, was contemporaneous but not cogenetic with the Acadian phase of the Caledonian Orogeny. The remaining three phases, Bretonian (Late Devonian to Early Carboniferous), Sudetian (Viséan to early Namurian) and Asturian (Westphalian to Early Permian), which together spanned an impressive 100 Myr in time, are assigned to the Variscan Orogeny.

Despite its complex tectono-thermal history, a relatively simple bilateral symmetry can be recognized across the European Variscides (Fig. 15.1a).

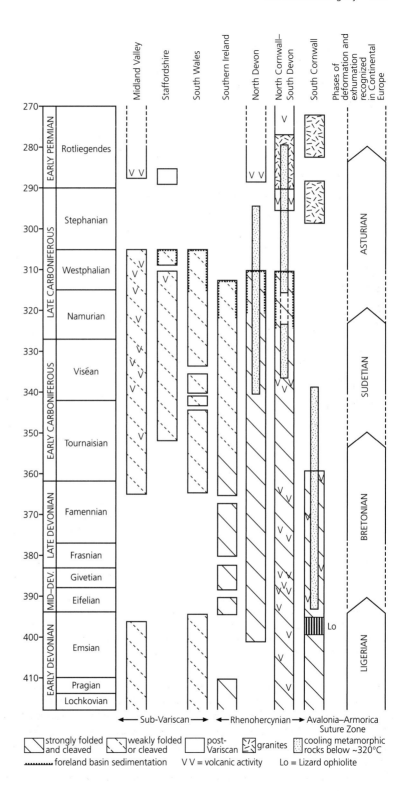

Fig. 15.2 Chronostratigraphic chart showing constraints on the timing of Variscan events in Britain and Ireland.

Two external fold-and-thrust belts, consisting of diagenetic-grade to low-temperature metamorphic Palaeozoic rocks of sedimentary and volcanic origin, are positioned on either side of an internal core of generally higher metamorphic grade. Although the direction of tectonic transport varies considerably around the belt, thrusting was typically directed outwards from the internal zones toward the external margins. The belt also contains a number of major strike-slip faults, such as those cutting Armorica. Although in general dextral displacements were concentrated along the eastern portion of the Variscan Belt, and sinistral displacements along the western portion, individual faults may show complex movement histories commonly involving both senses of slip.

To the north of the Variscan Belt lies the Subvariscan Zone. Here, unmetamorphosed Late Palaeozoic successions were deposited unconformably upon rocks deformed in the Caledonian Orogeny. These rocks were only weakly and heterogeneously strained at the end of the Late Carboniferous, during the Asturian phase of late Variscan deformation. The intensity of this deformation notably increased toward the Variscan Thrust Front, where coal-bearing foreland basin sediments were more intensely folded and thrust.

The basic features of the main Variscan geotectonic zones can be summarized as follows (from north to south).

1 *The South Portuguese–Rhenohercynian Zone* represents the northern external fold-and-thrust belt of the Variscan collision. The zone is comprised of Late Palaeozoic sedimentary and volcanic successions that formed during extensive thinning of the southern continental margin of Eastern Avalonia. The southern boundary of this zone is considered to represent a Variscan suture zone, with juxtaposition against Armorica to the south. A number of ophiolite complexes of Devonian to Carboniferous age are located along the suture, which show a long history of assemblage and cooling during the Bretonian phase of deformation. These successions were further deformed and metamorphosed at low temperatures during an intense and widespread Sudetian phase of folding and thrusting that migrated northwards across the foreland. Deformation continued with the Asturian phase, with incorporation of externally positioned foreland basins into the thrust belt. In south-west England, this event was shortly followed by the intrusion of a large granite batholith at depth and renewed basaltic volcanism at the surface.

2 *The Ossa Morena–North Armorica–Saxothuringian zones* consist of a number of complex terrane assemblages representing northern segments of the Armorican microplate. Low-grade metamorphic Devonian or Early Carboniferous sediments, volcanics and intrusives commonly rest unconformably on Early Palaeozoic or older successions that were variably metamorphosed up to high-grades by Early to Mid-Devonian times. A narrow belt of Late Devonian to Early Carboniferous calc-alkaline volcanics and intrusives runs along the Mid-German Crystalline Rise, representing remnants of a continental magmatic arc that formed along the northern margin of the Armorican microplate. These rocks, together with subduction-related high-pressure metamorphic rocks (eclogites, amphibolites and gneisses), were first deformed and rapidly uplifted during the Bretonian phase, before being further displaced by Sudetian thrusting. The southern boundary of this zone is also marked by ophiolites of both Early and Late Palaeozoic ages.

3 *The Central Iberian–Massif Central–Moldanubian zones* form the most internal units of the Variscan Belt and contain Gondwana-derived terrane assemblages of both the Iberian and Armorican microplates. They have many similarities with the northern segments of the Armorican microplate described above. Several windows of Cadomian basement are overlain by metamorphosed Palaeozoic successions that were affected by the Ligerian phase of deformation. Sudetian thrusting was widespread in this zone and was accompanied by high-grade metamorphism and intrusion of both I- and S-type syn- to post-collisional granites. High-level thrust nappes often contain an array of metamorphic rocks of variable age, including eclogites, granulites and high-pressure ultramafic rocks, such as mantle-derived garnet peridotites. The Asturian phase of deformation was largely absent as Late Carboniferous sedimentation continued uninterrupted into the Permian. High-potassium ignimbrite volcanism and limnic continental and coal-bearing sediments were deposited in a series of intramontane basins.

4 *The Cantabrian Zone* represents the southern external fold-and-thrust belt of the Variscan collision and preserves a near-complete weakly metamorphosed succession of Palaeozoic sedimentary rocks that were deposited upon the Iberian microplate. Late

Carboniferous foreland basins and localized pull-apart basins along narrow strike-slip zones also developed, contemporaneous with those formed within the South Portuguese–Rhenohercynian Zone. The dominant phase of thrusting occurred during the Late Carboniferous Asturian phase of deformation, which is also held to be responsible for the strong secondary oroclinal bending of the Ibero-Armorican arc, centred around northern Spain. An extension of the southern flank of Variscan Belt can also be found further east in southern France, and in Austria where less metamorphosed Palaeozoic units are preserved in the eastern Alps.

In general, the Variscan Belt provides an excellent opportunity to study the deeper roots of a collisional orogen in which the important crustal processes were: (i) oblique indentation of microplates; (ii) tectonic underplating; (iii) widespread high-temperature/low-pressure metamorphism; (iv) granite emplacement; (v) late orogenic extensional collapse. Also of particular economic interest are the world-class base- and lithophile-metal mineralizations of the South Portuguese–Rhenohercynian Zone, and the array of coal-bearing late orogenic foreland basins that developed along the external margins of the mountain belt.

15.3 Regional timing and character

Situated along the northern external margin of the European Variscides, Britain and Ireland experienced a range of Devonian–Carboniferous geological conditions, which varied greatly in time and space (Fig. 15.2). As a result, a diverse array of Variscan structures developed across the region, in association with localized syn-tectonic sedimentation, volcanism, metamorphism and the intrusion of a major granite batholith (Fig. 15.3). Understanding of the complex geological history along the northern Variscan margin has benefited greatly from the almost continuous coastal exposures of this region, the quality of which is unrivalled anywhere else in Europe.

During Early to Mid-Devonian times, the later stages of Caledonian deformation, uplift and erosion were affecting much of Britain and Ireland (see Chapter 12), as well as central Europe. At the same time, however, rifting along the southern continental margin of Eastern Avalonia (newly accreted to Laurentia to form Laurussia) led to the deposition of near-complete sequences of Devonian marine sediments and associated volcanics in south-west England and southern Ireland (see Chapter 13). Extensive thinning of continental crust occurred across Cornwall and Devon throughout the Devonian and continued into the Early Carboniferous. Sedimentation was characterized by complex facies variations and intracontinental (alkaline) volcanism.

Crustal thinning in south Cornwall was locally so intense that oceanic crust formed to floor a separate Devonian sedimentary basin lying to the south. An oceanic separation of restricted width, commonly referred to as the Rhenohercynian Ocean, is considered to have formed along the suture zone between the Avalonian and Armorican microplates. This suture zone represents the site of the consumed Rheic Ocean, which separated Avalonia from Armorica throughout much of the Early Palaeozoic before closing by the end of the Silurian (see also Fig. 2.6d, p. 29). The first onset of Variscan convergence can be recognized during the Mid- to Late Devonian (beginning of the Bretonian phase, Fig. 15.2), during which a fragment of oceanic crust, the Lizard ophiolite, was obducted along the northern continental margin of Armorica. These units were incorporated into advancing thrust sheets that shed their debris northward to form olistostrome deposits. The Early Palaeozoic high-grade crystalline rocks of Armorican affinity found in the English Channel area confirm that the suture zone between the Avalonian and Armorican microplates, the site of the former Rheic and Rhenohercynian oceans, runs through the southern extremity of Britain.

By Late Viséan–Early Namurian times, Sudetian phase north-directed thrusting had reached north Cornwall and Devon. Large crustal downflexuring in advance of the orogenic thrust load led to the development of new rapidly subsiding depocentres to the north, which hosted the Late Carboniferous clastic synorogenic successions in south-west England, south Wales and southern Ireland (see Chapter 14). While the metamorphic slates of Cornwall were cooling during exhumation within actively eroding thrust sheets, foreland basin sedimentation was occurring within the Culm Basin of Devon, within the Clare Basin of Ireland and in the South Wales Coalfield (Fig. 15.3). The progressively younger onset of this style of sedimentation within each basin towards the north records the steady migration of the flexural downwarp of the lithosphere in response to the northward propagating Variscan thrust load. This mechanism of

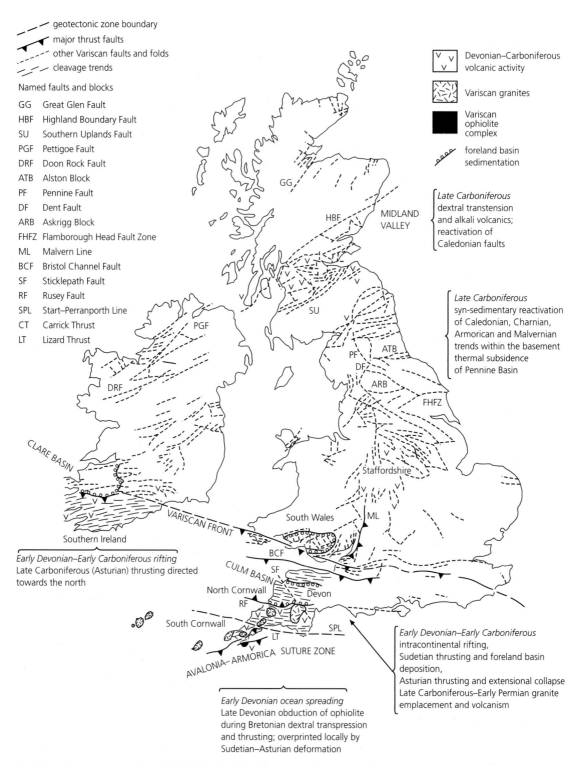

Fig. 15.3 Variscan structures of Britain and Ireland (based on the tectonic map of Britain, Ireland and adjacent areas published by the British Geological Survey 1996).

subsidence had a restricted influence on Late Carboniferous sedimentation further north, where rifting and thermal subsidence within a dextral strike-slip regime occurred in central England (such as Staffordshire) and the Midland Valley of Scotland.

Asturian thrusting propagated northwards across south-western Britain and southern Ireland during the late Westphalian to Stephanian time, progressively diminishing in intensity northwards. The most prominent northerly east–west-trending thrust marks the Variscan thrust front that divides the Subvariscan from the Rhenohercynian Zone. North of this front, the intensity of Variscan deformation greatly decreases. The gentle Variscan folding and faulting tended to adopt the trends of pre-existing structures inherited from underlying Lower Palaeozoic and Proterozoic rocks. Toward the end of Asturian thrusting, a phase of extensional collapse occurred within parts of the fold-and-thrust belt in south-west England. This was associated with the late to post-collisional intrusion of the Cornubian granites and with the regional uplift and exhumation that gave rise to the major Variscan unconformity across Britain and Ireland. Such activity provided the template for the deposition of the post-orogenic molasse deposits that characterize the beginning of the Permian (see Chapter 16).

The processes that occurred during the Variscan Orogeny will now be examined (see Sections 15.4–15.9), before considering the plate tectonic origin of the orogenic belt as a whole (see Section 15.10).

15.4 Ocean spreading and deformation along the Avalonia–Armorica Suture Zone

The Lizard and Start complexes represent some of the most internal and structurally highest tectonic units of the Variscan Orogen of mainland Britain, and are exposed at the southern most tips of south-western England. Both complexes contain basic igneous rocks of oceanic (mid-ocean ridge basalt; MORB) geochemistry envisaged to have formed during a Devonian phase of ocean spreading along the Avalonia–Armorica Suture Zone. The Lizard Complex preserves tectonically disrupted fragments of an ophiolite suite now shown to be of Early Devonian age (397±2 Ma). It represents one of the best examples of an ophiolite found in Britain. The pillow lava and pelagic sediments that form at the top of a complete

oceanic suite are not intact, as the unit has been dismembered by complicated polyphase folding and faulting. The ophiolite does, however, record a rare history of high-temperature, preobduction deformation that occurred within an oceanic environment. Mylonitized oceanic crustal rocks within low-angle, amphibolitic-facies ductile shear zones are considered to have formed close to, and dipping toward, an ocean spreading ridge axis or along transform faults. These characteristics are considered typical for slow spreading rates of <35 mm/year.

Three sets of basaltic–dolerite dykes are recognized in the ophiolite complex—in general, the older the intrusions, the shallower the dip of the dykes. In a tectono-magmatic reconstruction (Fig. 15.4a), the oldest set of dykes was intruded as early flat-lying

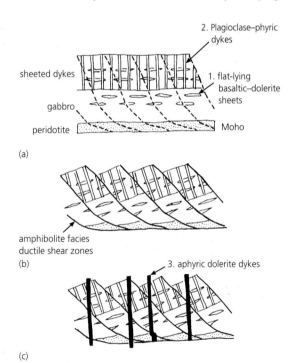

Fig. 15.4 Schematic tectono-magmatic evolution of the Lizard ophiolite (east coast section) showing three main stages of development during the formation of oceanic crust. (a) Initial generation of an ophiolite stratigraphy with early flat-lying basaltic–dolerite sheets cross-cut by plagioclase-phyric dykes; (b) activation of extensional structures in response to continued spreading during a period of low magma production; (c) renewed magmatic activity with the injection of a series of aphyric dolerite dykes (after Roberts *et al.* 1993, with permission from Elsevier Science (1999)).

basaltic–dolerite sheets in the gabbro, near the top of an early magma chamber. Rifting and ocean floor spreading gave rise to the main sheeted dyke complex, which was initially intruded at a high angle, but subsequently became rotated in blocks upon a series of listric extensional shear zones (Fig. 15.4b). These zones of displacement dipped toward the ridge axis and presumably flattened into a basal shear zone near to the petrological Moho. The amount of rotation (about 30%) suggests that as much as 40% thinning of the oceanic lithosphere may have taken place during this period of low magma production. Following rotation and cooling, a new phase of rifting injected the youngest set of dykes vertically through the tectonized peridotite, the shear zones and the rotated gabbro blocks at high crustal levels (Fig. 15.4c). These dykes are characterized by chilled margins and may represent late intrusions along the spreading axis, or could have propagated from a later gabbro magma chamber developed below the already thinned crust.

Other interesting features of the Lizard Complex are observed in the mantle peridotite. These rocks contain early subvertical mylonitic fabrics and steeply dipping mineral lineations that are cut by both gabbros and sheeted dykes. Such structures are suggestive of thinning of the upper mantle prior to the formation of oceanic crust, which led to exhumation of the peridotite during the initial stages of rifting.

The north-west–south-east strike of the sheeted dykes in the Lizard ophiolite formed oblique to the major Armorican-trending east–west faults suspected in the pre-Devonian basement. These relationships indicate either transtensional opening across a dextral shear zone (bulk extension across the suture zone), or more localized spreading within pull-aparts along an intracontinental strike-slip fault system. The restricted nature of the ocean is also shown by the apparent wander paths for Avalonia and Armorica, which reveal no significant separation in latitude between the two microplates during Late Silurian to Mid-Devonian time. A crude estimate of ocean width may be made by assuming a spreading rate of 10 mm/year (the accretion rate per ridge flank of the Red Sea) operated over a period of 20 Myr, the estimated hiatus between the generation of the Lizard ocean crust and its obduction. This would produce a maximum oceanic separation of just 400 km, a width not resolvable by palaeomagnetic methods.

Despite the restricted width of the ocean, it does appear that the oceanic crust that formed along the Avalonia–Armorican Suture Zone was consumed by southward-directed subduction. The detrital material within the Devonian sediments of the Gramscatho Basin has a forearc signature. This basin, and the occurrence of Early Palaeozoic gneisses (e.g. the Man of War Gneiss) that formed part of the Normannian High of the English Channel area, occur along strike from the Late Devonian to Early Carboniferous age continental forearc–arc assemblage of the Mid-German Crystalline Rise. These relationships suggest that the Lizard ophiolite was obducted within a south-dipping subduction zone complex that ran along the northern margin of Armorica.

The age of ophiolite obduction is well constrained by recent radiometric dating. At the basal tectonic contact of the uppermost ophiolite thrust sheet, mixed acidic and basic material, known as the Kennack Gneiss, was injected along the flat-lying tectonic contact between peridotite and amphibolite. The gneisses, probably derived from melting of Gramscatho flysch, appear to have been intruded and metamorphosed during thrusting, with cooling below 500°C during the Late Devonian (≈ 370 Ma). The final emplacement of the Lizard ophiolite occurred after about 360 Ma, because Late Devonian (Famennian age) palynomorphs occur in sedimentary rocks lying beneath the Lizard Boundary Fault. As no thermal effect has been recorded in the immediately underlying mudrock-dominated units, the ophiolite is considered to have cooled significantly during late stages of thrusting.

The emplacement of the ophiolite and the assemblage of tectonic units found along the Avalonia–Armorican Suture Zone in south Cornwall can be attributed to the Bretonian phase of deformation (Fig. 15.5). This phase was characterized by intense transpressive shortening and dextral shear accompanying closure of the Rhenohercynian oceanic separation, and reconvergence of the Avalonia and Armorican microplates. Collisional movements led not only to the northward thrusting of the Lizard ophiolite, but also to the inclusion and uplift of basement blocks of Armorican affinity (the Normannian High). South of the Start–Perranporth Line that marks the northern extent of the suture zone, the basement highs, together with a series of newly forming north-west–south-east-trending thrust sheets

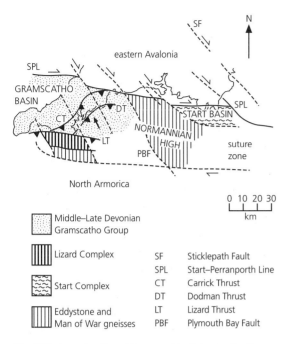

SPL – Start–Perranporth Line
SF – Sticklepath Fault
CT – Carrick Thrust
DT – Dodman Thrust
LT – Lizard Thrust
PBF – Plymouth Bay Fault

Fig. 15.5 Reconstruction of the assembly of structural units along the Avalonia–Armorican Suture Zone during the Bretonian phase of deformation (Late Devonian to Early Carboniferous). The Lizard and Start complexes were emplaced within a dextral transpressive shear zone following closure of the Rhenohercynian Ocean (modified after Holdsworth 1989, with permission from the Geological Society, London 1999).

(the Carrick, Dodman and Lizard nappes), shed their debris northward into the Gramscatho Basin to source both flysch and olistostrome deposits. The massive olistostromes found in the footwall of the Lizard Thrust, known as the Meneage, contain clasts of Ordovician quartzites bearing Gondwana faunas derived from the Armorican continent. The Bretonian thrust-system is correlatable with seismic reflectors dipping at 25–30° southwards beneath south Cornwall and the English Channel, which broadly correspond to the suture zone at depth.

15.5 Fold-and-thrust belt deformation in the Rhenohercynian Zone; basin inversion, dextral transpression and backthrusting

In comparison with many external fold-and-thrust belts of the world, the Rhenohercynian Zone of south-west Britain and southern Ireland is fascinatingly complex. In addition to multiple phases of deformation (the Devonian slates of south-west England may contain as many four rock cleavages), the belt shows examples of oblique thrust traces, lateral ramps, out-of-sequence thrusts and distinct areas of backthrusting. This complexity can be attributed to:

1 the rift–fault architecture of the Eastern Avalonia continental margin prior to collision;
2 the transpressive nature of the collisional deformation;
3 the multiple phases of Variscan folding and thrusting (Bretonian, Sudetian and Asturian phases);
4 the late orogenic extensional collapse of the thrust belt.

The transport directions of thrusting, folding and extensional faulting that have been mapped across south-west Britain and southern Ireland reflect these influences (Fig. 15.6). As the Variscan deformation propagated northward toward the foreland, thrusting interacted with east–west-trending fault-bounded sedimentary basins. The thick syn-rift sequences of the Gramscatho, Trevone, south Devon, north Devon (Exmoor) and Munster basins were successively inverted, and the style of deformation was strongly controlled by the position of pre-existing faults and the geometry of the basin-fill.

The principal direction of Variscan convergence was probably directed toward the north-west. This caused oblique compression and inversion of the extensional basins within a dextral transpressive regime. The predominant northerly direction of thrusting was clockwise oblique to the direction of orogen-wide convergence, so was accompanied by a component of dextral orogen-parallel shear. Reactivated steep faults, which presumably extended down into basement rocks, acted as lines where those oblique movements were focused. When accompanied by high strains, fold axes became rotated to lie oblique to the regional Variscan trend, as seen north of the Lizard ophiolite and in the Tintagel Strain Zone of north Cornwall. In other areas, such as along the Start–Perranporth Line, folds are strongly curvilinear about subhorizontal east–west extensional lineations, which are about 90° to the regional transport. These formed within zones of intense strike-slip ductile shear.

The faulted margins of some basins appear to have acted as buttresses to the northward-migrating compression and initiated zones of backthrusting.

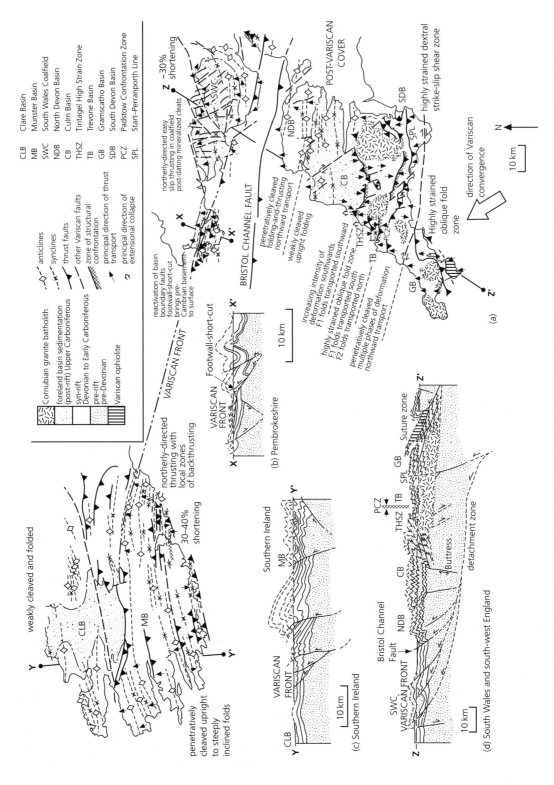

Fig. 15.6 (a) Structural map of south-west Britain and southern Ireland (compiled mostly from Sanderson 1984; Hanna & Graham 1988; Seago & Chapman 1988; Powell 1989; Warr 1993; Alexander & Shail 1995; Gayer et al. 1998). (b) Cross-section through Pembrokeshire. (c) Cross-section through southern Ireland. (d) Cross-section through south Wales and south-west England.

For instance, early backthrusts that influenced the northern shelf of the Trevone Basin propagated southward into the central part of the basin. Here they interacted with forethrusts in a zone of structural confrontation that can be found in both the Padstow and Plymouth areas (Fig. 15.6a,d). Although of less complexity, another intrabasinal confrontation zone can be found in the extreme south of Ireland (Fig. 15.6a). In Pembrokeshire, the reactivation of syn-sedimentary extensional faults truncated fragments of basement (footwall short-cuts), transporting them upward into higher thrust sheets (Fig. 15.6a,b).

The intense polyphase folding and thrusting across south-west England resulted in dramatic shortening of the upper crust, in excess of 50%. Although difficult to quantify, the degree of crustal extension prior to compression may have been equally high. Estimates of horizontal shortening in southern Ireland are similar to those of south Wales, with average values of around 30%. Shortening rapidly diminishes north of the Variscan Front, particularly along the northern crop of the South Wales Coalfield and in the Clare Basin.

The nature of the Variscan displacements at depth is still speculative. The recognition of shallow seismic reflectors has been used to propose thin-skinned models for the deformation, with a single regional décollement horizon flooring the Rhenohercynian fold-and-thrust system at about 10 km depth. Alternatively, displacements at depth may have occurred along a series of both shallow and deep detachment horizons that were linked to steeper faults in the basement (Fig. 15.6b–d). This model is more compatible with the strong precollision rifting, the significant components of strike-slip shear, and the steepness of structures observed in some parts of the belt (such as in the Munster Basin). The positions of detachments were likely to have been controlled by the geometry of the pre-, syn- and post-rift sedimentary layers, and distributed throughout the upper 15–20 km of the rethickened Variscan orogenic wedge. This style of tectonic deformation is considered to be typical for restacking and thickening of previously thinned continental lithosphere (Fig. 15.7).

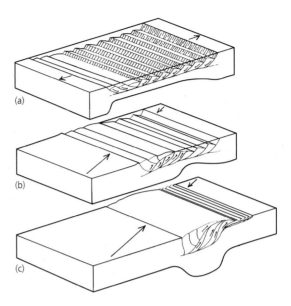

Fig. 15.7 Block diagram showing the thickening of previously thinned continental lithosphere: (a) extension of the upper continental crust by listric normal faulting; (b) onset of restacking with inversion of normal faults; (c) complete restacking and thickening of the lithosphere (after Dewey *et al.* 1986, with permission from the Geological Society, London 1999).

15.6 Late Carboniferous foreland basin development and deformation; underthrusting, rolling fold nappes and easy-slip thrusting

As the wave of Late Carboniferous deformation spread northward across the Rhenohercynian Zone, the thickening orogenic wedge loaded the lithosphere and flexed down the foreland. New sites of deposition formed between the Variscan thrust-wedge and the Laurussian continental craton to the north. The 3–4-km-thick clastic successions preserved in each of the Culm, South Wales Coalfield and Clare basins provide an indirect record of the developing mountain belt. The onset of foreland basin sedimentation occurred progressively later toward the north (Fig. 15.2), with rapid subsidence commencing in the southern part of the Culm Basin by the early Namurian, in south Wales by the early Westphalian and possibly in the Staffordshire Coalfield at the end of the Westphalian.

The Culm Basin shows a rather typical foreland basin infill (see Chapter 14). The distal turbidites at the base reflect the initial deepening of the basin due to high rates of subsidence, together with sediment starvation induced by rapid downflexure. With time the depocentre migrated northwards, in-phase with the propagating thrust-wedge (Fig. 15.8a). This led to the basin being infilled diachronously upward and towards the north, by a shallowing-upward sequence of deltaic facies, and then storm-influenced facies by Westphalian C times. An unusual feature of the Culm Basin is that much of the Westphalian sediment was derived from uplift to the north, in the area of the Bristol Channel, rather than from the orogenic wedge itself in the south. These sediments contained huge volumes of late Ordovician and older detritus (white mica $^{40}Ar–^{39}Ar$ ages, $\approx 450–470$ Ma) preserved in the diagenetic grade clastics. This source area may reflect the development of a flexural bulge that separated the Culm and South Wales basins, or transpressive uplift associated with strike-slip faulting. Another possibility is that the younger part of the Culm Basin fill developed as a piggy-back basin, whereby thrusting propagated through to the Bristol Channel Fault to form a landmass in the hangingwall of a thrust fault.

The intimate relationship between sedimentation and thrusting is well documented in the sedimentary record of the Culm Basin. Unstable slump horizons with soft-sediment deformation appear to have been directly utilized as detachment horizons during initial folding and thrusting. In contrast to many other foreland basins in the European Variscides, the southern margin of this basin shows a rare history of underthrusting (Fig. 15.8a). Boundary conditions initiated underthrusting of a metamorphic wedge and backthrusting of the Culm succession. This process may have been driven partly by density differences between the freshly deposited clastic sediments, and the denser, highly strained greenschist facies rocks of the underlying Tintagel High Strain Zone. As underthrusting proceeded, the fold nappes within the Culm hangingwall successively rolled over, increasing the length of overturned limbs at the expense of the right-way-up limbs. The last stage of structural development along the southern margin was the extensional collapse of the thrust pile toward the north, associated with the intrusion of the granite batholith (see Section 15.8).

The South Wales Coalfield also contains features that testify to the intimate relationship between sedimentation and deformation. Coal seams of early Westphalian A age were cannibalized and redeposited as coal clasts within late Westphalian C channel lag deposits. As these fragments were compacted and partially coalified before being incorporated into the fluvial channels, sedimentation, burial, uplift and resedimentation occurred within a short period of time (≈ 5 Myr). This rapid redistribution of material reflects the continued northward propagation of thrusting across the basin during the Westphalian. In contrast to the Culm Basin, the South Wales Coalfield shows a different style of thrusting reflecting the large number of coal seams that acted as easy-slip horizons. A variety of complex structures evolved, controlled by the position and geometry of the coal seams (Fig. 15.8b). Thrusts initially moved along the floors of the coal horizons until they locked up and produced hangingwall folds. Further displacements eventually caused the faults to break through the fold hinges, and to initiate new break-back thrusts in either the hangingwall or footwall segments. This process led to numerous repetitions of the coal seams, and although the average degree of shortening across the South Wales Coalfield is about 30%, localized shortening of up to 67% has been recorded.

15.7 Variscan metamorphism and thermal history

Regional metamorphism in the Rhenohercynian Zone ranged from deep diagenetic to the greenschist facies, and locally reached amphibolite facies in the Lizard and Start complexes. The pattern of metamorphic grade across south-west Britain and southern Ireland (Fig. 15.9) reflects a range of tectono-thermal conditions experienced during the Variscan development. In general, the Devonian rocks were more strongly metamorphosed than those of the Carboniferous.

The long period of lithospheric stretching that preceded the Variscan collision led to very high geothermal gradients along the southern continental margin of Eastern Avalonia. Sedimentary rocks in the deepest parts of the syn-rift sequences experienced temperatures probably as high as 300–350°C. Early stages of metamorphism in this extensional regime are suggested to have taken place in the north Devon, Trevone, south Devon and Munster basins. This style of metamorphism, characterized by low pressures,

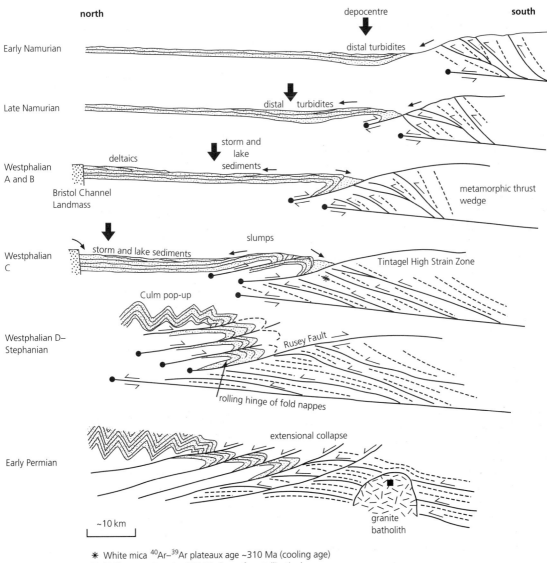

* White mica ^{40}Ar–^{39}Ar plateaux age ~310 Ma (cooling age)
■ U–Pb monazite age ~281 Ma (age of crystallization)

Fig. 15.8 (a) Early Namurian to Early Permian structural development of the Culm foreland basin of south-west England, showing the development of rolling fold nappes and the Culm pop-up during underthrusting by the Tintagel High Strain Zone. The area also underwent extensional collapse during emplacement of the granite batholith (based on the models of Seago & Chapman 1988 and Warr 1993; isotopic constraints provided by A. Clark, personal communication). (b) Progressive easy-slip thrusting in the South Wales Coalfield (Ffos Las).

Complex fold-and-thrust geometries are described in a four-stage model. Stage 1: Initial propagation of thrusts along floors of the coal seams. Stage 2: Formation of folds induced by the locking-up of faults. Stage 3: Initial break-back thrusting as the faults cut through the fold hinges. Stage 4: Continued out-of-sequence break-back thrusting in both hangingwalls (h) and footwalls (f) (after Frodsham et al. 1993, with permission from Viewegy (1999)).

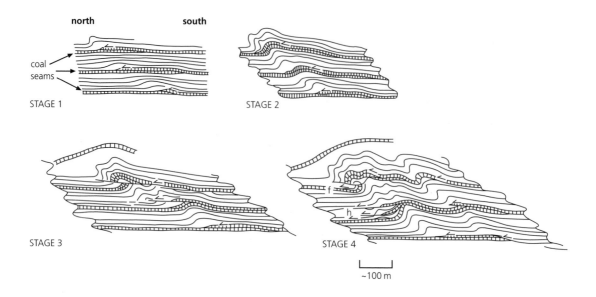

north south

coal seams

STAGE 1

STAGE 2

STAGE 3

STAGE 4

~100 m

Fig. 15.8 (*Continued.*)

has a sedimentary burial signature and is referred to as diastathermal metamorphism.

As a result of the enhanced temperatures, these rocks were particularly sensitive to recrystallization during the onset of deformation, and the mud-dominated lithologies developed an intense slaty cleavage during folding and thrusting (such as the Delabole slate of north Cornwall). Large volumes of fluids were released from the clay-rich pelitic rocks to form abundant syn-tectonic quartz–carbonate veins. The progressive thickening of the lithosphere from the south, caused by the Variscan collision, led to the tectonic burial of rocks and to thrust-related metamorphism. Examples of such metamorphism are documented from south and north Cornwall, where inverted metamorphic sequences exist. Although temperatures were not so different from the earlier diastathermal conditions, the pressure was notably higher. Polyphase deformation led to extensive syn-kinematic phases of recrystallization as well as the growth of metamorphic porphyroblasts.

Three periods of exhumation can be recognized in the K–Ar and ^{40}Ar–^{39}Ar white mica isotopic ages of the metapelites of south-west England, which correspond to the Bretonian, Sudetian and Asturian phases of deformation (Fig. 15.9). These ages are thought to record the time at which the rock cooled down through temperatures of ≈ 250–$320°C$. Bretonian ages (≈ 350–$385\,Ma$) are recorded only in the area between the Lizard and Dodman complexes. These old ages represent exhumation of the amphibolite and greenschist facies units during the early stages of obduction. Sudetian ages (≈ 320–$340\,Ma$) are found spread across much of south-west England and represent exhumation during, or immediately following, the first major phase of thrusting. The youngest cooling ages of the Asturian phase (≈ 260–$310\,Ma$) may represent areas of protracted burial, such as the underthrust rocks of the Tintagel High Strain Zone, or areas such as north Devon where slates remained deeply buried until Late Carboniferous time. As many of the Asturian ages lie in the vicinity of the granite batholith, hot hydrothermal fluids associated with intrusion could also be responsible for these younger cooling ages. Continued hydrothermal fluid activity during late to post-Variscan times is also evident from the extensive kaolinization of the granites, which gave rise to the world famous Cornish china clay deposits.

Crustal thickening across south-west Britain and Ireland during the Carboniferous led to the progressive lowering of the geothermal gradients from the south. Temperature gradients remained high in the foreland, with estimates of 40–$50°C$ per kilometre for

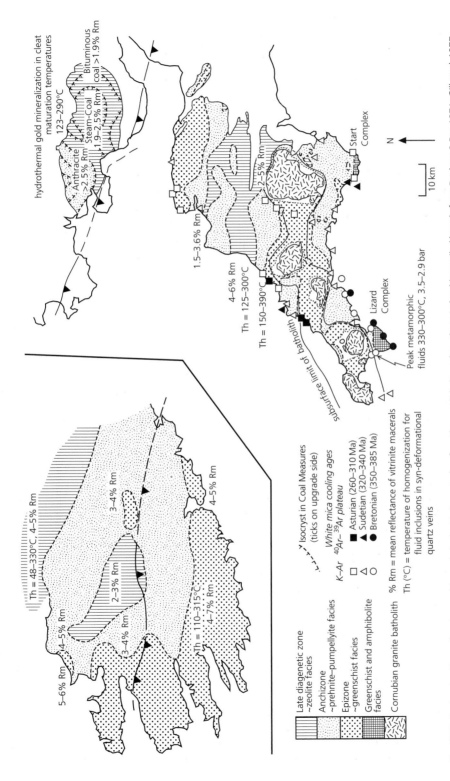

Fig. 15.9 Regional pattern of metamorphic grade in the Variscan rocks of south-west Britain and southern Ireland (compiled largely from Dodson & Rex 1971; Gill *et al.* 1977; Warr *et al.* 1991; Blackmore 1995) (unpublished Ar–Ar results were kindly supplied by A. Clark, Kingston, Ontario).

the Culm and South Wales Coalfield basins. These conditions reflect the high degree of crustal thinning that occurred prior to collision. Although the degree of coalification generally increases with burial depth, the South Wales Coalfield does show an interesting temperature anomaly in the north-west part of the basin. Higher temperatures there were responsible for the anthracite grades of coal, which have been of significant economic importance. One suggestion is that the anomaly reflects the upward surge of hotter Palaeozoic metamorphic fluids, pumped up by the advancing thrust-wedge, so high-temperature geotherms need not be solely due to earlier crustal thinning. These fluids infiltrated the coal seams along brittle fractures (cleat), causing hydrothermal mineralization and the precipitation of gold before the onset of easy-slip thrusting.

15.8 Granite plutonism and extensional collapse

The Cornubian Batholith is the most voluminous igneous body ($\approx$ 68 000 km^3) in the Variscan Belt of Britain and Ireland (Fig. 15.3). It consists of a broadly linear array of plutons extending northeast–south-west for over 220 km from the underwater platform of Haig Fras to the highland of Dartmoor. This extensive suite of dominantly coarse-grained biotite granites, containing both biotite–muscovite and biotite–cordierite varieties, are peraluminous, broadly S-type and anatectic in origin. The body is rather anomalous in the Rhenohercynian Zone of Europe, although compositionally similar Variscan batholiths of older ages are common in the more internal parts of the orogen.

The relatively young, less hydrous and higher-temperature biotite–cordierite granites were generated by partial melting of pelitic rocks in the lower or intermediate crust, and further modified by crystal fractionation at their level of emplacement. Anatexis has been estimated to have occurred at about 800°C and 5 kbar. Enclaves of diverse origin, consisting of both non-igneous pelitic and intermediate igneous compositions, are present in minor amounts (up to 5 vol%). Whereas the textures and compositions of the pelitic enclaves indicate variable degrees of melt extraction from metasedimentary source rocks, the igneous enclaves are considered to originate from magma mixing and anatexis caused by incursion of mafic melt from the mantle.

The batholith comprises six main granite plutons, which show a complex intrusive history. U–Pb and ^{40}Ar–^{39}Ar dating studies indicate that the plutons were emplaced successively over an interval of some 20 Myr, spanning the Late Carboniferous to Early Permian (Fig. 15.2), and cooled rapidly below $\approx$320°C within 4–5 Myr of intrusion (at their present erosion level). This protracted history of intrusion and cooling was accompanied by intense hydrothermal mineralization caused by the circulation of post-magmatic metalliferous brines. Tungsten (W)-rich (greisen-bordered) vein systems have been ascribed to fluid release during retrograde boiling that accompanied cooling of the outer zones of the intrusions. Main-stage Sn–Cu lodes resulted from the continued release of fluid that migrated through substantial volumes of the almost-solid granite. This world-class mineralization extended from 286 to $\approx$265 Ma and was controlled largely by extensional faulting during cooling of each intrusive centre. The east-north-east- or north-north-west-trending mineralized fault zones probably formed by reactivation of steep Variscan or older basement structures (Armorican and Charnian trends; see Section 15.9) and were associated with an important phase of extensional collapse that accompanied emplacement of the granites. Along the margins of the plutons, late magmatic quartz porphyry (elvans) and lamprophyre dykes commonly infiltrated extensional faults.

The crystallization ages of the granites, and the way the intrusions deflect and truncate Variscan thrusts, confirm the late orogenic origin of the Cornubian Batholith. The age of the earliest major intrusion (293±1 Ma for the Carnmenellis granite) indicates that the north-north-west-directed thrusting in south Cornwall had ended prior to the late Stephanian. Upwarping of adjacent and overlying rocks led to the exhumation of deeper metamorphic units (epizonal grades of white mica [illite] crystallinity) above the subsurface limit of the batholith (Fig. 15.9). Accompanying late orogenic extensional collapse produced a vast number of both low- and high-angle extensional detachments, with down-dip movements along pre-existing rock fabrics directed away from the core areas of the batholith (Fig. 15.6a). Extensional faulting and magmatic activity was also contemporaneous with the subaerial eruption of the lamprophyric and K-rich basaltic Exeter Volcanic Series during the Stephanian to Early Permian. Inclusions of mantle-derived mafic rocks within the Exeter volcanics show close chemical affiliation with the igneous enclaves found in the granites.

Horizontal seismic reflectors at crustal depths of about 8 and 20 km have led to the suggestion that the batholith formed as a shallow, flat-bottomed body that was injected along a regional décollement surface toward the base of the Variscan thrust-wedge. The individual granites then moved upwards, diapirically and by stoping, to form the separately exposed intrusions. Major anatexis probably occurred during the stage at which the orogenic wedge was undergoing abrupt and major thinning by extensional collapse of the Variscan mountain belt in a similar fashion to the Himalyan–Tibet Orogen. The igneous and hydrothermal character, however, show closer similarities to that of the Cordillera Oriental (Eastern Cordillera) of south-east Peru and north-west Bolivia, which are considered to represent a collisional but strictly intraplate setting. The intrusion of the Cornubian Batholith, and the lithophile and base metal hydrothermal activity, therefore appears to have spanned the late to post-collisional plutonic stages of orogenic development. The large volume of crustal melting, the invasion of mantle-derived mafic melt, the extensional collapse and exhumation of metamorphic rocks, as well as the coeval volcanic activity, together suggest the advection of subcrustal heat into regions of actively thinning crust.

15.9 Reactivation tectonics and basin inversion in the Subvariscan Zone

The intensity of compressive deformation rapidly decreased northward from the Variscan Front. Throughout much of the Subvariscan of Britain and Ireland, Late Palaeozoic rocks were only gently folded and faulted, with a large variety of structural trends reflecting the rejuvenation of the underlying basement grain. These structures are summarized as follows in order of their abundance (refer to Fig. 15.3):

1 *The north-east–south-west Caledonian trend.* Dominant across central and northern Ireland, north-ern England and Scotland. This trend reflects significant reactivation of pre-existing Caledonian faults with predominantly dextral strike-slip movements, such as that recorded along the Highland Boundary and Southern Uplands faults.

2 *The north-west–south-east Charnian trend.* Common in central England, consisting of normal faults with some component of sinistral strike-slip shear. This structural trend reflects the orientation of the north-eastern edge of Avalonia (parallel to the Tornquist Line of eastern Europe).

3 *The east–west Armorican trend.* Well developed just north of the Variscan Front, but also found scattered across northern England and the Midland Valley. These structures largely represent normal faults that reactivated old east–west lineaments. South of the Iapetus Suture, this east–west grain occurs within the Cadomian basement and also runs roughly parallel to the Avalonia–Armorican Suture Zone.

4 *The north–south Malvernian trend.* A distinctive zone of lineaments running from the Malvern Line, through the Midlands (e.g. Staffordshire), to the western border of the Alston Block. These are largely compressive structures such as folds and thrusts that also reflect an older Cadomian north–south grain. This regional trend parallels the suture between Laurentia and Baltica and formed the lines of weakness utilized by the younger Atlantic and North Sea rift systems.

In the Pennine and Midland Valley basins, these structures were active during Late Carboniferous sedimentation, producing a variety of syn-sedimentary folds and faults (Fig. 15.10a,b). A complex interaction of extension and compressional movements occurred synchronous with phases of rifting, thermal subsidence and basin inversion that overlapped in time.

In the Pennine Basin, the influence of extensional structures on basin evolution broadly decreased from Early Carboniferous to Westphalian C times. Growth folds and reverse faults developed along Caledonian and Malvernian trends, whereas syn-sedimentary normal faults were active along the lines of Armorican and Charnian trending structures (Fig. 15.10a). During this period, there was also a concomitant increase in the influence of compressive structures, which predominated between Westphalian C and

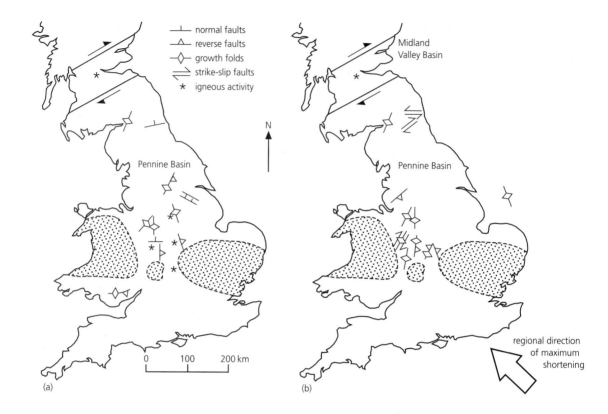

Fig. 15.10 Principal syn-sedimentary tectonic structures present in the Pennine Basin during: (a) late Namurian to early Westphalian C times; (b) late Westphalian C to Permian times. Stipple represents the extent of the emergent Wales–Brabant Massif (after Waters *et al.* 1994, with permission from the Geological Society, London 1999).

Permian times (Fig. 15.10b). Extensional faults were inverted to form thrusts or strike-slip faults and Caledonian, Charnian and Malvernian trends were marked by syn-sedimentary growth folds and reverse faults, along with components of oblique movement, that gave rise to en-echelon periclines and flower structures.

In the Midland Valley of Scotland, dextral strike-slip movements along Caledonian faults produced a zone of transtensional subsidence and the formation of a complex pull-apart basin. The predominance of crustal extension in this area is also supported by the widespread alkaline volcanic activity, which continued until the end of the Westphalian. Similar volcanism is also recorded in the Pennine Basin until Westphalian C times.

Overall, Late Carboniferous syn-sedimentary deformation across the Subvariscan Zone was characterized by pulses of north–south extension and east–west to north-west–south-east compression. This pattern can be attributed to the reactivation of major basement faults within a regional dextral strike-slip regime, which was superimposed upon a region undergoing extensive thermal subsidence. The geometry and intensity of structures developed during basin inversion toward the end of the Carboniferous was primarily controlled by the position and orientation of the basement grain in relation to a north-west–south-east to north-north-west–south-south-east regional direction of maximum shortening.

15.10 Geodynamic evolution and plate tectonic character

Having documented the history across Britain and Ireland, this section will consider the geodynamic evolution and plate tectonic origin of the Variscan

Orogeny. There is now general agreement that the Variscan Belt presents a mosaic of microplates and geological terranes that collided within an oblique (dextral) collisional regime throughout Mid- to Late Palaeozoic times. The Late Carboniferous constellation of welded continents that made up Pangaea at the end of the Variscan Orogeny produced a broad triangular zone of microplates sandwiched between Avalonia and Baltica, to the north, and Gondwana (Africa) to the south. However, the precise history of the continental interactions involved is not at all well constrained and is a subject of continued debate. A speculative plate tectonic reconstruction is here presented (Fig. 15.11a–c).

Based on palaeomagnetic pole positions, Armorica is suspected to have encroached upon Avalonia by the end of the Silurian, with closure of the Rheic Ocean. This convergence was, however, locally reversed during a phase of Early to Mid-Devonian transtensional shearing, with the development of a restricted oceanic separation along the Avalonia–Armorican Suture Zone (Fig. 15.11a). This Rhenohercynian

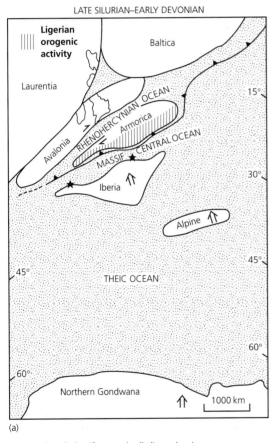

(a)

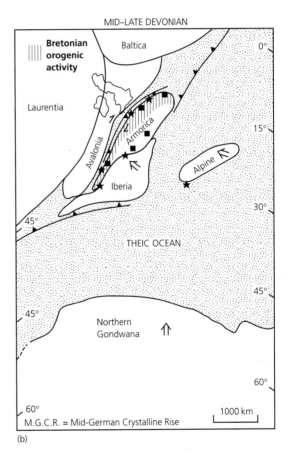

(b)

M.G.C.R. = Mid-German Crystalline Rise

■ Devonian–Carboniferous calc alkaline volcanics
★ Devonian–Carboniferous ophiolites

Fig. 15.11 Plate tectonic model for the construction of the Variscan Orogeny. Latitudes are based on available palaeomagnetic data (e.g. Tait *et al.* 1997). (a) Late Silurian closure of the Rheic Ocean. (b) Mid- to Late-Devonian southward subduction of the Rhenohercynian Ocean beneath the Mid-German Crystalline Rise. (c) Late Carboniferous final amalgamation of Pangaea, with indentation and formation of the Iberian–Armorican arc. (d) Tectonic collage of microplates welded to the southern margin of Asia. The Himalayan–Tibet Orogeny, caused by the indentation of India, shows many similarities to the Variscan development (based on Dewey & Burke 1973, with permission from the University of Chicago Press (1999); McElhinny *et al.* 1981).

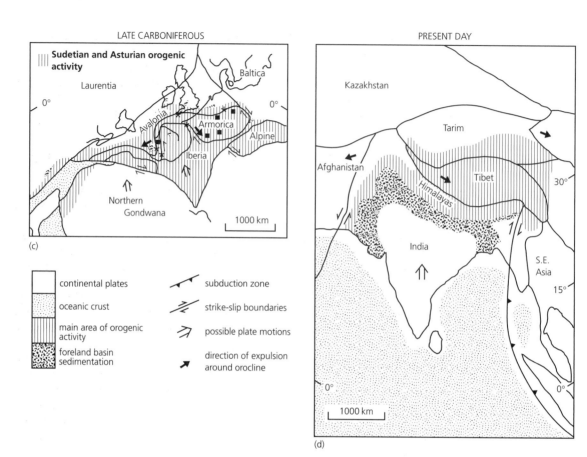

Fig. 15.11 (*Continued.*)

Ocean, part of which is preserved as the Lizard ophiolite, may have formed during oblique back-arc spreading above a subduction complex that dipped northward beneath the southern margin of the Armorican microplate (Massif Central). Northerly subduction and collision between Iberia and Armorica is documented by the Ligerian phase of orogenic activity in central Europe, which ended during the Mid-Devonian.

Towards the end of the Devonian, renewed convergence occurred across the Avalonia–Armorican Suture Zone, with transpressive closure of the restricted oceanic basin (Fig. 15.11b). The Bretonian phase of deformation appears to have been associated

with southward-directed subduction of Rhenohercynian oceanic crust and the formation of a continental arc along the Mid-German Crystalline Rise and Normannian High. During this phase, the Rhenoherycnian plate was underthrust beneath the Saxothuringian and was tectonically accreted to its base by underplating. The Gramscatho sediments could well in part represent deposition within the fore-arc basin associated with this subduction. Meanwhile, to the north of the subduction complex, extensive thinning of the passive continental margin of Eastern Avalonia continued with deposition of the typical Rhenohercynian basin and swell facies. This thinning may reflect renewed back-arc extensional forces associated with northerly subduction along the southern margin of the Iberian–Armorican continents that led to the closure of the remaining Theic Ocean.

Whereas rifting and thermal subsidence characterized much of the Subvariscan Zone throughout the Carboniferous, final closure and collision between the southern margin of the Iberian–Armorican microplates and the Gondwana continent led to widespread deformation across the orogenic zone during the Late Carboniferous (Fig. 15.11c). The Sudetian phase of deformation may have been initiated by the collision of an Alpine microplate, which is thought to now lie beneath the Alpine mountain chain, or it may reflect the first contact with promontories of the Gondwana continent itself. The resulting regional compression, which was transmitted across most of the belt, caused large-scale crustal thickening of the internal core areas and intrusion of both I- and S-type collisional granites. Both compressional and subsequent extensional deformation migrated outwards, from the internal core of the orogen, towards the forelands. Extensive folding, thrusting and downflexuring of the previous stretched lithosphere formed coal-bearing foreland basins along the external thrust front.

By the end of the Carboniferous, the Asturian phase of deformation marked the locking up and welding together of the Variscan mosaic as the collision with Gondwana reached its peak. Strong secondary oroclinal bending of the belt gave rise to its present complex curvilinear shape, with formation of the Iberian–Armorican arc (Fig. 15.11c). Indentation also caused lateral expulsion and widespread strike-slip movements. This terminating phase of orogenic activity was accompanied by the intrusion of large volumes of late to post-collisional S-type granites and late orogenic extensional collapse. Collapse within the internal zones of the mountain belt led to intramontane sedimentation and high-potassium volcanism during Late Carboniferous–Early Permian times.

Many of the characteristics described in the Variscan Orogeny can be studied at higher tectonic levels in southern Asia. The late collisional stages show particular similarities with the more recently formed Himalayan–Tibet mountain range (compare Fig. 15.11c and d). Southern Asia consists of a collage of microplates that have been accreting since the Permian. The youngest indentation of India over the last 30 Myr has formed a broad orogenic zone that, like the Variscan, is complexly curved and contains large volumes of granite. Other familiar aspects are the high-temperature, low-pressure regional metamorphism, the rapid changes in metamorphic grade across major steeply dipping shear zones, and the small volumes of locally obducted ophiolites. The more recent tectonic events of the last 10 Myr also show notable similarities with the late Variscan development, characterized by extensional collapse, the formation of intramontane rift basins and high-potassium volcanism.

Such a comparison aids understanding of the mountain-building processes that took place during the Variscan Orogeny. The combination of plate tectonic processes that appear to be important were the oblique nature of the collision, the irregular shape of microplates and the tectonic underplating that occurred along convergent margins. Central parts of the collisional zone also suffered significant lithospheric thickening, which presumably led to high topographic elevations during the Carboniferous. Thickening and uplift gave way to extensional collapse and crustal thinning, which induced significant anatexis of the lower crust during late and post-collisional stages. These events may have occurred in response to a sudden increase in buoyancy of the lithosphere, driven by the downward detachment of a part of its heavy roots (thickened upper mantle). Whatever the processes involved, the complexity of the Variscan Orogeny is in strong contrast to the more linear segments of the contemporaneous Alleghenian Belt of North America, which is attributed to a relatively straightforward collision normal to its strike.

15.11 Summary: the welding of Pangaea

The Variscan Orogeny produced a wide range of geological conditions through Devonian and Carboniferous times, which are summarized as follows (Fig. 15.12):
1 In the Early Devonian, a phase of transtensional rifting occurred along the suture zone between Avalonia and Armorica. This may reflect back-arc spreading above a northerly dipping subduction zone situated at the southern margin of Armorica. Ocean spreading gave rise to the Lizard ophiolite, which formed part of a restricted Rhenohercynian Ocean Complex. This phase of ocean rifting probably ceased

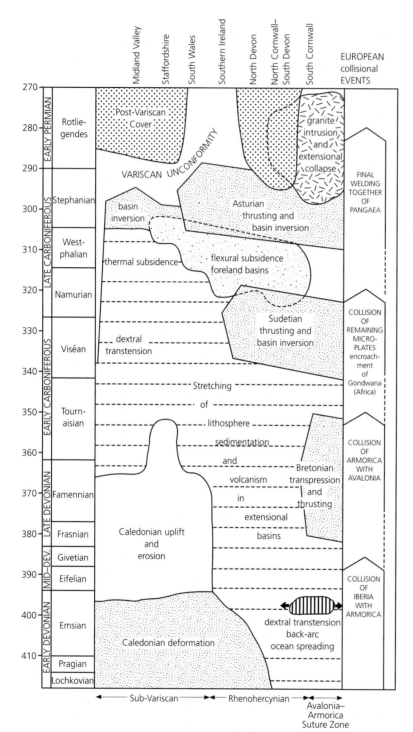

Fig. 15.12 Proposed chronology of Variscan events.

during the Ligerian collision of Iberia with Armorica.

2 The Mid-Devonian to Early Carboniferous Breton-ian phase of transpression and northerly directed thrusting was associated with southerly subduction of Rhenohercynian oceanic crust and convergence of Armorica with Avalonia. To the north of this subduction complex, lithospheric stretching continued along the passive continental margin of Eastern Avalonia. This stretching may have resulted from intermittent back-arc extension in association with renewed northerly directed subduction beneath the southern margin of the Iberian microplate.

3 During the Early to Mid-Carboniferous, the Sudet-ian phase of deformation caused north-north-west foreland-directed thrusting across the Rhenohercyn-ian Zone with tectonic inversion of sedimentary basins across Cornubia. A northward-propagating thrust-wedge initiated flexural subsidence along the external thrust front, with the formation of coal-bearing foreland basins.

4 In Late Carboniferous to Early Permian time, the Asturian phase of deformation was driven by the final assembly of remaining Gondwana-derived microplates, and the Gondwana continent itself, to form the supercontinent of Pangaea. This consolida-tion caused large-scale secondary bending of the oro-genic belt, and the intrusion of late to post-collisional granites over a period of 20 Myr in south-west England. Intrusive activity was associated with extensional collapse of the orogenic belt. Across the Subvariscan, thermal and flexural mechanisms of subsidence interacted with the intensifying Variscan deformation to produce a number of syn-sedimentary folds and faults along reactivated basement struc-tures. Pulses of dextral strike-slip shear deformation and a north-west-directed regional compression led to widespread basin inversion.

References

Alexander, A.C. & Shail, R.K. (1995) Late Variscan structures on the coast between Perranporth and St Ives, Cornwall. *Proceedings of the Ussher Society* **8**, 398–404.

Blackmore, R. (1995) Low-grade metamorphism in the Upper Palaeozoic Munster Basin, southern Ireland. *Irish Journal of Earth Sciences* **14**, 115–133.

Dewey, J.F. & Burke, K.C.A. (1973) Tibetan, Variscan, and Precambrian reactivation: products of continental collision. *Journal of Geology* **81**, 683–692.

Dewey, J.F., Hempton, M.R., Kidd, W.S.F., Saroglu, F. &

Sengör, A.M.C. (1986) Shortening of continental lithosphere: the neotectonics of Eastern Anatolia—a young collision zone. In: *Collision Tectonics* (eds M. P. Coward & A. C. Ries), Special Publication 19, pp. 3–36. Geological Society, London.

Dodson, M.H. & Rex, D.C. (1971) Potassium–argon ages of slates and phyllites from south-west England. *Quarterly Journal of the Geological Society, London* **139**, 371–412.

Franke, W. (1989) Tectono-stratigraphic units in the Variscan belt of central Europe. *Geological Society of America, Special Paper* **230**, 67–89.

Frodsham, K., Gayer, R.A., James, E. & Pryce, R. (1993) Variscan Thrust Deformation in the south Wales coalfield—a Case Study from Ffos–Las Opencast Coal Site. In: *Rhenohercynian and Subvariscan Fold Belts* (eds R. Gayer, R. O. Greiling & A. K. Vogel), pp. 315–348. Vieweg, Braunschweig/Wiesbaden.

Gayer, R., Hathaway, T. & Nemcok, M. (1998) Transpressionally driven rotation in the external orogenic zones of the Western Carpathians and the SW British Variscides. In: *Continental Transpressional and Transtensional Tectonics* (eds R. E. Holdsworth, R. A. Strachan & J. F. Dewey), Special Publication 135, pp. 253–266. Geological Society, London.

Gill, W.D., Khalaf, F.I. & Massoud, M.S. (1977) Clay minerals as an index of the degree of metamorphism of the carbonate and terrigenous rocks in the South Wales coalfield. *Sedimentology* **24**, 675–691.

Hanna, S.S. & Graham, R.H. (1988) A structural context of strain measurements on reduction spots in the Alpes Maritimes and the Hercynian fold belt of Southern Britain. *Annales Tectonicae* **2**, 71–83.

Holdsworth, R.E. (1989) The Start–Perranporth line: a Devonian terrane boundary in the Variscan orogen of SW England. *Journal of the Geological Society, London* **146**, 419–421.

McElhinny, M.W., Embleton, B.J.J., Ma, X.H. & Zhang, Z.K. (1981) Fragmentation of Asia in the Permian. *Nature (London)* **293**, 212–216.

Powell, C.M. (1989) Structural controls on Palaeozoic basin evolution and inversion of south-west Wales. *Journal of the Geological Society, London* **146**, 439–446.

Roberts, S., Andrews, A., Bull, J.M. & Sanderson, D.J. (1993) Slow-spreading ridge tectonics: evidence from the Lizard complex, UK. *Earth and Planetary Science Letters* **116**, 101–112.

Sanderson, D.J. (1984) Structural variations across the northern margin of the Variscides in NW Europe. In: *Variscan Tectonics of the North Atlantic region* (eds D. H. W. Hutton & D. J. Sanderson), Special Publication 14, pp. 149–165. Geological Society, London.

Seago, R.D. & Chapman, T.J. (1988) The confrontation of structural styles and the evolution of a foreland basin in central SW England. *Journal of the Geological Society, London* **145**, 789–800.

Tait, J.A., Bachtadse, V., Franke, W. & Soffel, H.C. (1997) Geodynamic evolution of the European Variscan fold belt: palaeomagnetic and geological constraints. *Geologische Rundschau* **86**, 585–598.

Warr, L.N. (1993) Basin Inversion and Foreland Basin Development in the Rhenohercynian Zone of South-west England. In: *Rhenohercynian and Subvariscan Fold Belts* (eds R. Gayer, R. O. Greiling & A. K. Vogel), pp. 198–224. Vieweg, Braunschweig/Wiesbaden.

Warr, L.N., Primmer, T.J. & Robinson, D. (1991) Variscan very low-grade metamorphism in south-west England: a diastathermal and thrust-related origin. *Journal of Metamorphic Geology* **9**, 751–764.

Waters, C.N., Glover, B.W. & Powell, J.H. (1994) Structural synthesis of S Staffordshire, UK: implications for the Variscan evolution of the Pennine Basin. *Journal of the Geological Society, London* **151**, 697–713.

Ziegler, P. (1989) *Evolution of Laurussia.* Kluwer Academic, Dordrecht.

Further reading

Chen, Y., Clark, A.H., Farrar, E., Wasteneys, H.A.H.P., Hodgson, M.J. & Bromley, A.V. (1993) Diachronous and independent histories of plutonism and mineralisation in the Cornubian Batholith, south-west England. *Journal of the Geological Society, London* **150**, 1183–1119. [A key paper presenting the latest isotopic dating of the Variscan Cornubian plutons and their related mineralizations.]

Corfield, S.M., Gawthorpe, R.L., Gage, M., Fraser, A.J. & Besly, B.M. (1996) Inversion tectonics of the Variscan foreland of the British Isles. *Journal of the Geological Society, London* **153**, 17–32. [A comprehensive case study of Late Carboniferous inversion tectonics in the Subvariscan.]

Gayer, R.A., Greiling, R.O. & Vogel, A.K. (eds) (1993) *Rhenohercynian and Subvariscan Fold Belts*, pp. 1–390. Vieweg, Braunschweig/Wiesbaden. [A compilation of research covering the northern external margin of the Variscan Orogeny. Particularly useful for its collection of papers concerning the Late Carboniferous foreland basin history of Europe.]

Gayer, R., Garven, G. & Rickard, D. (1998) Fluid migration and coal-rank development in foreland basins. *Geology* **26**, 679–682. [A short topical paper suggesting that the high coal-rank grades of the South Wales Coalfield were related to temperature anomalies caused by fluid discharge.]

Holder, M.T. & Leveridge, B.E. (1986) A model for the tectonic evolution of south Cornwall. *Journal of the Geological Society, London* **143**, 125–134. [A detailed tectono-stratigraphic model for the emplacement of the Lizard ophiolite and associated rocks, which conforms well with recent isotopic dating.]

Holdsworth, R.E. (1989) The Start–Perranporth line: a Devonian terrane boundary in the Variscan orogen of SW England. *Journal of the Geological Society, London* **146**, 419–421. [A short and readable paper presenting the idea of a E–W striking suture zone running through south-west England, which separated the Avalonian and Armorian microplates.]

Hutton, D.H.W. & Sanderson, D.J. (eds) (1984) *Variscan Tectonics of the North Atlantic region.* Special Publication 14, pp. 1–270. Geological Society, London. [Probably the best compilation of research dealing with controversial issues of Variscan tectonics of the North Atlantic region.]

Matte, P. (1991) Accretionary history and crustal evolution of the Variscan belt in Western Europe. *Tectonophysics* **196**, 309–337. [A good overview of crustal evolution during the Variscan Orogeny.]

Seago, R.D. & Chapman, T.J. (1988) The confrontation of structural styles and the evolution of a foreland basin in central SW England. *Journal of the Geological Society, London* **145**, 789–800. [An interesting contribution relating early backthrusting in south-west England to progressive underthrusting of the southern margin of the Culm foreland basin.]

Shail, R.K. & Wilkinson, J.J. (1994) Late- to post-Variscan extensional tectonics in south Cornwall. *Proceedings of the Ussher Society* **8**, 262–270. [A detailed case study of the extensional structures in south Cornwall, and their relationships to magmatism and accompanying main-stage phases of mineralization.]

Warr, L.N., Primmer, T.J. & Robinson, D. (1991) Variscan very low-grade metamorphism in southwest England: a diastathermal and thrust-related origin. *Journal of Metamorphic Geology* **9**, 751–764. [An extensive review of the tectono-thermal development of the Variscan fold-and-thrust belt of south-west England.]

Part 6
Post-Variscan Intraplate Setting

Illustration overleaf: View east of interbedded Upper Jurassic (Portlandian Stage) limestones and shales (Purbeck Beds) at Lulworth Cove, Dorset (reproduced by kind permission of Landform Slides). The folds formed above old normal faults in the Mesozoic Wessex Basin, reactivated as reverse faults during late Alpine compression in the Tertiary.

16 Permian to Late Triassic post-orogenic collapse, early Atlantic rifting, deserts, evaporating seas and mass extinctions

A. H. RUFFELL AND R. G. SHELTON

16.1 Plate tectonic framework; post-orogenic collapse and proto-Atlantic rifting

Variscan collision occurred in a zone of deformation from eastern Europe and northern Germany, through northern France and southern England and Wales to eastern North America and beyond. The resultant roughly east–west fold and thrust belt appears to have no geometric relation to the succeeding north–south proto-Atlantic rifting of Permian and Triassic times. Plate reconstructions of northern Europe show that, rotated back to its original position, the east-north-east–west-south-west trending Variscan foldbelt is not too dissimilar to the proto-Atlantic rift trend from the Arctic Ocean to north Africa (Fig. 16.1). The Variscan collision of Gondwana from the south with Laurussia to the north and the closure of the Rheic Ocean was but one part of the coalescence of the Permian–Triassic supercontinent, Pangaea. Very few continental fragments escaped incorporation into Pangaea, which was, by sheer size, unstable from its creation. Parts of the supercontinent began to break away, while in other areas collision continued. In the majority of areas around the globe, this coalescence occurred in the latest Carboniferous and earliest Permian, or from 300 to 296 Ma. The timing of break-up was varied: immediately north of the Variscan Orogen, Carboniferous–Permian syn-orogenic collapse occurred. In many places (e.g. Germany and Belgium), extension to the north of the Variscides occurred at the same time as thrusting and strike-slip faulting further south. The previous chapter showed how the Asturian phase of deformation continued into the Early Permian in Iberia, southern France and Germany. Following the main phases of compression, thermal relaxation of the crust occurred in Early Permian times, creating the rifts and graben that allowed accumulation of the first phase of sedimentation we are concerned with here. Later, there was renewed intracontinental rifting in the Triassic. Very Early Permian rocks are not widespread in Northern Europe, partly as a consequence of later erosion, partly because of their originally limited areal extent.

The late Carboniferous Variscan Orogeny caused the simultaneous uplift, intermontane rifting and foreland post-orogenic collapse that produced a widespread unconformity between the Permian–Triassic and pre-Permian strata. Above this unconformity and in the few areas where a conformable late Carboniferous–Early Permian succession is preserved, deep sedimentary basins formed. There is very little inheritance of Devonian–Carboniferous basins in the Permian–Triassic; instead, many of the later basins are situated in new rifts away from earlier Variscan syn-tectonic basins. Seismic data and cross-sections from boreholes and outcrop maps show these basins as having either a roughly symmetrical 'sag'-like form (Southern North Sea Basin; Plymouth Bay Basin) or as graben and half-graben (Worcester Graben, Minches Basin, Western Approaches Basin). These basin forms probably relate to their origin as post-collisional thermal collapse (Southern North Sea) or mechanical rift (Worcester Graben), respectively. Basins in the western British Isles that preserve a Permian succession are clearly parallel to the present-day Atlantic rift (Fig. 16.2), suggesting a common origin.

The timing and extent of individual phases of extension and rifting throughout the North Atlantic (and associated) rift systems are still subject to some debate, as the dating of Late Carboniferous and Early

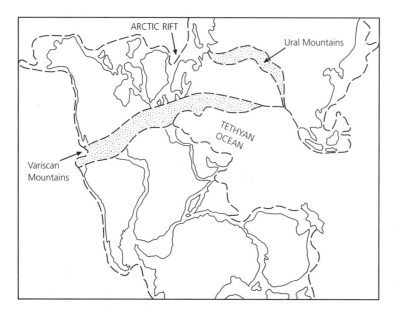

Fig. 16.1 Mid-Triassic reconstruction of Pangaea. It is unclear what height the Variscan Mountains would have been at this time: some workers consider them to have been an Early Permian feature, whereas more recent workers suggest uplands in this area. Dashed line = presumed shorelines; stipple = mountain chains. After Smith and Taylor (1992), with permission from the Geological Society, London (1999).

Permian red-beds is imprecise. Thus, it is unclear whether the Southern North Sea Basin began its subsidence in the Carboniferous or whether the main phase, now superimposed on an older rift, was of Permian age. Consequently, the area may be a mechanical rift, a Variscan foreland flexure (with or without post-orogenic collapse) or the result of pure thermal subsidence. The symmetrical nature of the basin, with its depocentre over the former Carboniferous thermal dome, suggests thermal collapse. The debates over the origin of the Southern North Sea Basin can be contrasted with the more fundamental question of whether any Permian at all is preserved in the North Atlantic basins. Many of these, such as Rockall, Porcupine and the Celtic Seas, are overlain by thousands of metres of Mesozoic–Tertiary sediment and occasionally thick volcanics, impeding our geophysical view of their geometry and making such places currently uneconomic to drill for hydrocarbons. When Permian and Triassic strata are observed in seismic surveys or at outcrop, then there is often evidence for syn-sedimentary fault activity and thus a likelihood that rifting was initiated by east–west and north-west–south-east mechanical stretching. What is less clear in these areas is the age of such extension. Some exposed (onshore) basins possess a dated Permian rift section (Cheshire Basin or Worcester Graben and

Wessex Basin), whereas other basins that are closer to the Atlantic rift (offshore Newfoundland, Parentis, Jeanne d'Arc) contain evidence of a Triassic episode of rifting. The preservation of an Early Triassic succession in some rift basins has been interpreted as reflecting regional uplift and intracratonic rifting. In many areas the first Triassic sediment deposited was sand; in the southern British Isles sand continued to be deposited (now Sherwood Sandstone Group) until the Mid-Triassic. In northern England and Northern Ireland the Sherwood Sandstone Group passes laterally into a mudstone and siltstone facies, or Mercia Mudstone Group (Fig. 16.3). Many of the basins fringing the North Atlantic show no Early Triassic rift but instead a Late Triassic–Early Jurassic break-up unconformity, followed by the deposition of evaporites and mudstones. In some areas, fault movement may have controlled facies; generally, however, Early Triassic sands are followed diachronously by Mid- to Late Triassic mudstones, siltstones and evaporites. Marine transgression in the latest Triassic was followed by more or less continuous marine deposition in the succeeding Jurassic.

16.2 Climate and sea-level changes

The essentially marginal marine and continental environments of the latest Carboniferous and Permian

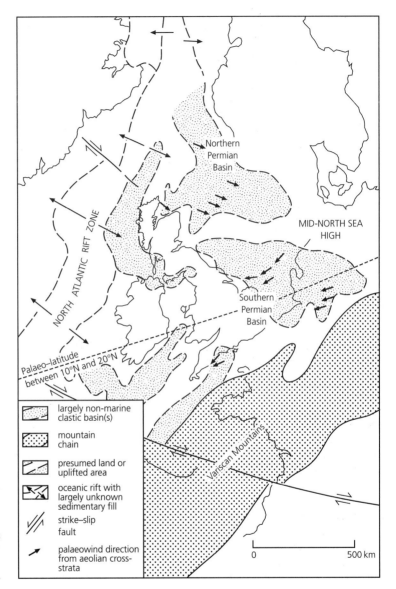

Fig. 16.2 Plate tectonic and palaeogeographic reconstruction of the North Atlantic area in the Early Permian. Note that when the plates are rotated back to this Permian position, then the proto-Atlantic and Variscan Mountains are not as dissimilar in their orientation as at the present day, suggesting a common origin of tectonic inheritance (data from text; Smith & Taylor 1992; Coward 1995).

form one of the clearest examples in the rock record where palaeoclimatic change can be demonstrated. Westphalian coal measures required high humidity and elevated water-tables for vegetation to grow and be preserved. The change to Stephanian and then Permian red-beds (still essentially non-marine to marginal marine) suggests that an arid climate developed gradually and coincidentally with the formation of Pangaea and the northward drift of the continents from a tropical to arid climate belt. In

southern Europe the Tethyan Ocean (some workers use Tethys Ocean) provided a maritime influence, whereas throughout the rest of the Pangaean super-continent an extreme continental monsoonal climate occurred. Aridity in northern Europe continued throughout the Permian with little change; Early Permian desert sands gave way to later marine limestones and evaporites, but both were formed under essentially arid conditions. The influx of siliciclastic detritus in the Early Triassic has been

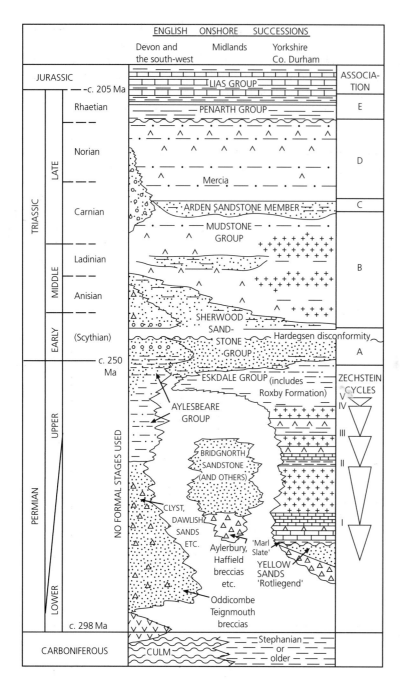

Fig. 16.3 Comparison of Permian–Triassic chronostratigraphy in the three most complete successions of onshore England. The Scottish, Welsh and Irish successions are more incomplete. Standard lithological symbols are used; blank areas represent times of apparent non-deposition (data largely from Smith and Taylor 1992; Warrington & Ivimey-Cook 1992). Associations are used in text (see Section 16.2).

suggested as evidence of renewed uplift and rifting. Conversely, the abundant evidence of fluvial deposition and/or raised water-tables in the deposits is suggestive of greater volumes of water being available for transport than in the Permian and thus possibly a wetter climate. There is evidence from evaporites and calcrete soils preserved in the Mercia Mudstone Group of a return to arid climates in the Mid- to Late Triassic. Whatever the precise nature of changes in humidity, there is separate evidence for a change in temperatures through the Permian and Triassic. Carboniferous–Permian glaciations were common in the southern hemisphere, their demise being coincident with a change from an icehouse to a greenhouse climate in the Late Triassic. The conclusion is that there was a general cooling of the world's climate through the Carboniferous, ending with the glaciations and reversed by a warming trend through the Permian and Triassic (Fig. 16.4). This warming initiated the greenhouse climate that was roughly coincident with the time (Late Triassic to Late Cretaceous) when dinosaurs existed.

Fig. 16.4 Comparison of suggested changes in sea level and climate through the late Carboniferous to Early Jurassic. The evidence for each curve is summarized in this chapter.

The long-term global (eustatic) sea-level trend for the Carboniferous–Permian–Triassic is a gradual fall, changing in the latest Triassic and Early Jurassic to a rising trend that continued into the Late Cretaceous. This eustatic fall in the Permian–Triassic is considered to have led to the lowest stand of global sea level experienced in the whole Phanerozoic and is coincident with the formation of Pangaea. Short-term and localized sea-level changes show significant variation from this overall trend. Throughout on- and offshore Britain and Ireland the Carboniferous–Permian succession of early carbonates, followed by later deltaic, swampy and often non-marine red-bed sediments, reflects a falling trend in sea level. Within the Early Permian, a eustatic sea-level lowstand, caused by the Carboniferous–Permian glaciations on

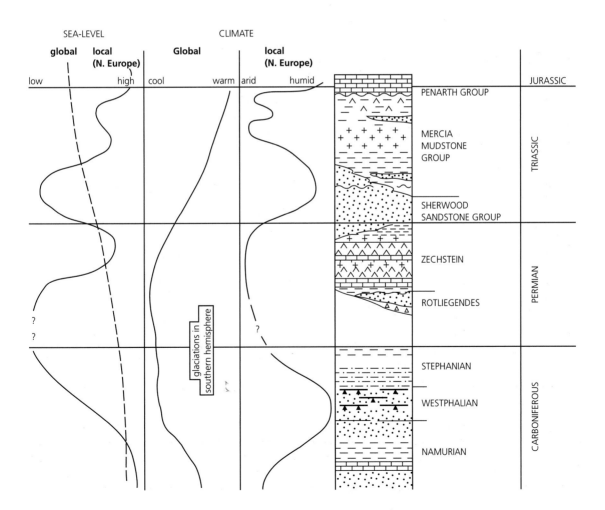

Gondwanaland, was accentuated by Variscan uplift, the end result being an almost complete lack of marine influence on earliest Permian sedimentation in many areas. Later in the Permian, rising sea levels, coupled with the incipient rifting in the North Atlantic or North Sea areas and erosion of the Variscan foreland uplifts, allowed marine transgression into the Irish Sea and North Sea regions. These two marine basins were separated by the proto-Pennines and are known as the Bakevellia Sea and Zechstein Sea basins, respectively; the two seas may have been connected north of Scotland as they show similarities in the development of limestone and evaporite facies.

Marine withdrawal from the Bakevellia and Zechstein basins in the latest Permian to Early Triassic was coincident with the global sea-level lowstand, final coalescence of Pangaea and more humid climate (see above). The resulting Early Triassic sandstones contain few primary marine indicators (fossils, sedimentary structures) in the British Isles and may reflect this global sea-level lowstand. The Triassic of the type locations in central Germany derives its name from the lithostratigraphic division into Bunter, Muschelkalk and Keuper. The German succession has some equivalent strata in Britain, there being a lower arenaceous Sherwood Sandstone Group and upper argillaceous Mercia Mudstone Group of very roughly the same facies and similar age. The German Middle Triassic Muschelkalk facies of marine fossiliferous limestones is absent in the British Isles and is represented throughout most of the onshore British Isles by a lithostratigraphic transition from sandstones to mudstones with the local development of intertidal marine sandstones and mudstones as in the English Midlands. Such intertidal deposits may represent the 'feather edge' of the Muschelkalk transgression as it penetrated north. Offshore in the North Sea the middle parts of the Triassic comprise salts (e.g. Rot Halite). The Mercia Mudstone Group succession can be divided into a lower halitic unit (Association B: Fig. 16.3), a middle arenaceous unit (Association C: Fig. 16.3) and an upper anhydritic unit (Association D: Fig. 16.3) that may chronicle the shallowing of the north European Triassic sea. Fossil soils and desiccation cracks are abundant near the top of the succession.

16.3 Correlation

Comparison of Permian or Triassic successions across on- and offshore Britain and Ireland is hampered by a lack of fossils or time-significant surfaces and lithostratigraphic units. In the thick (onshore, up to 2000 m) non-marine successions of the western Wessex Basin (Devon–Dorset) and south-west Scotland (Dumfries and Galloway), macrofossils are rare and restricted to vertebrates, trace fossils and exceptionally rare shelly fauna. Microfossils have been recovered and current work on extraction from redbeds and on magnetostratigraphy is providing a means of correlation. Few radiometric dates exist for Permian and Triassic strata. The absolute dates we have for the Carboniferous–Permian show how little time there was between events such as Variscan thrusting or granite intrusion and Permian rifting (see Chapter 15 for discussion). Some measure of correlation within the thinner (200–400 m) successions of the Zechstein and Bakevellia basins is possible through the regular cycles of deposition and fossils in the lower beds. Very few workers have attempted correlation of the Zechstein–Bakevellia deposits with the non-marine successions. The Zechstein succession may have been deposited at about the same time as the lower Aylesbeare Group red-beds in Devon (Fig. 16.3). The base of the Sherwood Sandstone Group, as well as the internal divisions of the Sherwood Sandstone Group and Mercia Mudstone Group, are lithostratigraphic rather than chronostratigraphic boundaries. The only surface currently identified as being of chronostratigraphic significance is the Hardegsen Disconformity. Sometimes also referred to as a disconformity, this break in sedimentation is believed to be widespread (extending from Germany to the East Irish Sea); it occurs in the middle and upper parts of the Sherwood Sandstone Group (suggesting diachroneity of the group) and may be the result of a phase of tectonic activity. The Hardegsen Unconformity separates fluvial and aeolian sandstones below from possibly more marine, argillaceous and fossil-bearing strata above. In southern England there is an influx of coarse material following the unconformity, whereas in the East Irish Sea there is a distinct change in diagenesis from illite-cemented to clay- and carbonate-cemented beds.

The Sherwood Sandstone Group–Mercia Mudstone Group boundary is believed to be diachronous

(Fig. 16.3): in north-west and north-east England, the Southern North Sea and Northern Ireland it is of (earlier) Scythian to Anisian age (Warrington *et al.* 1980), but in southern England, the Celtic Sea basins and the Western Approaches basins, it may be as young as Ladinian (possibly 10 Myr later). This diachroneity has implications for the timing of palaeoenvironmental change, whether it be from humid to more arid or continental to more marine. In terms of possible explanations for this diachroneity, it is likely that the overall climatic conditions would have been similar throughout Britain, as little variation would have been caused by latitudinal differences between the two areas (palaeo-position with respect to the rest of Pangaea or palaeoequator). There is no evidence of an extensive missing section from biostratigraphy or seismic data, which suggests true diachroneity of the boundary and not onlap. There are a number of interrelated regional effects that may have been responsible for the diachronous boundary between the Sherwood Sandstone and Mercia Mudstone groups. Highlands may have been preserved or created in the south of Britain, forming a barrier to the southward-transgressing Triassic seaway, or had the effect of raising base-level in that area. The latter in particular would have made the Sherwood–Mercia transition young to the south-west. This same topography may have had an orographic effect, which resulted in conditions of higher rainfall–runoff being restricted to the south and west. Such topography may have been either a local tectonic response to approximately east–west regional extension (i.e. early Atlantic opening), or a remnant of the Variscan highlands, with basement reactivation providing a control on resulting structures in both cases. Such basement controls throughout Britain may have been significant in partitioning basins through the creation of minor graben and half-graben. Such hangingwall sub-basins may be the sites of thicker sediment accumulation and preserved halite deposits.

16.4 Permian and Triassic thermal activity

Volcanic activity began in the Carboniferous, continuing like its associated tectonics into the Early Permian with a peak around the period of Variscan deformation (around 300 Ma). Throughout northern Europe, these volcanics are dominantly potassium rich.

Thermal events are known from Cheshire and Ireland in the Late Triassic (Fig. 16.5). The main centre of volcanic activity in the British Isles was the Midland Valley of Scotland, with numerous intrusive and extrusive episodes recorded, including the volcanic plug now preserved at Arthur's Seat in Edinburgh. Intrusive igneous events were also common, especially in the Variscan zone of deformation now exposed in Cornwall–Devon and the Scilly Isles (Cornubian granites). Other, minor areas of volcanic activity are likely to be rift related like the Midland Valley. These include minor intrusives and lavas on Arran; Islay–Jura; in the southern Uplands (Thornhill and Mauchline basins near Dumfries); in the subsurface of Northern Ireland and the east Irish Sea and in the western margins of the Wessex Basin close to the Cornubian granites around Exeter. In the Southern North Sea, volcanism and intrusive igneous activity (Lower Rotliegendes Volcanics) was common at 290–295 Ma. Evidence for similar igneous activity extends along the flanks of the Central Graben to the mid North Sea High and south into Germany as far as the Rhenish Massif. The limits of the northern and southern Permian basins of the North Sea (separated by the mid North Sea High) are roughly coincident with the limits of Permian igneous activity, that may be interpreted as a thermal dome of uplift and intrusion, followed by thermal relaxation which created the North Sea basins. In addition, the high concentration of Permian dykes and sills (e.g. the Whin Sill) along the present-day mid North Sea High suggests that this area may have had its origins as one of buoyant relief as far back as the Carbo-Permian and may have been the core of the dome. Such dykes and sills are earlier (301 + 4 Ma) than the Hebridean–western Scottish–Northern Ireland volcanics (285 + 6 Ma), suggesting a close relationship between the thermal dome and the dykes. The limit of igneous activity that can be related to North Sea subsidence extended westwards into the Irish Sea and Inner Hebrides, just incorporating the minor centres mentioned above. The two Carboniferous–Permian centres of igneous activity in the British Isles comprised: (i) the North Sea basins, Midland Valley and Hebrides (extensional in origin); and the (ii) Cornubian Granites, which can include minor lamprophyres and the nearby Exeter Volcanics. Most of the Cornubian granitic material is late syn-orogenic to post-orogenic and with little extrusive material preserved. The location of some of the Exeter Volcanics, being

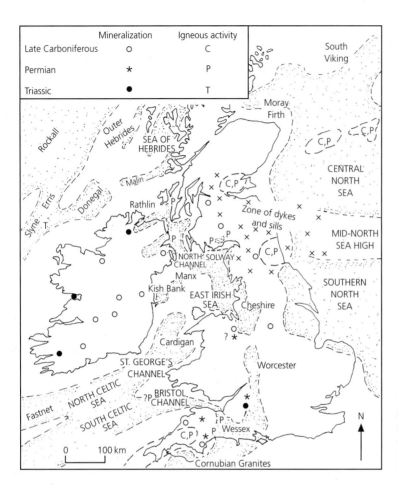

Fig. 16.5 Major sedimentary basins and sites of Late Carboniferous–Early Permian and Triassic igneous and mineralizing activity in Britain and Ireland (data from Smith & Taylor 1992; Warrington & Ivimey-Cook 1992; Coward 1995).

adjacent to extensional faults, may also provide evidence for tectonic control and post-granite extrusion. Industrially significant phases of mineralization (associated with Variscan tectonics and igneous activity) occurred in the late Carboniferous and Permian. The resultant mineral deposits include base metal veins in south-west England, South Wales, Cheshire, south Scotland and Ireland. A less significant phase of mineralization in the Late Triassic caused the partial remobilization of the earlier deposits. This later event coincided with renewed crustal extension and alkali volcanism in eastern North America and the Bay of Biscay.

16.5 Early Permian deserts

Carboniferous climate change, northward continental drift and Variscan uplift set the scene for the devel-

opment in the Early Permian of what Smith and Taylor (1992) described as 'one of the great deserts of world history'. The final preservation of Permian desert sediments was controlled by the location of sedimentary basins. Such basins were situated in one of two locations: the thermal foreland collapse of the Southern North Sea and Plymouth Bay and in the North Atlantic extensional graben of the Worcester and Minches basins. In some areas there is a complex mixture of these two symmetrical and asymmetrical basins (e.g. in the Manx–east Irish Sea–North Channel regions). Ultimately these basins preserve the record of desert conditions that also existed on the neighbouring Permian landmasses, for which there is now no sedimentary record. Although very little is known about the timing of events in the development of these deserts, the modern analysis of the Sahara (in terms of palaeolatitude) and the Asian and American deserts (in terms of environment) can be used as analogues for the Permian.

The base of the Permian succession comprises a widespread unconformity (often on reddened Carboniferous) overlain by lavas or breccias or desert sands (Rotliegendes and equivalents). These early sediments may be interpreted as the products of Variscan volcanism and of extensive fluvial erosion and transport (e.g. as flash floods). The landscape would have been one of actively eroding fault scarps (giving rise to screes or mudflows) with few areas of sand or pediplain in existence. The environment was unstable, but not harsh, as evidence of vegetation (fossil wood fragments in clays) is known from a few metres above the basal deposits. Reptile fossils are known from the Kenilworth Breccias of the Midlands. The basal Permian topography and infill is preserved in north-east and north-west England, south-west Scotland and south-west England by the subsidence discussed above. The evolution of this Early Permian environment was perhaps controlled by erosion of the Variscan uplifts in an arid climate characterized by flash floods. Thus in nearly all areas where a Permian succession is preserved, there is a common succession of coarse clastic sediments (alluvial fans), passing up and laterally into sands, silts and clays of outwash fan and playa lake origin. Such playas are thought to have developed in basin depocentres, synchronously with the aeolian conditions of the basin margins. Sand deposition was either fluvial–lacustrine in areas adjacent to the uplands or aeolian on upland intermontane basins and in the broad basins of the Southern North Sea. Such sands (e.g. Rotliegendes) display the classic features of desert sedimentation, including millet-seed quartz grains with iron–manganese oxide desert coating and preserved dune-forms. These features, together with the stratigraphic position of the Rotliegendes (above Carboniferous coal measures gas source-rocks and below Zechstein evaporite topseals), make the Early Permian sandstone reservoir one of the most economically important successions in Northern Europe.

Whether the switch from earliest Permian fluvial activity to later aeolian sedimentation was due to a significant climate change or the result of a decline in the relief of the surrounding uplands (and thus the rainfall such hills created) is not clear. Whatever the cause, there is no doubt that by the end of the Early Permian, prior to marine transgression, extensive dunefields occupied the basin margins with saline lakes (playas) in the basin depocentres. The dunes were created by the dominant winds of the time and their internal cross-stratification is remarkably consistent: those in the Northern North Sea show that in this area winds came mostly from the west. In the Southern North Sea and Wessex Basin the Early Permian winds came from the east (Fig. 16.2). The boundary between the two wind directions is roughly coincident with the mid-North Sea High, and the overall pattern is of peri-equatorial 'trade winds'. This is consistent with the palaeomagnetic reconstruction of the Permian British Isles lying in the trade wind belt, north of the Equator.

16.6 Late Permian Zechstein carbonate platforms and evaporites

A regional, possibly glacioeustatic, rise in sea level in the Permian (Zechstein) caused the rapid flooding (from the north) of the sand-rimmed and playa lake basins of the North Sea and east Irish Sea. What is remarkable about this transgression is the succession it helped preserve, in that there is very little evidence of extensive energetic shoreface reworking, there are no shallow marine sands and the original (underlying) dune topography was preserved. Such 'fossilized' dunes are now exposed, cropping out in parts of Yorkshire and Durham. The preservation of dune topography when the sands were probably still friable is so good that a very rapid transgression (a few centimetres per day) by rather quiet marine waters is the only plausible explanation. The dune sands contained sufficient early clay cement to have formed a contemporaneous crust, allowing significant stability during marine working to be achieved. No similar Permian transgression is recorded in the Wessex or Western Approaches basins, suggesting a marine connection to the North Sea and North Atlantic but not to the south, probably as a result of land barriers. Late Permian tectonics in the western basins of Britain and Ireland were similar to the Early Permian with continued (if slower) growth on extensional faults. In the North Sea basins, tectonic events are less well understood, largely because of the masking effect of halokinesis. The regional extent of Southern North Sea Basin Zechstein cycles suggests tectonic quiescence and a phase of regional subsidence prior to the Triassic.

The first marine deposit of the Zechstein transgression is no less remarkable: throughout the North Sea a carbonate-rich shale was deposited, being thicker in depressions (e.g. between dunes) and thinner or absent across intervening 'highs'. In Germany this shale is copper enriched (Kupferschiefer), whereas in eastern England the shales (Marl Slate) contain locally abundant fossil fish, along with plant debris and brachiopods. The Late Permian flooding of the East Irish Sea and North Channel (like the North Sea) came initially from the north, and thence from the west between Scotland and Northern Ireland (Fig. 16.6). This, the Bakevellia Sea, was separated from the Zechstein Sea by the proto-Pennines and Southern Uplands. Consequently, there are broad similarities between the two successions in terms of facies types, yet also significant differences in possible cor-

relations. For example, both the North Sea and the Bakevellia Sea basins contain a lower clastic unit (breccias and sands) followed by an upper carbonate unit (Magnesian Limestone). Conversely, the Bakevellia Sea succession contains no Marl Slate or Kupferschiefer, only silty dolomites.

The economic significance of the Zechstein–Bakevellia succession (salt–potash source and reservoir topseal) plus the cyclic variation in lithofacies has resulted in a better understanding of the correlation of this unit (within the Zechstein Basin) compared with underlying and overlying red-bed successions of the Permian and Triassic. In the Zechstein succession of eastern England there are four lower cycles and one upper minor cycle of deposition compared with three or four cycles in the Bakevellia succession in the east Irish Sea. Each cycle comprises varying proportions of mudstone, limestone, dolomitic limestone and evaporite deposited in a

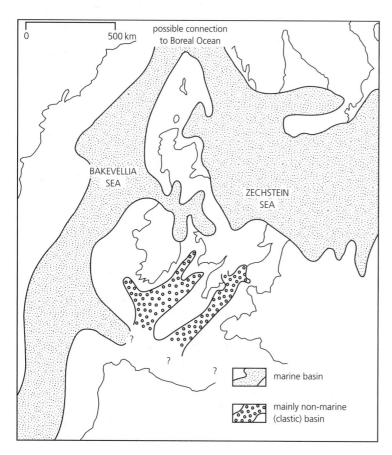

Fig. 16.6 Late Permian palaeogeography of the North Atlantic region (after Smith & Taylor 1992, with permission from the Geological Society, London, 1999; Coward 1995).

number of discrete phases (Fig. 16.7). The cycles are dominated by evaporites or carbonates or are a mix of siliciclastics and evaporites. The concentration of salts through each cycle of the Zechstein results from the desiccation of the surrounding playas and salinas with continued evaporation in the basin; occasionally this was still submarine with minor emergence of the evaporite beds in a sabkha-type environment. The upward concentration of salts and increasing 'harshness' of environment through each cycle is mirrored in the complete Zechstein succession, with the uppermost cycles showing thicker, more concentrated or more laterally extensive salts than the cycles below, culminating in the third and fourth cycles. This interpretation of the cycles is reflected in the fossil record of changing palaeoenvironments. In the lowest Zech-

stein cycle, although the fauna is restricted (i.e. no goniatites or corals), the Cadeby Formation limestones contain brachiopods, bivalves, gastropods, bryozoans and crinoids. In the succeeding cycles this diverse fauna becomes progressively more limited until only a single species of bivalve is found. Similar harsh conditions are also envisaged for the inhospitable Bakevellia Sea, where in places only one bivalve (*Bakevellia binneyi*) occurs in abundance.

The conventional view of Zechstein cycles as reflecting progressively greater evaporation and the shallowing of either the whole basin or the margins of the basin may be reconsidered using the analysis of depositional sequences. In this, thick gypsum deposits may be interpreted as accumulating directly above the sequence boundary as a lowstand wedge, surrounded by sabkhas and scattered carbonate shoals–reefs. These pass up into lowstand halite and potash deposits. The transgressive and highstand systems tracts are represented by shallow-water carbonate platforms that fringe some deeper-water areas in which muds and evaporites may still accumulate (Fig. 16.7). Significantly, this sequence stratigraphic model of Zechstein cycles is the complete reverse of our traditional view.

16.7 Permian–Triassic extinctions

'The end-Permian mass extinction is generally recognized as the most important biotic event in the Phanerozoic record' (Raup & Sepkoski 1986). It has been attributed to shifts of sea level, salinity or temperature, to mega-volcanism, anoxic sea water or extraterrestrial impact. For the first three possible causes, there is some supporting evidence in the British succession. The Zechstein transgressions were followed by a Late Permian regression: thus there is evidence of significant sea-level change. The development and cyclicity of Late Permian evaporites found throughout northern Europe suggest widespread changes in salinity, another possible cause of extinction. Late Carboniferous humid conditions and Permian aridity are suggestive of climate change, although this may have a regional cause in the northward drift of Pangaea. The end-Carboniferous glaciation indicates a more widespread change of temperature that may well be linked to extinctions, albeit too early to account for the end-Permian event. Conversely, most volcanics are of late Carboniferous

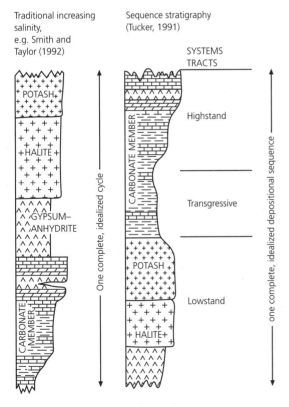

Fig. 16.7 Alternative interpretations of typical Zechstein cycles. The similar cyclicity of the Bakevellia Sea succession has yet to be the subject of similar analyses but could presumably be interpreted using either method.

and Early Permian age, making them too old to be linked (in the British and Irish successions at least) to mass extinction. Equally, there is as much evidence of palaeoclimate and sea-level change in the Carboniferous and Permian as there is in the Triassic. None the less, some workers have cited elements of the British and Irish Permian succession in support of their particular hypotheses.

A mass extinction must affect the higher orders of the biota (i.e. above species level) on a global scale (i.e. total extinction), but most importantly the event must be more or less synchronous, otherwise it is an episode of gradual extinction. The record of the Zechstein fossils suggests a gradual Late Permian extinction, although the fossil record here may reflect the regional environment (especially salinity) more than the global biota. To try to quantify the magnitude of the Permian biotic crisis, the number of orders and families lost may be compared with that number from the end-Cretaceous extinction (Fig. 16.8). Published estimates vary, but it seems likely that between 30% and 50% of all orders and families were lost around the end of the Permian. By contrast, the end-Cretaceous extinction resulted in 5–10% loss. The details reveal a truly cataclysmic event: in the ecologically (and palaeontologically) important shallow seas, of the 377 families known, only 181 survived into the Triassic; as Van Andel (1994) puts it 'like scattered shipwrecked immigrants struggling to the shores of an uninhabited island'. It may be that an assembling Pangaean supercontinent, with its monsoonal offshore winds, produced a highly unstable food supply. This unstable nutritional supply supported a shallow marine fauna that had very little geographical or ecological space in which to live because of the reduction in marine seaboard around the margins of Pangaea. In the Late Permian, as Pangaea coalesced, so the food supply would have become most unstable and the total area available for shallow marine ecological environments would be lowest. When, at this point, any change in sea level or palaeoclimate occurred (whether it was particularly unusual or not), then cataclysmic extinction of the shallow marine biota followed. The lifeless, anoxic oceans that followed have been likened to a post-holocaust world and the earliest Triassic marine environment has been dubbed the 'Strangelove Ocean'.

At first sight, this great event in the history of life on Earth seems hardly to have mattered in northern Europe. Permian limestones and mudstones gave way to Triassic sandstones and mudstones, and in the absence of any continual biostratigraphic record it is hard to know if any hiatus exists, or indeed just where the Permian–Triassic boundary should be placed in any given rock succession. The lack of evidence for any great changes in the northern European situation belies the fact that depositional changes did occur and that elsewhere (for example) the structure of reef communities altered radically. When more permanent marine deposition did return in the Jurassic of Britain and Ireland, the shelly fauna now found as fossils was rather different. Gone were the Palaeozoic trilobites and goniatites, many brachiopods, bivalves, corals and algae. In their place, new brachiopods and bivalves were now abundant, along with diverse molluscan predators such as ammonites and belemnites. This change is only partly real, the absence of marine fauna from many Triassic sections around the world, and in Britain and Ireland, obscuring the fact that a significant extinction also occurred in the Late Triassic. Far less is known about this event: we know from the Alpine successions that reef communities suffered, along with the conodonts, crinoids, fish, tetrapods and plants. Some workers favour an end-Triassic mass extinction, others suggest the event was earlier or that there was more than one major reorganization of the biota.

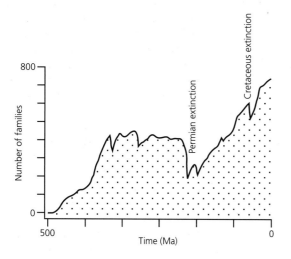

Fig. 16.8 The pattern of mass extinctions during the Phanerozoic (after Raup & Sepkoski 1986).

16.8 Triassic rifting

Post-compressional extension in the Variscan foldbelt of southern England (and parts of Germany) created the Permian tilted fault-block and half-graben intermontane basins that continued their development in the Mid-Triassic to Early Jurassic. Throughout southern England and the Irish Sea, the largely arenaceous Early and Mid-Triassic Sherwood Sandstone Group has been interpreted as deposition during a new syn-rift phase of crustal extension. Outcrop or borehole sections through the Permian to Triassic rarely show evidence of unconformable relationships, suggesting a rather more subtle change from Permian mudstones to Triassic sandstones. In basin margin successions, an unconformity may be observed between the Triassic sandstones and Palaeozoic strata. The presence of this unconformity has been interpreted as representing a phase of uplift, for which there is little evidence in the more complete, basinal locations. Instead, the basin-margin unconformity may be the result of onlap, aided by high sediment loads during time of flooding. Seismic data from some basins of the British Isles (e.g. the Cheshire Basin) show clear evidence of syn-sedimentary growth on faults during deposition of the Sherwood Sandstone Group. Conversely, other areas (e.g. the North Channel between Scotland and Northern Ireland) show no such growth faulting in such Early Triassic sandstones. The overlying Mercia Mudstone Group has been interpreted as representing the post-rift phase of thermal subsidence, following rifting in the Early Triassic. Conversely, many recent studies using offshore seismic and borehole evidence show significant extension in the middle to upper beds of the Mercia Mudstone Group rather than the underlying sandstones, casting doubt on the models in which the latter represent the syn-rift phase. Data from rift basins world-wide show that during maximum rift extension and subsidence, basins may be deep and restricted, commonly leading to the deposition of claystones and evaporites; periods of lesser tectonic activity are often reflected by coarse-grained sedimentation.

The findings from southern England and the Irish Sea (above) are consistent with events described from other North Atlantic basins. Many of the North Atlantic break-up unconformities appear to represent a Mid- to Late Triassic and Early Jurassic phase of rifting, rather than an Early Triassic one. There is evidence for Late Triassic (Late Carnian to Rhaetian) and probably Early Jurassic tectonic activity adjacent to the southern margin of the mid North Sea High. Extensional faults in the Southern North Sea and Danish Central Graben show syn-sedimentary growth on extensional faults from Scythian to Carnian times. The problem in many areas is differentiating between intra-Triassic thickening from extensional tectonics and post-depositional thickening from withdrawal of the underlying Permian salts. By subtracting any halokinetic effect Late Triassic through to Early Jurassic extension can be demonstrated. Passing north into the Central Graben and Norwegian Viking Graben there is important evidence from borehole and seismic data, data to demonstrate a post-Scythian, intra-Triassic unconformity. Many North Sea workers favour a Mid-Triassic phase of extension on the western platform of the Norwegian North Sea, in common with rift phases to the south and in the Atlantic.

The timing and style of Permian to Triassic rifting in north-western Europe may be compared with events in the marginal basins to the North American seaboard of the Atlantic Ocean. The syntheses of large volumes of borehole data with reliable bio- and magnetostratigraphy made by many North American workers demonstrate the preponderance of syn-rift extension and alkali volcanism beginning in the Carnian stage of the Late Triassic. A similar age for the onset of early syn-rift sedimentation is observed for offshore Portugal and Morocco, suggesting a synchronous and symmetrical rifting event. In short, the renewed, post-Variscan extensional phase detected in the Irish Sea (and Northern Ireland), southern England and the Southern to Central North Sea (above) is also observed as a break-up unconformity in some of the adjacent basins of the North Atlantic. Yet this widespread phase of rifting was associated with the deposition of evaporites and mudstones in the syn-rift, with some exceptions. Rift-margin uplift and the corresponding lowering of hangingwall blocks would be expected to have (together) increased hydraulic gradients and thus the input of coarse siliciclastics to the basin. Instead, this phase of rifting appears to be observable only in rotated syn-depositional fault blocks and by unconformable relationships, with coarse detritus very much restricted to hangingwall areas immediately adjacent to active faults. The overall lack of coarse detritus requires explanation that may be found through examination of Triassic palaeoclimates and sea-level change.

16.9 Early Triassic rivers and dunes

Earliest Triassic environments in many areas of the present British Isles are represented by conglomerates formed in braided streams, and largely comprise channel deposits. They pass laterally north-east and north-west from southern Britain into dominantly arenaceous sequences (such as the Bunter of the Southern North Sea) that were deposited in more distal fluvial environments. In southern and central Britain the Anisian saw the deposition of the Otter, Bromsgrove, and Helsby Sandstone formations, which represent continental fluvial environments with northward-flowing rivers. They comprise a complex of upward-fining sedimentary cycles that indicate deposition initially in predominantly braided stream environments and latterly in more mature meandering river systems with aeolian dunefields and calcrete soils between. Such meandering fluvial environments, which incorporate more vertical accretion (overbank) sediments, are characteristic of the distal (northerly) and younger parts of the fluvial succession, which diminish in importance southwards. The fluvial facies pass into interbedded sandstones, siltstones and mudstones, representing the fresh to brackish water, estuarine and intertidal environments that bordered areas in which evaporite-bearing mudstones of the Mercia Mudstone Group accumulated further north-west. The Wessex Basin of southern England has been intensively explored for hydrocarbons and the presence of Europe's largest onshore oilfield (beneath Poole Harbour), with Early Triassic sandstones as one reservoir and the Late Triassic–Early Jurassic mudstones as its topseal, have both provided abundant published information on the Triassic of this area. Southern England may be used as a typical area with Triassic successions known at outcrop and in the subsurface.

The Triassic succession of southern England is well documented, being both exposed on the south Devon coast and frequently penetrated in boreholes drilled for hydrocarbons. Many published works on the Sherwood Sandstone Group suggest a predominance of fluvial processes, with some aeolian and lacustrine environments. The lower, conglomeratic formations (e.g. Budleigh Salterton Pebble Beds and equivalents)

were deposited by northward-flowing braided rivers that in turn drained the alluvial fans surrounding the Variscan highland to the south. This general reconstruction confirms that derived from the chronostratigraphy (see above), wherein earlier Mercia Mudstone Group deposition was accommodated by lower palaeotopography in the northern British Isles. The arenaceous Sherwood Sandstone Group was deposited in a variety of fluvial (channel bars, braided streams) and aeolian environments. Few published studies have analysed the structure of the Triassic or the detailed tectonic controls on its deposition.

16.10 Mid-Triassic transgression

Throughout much of Europe and North America, Middle and Upper Triassic successions are dominated by facies typical of arid and semiarid environments. In northern Europe, this comprises the Mercia Mudstone Group and contiguous deposits, a succession of red and green mudstones and dolomitic siltstones with halites preserved in graben in the subsurface. Subordinate sandstones or anhydrite horizons also occur. Exact depositional environments for much of these Late Triassic calcareous mudstones are still poorly defined, having variously been described as 'evaporitic' or 'playa lakes' or 'epeiric seaways'. The group could also have been deposited in a predominantly lacustrine environment of varying depth with occasional episodes of emergence. This interpretation may require slight modification in order to de-emphasize the lacustrine aspect and highlight the evidence for a marine connection in order to allow salt deposition. Most recently, the concept of a shallow marine seaway has been supported by geochemistry and finds of marine fossils. Further support for a mid-Triassic transgression comes from the German and French successions where the tripartite stratigraphy was erected. Here, the lower, Sherwood Sandstone Group facies equivalent (Bunter Sandstone) is separated from the Mercia Mudstone Group facies equivalent (Keuper Marls) by marine limestones of the Muschelkalk, which are absent from onshore Britain and Ireland. The Muschelkalk carbonate platform limestones do, however, extend into the Southern North Sea and their onlapping transition into the successions of the English Midlands is thought to be represented by intertidal sandstones. Because the mudstones show evidence of marine deposition, these transitional intertidal sandstones are now considered to represent the initial transgression rather than the transgression–regression couplet or 'whiff of the sea' as was once thought.

The influx of Budleigh Salterton Pebble Beds and Otter Sandstone, following deposition of the fine-grained Aylesbeare Group, has previously been interpreted as the result of renewed hinterland uplift. Footwall uplift could have occurred during renewed, Early Triassic extension. Conversely, other workers (using the same evidence) have suggested that the influx of exotic basement-derived clasts indicated only minor uplift of remote areas, whereas the Sherwood Sandstone Group was probably deposited during a period of tectonic inactivity within the Wessex Basin. An alternative interpretation allows for the influx of coarse siliciclastics to be driven by a change to temporarily more humid climates, contrasting with the arid Aylesbeare Group and Mercia Mudstone Group below and above (respectively) and allowing a uniform thickness of Sherwood Sandstone across many syn-sedimentary faults. The fine-grained, onlapping nature of the Late Triassic and Early Jurassic mudstones may be interpreted as marking the onset of thermal subsidence in the Wessex Basin. Conversely, rapid thickening of the upper half of the Mercia Mudstone Group in boreholes penetrating subsurface graben and half-graben suggests fault-controlled subsidence (Fig. 16.9). Syn-sedimentary fault activity may have continued through the Triassic–Jurassic transition as local growth in the Penarth Group and Lias and evidence of contemporaneous slumping is recorded in Somerset and Devon–Dorset.

The change from arenaceous Early Triassic to argillaceous and halitic Mid- to Late Triassic reflects a number of changes, including provenance, length of the drainage system and climate. Arenaceous Early Triassic sediments in and around the UK, Ireland and Germany–France were sourced either from nearby fault-scarps or from long river drainage systems. In the UK such a drainage system in the Early Triassic was the successor to an earlier, south to north flowing fluvial system named the Budleighensis River; along which sand-rich (Devonian–Carboniferous) parent rocks were eroded. The drainage systems of the Late Triassic were more localized, perhaps because the area may have been broken up into separate horsts and graben. The fine-grained nature of the sedimen-

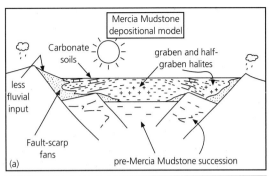

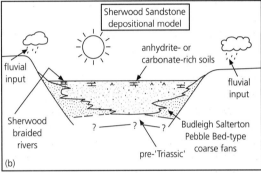

Fig. 16.9 Models of Triassic deposition in the rift basins of Britain & Ireland (after Ruffell & Shelton 1998, with permission of the Geological Society, London (1999)).

tary record at this time was also influenced by the prevailing arid climate. Marine influx resulted in the deposition of locally thick halite beds allied to fault-controlled thickening (Fig. 16.9). Major rivers are not in evidence and thus the long-distance movement of sediments from local footwall crests is not observed. Thickening of the Sherwood Sandstone Group into faults may have been produced through a combination of minor extension in the Early Triassic superimposed on thermal subsidence inherited from the important regional phase of extension in the Early Permian.

16.11 Summary

The paucity of fossils or recorded events of a widely correlatable nature, coupled with isolated or poor exposures, has resulted in less being written about related aspects of Permian–Triassic stratigraphy than the preceding and succeeding systems. Conversely, the enormous economic significance of the Permian or Triassic reservoir sands and later Permian and Triassic evaporite–mudstone topseals has resulted in a large number of publications, especially utilizing data from offshore exploration. Many of the new ideas to come forth in the past 10 or 15 years concerning the Permian and Triassic evolution of the British Isles relate to the tectonics of the area; the geochemistry–palaeoenvironment and sequence stratigraphy. Some of the major points to be drawn from the preceding discussion are listed below: some of these were well established 15 years ago and have merely been added to here; others are entirely new concepts and arguments and may not be included in a book such as this in another 20 years time!

1 The Variscan Orogeny left a juvenile topography of faulted uplands and post-orogenic volcanics being actively eroded in a semiarid climate.

2 Post-orogenic collapse and thermal subsidence occurred in areas of the Western Approaches and Southern North Sea. Proto-Atlantic rifting began.

3 Climate and sea-level change were important in influencing the facies of the Carboniferous and Permian–Triassic.

4 The analogy to the Himalayan–Tibet collision, developed in Chapter 15, is continued, as the Permian climate would also be similar to the monsoon of the Tibetan Plateau.

5 Humid Carboniferous coal swamps gave way to arid Permian deserts with flash floods and aeolian sedimentation of red-beds.

6 A glacio-eustatic rise in sea level may have caused the rapid Zechstein transgression.

7 The cyclic Zechstein deposits may be interpreted in terms of desiccation–concentration cycles or (more recently) their sequence stratigraphy.

8 Early Triassic sand deposition was probably caused by a regression, coincident with a humid climate and minor renewed rifting.

9 Mid- to Late Triassic mud and salt deposition was caused by transgression, coincident with a more arid climate and active rifting.

References

Coward, M.P. (1995) Structural and tectonic setting of the Permo-Triassic basins of northwest Europe. In: *Permian and Triassic Rifting in Northwest Europe* (ed. S. A. R. Boldy), Special Publication 91, pp. 7–39. Geological Society, London.

Raup, D.M. & Sepkoski, J.J. Jr (1986) Periodic extinction of families and genera. *Science* **231**, 833–836.

Ruffell, A. & Shelton, R.G. (1998) The control of sedimentary facies by climate during phases of crustal extension: examples from the Triassic of onshore and offshore England and Northern Ireland. *Journal of the Geological Society, London* **156**, 779–789.

Smith, D.B. & Taylor, J.C.M. (1992) Permian. In: *Atlas of Palaeogeography and Lithofacies* (eds J. C. W. Cope, J. K. Ingham & P. F. Rawson), Memoir 13, pp. 87–96. Geological Society, London.

Van Audel, T.H. (1994) *New Views on an Old Planet.* Cambridge University Press, Cambridge.

Warrington, G.W. & Ivimey-Cook, H.C. (1992) Triassic. In: *Atlas of Palaeogeography and Lithofacies* (eds J. C. W. Cope, J. K. Ingham & P. F. Rawson), Memoir 13, pp. 87–96. Geological Society, London.

Further reading

Anderton, R., Bridges, P., Leeder, M. & Sellwood, B.W. (1979) *A Dynamic Stratigraphy of the British Isles.* George Allen & Unwin, London. [Shows that many concepts have stood the test of time whereas others (tectonic controls; sequence stratigraphy) provide much new information.]

Benton, M.J. (1986) More than one event in the Late Triassic extinction. *Nature* **321**, 857–861. [Discusses the various patterns of Late Triassic extinction.]

Coward, M.P. (1995) Structural and tectonic setting of the Permo-Triassic basins of northwest Europe. In: *Permian and Triassic Rifting in Northwest Europe* (ed. S. A. R. Boldy), Special Publication 91, pp. 7–39. Geological Society, London. [An excellent introduction to the plate tectonic setting and structural controls on sedimentation.]

Glennie, K.W. (1995) Permian and Triassic rifting in northwest Europe. In: *Permian and Triassic Rifting in Northwest Europe* (ed. S. A. R. Boldy), Special Publication 91, pp. 7–39. Geological Society, London. [An up-to-date review of all that is happening in our thinking, especially on the Permian.]

Goodall, I.G., Mckie, T., Harwood, G.M. & Kendall, A.C. (1992) Discussion on sequence stratigraphy of carbonate evaporite basins—models and application to the Upper Permian (Zechstein) of Northeast England and adjoining North Sea. *Journal of the Geological Society, London* **149**, 1050–1054. [Gives an alternative opinion to that expressed in Tucker's (1991) work (see below).]

Manspeizer, W. (1988) Triassic–Jurassic rifting and the opening of the Atlantic: an overview. In: *Triassic–Jurassic Rifting; continental breakup and the origin of the Atlantic Ocean and passive margins* (ed. W. Manspeizer), pp. 41–79. Elsevier, New York. [Essential reading for those requiring a North Atlantic perspective on our islands.]

Musgrove, F.W., Murdoch, L.M. & Lenehan, T. (1995) The Variscan fold–thrust belt and its control on early

Mesozoic extension and deposition: a method to predict the Sherwood Sandstone. In: *The Petroleum Geology of Ireland's Offshore Basins* (eds P. F. Croker & P. M. Shannon), Special Publication 93, pp. 81–100. Geological Society, London. [An example of how our conservative views on tectonics may be overturned.]

Raup, D.M. & Sepkoski, J.J. Jr (1986) Periodic extinction of families and genera. *Science* **231**, 833–836. [A definitive short work on this contentious issue.]

Ruffell, A. & Shelton, R.G. (1998) The control of sedimentary facies by climate during phases of crustal extension: examples from the Triassic of onshore and offshore England and Northern Ireland. *Journal of the Geological Society, London* **156**, 779–789. [A recent work by the authors on the Triassic.]

Smith, D.B. & Taylor, J.C.M. (1992) Permian. In: *Atlas of Palaeogeography and Lithofacies* (eds J. C. W. Cope,

J. K. Ingham & P. F. Rawson), Memoir 13, 87–96. Geological Society, London. [Contains much of the current information used in this chapter.]

Tucker, M.E. (1991) Sequence stratigraphy of carbonate–evaporite basins: models and applications to the Upper Permian (Zechstein) of northeast England and adjoining North Sea. *Journal of the Geological Society, London* **148**, 1019–1036. [An excellent example of how traditional views of stratigraphy can be simply redefined to provide a modern explanation for the origin of carbonate–evaporite cycles.]

Warrington, G.W. & Ivimey-Cook, H.C. (1992) Triassic. In: *Atlas of Palaeogeography and Lithofacies* (eds J. C. W. Cope, J. K. Ingham & P. F. Rawson), Memoir 13, 97–104. Geological Society, London. [Contains much of the current information used in this chapter.]

17   Late Triassic and Jurassic: disintegrating Pangaea

S. P. HESSELBO

17.1 Background

Late Triassic and Jurassic geological history is set in the context of progressive supercontinental break-up. Over this time Pangaea had an almost symmetrical distribution about the equator and, in the Early Jurassic at least, no land was present at either geographical pole. Furthermore, with the docking of southern China to Laurasia in the Pliensbachian, the Early Jurassic configuration of Pangaea has been regarded by some as the most complete in the Phanerozoic. A degree of uncertainty does remain about Early Jurassic palaeogeography, particularly regarding the relative longitudinal position of northern versus southern continental masses. What is clear, however, is that complete continental assembly was ephemeral, and in the Early Jurassic the north-west European area was criss-crossed by a network of shallow seaways related to widespread lithospheric extension (Fig. 17.1; see Chapter 16). Ocean floor was created in the central Atlantic region by the Mid-Jurassic, paving the way for separation of Gondwana and Laurasia.

Long-term and profound changes in the palaeoclimate of the area of Britain and Ireland ensued from these palaeogeographical reorganizations over the course of the Jurassic. Instead of the 'mega-monsoonal' climate that had characterized earlier times, a latitudinally zoned climatic pattern, more akin to that of the present day, began to develop. Also, periodically over this time, massive outpourings of basaltic igneous rocks occurred, most notably at the Triassic–Jurassic boundary in the areas bordering what is now the Central Atlantic (North Africa, Brazil and eastern North America), and in the Early Toarcian in the area of southern Africa and Antarc-

tica (Fig. 17.2). The effects of these volcanic events on climate through the emission of carbon dioxide and other gases may have been important, but remain poorly known at present. Volcanic centres appeared also in our 'backgarden', during the Mid-Jurassic, clustered around the mid North Sea area, and these were associated with uplift of some considerable regional importance.

Across north-west Europe a pattern of sedimentary basins developed—summarized in palaeogeographic maps—which show some relation to the structural grain inherited from previous geological events. So, for example, the sedimentary basins of the south (the Wessex and Bristol Channel basins) are bounded by faults with a predominant east–west orientation, which follow precisely deep-seated Variscan thrusts in the Palaeozoic basement. Further north, the orientation of bounding faults was defined by Caledonian structures with a north-east–south-west orientation (e.g. in the Cardigan Bay Basin and the Hebrides Basin). However, many fault patterns appear to be newly formed in the Mesozoic. In the North Sea rift system, sedimentation took place primarily in axial rifts known as the Central Graben, the Viking Graben and the Moray Firth Basin. Analysis of deep seismic reflection profiles shows that fault systems in the Central Graben are orthogonal to the major Caledonian structures such as the Iapetus Suture or the Highland Boundary Fault.

Similarly, not all major fault systems in Jurassic basins extend into the basement. In regions underlain by Permo–Triassic evaporites, for example the western Wessex Basin, the Cleveland Basin, parts of the Cardigan Bay Basin and extensive areas of the North Sea, syn-sedimentary faults commonly 'sole out' into salt deposits, because these provided weak

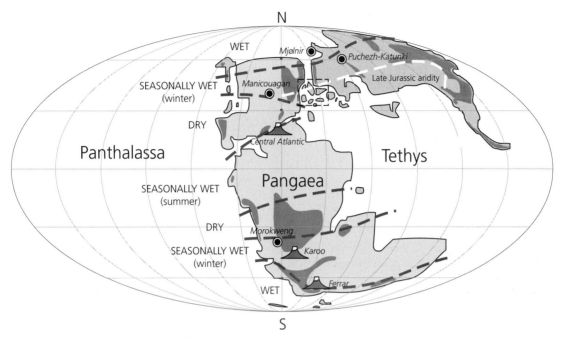

⛰ Continental flood basalt province   ◉ Impact crater (>30 km)

Fig. 17.1 Global palaeogeography and climate for the Mid-Jurassic (Bajocian). Continental configurations are based on Smith *et al.* (1994, with permission from Cambridge University Press 2000). The Mjølnir and Morokweng impact craters are at the Jurassic–Cretaceous boundary, but have been superimposed on this map to indicate their approximate palaeogeographic locations with respect to the area of Britain and Ireland. Refer to Fig. 17.2 for precise ages of marked features. The square box shows the approximate position of more detailed palaeogeographic maps of Figs 17.15 (see p. 335), 16.11 (see p. 324) and 17.7. The climatic belts are very generalized and are based on Hallam (1985) and Ziegler *et al.* (1993). Thick white dashed lines = spread of the arid belt across southern Asia in the Late Jurassic which may have been largely due to continental rotation; dark shading = high-altitude land.

horizons along which extensional deformation could be easily accommodated. In still other cases, fault systems developed because of gravitational collapse of sea-floor topography: many small-scale examples of these types of structures are known from the North Sea rifts, but it is also possible that much larger-scale gravitational-collapse faulting occurred in other basins starved of sediment, such as the Cardigan Bay Basin. Regional lithospheric extension caused fault systems in all the basins around Britain and Ireland to be active with varying degrees of strike-slip as well as dip-slip motion, and the importance of these components varied significantly during the course of the Jurassic.

The very localized spatial distribution of sandy sediment in many north-west European Jurassic basins suggests that a limited number of point sources were important in the sediment dispersal systems. Although the region is dotted with many small weakly subsiding or even uplifted areas, it is unlikely that these were major providers of sediment to the surrounding basins. The very large volumes of clay-sized sediment that were deposited over much of the region may instead have had source areas that were much further away, possibly the Fennoscandian, East European and Laurentian–Greenland areas. At particular times, new and important source regions were undoubtedly created, the prime example being the mid North Sea area, whose uplift created a radial pattern of sediment dispersal to neighbouring basins throughout Mid-Jurassic times.

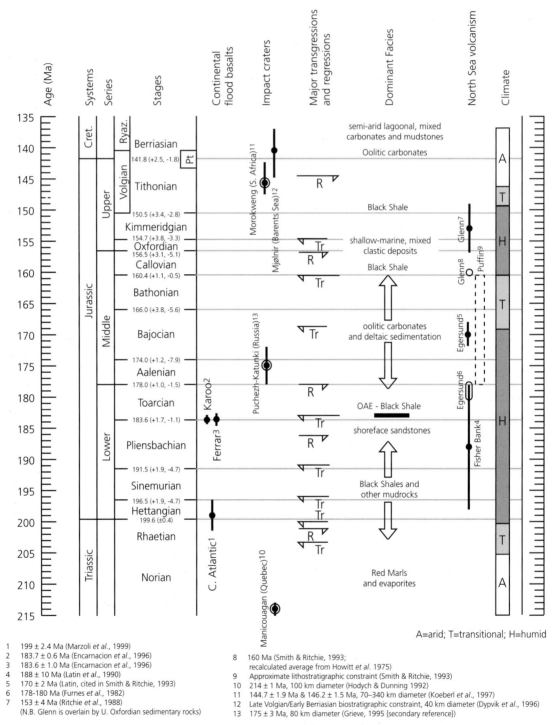

1	199 ± 2.4 Ma (Marzoli et al., 1999)
2	183.7 ± 0.6 Ma (Encarnacion et al., 1996)
3	183.6 ± 1.0 Ma (Encarnacion et al., 1996)
4	188 ± 10 Ma (Latin et al., 1990)
5	170 ± 2 Ma (Latin, cited in Smith & Ritchie, 1993)
6	178-180 Ma (Furnes et al., 1982)
7	153 ± 4 Ma (Ritchie et al., 1988) (N.B. Glenn is overlain by U. Oxfordian sedimentary rocks)
8	160 Ma (Smith & Ritchie, 1993; recalculated average from Howitt et al. 1975)
9	Approximate lithostratigraphic constraint (Smith & Ritchie, 1993)
10	214 ± 1 Ma, 100 km diameter (Hodych & Dunning 1992)
11	144.7 ± 1.9 Ma & 146.2 ± 1.5 Ma, 70–340 km diameter (Koeberl et al., 1997)
12	Late Volgian/Early Berriasian biostratigraphic constraint, 40 km diameter (Dypvik et al., 1996)
13	175 ± 3 Ma, 80 km diameter (Grieve, 1995 (secondary reference))

Fig. 17.2 Time-scale and summary of major features in the global and local stratigraphical record. The strikingly short duration for the Oxfordian stage is probably due to a scarcity of reliable ages. Some workers relegate the Rhaetian stage to be a substage of the Norian. The Kimmeridgian stage is used in its more restricted sense, i.e. 'sensu gallico'; the Portlandian stage (Pt) is also in its most restricted sense, i.e. 'sensu anglico' (cf. Cope 1993) (time-scale from Pálfy et al. in press).

17.2 Correlation and time-scale

Subdivision of the Jurassic into stages has remained reasonably stable since their introduction by d'Orbigny in the middle 19th century. These stages and their groupings into Lower, Middle and Upper Jurassic series are shown in Fig. 17.2. Approximately isochronous biostratigraphic zones and subzones have been defined using ammonites. For historical reasons, the zonal and subzonal schemes are best developed for the European area, and the average duration of an ammonite zone here is about 1 Myr. Further subdivision into more numerous subzones is unlikely to occur, simply because the more finely discriminated units are probably more geographically limited, and therefore less useful. A more recent trend in biostratigraphical work has been the recognition of faunal horizons based on ammonites. A faunal horizon has been defined as 'a bed or series of beds characterized by a fossil assemblage within which no further stratigraphical differentiation of fauna can be distinguished'. In a rock section, the time represented by a horizon could be very short compared with the duration represented by the interval between the horizons. Ammonite horizons have been defined (within the Middle Jurassic for example) in which the average time intervals between horizons may be as little as 70 kyr.

Zones and subzones based on micropalaeontological investigations are highly important and complementary to the ammonite-based schemes. Their particular advantages are that only small samples are needed so that they are useful for correlating borehole successions (whether from cores or chippings), something rarely possible using macrofossils. An important disadvantage lies in the relatively long ranges of guide taxa, which makes correlation based on Jurassic microfossils substantially less precise than ammonites. Problems inherent with all biostratigraphic correlations are potential diachroneity of the index species over long distances, faunal provinciality, palaeoenvironmental restriction, and the unsuitability of some lithofacies for preservation. Faunal provinciality, for example, is a problem for much of the Upper Jurassic, such that different stage names have had to be used for the terminal Jurassic stages in different parts of north-west Europe: in addition to the widely used primary standard, the Tethyan Tithonian Stage, a Boreal Volgian Stage and a Sub-boreal Portlandian Stage extend into the lowermost Cretaceous (Fig. 17.2).

Other methods of correlation may be used in the Jurassic on a global basis for parts of the column, and these can be superior to biostratigraphy. One newly established method is strontium-isotope stratigraphy. This works on the basis that the ratios of ^{87}Sr to ^{86}Sr change through geological time, but at any geological instant are homogeneous through the world's oceans. Strontium derived from continental weathering is relatively rich in the radiogenic isotope ^{87}Sr, whereas strontium derived from mantle sources (e.g. mid-ocean ridges) is relatively depleted. By determining $^{87}Sr/^{86}Sr$ ratios in marine calcitic fossils, which record the ocean chemistry in their shells, an age can be established by comparison with a standard curve constructed from biostratigraphically well-calibrated sections. The $^{87}Sr/^{86}Sr$ curve shown in Fig. 17.3 is based principally on oyster and belemnite samples from the ammonitiferous and much studied sections exposed on the coasts of Dorset and Yorkshire. The curve can only be used for correlation where changes in $^{87}Sr/^{86}Sr$ are rapid, and thus the technique has little application in the early Mid-Jurassic. Large-scale fluctuations are likely to reflect changing dominance of mid-ocean ridge and continental sources of strontium through time, which may link closely to phases of rifting and sea-floor spreading.

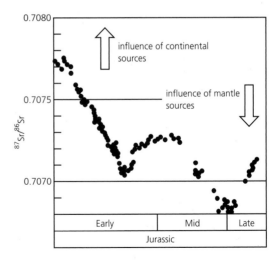

Fig. 17.3 Evolution of strontium-isotope composition of Jurassic sea water (modified from Jones *et al.* 1994 and reprinted with permission from Elsevier Science).

Another isotopic technique that is of value over more restricted intervals is carbon-isotope stratigraphy. Major perturbations of the ratios of ^{12}C to ^{13}C occur in the world's oceans primarily because of burial and release of light-carbon from organic matter and, probably, a number of additional factors such as volcanic outgassing. During the Jurassic there were several episodes of global carbon burial, particularly in the Toarcian and, to a lesser extent, the Bajocian and the Oxfordian. Associated with these carbon-burial events are carbon-isotope excursions, recorded in both marine calcite and marine organic matter, which may be used for very precise correlation of specific points in time.

Magnetostratigraphy, a very important technique for later periods of geological time (see following chapters), is of less use in the Jurassic currently. The principal reason for this is that the sea-floor magnetic anomaly record, one of the best data sources for linking magnetic reversals to a numerical time-scale, extends back only to the Callovian, older oceanic crust having been largely destroyed. However, magnetostratigraphic zonations can be based solely on continuous sedimentary successions; schemes of this kind are at present being constructed, so that magnetostratigraphy may be more often used in future studies.

Radiometric ages have been derived by analysis of isotopic ratios in magmatic intrusions, bentonites, glaucony and, most recently, black shales; time-scale compilations have treated Mesozoic glaucony dates as either minimum ages or have excluded them altogether, as there is clear evidence that analysis of multiple glaucony grains produces dates that are too young. Until very recently, radiometric dates from Jurassic rocks have been too sparse and biostratigraphically too poorly constrained to use directly to date stage boundaries. The time-scale shown in Fig. 17.2, however, is based on high-quality dates obtained from fossiliferous marine strata, particularly for the Early and Mid-Jurassic. Numerical ages have been assigned to the Late Jurassic boundaries mainly by a variety of geological and mathematical interpolation methods. The primary interpolation method used for the younger parts of the Mesozoic (and Cenozoic) time-scale employs the concept of constant sea-floor spreading rates. This assumption allows ages to be assigned to stratigraphically significant sea-floor magnetic anomalies (those correspond-

ing to stage boundaries) that lie between other sea-floor magnetic anomalies for which radiometric dates have been obtained; a method that can be used back to the Callovian.

17.3 Climate and sea-level change

In the latest Triassic and through the Jurassic, strata in the area of Britain and Ireland, at palaeolatitudes of between 30° and 40°N, were laid down in a region that appears to have been on the borderline between a mid- to low-latitude, continuously dry climatic belt, and a more northerly, winter-wet climatic belt (Fig. 17.1). In contrast to the present-day pattern, in which equatorial areas are characterized by warm aseasonal and humid conditions, the Late Triassic–Jurassic equator was a relatively dry, high-pressure region, because of strong preferential heating of the higher-latitude continental masses in the summer and their dramatic cooling in the winter. The effect of summer heating was to draw in humid air by convection; precipitation in upwardly mobile air led to further release of latent heat and enhanced warming. This is the basis of the so-called 'mega-monsoonal' climate that affected Pangaea through much of its history (see Chapter 16).

Because of its palaeogeographical setting, the sedimentary succession of the region of Britain and Ireland records rather well some of the major features of climatic change that affected the Earth during the Jurassic. For example, the marked spread of the equatorial arid zone across southern Eurasia during the Late Jurassic (Fig. 17.1) appears to be well expressed in the development of latest Jurassic (Purbeck Group) evaporitic carbonate facies in southern England. This apparent large-scale climatic change has been attributed to three alternative mechanisms:

1 building of a mountain belt, the 'Cimmerides', along the north-west coast of Tethys, creating an effective rain-shadow over continental areas;
2 progressive weakening of monsoonal circulation during the Jurassic caused by the more dispersed and increasingly asymmetrical continental configurations;
3 southward movement of continental masses through stationary climatic belts.

One major issue concerning Jurassic climate is the extent to which ice accumulation was important in polar regions. This is a key topic, because the waxing and waning of land-based ice is one of the few known mechanisms capable of generating high-amplitude, high-frequency global sea-level changes (i.e. 100–150-m rise and fall over 10–100-kyr periods). The rather patchy evidence for 'tillites' from northern Siberia and Kamchatka, comprising matrix-supported, very poorly sorted, pebbly siliciclastic sediments, indicates that at least some polar seasonal ice existed in the Mid-Jurassic. There is, however, little strong evidence for the existence of any major polar ice-caps.

Palaeotemperature estimates, based on oxygen-isotope studies of marine invertebrates and vertebrates, are in the region of 12–29°C for the British Jurassic. The temperature ranges in some cases clearly reflect the life habits of the invertebrates analysed, as discussed in more detail below. Because of its distinctly maritime palaeogeographical setting within shallow seaways linking Tethys to the Boreal ocean, the north-west European area suffered less extreme seasonal change than has been inferred for areas within the continental interiors. Nevertheless, the ubiquitous presence of annual tree rings through fossil wood in many of the local Jurassic successions does indicate some marked seasonal changes even here.

On a longer time-scale, high-frequency climatic fluctuations, probably forced by the variations in insolation resulting from cyclicity in the character of the Earth's orbit (Milankovitch cycles), are a prominent feature of Jurassic successions, as they are for other periods. Milankovitch cyclicity, as opposed to irregular, non-periodic cyclicity, has been established from length series (sometimes referred to as 'time series') of stratigraphic data using a statistical technique known as power spectral analysis, by estimation of cycle durations, and by recognizing characteristic wavelength ratios between superimposed cycles. Orbital forcing as expressed in lithological cyclicity has changed its dominant mode through the Jurassic of the region: at some times obliquity has been paramount (≈40 kyr cyclicity), whereas during others precession (≈20 kyr) or eccentricity (≈100 kyr and 400 kyr) have been more important. Climate models have been used to demonstrate, using a simplified Pangaean palaeogeography, that precession-related changes in insolation would have had a marked effect on monsoonal circulation, such that very large changes in seasonal rainfall might be expected in circum-Tethyan coastal regions like that that occupied by the area of British and Ireland. Obliquity can also be expressed in low-latitude settings, but this requires some amplifying mechanism or means to transfer the effects of high-latitude climatic processes down to low latitudes.

There is broad agreement that a general increase in global sea level occurred through the Jurassic (Fig. 17.4). It has been guestimated that this long-term rise was in the region of 100 m, based principally on compilation of the areal extent of marine sediments, modified on the basis of assumptions concerning continental hypsometry (i.e. the distribution of area with respect to height). Rather than a continuous slow rise through the Jurassic, a stillstand, or even fall, during the Mid-Jurassic (Bajocian–Bathonian) has also been

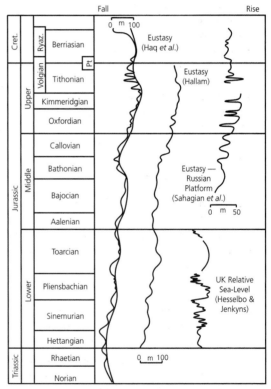

Fig. 17.4 Proposed global sea-level curves for the Jurassic all based on analysis of large-scale sedimentary cycles ('depositional sequences') (from Haq et al. 1987; Norris & Hallam 1995; Hallam 1996; Sahagian et al. 1996). Also shown is a relative sea-level curve for the British area (Hesselbo & Jenkins 1998). Relative length of stages reflects the time-scale of Gradstein et al. (1994), which differs slightly from the more recent scale used in Fig. 17.2.

suggested, although it is difficult to tell to what extent this is strictly a north-west European phenomenon. In view of the lack of evidence for substantial polar ice, it is likely that the overall slow rise in sea level is related to increasing production of oceanic crust consequent upon the break-up of Pangaea. Shorter-term major eustatic sea-level fluctuations also undoubtedly occurred, the most dramatic of which appears to be the rise in the early Toarcian, but global rises also occurred in the Sinemurian, Pliensbachian, Bajocian, Callovian and Tithonian, and major global falls have been suggested for the Late Rhaetian, Late Sinemurian, Aalenian, Early to Mid-Oxfordian and mid-Tithonian. Quantification of these more rapid changes, and identification of their causes, has been elusive. Estimates of tens of metres of sea-level change have been suggested based on interpretation of relative water depths from biofacies and lithofacies, most recently from the Russian Platform, which has been considered to provide a 'stable' reference.

17.4 Catastrophic events

The Earth in the Late Triassic and Jurassic was subject to extreme geological events, just as at any other time period, namely massive extraterrestrial impacts and episodic flood basalt volcanism. The effects of these events are either masked in the geological record of Britain and Ireland or have not yet been recognized, because the epicentres are located some distance from the area. Their effects are, however, likely to have been global.

Impacts of extraterrestrial bodies, asteroids or comets, are a normal aspect of planetary development, particularly during the early stages of solar system evolution. Evidence can be in the form of craters or strewn fields of impact ejecta or their geochemical traces (e.g. anomalous iridium concentrations). Several large (>30 km) impact craters of Late Triassic to Jurassic age are now known and are mostly well dated by radiometric or biostratigraphic means (Figs 17.1, 17.2). Because most Jurassic ocean floor has been subducted, it is likely that other impacts, with a more cryptic geological record, also occurred.

Three craters are close enough to north-west Europe that some expression in the sedimentary record of Britain and Ireland might be anticipated (the Chicxulub Crater at the Cretaceous–Tertiary boundary, Mexico, has an estimated diameter of 180 or 310 km, depending on how interpreted, and undoubtedly left a global stratigraphical imprint). The following large craters have been identified as end-Triassic to Jurassic age (Figs 17.1, 17.2).

1 Manicouagan Crater in Quebec has a diameter of 100 km and is dated radiometrically at 214±1 Ma, which may be as much as 15 Myr older than the Triassic–Jurassic boundary.

2 Puchezh–Katungi Crater in Russia, with a diameter of 80 km, is dated radiometrically as Aalenian–Bajocian.

3 Mjølnir Crater, with a 40-km diameter, had an impact site in a marine basin of the Barents Sea, and is constrained biostratigraphically to the Volgian–Ryazanian boundary interval, i.e. just above the Jurassic–Cretaceous boundary.

4 Morokweng Crater in South Africa is also dated radiometrically as approximately coincident with the Jurassic–Cretaceous boundary, with a diameter as yet poorly constrained as between 70 and 340 km.

The potential relationship between extraterrestrial impact and mass extinction is very well explored for the Cretaceous–Tertiary boundary (see following chapters), but fewer data are available for the Late Triassic to Jurassic interval, and apart from some suggestions of coincident timing, there is not yet any strong evidence for a similar linkage in the examples cited above. Intriguingly, the Manicouagan Crater lies on the same palaeolatitude, 22.8°, as two other substantial craters thought to be the same age, and may thus form part of a linear crater chain resulting from the preimpact break-up of a particularly large extraterrestrial body.

Three major continental flood basalt provinces have been recognized as being of Late Triassic–Jurassic age and appear to pre-date, by a few million years, major phases of continental break-up and sea-floor spreading in what was to become the Central Atlantic area and in south-west Gondwanaland (Figs 17.1, 17.2). As with the cratering, it is possible to make a case for linkage to mass extinctions (this has certainly been claimed, via volcanic emissions into the atmosphere or stratosphere), but as yet there is too little agreement as to the ages of the flood basalt events, and their relative timing with respect to faunal crises.

Three important mass extinctions took place over latest Triassic–Jurassic time:

1 The first, and greatest, was at the Triassic–Jurassic boundary and was one of the five largest mass extinctions in the Phanerozoic, affecting organisms in marine (ammonites, bivalves and reef ecosystems) and terrestrial (tetrapods, floras) environments. Conodonts and strophomenid brachiopods, both taxa that are important constituents of the Palaeozoic fauna, finally became totally extinct at the Triassic–Jurassic boundary. The evidence suggests that the extinction was catastrophic in that it apparently took place in less than 40 kyr, but careful examination of boundary beds, including those of south-west England (Bristol Channel Basin), has failed to reveal definitive evidence of impact debris, which has led some workers to rule out an impact origin for the extinction. Recent dating of the flood basaltic volcanism in the Central Atlantic region makes this a more plausible trigger to extinction on the basis of 'guilt by coincident timing'.

2 A second extinction event was in the early Toarcian and coincided with the development of poorly oxygenated bottom waters, described below, with which it was probably causally linked. In this case, extinctions affected principally benthos rather than plankton, and most clearly affected organisms in the European region, which were replaced through immigration from elsewhere. Nevertheless, the extinction was certainly global.

3 The third extinction event took place in the Tithonian and was probably the least dramatic of the three, affecting principally European bivalves and some other benthic organisms.

17.5 Sedimentary basins and syn-sedimentary faulting

Many of the faults that now divide up the Jurassic sedimentary basins were active, at least episodically, during sediment deposition. Pinpointing times of actual fault movements can be difficult, but many of the major faults would have had at least a passive effect on deposition, even when there was no slippage on the fault plane, through mechanisms such as differential compaction of contrasting lithology either side of the fault. Evidence for fault movements in the Early Jurassic includes:

1 thickness changes across faults;
2 sedimentary facies genetically related to fault-crests or fault-scarps; and
3 fault-related syn-sedimentary neptunian intrusions.

Studies of basins in the south and west of the region considered here all indicate the likelihood of rapid Early Jurassic subsidence, possibly associated with rifting. This is compatible with demonstrable rift-related subsidence that occurred along the northern margin of Tethys at the same time. In southern basins, Early Jurassic faults can be inferred on the basis of borehole records of thickness changes in Lower Jurassic strata traced across major fault zones, such as the Isle of Wight–Purbeck system of the Wessex Basin. Smaller-scale thickness changes can also be observed across the smaller faults exposed cutting the coast of Dorset, southern England. Some of the Lower Jurassic rocks adjacent to these faults contain very fine-grained, creamy-coloured, complexly bedded and laminated limestones in dykes and sills intruded into other limestones or sandstones. These sedimentary intrusions contain marine fossils, which include ammonites that are commonly two or three zones younger than the host material. The close spatial relationship to faults has been used to infer a fault-slip origin for the intrusions, dated by means of the included fossils.

In areas such as the North Sea, where seismic reflection data have been gathered in greatest abundance, lateral thickness changes resulting from differential subsidence across faults are most dramatically evident in the Middle to Upper Jurassic, and are a result of crustal extension. Nevertheless, even during the deposition of 'post-rift' Early Jurassic aged strata in the North Sea (which post-date Triassic rifting), evidence of some significant syn-sedimentary fault movements is commonplace.

The Cardigan Bay Basin was probably also predominantly an extensional basin, at least initially in the Jurassic, but the basin may also have developed a very large-scale, gravitationally driven system of faults during the later Early Jurassic (Fig. 17.5). Seismic evidence indicates that failure occurred of the Bala Fault footwall (incidentally, an important Caledonian fault), probably because the basin to the west was subsiding so rapidly that sedimentation was unable keep pace and a very great bathymetry developed (although the estimated 2 km is hard to credit). Certainly, we know that the Lower Jurassic succession in the Cardigan Bay Basin, at about 2.5 km thick, is the thickest and most complete in the area of

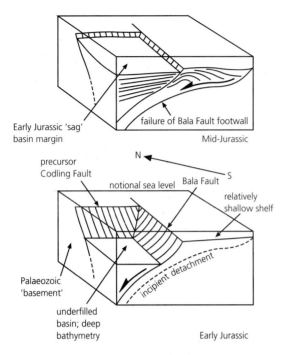

Fig. 17.5 Schematic illustration of possible gravity-driven tectonics operating in the Early to Mid-Jurassic Cardigan Bay Basin (from Turner 1996).

Britain and Ireland yet discovered. This kind of structural evidence presents some challenges to the long-held view that Early to Mid-Jurassic sedimentary basins of the mainland British region rarely developed very significant sea-floor relief.

17.6 The early seaways

The later Jurassic erosion of much of the Upper Triassic and Lower Jurassic section around Britain and Ireland results in some uncertainty in reconstruction of the geological history of the region, particularly in the vicinity of the middle North Sea area. Nevertheless, away from the areas of later uplift, around the present-day coasts and across much of the British mainland, some excellent and stratigraphically quite complete exposures of Upper Triassic to Lower Jurassic can be seen. Extensive preservation of these strata also occurs in offshore basins in the west of the area, such as those of Fastnet, the Celtic Sea and Cardigan Bay.

Areas of upstanding basement undoubtedly existed

in Late Triassic–Early Jurassic time (Figs 17.6, 17.7), but their nature and extent is debatable (as it is for much of the later Jurassic), because there are only rarely extensive coarse clastic sediment fringes associated with them. Heavy-mineral assemblages indicate that at times the Scottish landmasses fed siliciclastic sediment into adjacent basins. In the south-west, the existence of the land areas has generally been inferred on the basis of lack of preservation of Jurassic strata, but there is no positive evidence for their existence in the Early Jurassic. Perhaps the best established land area in the southern region was at the western end of the Anglo–Brabant Massif (Fig. 17.7), otherwise known as the London Platform, across which Lower Jurassic strata preserve a pattern of progressive onlap when mapped using geophysical log data. Even in this case, definitive shoreline deposits are unknown. Much smaller areas of sharp relief existed in a belt along the northern margin of the Bristol Channel Basin and, possibly, in parts of the Hebrides Basin, but these islands probably had areas of only a few tens of square kilometres (Fig. 17.6).

In the principal sedimentary basins of the British mainland area, sediment deposition from the Late Triassic to the Early Jurassic was mostly argillaceous. There was, however, a very striking change from non-marine 'red-bed' deposition through to marine organic-rich strata. This overall transgression, characteristic of the Early Jurassic, was not a gradual drowning, but was pulsed in nature.

The initial latest Triassic transition from non-marine to marine strata is recorded over much of the British mainland in a remarkable series of deposits, well exposed in south-west England and South Wales (Fig. 17.6). The uppermost red-beds of the *Mercia Mudstone Group* are strongly argillaceous with only minor (though spectacular) deposits of coarse clastic sediment restricted to the fringes of the localized upstanding basement blocks. The upper beds of the Mercia Mudstone contain only sulphate evaporites (as opposed to halite, common in some lower levels), which were deposited by interstitial brines, and in coastal sabkhas bordering hypersaline water bodies of marine origin. Upwards through the group the colour changes from red, green and grey, to green and grey (this multicoloured formation is rather incongruously called the *Blue Anchor Formation*). These rhythmic beds show very strong similarities to Milankovitch cycles described from similar-aged

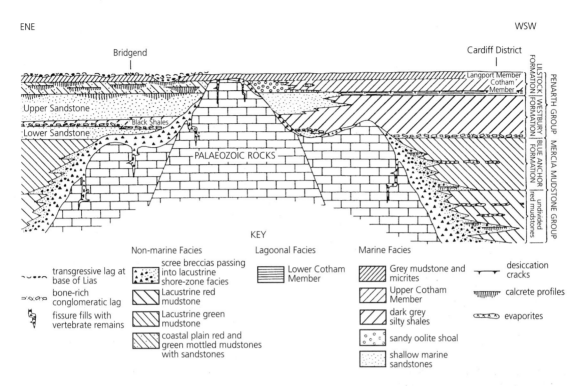

ENE   WSW

Fig. 17.6 Schematic section through the Late Triassic succession of the Bridgend district, South Wales (from Wilson *et al.* 1990, with permission of the British Geological Survey © NERC. All right reserved).

deposits in the eastern USA, and may reflect a particular sensitivity of these transitional environments to palaeoclimatic fluctuations.

A marked expansion of the marine area in a northward direction is represented by the basal formation of the *Penarth Group*, a thin black shale unit known as the *Westbury Formation*, which is the precursor to the much more persistent organic-rich shales of the Jurassic. This facies is evident as far north as the Hebrides and also in basins around the supposed Irish landmass and Antrim. The formation has yielded one specimen of a diminutive psiloceratid ammonite, of Jurassic affinity, but otherwise contains abundant reworked terrestrial- and marine-vertebrate bones and numerous bivalves. Fully marine conditions were, however, not well established, because the overlying beds represent a widespread lagoonal deposit; in the area of Bristol, one stromatolitic horizon is particularly well known (*Cotham Marble*) and other indicators of low water levels include the local occurrence of very deep

(≈ 50 cm) polygonal mudcracks. Marked interbedding of different faunas and floras is taken as evidence of rapid and frequent salinity changes from fresh to brackish to normal marine, and to hypersaline. At the top of the Penarth Group, in the area of southern England, a peculiar limestone facies known as the 'White Lias' is well developed. Lying stratigraphically between underlying lagoonal deposits and overlying open-marine strata, this interval might be expected to represent some intermediate depositional environment. Although it does contain shallow-marine fauna, it also bears the hallmarks of slope-related deposition, such as slumps and slump scars, flow banding, and matrix-supported pebbles (debris flows), and indicates a major and abrupt environmental change, possibly deepening.

In addition to the rather more conventionally preserved fauna from these times, fissures within the basement limestones adjacent to the Bristol Channel Basin are a rich store of more unusual fossils, which include Late Triassic reptiles and early mammals. Whereas some of these fissures may be dissolution modified, it is likely that others have a purely tectonic origin, and opened up in the basement rocks under a cover of later sediment.

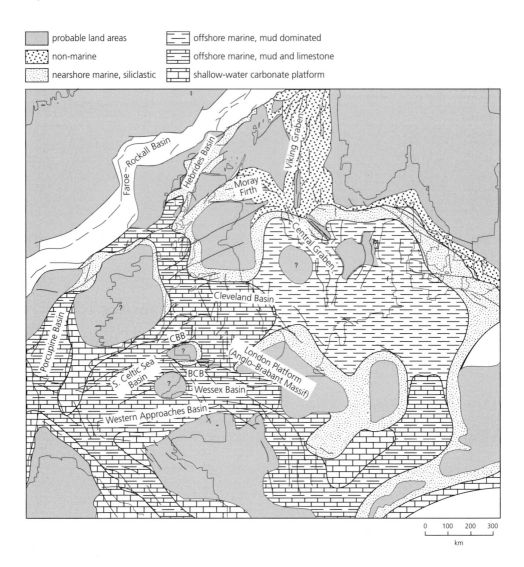

Fig. 17.7 Palaeogeographic map for the early Hettangian of the area of Britain and Ireland (based on Ziegler 1990, modified in region of the British Isles on the basis of Bradshaw *et al.* 1992). Facies patterns in the graben area north-west of the Britain and Ireland are relatively poorly known at present.

17.7 Early Jurassic marine environments

Perhaps the best known of the Lower Jurassic mudstone formations is the Hettangian–Sinemurian aged *Blue Lias* (Fig. 17.8), a unit of interbedded limestones, marls and shales of very wide distribution. This facies is prevalent in the Wessex and Bristol Channel basins in the south, the Cardigan and Celtic Sea basins in the west, through to the Cleveland and Hebrides basins in the east and north. Because of the abundance of layers with high carbonate content, cliff exposures of this formation appear strikingly banded. The rhythmic, decimetre-scale interbedding is a reflection of extensive and geologically frequent changes in the degree of bottom-water oxygenation. The dark grey, highly organic-rich shales containing, if anything, only limited diversity, small trace fossils, were lain down in very poorly oxygenated bottom waters. Such conditions required at least seasonal density stratification of the water column, which may have been related to variations in, for example,

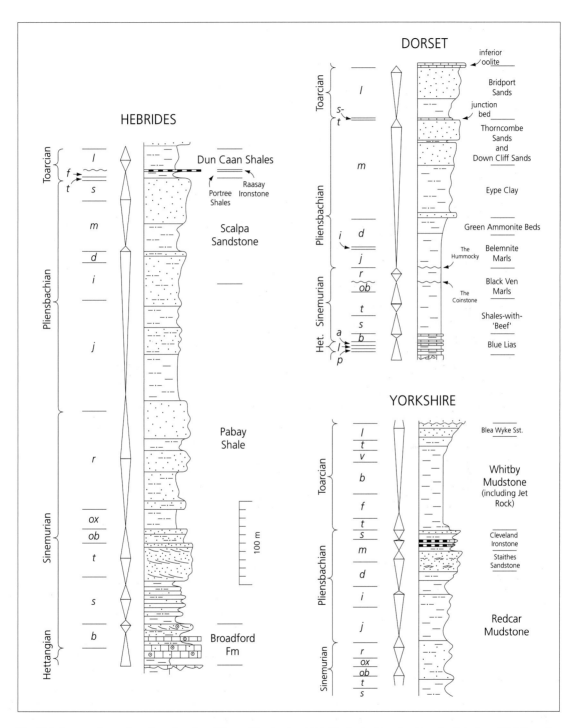

Fig. 17.8 Summary Lower Jurassic sections from Wessex, Cleveland and Hebrides basins (based on Hesselbo & Jenkyns 1995; Hesselbo *et al.* 1998). See Fig. 17.12 for key.

surface water runoff. However, the extent to which black shales, such as these, are a result of enhanced organic-matter preservation or enhanced organic-matter productivity has been a matter of heated debate, not least because of their importance as hydrocarbon source rocks. As yet there is no consensus. Part of the problem is that productivity and preservation are rarely unrelated; enhanced marine organic production undoubtedly leads to enhanced oxygen consumption in bottom waters by way of bacterial degradation. Conversely, development of poorly oxygenated bottom waters may lead to enhanced nutrient recycling, for example of phosphorus, thus stimulating productivity. At present, no one palaeoceanographic explanation for the origin of black shales can account for the genesis of all the Jurassic examples from the area of Britain and Ireland: all are to some extent special cases.

Another well-known, highly organic-rich formation of the Lower Jurassic is the Toarcian-aged *Whitby Mudstone*, exposed on the coast of Yorkshire in the Cleveland Basin, which includes a member known as the '*Jet Rock*', so named because of the abundant polishable fossil wood it contains, the 'jet' much used for jewellery during Victorian times. The Whitby Mudstone contrasts with the Blue Lias in that it is not rhythmically bedded on a decimetre scale, but includes an approximately 10-m-thick interval that is millimetrically laminated black shale containing up to 15% (typically 5–10%) total organic carbon (TOC). Carbon burial was a global phenomenon at this time, and widespread burial of relatively light (^{12}C-rich) organic carbon led to an enrichment of heavy carbon (^{13}C) remaining dissolved in the water column. Thus, early Toarcian marine carbonates commonly record a 'positive' (i.e. heavy) carbon-isotope excursion (Fig. 17.9). However, carbon-isotope values from organic matter from the lower Jet Rock (and other time-equivalent strata in the northwest European area) show a sharp 'negative' excur-

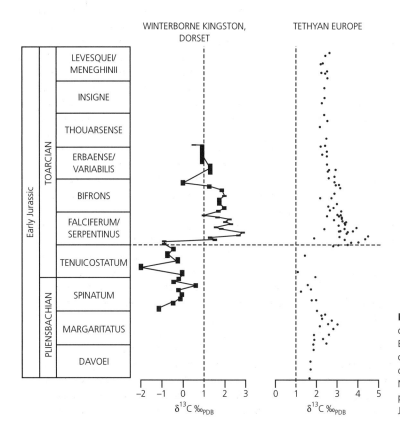

Fig. 17.9 Carbon-isotope data for a carbonate-rich succession in the Wessex Basin compared with Tethyan pelagic carbonates. '$\delta^{13}C$' refers to deviation of carbon-isotopic ratios from a standard. Negative values mean relatively ^{12}C rich; positive mean relatively ^{12}C poor (from Jenkyns & Clayton 1997).

sion interrupting the 'positive' excursion, which may have originated from addition to the oceans of isotopically light carbon derived from previously buried organic matter.

The phenomena exemplified by the Jet Rock have been termed Oceanic Anoxic Events (OAEs), and they are an important feature also of Cretaceous palaeoceanography (see Chapter 18). Relatively light oxygen in carbonates (i.e. relatively ^{16}O-enriched) coincident with the level of the OAE suggests that highest Early Jurassic water temperatures were achieved in early Toarcian times, although the values cannot yet be quantified with any confidence. The early Toarcian OAE in epicontinental north-west Europe has been explained in terms of enhanced planktonic productivity, producing intense oxygen demand in bottom waters (or, in southern Europe, a mid-water oxygen minimum zone). It is likely that a chain of events occurred to bring this about, which included global warming, reorganization of oceanic circulation, and liberation and oxidation of methane hydrate, and whose ultimate cause was flood-basalt volcanicity.

In the Wessex Basin, much of the Toarcian is represented by fine-grained pink and yellow limestone, the *Junction Bed*, commonly not much more than 0.5 m thick. In other words, the succession is extremely condensed. The fossil content of this rock is dominated by nekto-pelagic organisms, such as ammonites and belemnites and, like the mudrocks, the limestone probably represents an undisturbed, deep, dark marine environment. In this case, however, the limestones developed on a sea floor that was severely starved of fine-grained siliciclastic sediment, so that all organic matter falling to the bottom was oxidized before it could be buried, particularly on the less subsident, fault-bounded intrabasinal 'swells'.

Many basins in this region also accumulated sediments in relatively shallow marine environments during particular times in the Early Jurassic. One of the most prominent and widespread episodes of shallow-water deposition of sand took place during the late Pliensbachian, for example the *Staithes Sandstone* of the Cleveland Basin (Fig. 17.8). These sandstones are commonly characterized by the occurrence of hummocky cross-stratification, beds of transported shells, and other features together indicating storm deposition in shoreface environments. Indeed, evidence of storm processes, as opposed to

tides, dominates the outcropping sandy Lower Jurassic in all the basins of the region except the Hebrides. Other shallow-water facies developing at this time include oolitic ironstones, containing the mineral berthierine [$\sim$(Fe^{3+},Mg,Fe^{2+})$_5$Al(Si$_3$,Al)O$_{10}$(OH,O)$_8$] as matrix and as ooids. The concentrations of iron, relative abundance of other insoluble elements such as thorium, and association with the clay mineral kaolinite, support an origin for these deposits related to intense humid weathering of the hinterland.

17.8 Mid-Jurassic Uplift: the 'North Sea Dome'

Regional sedimentation patterns changed dramatically at the beginning of the Mid-Jurassic (Aalenian). Of particular importance to much of the British area was the updoming and volcanic activity of the central North Sea region (Figs 17.2, 17.10, 17.11). Partly synchronous with this was the creation of oceanic crust in the central Atlantic, probably starting in the Bajocian. As a consequence, the north-west European region experienced a rejuvenation of clastic source areas, and deposition of a broad range of arenaceous clastic sedimentary rocks. Marine connections between the Arctic and Tethys oceans must at this time have been restricted to a narrow seaway west of the British Isles, including the Hebrides Basin, which became more strongly tidal. Even this seaway became occluded in the Bathonian. Southern shallow-water areas developed extensive carbonate platforms dominated by oolite deposition on apparently gently sloping ramps. The northward extension of carbonate deposition has been variously attributed to dominance of warm Tethyan waters or restriction of supply of fine clastic sediment.

Uplift of the central North Sea has been described as a broad arch transected by the Central Graben and Moray Firth Basin, perhaps with a north-east (Norwegian) flank that was steeper than the others. Lower Jurassic and older sedimentary rocks and, locally, even Caledonian basement, are truncated. Erosion at this major unconformity increases progressively when traced in from the edges of the dome towards the margins of the Central Graben and Moray Firth Basin. A much more complete record of earlier deposition has been preserved within the grabens themselves. Significant volumes of sediment were eroded from the shoulders of the rift blocks,

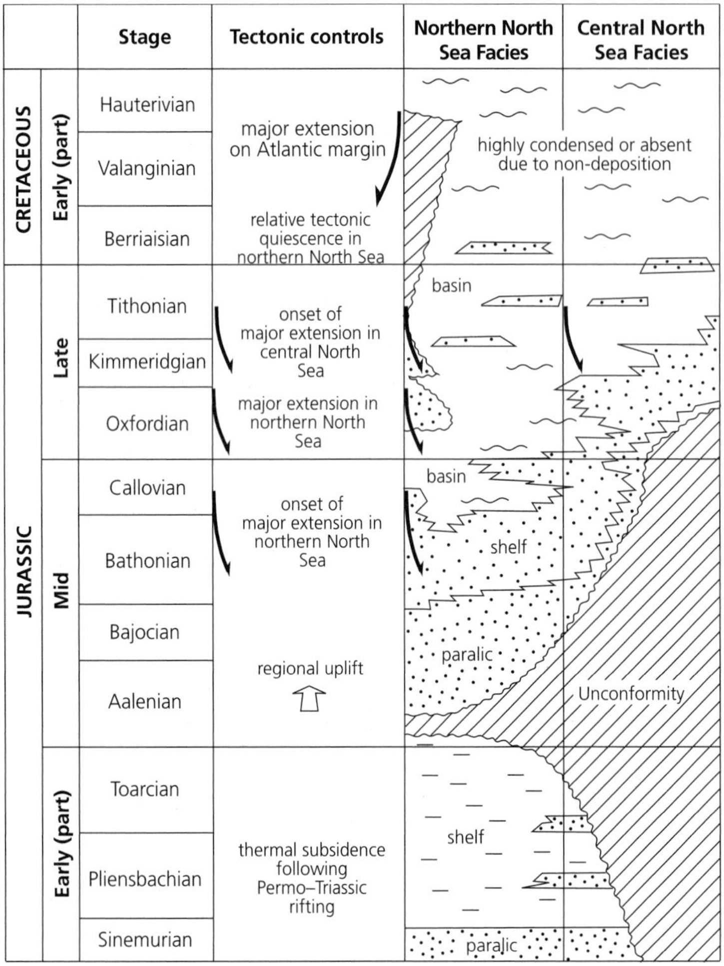

Fig. 17.10 Summary of Jurassic to earliest Cretaceous stratigraphy of the central and northern North Sea in relation to major tectonic controls on sea level (modified from Rattey & Hayward 1993).

but the topography of the dome may never have amounted to much because of rapid erosion of the generally poorly consolidated sediments. Estimates of maximum eroded thicknesses range from 250 m to in excess of 2 km; estimates dependent on the extent to which the region is thought to have been subject to prior Jurassic rift-related subsidence. Some of the erosional products were deposited in neighbouring basins in major deltaic complexes such as the *Brent Group* of the North Sea region, which was supplied from the south and east into the Viking Graben, and the *Ravenscar Group* now exposed in north-east Yorkshire, which prograded southward from the flanks of the dome, down north–south-orientated rift systems of the Cleveland Basin. Extensive deltaic or estuarine complexes accumulated in other northern British basins at this time, such as the Moray Firth and Hebrides Basins (e.g. the *Bearreraig Sandstone* and *Great Estuarine Groups*), the latter with sediment sources that were unrelated to North Sea-centred uplift.

Uplift of the North Sea Dome is connected with an important episode of volcanism. The igneous rocks, which accumulated mostly in the vicinity of the triple junction of the North Sea rift system, may be up to 2 km thick, are predominantly of alkaline affinity, and include alkali–olivine basalts. Their geochemistry suggests derivation from an 'enriched' lithospheric mantle source, with possibly an asthenospheric component. An explanation for both

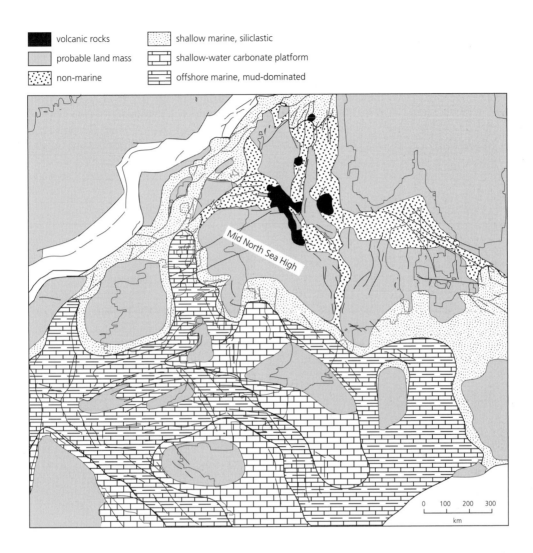

■	volcanic rocks		⋯	shallow marine, siliclastic
▢	probable land mass		▦	shallow-water carbonate platform
▨	non-marine		▤	offshore marine, mud-dominated

Mid North Sea High

0 100 200 300
km

Fig. 17.11 Palaeogeographic map for the early Bajocian of the area of Britain and Ireland (based on Ziegler 1990, modified in region of British Isles on the basis of Bradshaw *et al.* 1992). CBB, Cardigan Bay Basin; BCB, Bristol Channel Basin.

the dome and the volcanism is that renewed Mid-Jurassic lithospheric extension resulted in decompression of previously melted mantle lithosphere to form new melts that rose diapirically and spread out laterally along the crust–mantle boundary. Alternatively, the same features have been explained by a short-lived mantle plume head that impinged on the base of the crust in the Aalenian. The key difference between these two hypotheses lies in the extent to which melt

generation was a consequence of extension, or extension a consequence of mantle upwelling. Concentrations of smectitic clays in several onshore British Bathonian sections have been interpreted as the products of weathering of volcanic ash with a possible North Sea source.

By mapping the distribution of coarse-grained siliciclastic sediment around the North Sea region it has been possible to reconstruct the history of 'inflation' and 'deflation' of the dome. The oldest coarse-grained sediments in the neighbouring basins appear in the Late Toarcian—for example the *Grey Sandstone* and *Yellow Sandstone* of the Cleveland Basin, and the Bearreraig Sandstone of the Hebrides Basin

(Fig. 17.12). Shallow-water sandstone deposition also became important in the Wessex Basin of southern England in the Late Toarcian (the *Bridport Sands*), whose source area is still somewhat uncertain. Because enhanced late Toarcian siliciclastic deposition is a common feature of north-west European rift basins, for instance the Lusitanian Basin of Portugal, it may be asked whether these deposits are characteristic of some larger-scale tectonic and/or sea-level process, rather than the North Sea Dome *per se*. All these occurrences are of shallow-marine, commonly shoreface-deposited, sandstones and bear strong similarities in their distribution and facies to the widespread late Pliensbachian episode of coarse-clastic sediment deposition, suggesting a similar origin. Throughout the

Fig. 17.12 Summary Middle Jurassic sections for the Cleveland and Hebrides basins (based on field observations and references cited in Hesselbo & Jenkyns 1995; Morton & Hudson 1995).

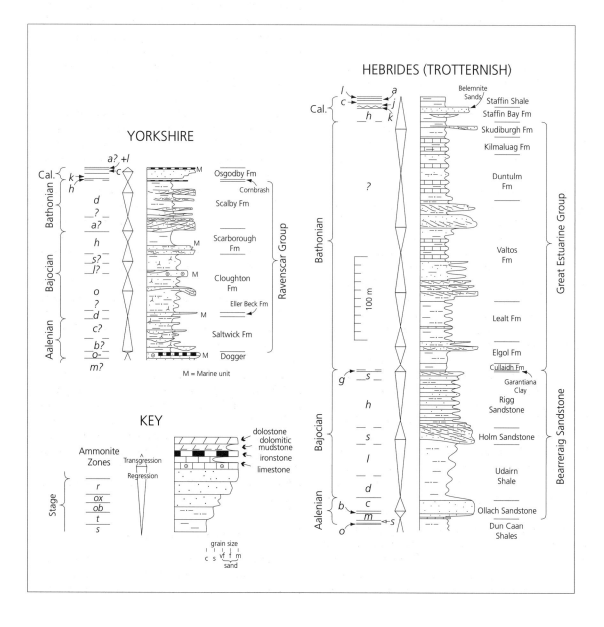

Mid-Jurassic, and more clearly restricted to the North Sea region, episodically more shallow-water deposition occurred through successive ages, with maximum regression possibly at about the Bajocian–Bathonian boundary (although the paucity of biostratigraphically useful taxa makes these non-marine strata difficult to date and correlate accurately).

17.9 Mid-Jurassic non-marine and marginal marine depositional environments

The extensive development of non-marine and marginal marine siliciclastic settings in the Mid-Jurassic provides an opportunity for understanding diverse aspects of coastal and terrestrial environments at this time. Sedimentary accumulations of this age are also economically important, as they form a reservoir interval for North Sea oil.

In the Cleveland Basin of north-east Yorkshire the fluvial and deltaic sediments deposited by rivers flowing off the flanks of the North Sea Dome interfinger with marine strata, and together they make up the Ravenscar Group (Fig. 17.12). The sedimentary facies represent a great range of river-related depositional environments, from flood plains, interdistributary bays and lagoons, to braided and meandering river-channel systems. On at least three occasions the area was inundated by the sea. From fossil evidence it is known that a diverse and abundant flora grew in the Yorkshire terrestrial environments, which included at least two hundred species. Over the course of this century, so much work has been completed on the Yorkshire flora that it is now one of the best characterized Jurassic floras in the world.

The Bathonian-aged *Great Estuarine Group* of the Hebrides Basin shows some similarities to the Ravenscar Group, but (despite its name) may be regarded as less fluvially influenced and more lagoonal. Numerous studies of the assemblages of biota preserved in the Great Estuarine Group lagoons have provided much information about the relationship of fossil faunas to palaeosalinity, which ranged here from near freshwater to hypersaline (Fig. 17.13). In particular, bivalves whose nearest living relatives have a specific salinity preference have been used to infer palaeosalinities for the fossil communities. In the broadly brackish range, the inferred salinities based on bivalve communities receive support from study of other fossil groups from the same horizons (e.g. ostra-

cods and choncostracans), which also show definite preferences for particular ranges of salinity. Some palynomorphs (e.g. dinoflagellate cysts and *Botryococcus*) are sensitive to salinity and, unlike the invertebrates, are relatively insensitive to substrate conditions. Stable isotopic data may also be used to support the palaeosalinity inferences based on the fossil communities: the light isotopes of oxygen and carbon (^{16}O and ^{12}C) are concentrated in fresh water compared with sea water, through preferential evaporation in the case of oxygen, and through derivation of dissolved bicarbonate from organic matter in soils in the case of carbon. The isotopic characteristics of the waters are reflected in the $^{18}O/^{16}O$ and $^{13}C/^{12}C$ composition of the biominerals—if precipitation is in isotopic equilibrium with water and is not diagenetically altered. Oxygen-isotopic compositions also monitor palaeotemperatures, because in carbonate light oxygen is preferentially incorporated at higher temperatures: $^{18}O/^{16}O$ from bivalves in the Great Estuarine Group have been used to bracket temperature ranges as 15–25°C.

At times in the Mid-Jurassic, particularly in southern locations, extensive shallow-water carbonate platforms developed in place of siliciclastic shoreline sediments. These accumulated in ooid- and peloid-dominated shoals, which are commonly preserved with metre-scale cross-bedding, signifying strong, possibly tidal current action, and in quiet lime-mud-dominated 'lagoonal' settings. The earlier of these thick carbonate platforms (*Inferior Oolite*) passes southwards (southern England and north France) into strongly condensed facies reminiscent of the older Junction Bed—thin limestones, commonly conglomeratic, containing an abundant nekto-pelagic and benthic fauna. It is likely that these deposits represent relatively sediment-starved distal environments, lying beyond the reach of North Sea-derived coarse-grained siliciclastic sediment during much of the Aalenian–Early Bathonian. Oolite deposition continued into the Bathonian (*Great Oolite*) but the ooid shoals passed offshore southwards into mud-clouded waters.

17.10 The Late Jurassic seaways

Dramatic palaeogeographic changes took place once again towards the close of the Mid-Jurassic, and involved mostly the regional drowning of paralic

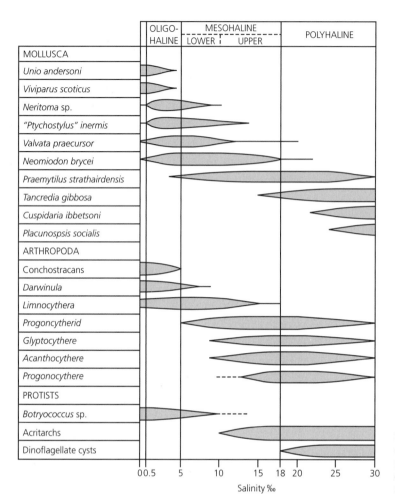

Fig. 17.13 Salinity ranges of a variety of organisms determined from palaeoecological and geochemical studies of the Great Estuarine Group (from Hudson *et al.* 1995).

sedimentary environments. In many areas of Britain and Ireland, this major transgression took place across the Bathonian–Callovian boundary, when a blanket of marine mudstone was deposited, with subsequent intermittent deposition of marine sands, commonly glauconitic. Better connections between seaways allowed the migratory spread southwards of Boreal ammonites such as the kosmoceratids. The overall transgression was accompanied by the initiation of major block-fault movements across much of the North Sea area. On the basis of the occurrence of volcaniclastic debris in the Hebrides, phases of volcanic activity may have peaked in the Faeroe–Rockall basins at the Mid-Jurassic–Late Jurassic boundary. Volcanicity may also have been important in the Netherlands area at this time (the Zuidval volcanic

complex has been assigned to the Oxfordian–Kimmeridgian but its age is unclear and it has also been regarded as Cretaceous). The overall deepening that occurred through the early Late Jurassic was interrupted in the Mid-Oxfordian by a marked spread of shallow-water sandstones and limestones over much of the region, exemplified by the *Corallian Group* of the Wessex and Cleveland Basins (Fig. 17.14), and uplift of regions of the northern Netherlands and the Celtic Sea. Renewed deepening took place over the area of Britain and Ireland in the late Oxfordian, and then more markedly through the Kimmeridgian, but was followed closely by a Volgian (Tithonian) shallowing that affected most of Europe. Some marked contrasts in sediment supply to northern and southern areas are evident by the end of the

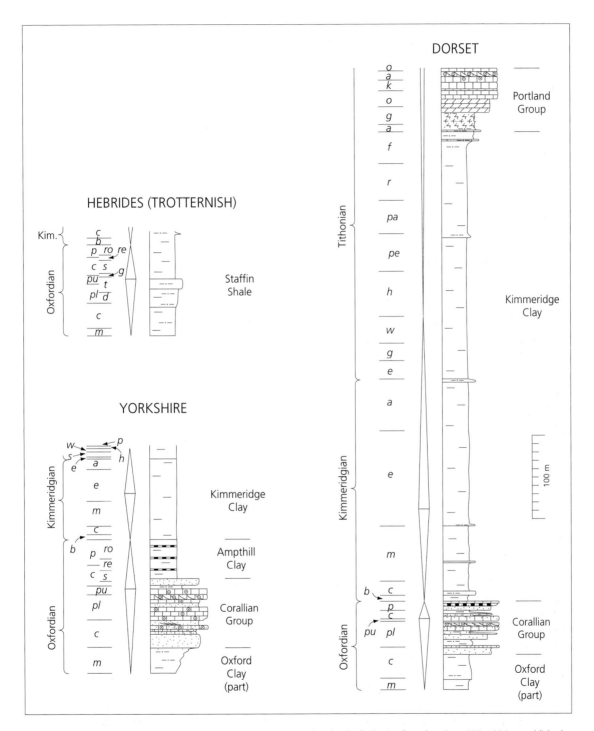

Fig. 17.14 Summary Upper Jurassic sections for the Wessex, Cleveland and Hebrides basins (based on Coe 1995, 1996, unpublished data).

Jurassic: southern basins continued to be supplied with abundant sediment, whereas northern basins became sediment starved, a pattern that continued in the Early Cretaceous. Reasons for these major changes in sediment supply have not been identified with certainty, but could include drowning of source regions, diversion of sediment transport paths and increasing aridity. As a result of regression in the intervening areas, Boreal and Tethyan marine biota were once again separated.

One of the first products of the Late Jurassic drowning is the *Oxford Clay* (Fig. 17.14). This is the 'mud blanket' that characterizes much of the southern onshore outcrop of our area of interest and it passes laterally into sandy strata in much of the North Sea region. The Callovian part of the Oxford Clay is particularly organic rich. The Oxford Clay is exceptional in having yielded spectacular fossil faunas which, in areas where not deeply buried, retain intact much of the original biomineralogy. As a result the Oxford Clay faunas have become the focus of efforts to determine, by combined palaeobiological and geochemical means, much detail about Jurassic open-marine, mud-dominated environments. Stable isotope (C and O) analyses of ammonites, belemnites, brachiopods, bivalves and diverse vertebrates have indicated a wide range of palaeotemperatures. Benthic bivalves and belemnites indicate 12–19°C, whereas the ammonite *Kosmosceras* and the vertebrate faunas suggest temperatures from 16°C to as high as 29°C. Of course, there is always the possibility of diagenetic alteration of original isotopic values, or taxon-specific fractionation, but the data from the Oxford Clay rather suggest that these temperatures reflect real differences between bottom and surface waters. In a stratified water column, some difference in $^{12}C/^{13}C$ ratios would be expected at different depths (because of oxidation of descending organic matter rich in light carbon). In the case of the Oxford Clay no such differentiation between bottom-living and surface-dwelling forms has been discovered, indicating a well-mixed water column, at least seasonally.

In the area of Britain and Ireland, one of the most economically important Upper Jurassic formations is the *Kimmeridge Clay* (Fig. 17.15 shows the palaeogeography at an instant during the deposition of this unit). The Kimmeridge Clay, or its lithological equivalent, is present in the North Sea, in the Porcupine Basin to the west of Ireland, and areas east of Newfoundland and in East Greenland. Where best exposed, in the area around Kimmeridge Bay in south Dorset, the Kimmeridge Clay bears some superficial resemblance to much of the Lower Jurassic mudrock succession, in that these argillaceous deposits are rhythmically bedded, decimetre-scale, alternations of relatively organic-rich and relatively carbonate-rich mudstones. However, the total organic carbon values in the Kimmeridge Clay can be exceptionally high, in the case of the millimetre-laminated 'Blackstone Band' rising above 50%. This unique horizon was used as long ago as the Bronze Age for the manufacture of decorative bracelets, perhaps because of its attractive colour, unusually low density and ease of working. Limestones almost entirely composed of coccoliths are also found, for example the 'White Stone Band', which is likewise millimetrically laminated.

There is little variation in the type of organic matter present in the formation, most of it being amorphous material of marine algal or bacterial origin. Palaeoecological evidence shows that bottom waters were either poorly oxygenated or entirely lacking in free oxygen, at least on a periodic or seasonal basis. The occurrence of horizons interpreted as storm-reworked sea floor, mostly thin mudchip conglomerates, has been used to argue that water depths may not have been much more than about 50 m. In the North Sea, the Kimmeridge Clay is a very widespread formation. The mudstones there are also organic rich, and although not normally above about 15% total organic carbon, are the major source for North Sea oil. In the North Sea the Kimmeridge Clay grades up not into shallow-water facies, as it does in southern basins, but passes up rather abruptly into calcareous mudstones deposited under oxic conditions of Early Cretaceous age (Fig. 17.10).

The development of the Kimmeridge Clay, and the basic patterns it exhibits, have been explained by palaeoceanographic models, involving currents generated in the epicontinental seaways between Tethys and the Boreal Ocean. Two possible scenarios have been proposed.

1 Tethyan surface waters were driven northwards by a hypothetical seasonal wind from the south and south-west, and upwelling cold water generated in the arctic regions moved southwards as bottom water, thus forming a two-layer system. In this expla-

▨ probable land area	
⬚ non-marine	⊟ shallow-water carbonate platform
⬚ near shore-marine, siliciclastic	☰ offshore marine, deep water, mud dominated
⊞ offshore marine, shallow water, mud dominated	ⱽⱽ evaporitic

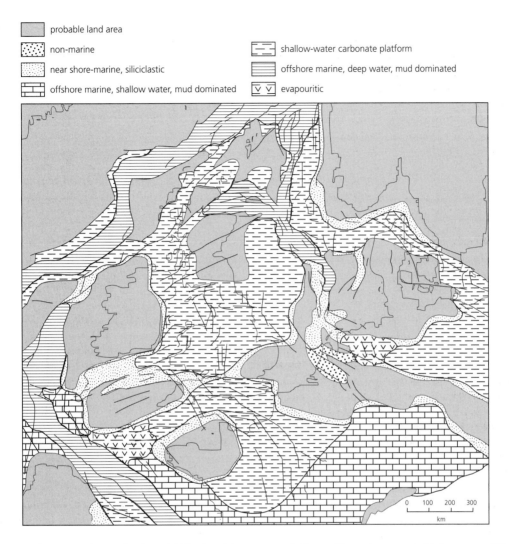

Fig. 17.15 Palaeogeographic map for the early Tithonian of the area of Britain and Ireland (based on Ziegler 1990, modified in region of British Isles on the basis of Bradshaw *et al.* 1992).

nation, the cool bottom waters became progressively deoxygenated as they flowed south, and the model explains the general pattern of increasing organic richness in a southerly direction.

2 Alternatively, high rates of evaporation in poorly connected southern basins caused warm saline bottom waters to form, which migrated and ponded in the deepest water areas. The northern affinities of the preserved macrofossils in the Kimmeridge Clay are used as a basis for proposing a general south-

ward-directed boreal current, which was amplified by Coriolis effects along the most westerly seaways, and would have been most sluggish across the North Sea region and southern England. This model is supported by the clear evidence of evaporitic conditions, which developed towards the close of the Jurassic in southern areas such as the Wessex Basin, and by the lack of evidence for significant fluvial input.

In northern basins, the strong rift-related subsidence during the Late Jurassic resulted in deep-water basins forming adjacent to active faults. Coarse-grained sediment was shed off the fault scarps into these basins. One of the best known examples is from the Moray Firth Basin in north-east Scotland where, throughout the Kimmeridgian and Volgian, spectacu-

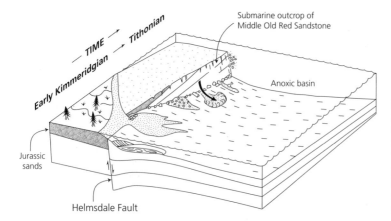

Fig. 17.16 Spatial–temporal reconstruction of Kimmeridgian–Tithonian depositional conditions adjacent to the Helmsdale Fault, north-east Scotland. Through time a significant submarine scarp developed, down which sediment was transported by rock fall, debris flow and turbidity current processes (modified from Wignall & Pickering 1993).

lar rock-fall breccias, debris flows and turbidites were generated (Fig. 17.16).

17.11 Late Jurassic aridity, carbonate ramps and the Purbeck lagoons

The first evidence of increasing aridity in the region towards the end of the Jurassic comprises a change in the dominant detrital clay mineral from kaolinite to illite, and an associated increase in the abundance of the very particular pollen genus *Classopollis* within the middle parts of the Kimmeridge Clay. Mudstones give way upwards in southern England to fine-grained dolostones and bioclastic, peloidal and oolitic limestones in a carbonate ramp succession (the *Portland Group*, parts of which are still used as a monumental building stone). This unit in turn passes into mixed limestones and shales deposited in lagoons containing evidence of evaporite deposition (the *Purbeck Group*). The abundance of dolomite in marine strata of the lower part of the Portland Group has also been attributed to the generation of heavy, Mg-rich brines through sea-surface evaporation. Evaporite minerals known from the Purbeck beds include celestite, gypsum, anhydrite and halite. Offshore from south-west Ireland, in the Porcupine Basin, appreciable quantities of halite were deposited. In southern England, the Purbeck 'fossil forest' has long been known for the spectacular silicified trees: analysis of the rings from Purbeck trees reveals great irregularity in growth throughout the year, and from year to year, demonstrating the highly stressed environment that they inhabited. By comparison with the modern 'Mediterranean-type' climates, for example,

that of South Australia, it is likely that Purbeck tree growth occurred in a warm wet winter season (when, also, the lagoons would have been filled with fresh water in which short-lived arthropods were abundant). Tree growth ceased in the hot dry summers, and that would have been the season of evaporite formation. The pattern is of a seasonality stronger than that inferred from either older Jurassic or younger Cretaceous fossil wood and represents a regional palaeoclimatic extreme.

17.12 Summary

The geological history of the area of Britain and Ireland during the Late Triassic and Jurassic can be encapsulated as three major phases.
1 A major pulsed transgression occurred, caused by continued lithospheric extension and associated rifting, combined with global sea-level rise. This brought about a change from non-marine to marine deposition over much of the region, and resulted in a regional marine connection between the Arctic and Tethys oceans. At the same time, the local climate changed from semiarid to humid as a result of northward drift of the continents and increasing maritime influence. The overall transgression resulted in the widespread deposition of mudrocks, some highly organic rich. However, this pattern was interrupted by phases of regional regression, when shallow-marine sands accumulated, probably from localized sediment sources. During major phases of sea-level rise, poor sediment supply to southern regions resulted in the deposition of highly sediment-starved limestones, whereas expanded 'black shales' accumu-

lated where abundant clay-sized sediment was available, probably derived from more distant landmasses.

2 Regressive sedimentary facies (carbonate ramps, deltas and coastal plains) and widespread erosion of pre-existing strata characterized much of the Mid-Jurassic, particularly well known for the central North Sea area, where these events were accompanied by an important phase of alkaline igneous activity. Uplift of a central North Sea 'dome' may have been brought about by limited mantle upwelling, or renewed crustal extension, or a combination of these. At the peak regression, marine connections between the Arctic and Tethys were once again severed, leading to marked faunal provinciality.

3 In the later Jurassic, crustal extension and rifting, centred on the northern North Sea region, was the dominant tectonic pattern. Whereas sediment supply to the more northerly regions of the area of Britain and Ireland progressive decreased, resulting in sediment condensation in deep marine environments, supply of sediment continued abundantly in many southern areas. Rifting in the northern North Sea ended in the latest Jurassic, but continued on the Atlantic margins and in southern basins into the Early Cretaceous. A notable widespread regressive phase occurred in the Mid-Oxfordian, resulting in the deposition of the 'Corallian' strata; towards the close of the Jurassic and in the earliest Cretaceous a second major regressive phase occurred in southern basins. Poor marine connections during regressive episodes resulted in the enhancement of faunal provinciality, particularly leading up to the Jurassic–Cretaceous boundary. A pronounced climatic change also took place in the early Tithonian and lasted into the earliest Cretaceous: evidence of increasing seasonal aridity is apparent, and this culminated in the accumulation of minor evaporites in southern areas.

References

Bradshaw, M.J., Cope, J.C.W., Cripps, D.W. *et al.* (1992) Jurassic. In: *Atlas of Palaeogeography and Lithofacies* (eds J. C. W. Cope, J. K. Ingram & P. F. Rawson), Memoir 13, pp. 107–129. Geological Society, London.

Coe, A.L. (1995) A comparison of the Oxfordian successions of Dorset, Oxfordshire, and Yorkshire. In: *Field Geology of the British Jurassic* (ed. P. D. Taylor), pp. 151–172. Geological Society, London.

Coe, A.L. (1996) Unconformities within the Portlandian Stage of the Wessex Basin and their sequence-

stratigraphical significance. In: *Sequence Stratigraphy in British Geology* (eds S. P. Hesselbo & D. N. Parkinson), Special Publication 103, pp. 109–143. Geological Society, London.

Cope, J.C.W. (1993) The Bolonian Stage: an old answer to an old problem. *Newsletters on Stratigraphy* 28, 151–156.

Gradstein, F.M., Agterberg, F.P., Ogg, J.G. *et al.* (1994) A Mesozoic time scale. *Journal of Geophysical Research* 99, 24051–24074.

Hallam, A. (1985) A review of Mesozoic climates. *Journal of the Geological Society, London* 142, 433–445.

Hallam, A. (1996) Phanerozoic sea-level changes. *Perspectives in Paleobiology and Earth History.* Columbia University Press, New York.

Haq, B.U., Hardenbol, J. & Vail, P.R. (1997) Chronology of fluctuating sea levels since the Triassic (250 millon years ago BP). *Science* 235, 1156–1167.

Hesselbo, S.P. & Jenkyns, H.C. (1995) A comparison of the Hettangian to Bajocian sections of Dorset and Yorkshire. In: *Field Geology of the British Jurassic* (ed. P. D. Taylor), pp. 105–150. Geological Society, London.

Hesselbo, S.P. & Jenkyns, H.C. (1998) British Lower Jurassic sequence stratigraphy. In: *Mesozoic and Cenozoic Sequence Stratigraphy of European Basins* (eds P.-C. de Gracianisky, J. Hardenbol, T. Jacquin & P. Vail), pp. 561–581. SEPM.

Hesselbo, S.P., Oates, M.J. & Jenkyns, H.C. (1998) The lower Lias Group of the Hebrides Basin. *Scottish Journal of Geology* 34, 1–38.

Hudson, J.D., Clements, R.G., Riding, J.B., Wakefield, M.I. & Walton, W. (1995) Jurassic palaeosalinities and brackish-water communities — a case study. *Palaios* 10, 392–407.

Jenkyns, H.C. & Clayton, C.J. (1997) Lower Jurassic epicontinental carbonates and mudstones from England and Wales: chemostratigraphic signals and the early Toarcian anoxic event. *Sedimentology* 144, 687–706.

Jones, C.E., Jenkyns, H.C., Coe, A.L. & Hesselbo, S.P. (1994) Strontium isotopic variations in Jurassic and Cretaceous seawater. *Geochimica et Cosmochimica Acta* 58, 3061–3074.

Morton, N. & Hudson, J.D. (1995) Field guide to the Jurassic of the Isles of Raasay and Skye, Inner Hebrides, Scotland. In: *Field Geology of the British Jurassic* (ed. P. D. Taylor), pp. 105–150. Geological Society, London.

Norris, M.S. & Hallam, A. (1995) Facies variations across the Middle–Upper Jurassic boundary in Western Europe and the relationship to sea-level changes. *Palaeogeography, Palaeoclimatology, Palaeoecology* 116, 189–245.

Pálfy, J., Smith, P.L. & Mortensen, J.K. (2000) A U–Pb and ^{40}Ar–^{39}Ar timescale for the Jurassic. *Canadian Journal of Earth Sciences*, 37.

Rattey, R.P. & Hayward, A.B. (1993) Sequence stratigraphy of a failed rift system: the Middle Jurassic to early Cretaceous basin evolution of the Central and Northern North Sea. In: *Petroleum Geology of Northwest Europe: Proceedings of the 4th Conference* (ed. J. R. Parker), pp. 215–249. Geological Society, London.

Sahagian, D., Pinous, O., Olferiev, A. & Zakharov, V. (1996) Eustatic curve for the Middle Jurassic–Cretaceous based on Russian Platform and Siberian stratigraphy: zonal resolution. *American Association of Petroleum Geologists Bulletin* 80, 1433–1458.

Smith, A.G., Smith, D.G. & Funnell, B.M. (1994). *Atlas of Mesozoic and Cenozoic Coastlines.* Cambridge University Press, Cambridge.

Turner, J.P. (1996) Gravity-driven nappes and their relation to palaeobathymetry: examples from West Africa and Cardigan Bay, UK. In: *Modern Developments in Structural Interpretation, Validation and Modelling* (eds P. G. Buchanan & D. A. Nieuwland), Special Publication 99, pp. 345–362. Geological Society, London.

Wignall, P.B. & Pickering, K.T. (1993) Palaeoecology and sedimentology across a Jurassic fault scarp, NE Scotland. *Journal of the Geological Society, London* 150, 323–340.

Wilson, D., Davies, J.R., Fletcher, C.J.N. & Smith, M. (1990) Geology of the South Wales Coalfield, Part VI, the country around Bridgend. In: *British Geological Survey Memoir, Sheets 262 and 262 (England and Wales).* HMSO, London.

Ziegler, P.A. (1990) *Geological Atlas of Western and Central Europe.* Shell International Petroleum Maatschappij BV, The Hague.

Ziegler, A.M., Parrish, J.M., Jiping, Y. *et al.* (1993) Early Mesozoic phytogeography and climate. *Philosophical Transactions of the Royal Society of London* B341, 297–305.

Further reading

Anderson, F.T., Popp, B.N., Williams, A.C., Ho, L.Z. & Hudson, J.D. (1994) The stable isotopic records of fossils from the Peterborough Member, Oxford Clay Formation (Jurassic), UK: palaeoenvironmental implications. *Journal of the Geological Society, London* 151, 125–138. [This, and companion papers in the same issue, give a very thorough account of just how much environmental information can be extracted from focused multidisciplinary studies.]

Bradshaw, M.J., Cope, J.C.W., Cripps, D.W. *et al.* (1992) Jurassic. In: *Atlas of Palaeogeography and Lithofacies* (eds J. C. W. Cope, J. K. Ingram & P. F. Rawson), Memoir 13, pp. 107–129. Geological Society, London [An important compilation of palaeogeographic information for the region.]

Chandler, M.A. (1992) Pangaean climate during the Early Jurassic: GCM simulations and the sedimentary record of palaeoclimate. *Geological Society of America Bulletin* 104, 543–559.

Francis, J.E. (1984) The seasonal environment of the Purbeck (Upper Jurassic) fossil forests. *Palaeogeography, Palaeoclimatology, Palaeoecology* 48, 285–307.

Hallam, A. (1993) Jurassic climates as inferred from the sedimentary and fossil record. *Philosophical Transactions of the Royal Society of London, B* 341, 287. [A simple and readable review of global climatic patterns during the Jurassic as inferred from the geological record.]

Hesselbo, S.P., Gröcke, D.R., Jenkyns, H.C., Bjerrum, C.J., Farrimond, P., Morgans Bell, H.S. & Green, O.R. (2000) Massive dissociation of gas hydrate during a Jurassic oceanic anoxic event. *Nature* 406, 392–395.

Hudson, J.D., Clements, R.G., Riding, J.B., Wakefield, M.I. & Walton, W. (1995) Jurassic palaeosalinities and brackish-water communities—a case study. *Palaios* 10, 392–407.

Taylor, P.D. (ed.) (1995) *Field Geology of the British Jurassic.* Geological Society, London. [A key source of detailed sedimentological and stratigraphic information for the Jurassic of the British Isles.]

Underhill, J.R. & Partington, M.A. (1993) Jurassic thermal doming and deflation in the North Sea: implications of the sequence stratigraphic evidence. In: *Petroleum Geology of Northwest Europe: Proceedings of the 4th Conference* (ed. J.R. Parker), pp. 337–345. Geological Society, London.

Wilson, M. (1997) Thermal evolution of the Central Atlantic passive margins: continental break-up above a Mesozoic super-plume. *Journal of the Geological Society, London* 154, 491–495.

Ziegler, P.A. (1990) *Geological Atlas of Western and Central Europe.* Shell International Petroleum Maatschappij BV, The Hague. [A very broad summary of the tectonic and palaeogeographic changes that characterised the geological evolution of western and central Europe put in a global context.]

18 Early Cretaceous: rifting and sedimentation before the flood

A. S. GALE

18.1 The Cretaceous Earth

The Cretaceous was a time of greenhouse climate, characterized by low pole-to-equator thermal gradients and absence of polar ice at sea level. Limited oxygen-isotope data provide evidence that global warming began in the Berriasian and continued through the Early Cretaceous (Fig. 18.1j). The mid-Cretaceous (Aptian–Cenomanian) saw a huge pulse of submarine volcanism, probably caused by a mantle plume, which resulted in the formation of oceanic plateaux and increased ocean ridge production centred in the Pacific Ocean (Fig. 18.1a). This event began in the Aptian and continued through to the Santonian. One result was the high sea level in the Cenomanian–Coniacian interval as increased spreading ridge volume displaced sea water (Fig. 18.1f). Carbon dioxide, generated by the volcanic event, elevated atmospheric concentrations to as much as 10 times those at present, and was the direct cause of a mid-Cretaceous (Cenomanian–Santonian) ultrathermal period. The Late Cenomanian–Campanian (Late Cretaceous) interval saw the development of a mid-latitude arid zone in Europe and central Asia (Fig. 18.1k). In the epeiric sea that stretched east–west across Europe, the white chalk facies was deposited. A high-latitude humid zone in which coals formed was present in the northern hemisphere for much of the Cretaceous.

Cretaceous deep ocean water was dominantly warm and saline and formed at low latitudes by evaporation. Because of its warmth, bottom water had a low oxygen concentration, which was partially responsible for periodic sea-floor anoxia in the ocean basins and over the outer shelves. These oceanic anoxic events coincide with major transgres-sions and occurred in the Aptian, Cenomanian and Coniacian–Santonian. The events resulted in locally high levels of organic preservation in marine sediments concentrated within areas of predicted marine upwelling (Fig. 18.1h). The organic burial caused drawdown of atmospheric carbon dioxide, which in turn caused cooling. Global cooling continued throughout the Campanian and accelerated in the Maastrichtian (Fig. 18.1j), resulting in decreased marine organic burial. The cooling may be related to rising mountain chains in western South and North America and in Asia, where increased chemical weathering induced drawdown of atmospheric carbon dioxide. The Maastrichtian in particular was a time of considerable tectonic and environmental change, and the rapid extrusion of the Deccan Traps (continental flood basalts) and the initial collision between India and Asia commenced at the end of this stage (Fig. 18.1b,c,e).

18.2 Lower Cretaceous palaeogeography and general setting

The Early Cretaceous saw the continued continental break-up of Pangaea and Gondwana, which had begun in the Triassic. The South Atlantic began to rift in the Valanginian–Aptian interval and by the late Albian there is evidence of marine connection between the North and South Atlantic Oceans. Opening of the North Atlantic had a direct and major effect on north-west Europe, because tensional stresses caused by spreading resulted in extensional, syn-rift reactivation of pre-existing basement structures and the formation of horsts and half-grabens. In the North Sea, volcanism had mostly ceased by the Early Cretaceous, but the Central Graben (formed by

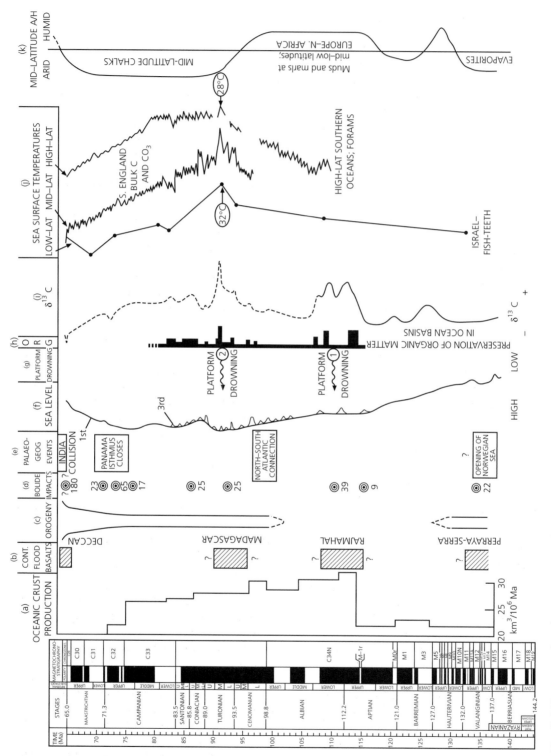

Fig. 18.1 The patterns through Cretaceous time of various indicators of global change.

Jurassic rifting, Chapter 17) continued to act as a focus for rapid sedimentation.

In the interval from 120 to 80 Ma (Aptian–early Campanian) there is evidence of a doubling of the rate of ocean crust formation, both by increased mid-ocean ridge spreading rates and by plateau formation in the Pacific Ocean (Fig. 18.1a). This magmatic event coincides with the long normal Cretaceous magneto-zone (34N) and has been interpreted as evidence for a mantle superplume that caused large volumes of basaltic magma to erupt on the Pacific sea floor. This superplume episode probably provided the carbon dioxide responsible for warming the Cretaceous greenhouse, which became hotter from the Barremian onwards, reaching a temperature maximum in the late Cenomanian. It also accounts for the progressive rise in overall sea level that took place through the latter part of the Early Cretaceous (Fig. 18.1f). Volcanism may explain another curious phenomenon of the Cretaceous oceans, the tendency for episodes of stagnation and oxygen depletion of the ocean basins and outer shelves, resulting in deposition of shales rich in organic matter. Although organic-rich shales of Aptian age are not found in north-west Europe, the effect of their deposition on the ratios of stable carbon isotopes (positive $\delta^{13}C$ excursion) is recorded in the succession in southern England (Fig. 18.1i). It appears that there is a relationship between individual episodes of marine organic deposition (anoxic events) and volcanism. For example, formation of the Ontong–Java Plateau in the equatorial mid-Pacific is exactly coincident with the maximum development of Aptian black shales.

18.3 Correlation and stratigraphical framework

As with much of the Mesozoic, the highest-resolution correlation of Cretaceous marine sediments is achieved with the use of ammonites. The succession of Aptian and Albian ammonites found in the marine Cretaceous sediments (Lower Greensand, Gault) of southern England has been important in the development of zonal and subzonal schemes, which are now applied more widely across Europe and into central Asia (Fig. 18.2). The marine successions in north-east England also yield abundant ammonites of the Ryazanian, Valanginian, Hauterivian and Barremian Stages, which facilitate correlation with northern

Europe (Germany, Russia). Where ammonites are absent or rare (as in the Late Aptian to Early Albian part of the Lower Greensand Formation), there are considerable uncertainties as to the precise ages of rock units. Calcareous nannofossils (coccoliths, etc.) are superbly preserved in Early Cretaceous clay successions and provide an increasingly refined and widely applicable means of correlation.

Within the non-marine Wealden facies, ostracods and plant macro- and microfossils afford useful means of correlation both within the UK and with comparable facies across western Europe (Fig. 18.2). However, Anderson's very detailed scheme of fauni-cycles, based on ostracods and salinity changes, has recently been shown to be largely mythical. The major correlation problems are those between the non-marine Wealden facies and the marine record, which has been based largely on palynology (spores and pollen and some dinoflagellates). This correlation is very important in understanding the facies shifts in relation to both sea-level and climatic change.

Magnetostratigraphy offers some potential in the resolution of these problems. The earlier part of the Early Cretaceous succession contains a number of reversed and normal magnetozones that have been identified in the deep marine pelagic successions in southern Europe. One of these (MO, Fig. 18.1), taken as marking the base of the Aptian Stage, has been found in the highest part of the Wealden of the Isle of Wight (Fig. 18.2). However, even if normal and reversed polarity intervals are identified in the UK, these can be notoriously difficult to identify without the assistance of good-quality biostrati-graphical data.

Stable carbon isotopes ($\delta^{13}C$) provide a valuable means of correlation, because changes are more or less instantaneous, global in distribution and can be identified in both marine and terrestrial carbon. Curves have been established for marine carbonate successions, particularly in southern Europe, which show the existence of several major positive excursions within the Early Cretaceous. For example, a large double-peaked positive $\delta^{13}C$ excursion is now known in the Aptian. This has been identified in southern England using carbon from marine drift-wood in the Lower Greensand (Fig. 18.2).

For the future it is hoped that both magnetostratigraphy and carbon-isotope stratigraphy will improve the resolution of correlation between Wealden facies

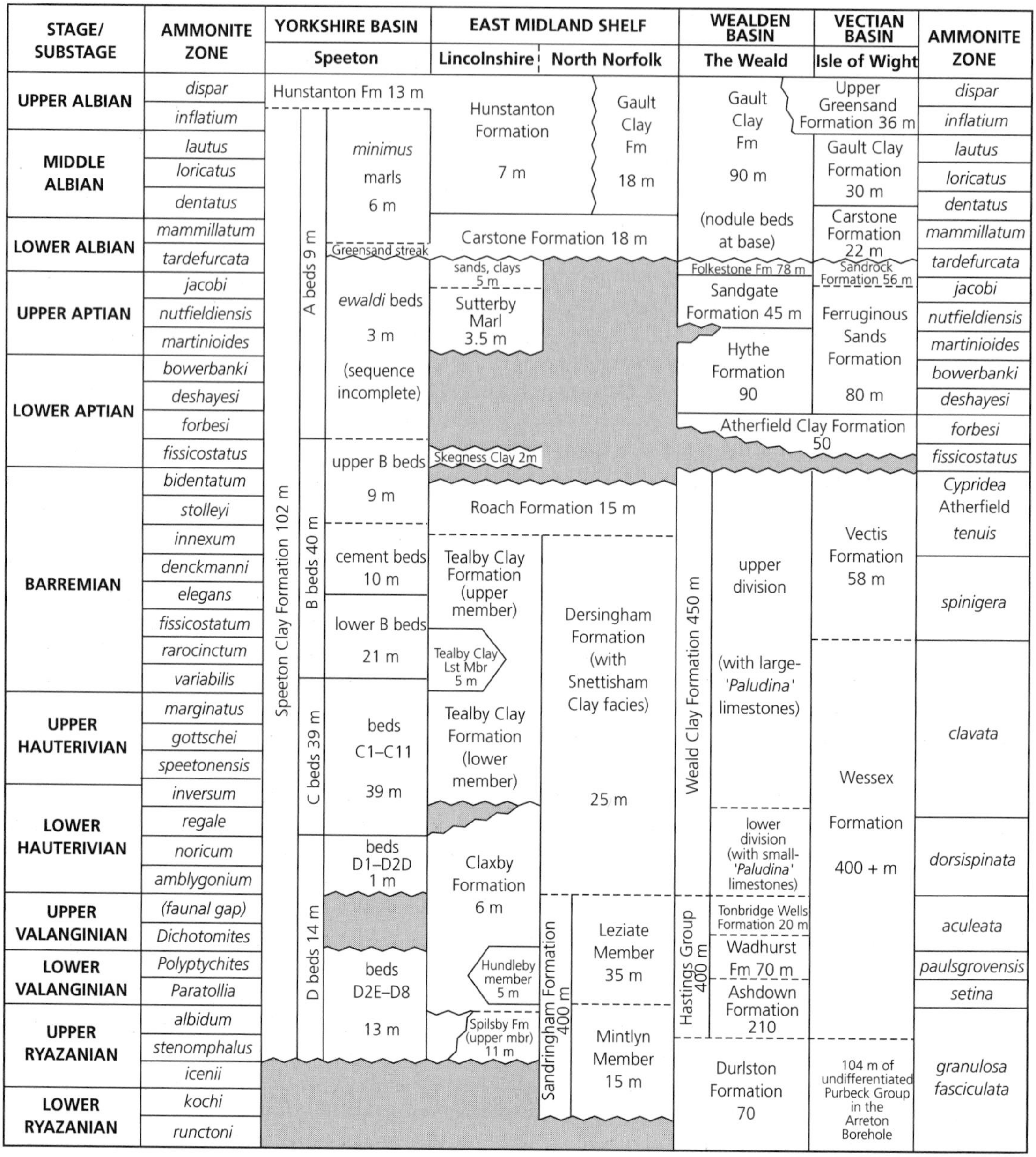

STAGE/ SUBSTAGE	AMMONITE ZONE	YORKSHIRE BASIN — Speeton	EAST MIDLAND SHELF — Lincolnshire	EAST MIDLAND SHELF — North Norfolk	WEALDEN BASIN — The Weald	VECTIAN BASIN — Isle of Wight	AMMONITE ZONE
UPPER ALBIAN	dispar	Hunstanton Fm 13 m	Hunstanton Formation 7 m	Gault Clay Fm 18 m	Gault Clay Fm 90 m	Upper Greensand Formation 36 m	dispar
	inflatium						inflatium
MIDDLE ALBIAN	lautus	minimus marls 6 m				Gault Clay Formation 30 m	lautus
	loricatus						loricatus
	dentatus				(nodule beds at base)	Carstone Formation 22 m	dentatus
LOWER ALBIAN	mammillatum	A beds 9 m — Greensand streak	Carstone Formation 18 m				mammillatum
	tardefurcata				Folkestone Fm 78 m	Sandrock Formation 56 m	tardefurcata
UPPER APTIAN	jacobi	ewaldi beds 3 m	sands, clays 5 m	Sutterby Marl 3.5 m	Sandgate Formation 45 m	Ferruginous Sands Formation 80 m	jacobi
	nutfieldiensis						nutfieldiensis
	martinioides	(sequence incomplete)			Hythe Formation 90		martinioides
LOWER APTIAN	bowerbanki						bowerbanki
	deshayesi						deshayesi
	forbesi				Atherfield Clay Formation 50		forbesi
	fissicostatus	upper B beds 9 m — Skegness Clay 2 m					fissicostatus
BARREMIAN	bidentatum		Roach Formation 15 m			Vectis Formation 58 m	Cypridea Atherfield tenuis
	stolleyi						
	innexum	cement beds 10 m	Tealby Clay Formation (upper member)		upper division		
	denckmanni			Dersingham Formation (with Snettisham Clay facies)			spinigera
	elegans						
	fissicostatum	lower B beds 21 m — Tealby Clay Lst Mbr 5 m					
	rarocinctum				(with large-'Paludina' limestones)		
	variabilis						
UPPER HAUTERIVIAN	marginatus	beds C1–C11 39 m	Tealby Clay Formation (lower member)	25 m		Wessex Formation	clavata
	gottschei						
	speetonensis						
	inversum						
LOWER HAUTERIVIAN	regale	beds D1–D2D 1 m	Claxby Formation 6 m		lower division (with small-'Paludina' limestones)	400 + m	dorsispinata
	noricum						
	amblygonium						
UPPER VALANGINIAN	(faunal gap)	beds D2E–D8 13 m	Hundleby member 5 m	Leziate Member 35 m	Tonbridge Wells Formation 20 m		aculeata
	Dichotomites				Wadhurst Fm 70 m		
LOWER VALANGINIAN	Polyptychites				Ashdown Formation 210		paulsgrovensis
	Paratollia		Spilsby Fm (upper mbr) 11 m	Mintlyn Member 15 m			setina
UPPER RYAZANIAN	albidum				Durlston Formation 70	104 m of undifferentiated Purbeck Group in the Arreton Borehole	granulosa fasciculata
	stenomphalus						
	icenii						
LOWER RYAZANIAN	kochi						
	runctoni						

Other labels on the chart: Speeton Clay Formation 102 m; A beds 9 m; B beds 40 m; C beds 39 m; D beds 14 m; Weald Clay Formation 450 m; Hastings Group 400 m; Sandringham Formation 400 m.

Fig. 18.2 Stratigraphical nomenclature of the Lower Cretaceous in England (after Rawson 1992).

and marine successions, which is also uncertain at present. The paucity of good-quality radiometric dates remains a problem and considerable margins of error exist on ages for all the Lower Cretaceous stages, especially the earlier ones. The development of an astronomical time-scale for the Early Cretaceous, based on the rhythmically bedded deep-water successions in southern Europe, should greatly improve the resolution and calibration of Early Cretaceous time.

18.4 Early Cretaceous tectonic framework

The Palaeozoic massifs that had exercised such an important control on Jurassic sedimentation continued to act as upland sources of sediment through the earlier part of the Early Cretaceous. However, many were at least partially submerged by the end of the Albian (Fig. 18.3). In south-west England (Devon and Cornwall), the Cornubian Massif was made up of dominantly Late Palaeozoic rocks and existed as land through into the Late Cretaceous. Another region of older rocks, the East Anglian Massif—part of the London–Brabant Massif—was just submerged by the end of the Early Cretaceous (Fig. 18.3d). In north-east England, the East Midlands Shelf and the Market Weighton High on its northern flank continued to act as positive structures and controlled sedimentation well into the Late Cretaceous.

The structural history of southern England during the Early Cretaceous was thus dominated by structures developed within the Devonian–Carboniferous basement underlying the region and bordering the massifs. These structures were originally Variscan thrusts, which were reactivated as normal faults during the extensional movements associated with the early opening history of the North Atlantic. The east–west structures that were active through the Permian, Triassic and Jurassic defined the boundaries of many sedimentary basins such as the Wessex Basin and the northern boundary of the Weald Basin.

The structural history of the Early Cretaceous was marked by episodic normal movements on east–west basement structures during a syn-rift phase that extended up to the Late Aptian. These movements had a profound effect on sedimentation, by rejuvenating source areas and probably thereby affecting climate. The basement structures became relatively quiet in the Early Albian, perhaps marking the change from the syn-rift to post-rift phases, and had little effect on Gault and Upper Greensand successions other than causing local changes in thickness.

18.5 Rivers, lakes and soils: the Wealden facies

The Wealden facies comprises sands, silts, clays and a few conglomerates deposited in river, lake and brackish coastal plain environments. Palaeosols are locally abundant, and have led to the distinctive colour mottling of many Wealden deposits in red, purple and orange hues. Wealden sediments contain a distinctive fauna and flora including dinosaurs (most commonly *Iguanodon*, which also left numerous footprints), crocodiles, fishes, diverse ostracods and molluscs, and plants including ferns, horsetails, cycads and abundant coniferous wood. Rocks of Wealden facies are widely represented across north-west Europe at approximately the latitude of the southern UK; they extend from the Celtic Basin (south of Ireland), across southern England and northern France and into north-west Germany.

The transition from the interbedded limestones and shales of the Purbeck Limestone to the clastic sands of the Wealden Group was a dramatic facies change. The Purbeck was deposited in a semiarid to hot-arid marginal marine environment of lagoons, bitter lakes and sabkhas in which high salinities led to the local formation of evaporites. The Wealden represents a period of considerable precipitation (including storms and flash floods) and reduced seasonality. The precise causes of this change of climate are uncertain, but it was coincident with a major uplift of the massifs surrounding the Wealden basins and the new uplands may have generated orographic rainfall.

Owing to extensive outcrops and boreholes in the Weald and Wessex Basins, Wealden palaeoenvironments have been reconstructed in considerable detail (Fig. 18.4). Evidence has come from sedimentology, the provenance of heavy mineral grains and larger clasts, soils, clay mineralogy, and fauna and flora.

The Wealden of the Weald is divided into a lower Hastings Group (about 400 m in thickness) and a muddier upper Weald Clay Group (Fig. 18.2). The Hastings Group comprises three major sedimentary cycles in which dominantly arenaceous phases (mostly outwash fans and sandy braided stream deposits) alternate with mostly argillaceous ones (mud-plains, lakes, lagoons). The argillaceous formations were deposited on a mud-plain occupied by ephemeral lakes, in which numerous freshwater

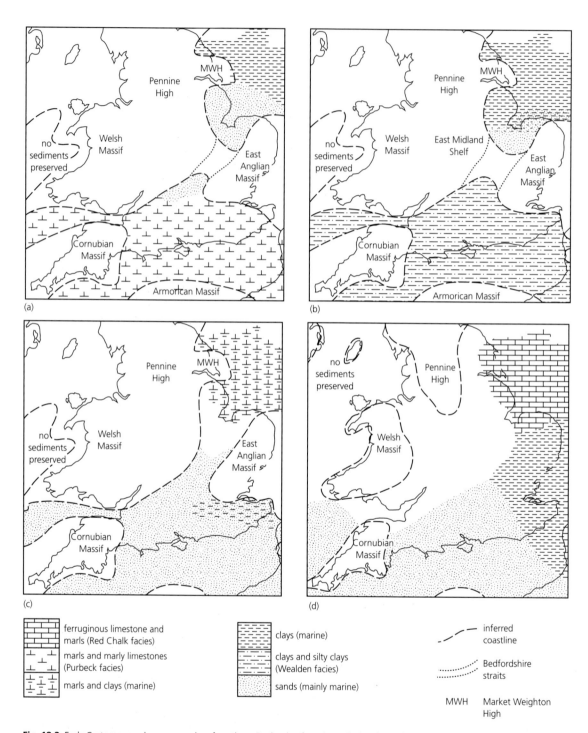

Fig. 18.3 Early Cretaceous palaeogeography of southern England at four times during the Early Cretaceous: (a) mid-Ryazanian; (b) mid-Hauterivian; (c) late Aptian; (d) late Albian (after Rawson 1992).

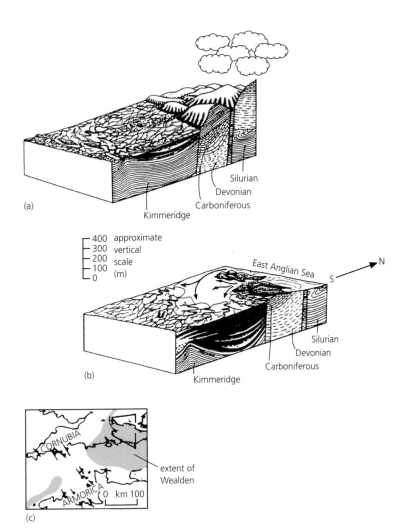

Fig. 18.4 Reconstruction of the London Platform during deposition of arenaceous (a) and argillaceous (b) formations of the Wealden Group (after Allen 1975).

ostracods and molluscs (e.g. *Unio, Viviparus*) lived and in which horsetails thrived. Towards the north-west, where there was a connection with the northern Boreal Sea (the 'Bedfordshire Straits' see below)—the marine influence increased, as shown by increasing abundance of brackish-water molluscs.

The arenaceous formations (Figs 18.4, 18.5) record the successive advances of alluvial braidplains south-wards from the London–Brabant Massif into the Weald. The Ashdown and Lower Tunbridge Wells Sands (Fig. 18.2) both display progressive coarsening up, reflecting initial meander-plain development followed by a later phase of coarser braid-plain development.

The cycles have been attributed to eustatic sea-level changes. Rising base levels during transgression and highstand may have caused the formation of exten-sive lakes and lagoons. To prove that this control was truly eustatic it is necessary to demonstrate a precise correlation with sea-level changes in adjacent marine successions, which has not yet proved possible. Fur-thermore, the changes from arenaceous to argilla-ceous sediments (the 'transgressive' parts of the cycles) do not seem to correlate between adjacent basins. Thus the sharp boundary between the muddy Vectis Formation overlying the sandy Wessex Forma-tion in the Wessex Basin does not appear to have any equivalent in the Weald Basin only 50 km away (Fig. 18.2). It is likely therefore that arenaceous units cor-respond to periods of maximum uplift of source areas

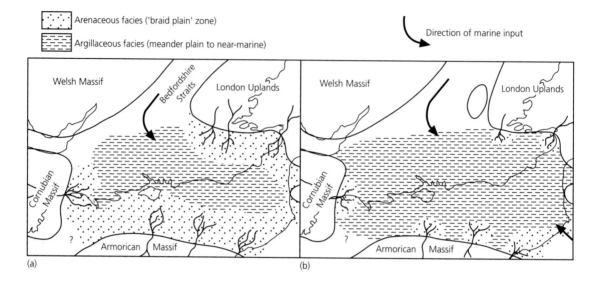

Arenaceous facies ('braid plain' zone)

Argillaceous facies (meander plain to near-marine)

Direction of marine input

Fig. 18.5 Palaeogeography during the formation of arenaceous (a) and argillaceous (b) deposits in the Wealden of southern England (after Allen 1975).

and argillaceous ones to times when those areas were most degraded. The highest part of the Wealden of the Weald, the Weald Clay Group (up to 400 km thick, Fig. 18.2), probably formed at a time when the London–Brabant Massif was at its lowest. It contains a number of marine horizons, which record short-lived breaching of the Weald mud-plains by the Boreal Sea.

Detailed provenance studies in the Weald, using heavy mineral grains, have shown that most of the detritus was derived from Jurassic and to a lesser extent Palaeozoic rocks of the London Platform to the north. However, in the upper part of the Hastings Group, detritus originating in Cornubia spilled over from the Wessex Basin. The general abundance of detritus derived from the west, which includes material from the Iberian Peninsula, points to doming on the continental margin west of the UK during Early Cretaceous times.

The Wealden of the Wessex Basin (Isle of Wight, Dorset) has a rather different sedimentary history from that of the Weald. The Wessex Formation comprises pedogenically altered, strongly mottled silts and clays, which represent muddy floodplain

deposits. Subsidiary sandstones, 5–10 m thick, were deposited as point bars within a major east–west river, flowing just south of the fault-bounded north basin margin. The regular vertical distribution of these sands suggests that the river development was controlled by long-term climatic factors. Units of mud, sand, coarse detritus and plant debris represent mudflows following heavy seasonal rainfall. Westwards into Dorset the Wessex Formation thins rapidly and coarse quartz gravels are locally developed, formed as proximal braided fan deposits. The Vectis Formation, which overlies the Wessex Formation (Fig. 18.2), comprises dark silty clays and was deposited in a large standing body of water of varying but generally low salinity (Fig. 18.6). In the eastern Isle of Wight, the Vectis Formation contains Jurassic fossils and clasts derived from the adjacent fault-bounded basin margin. The source of detritus in the Wessex Basin was Cornubia to the west, and to a lesser extent, Armorica to the south.

18.6 Transgressive cycles and tidal influences: the Lower Greensand

To the south and west of the London Platform, a succession of more and less glauconitic sands with subsidiary clays was deposited in shallow seas in the Weald and Wessex basins. Deposition periodically extended onto the western part of the London Plat-

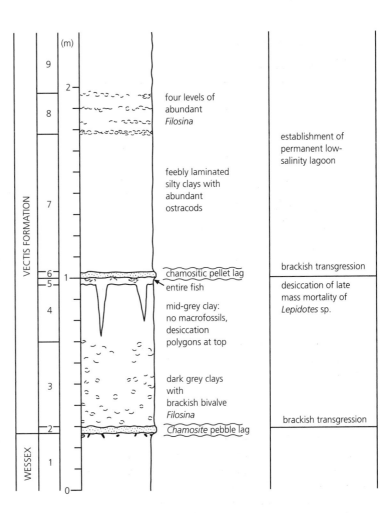

Fig. 18.6 Basal part of an argillaceous unit (Vectis Formation) in the Wealden Group of the Isle of Wight.

form where sediments are preserved as scattered outliers. The high-resolution ammonite stratigraphy applied to these deposits has made detailed correlations possible both between separate depositional areas and on an international scale. The development of a sequence stratigraphic framework, tied to the biostratigraphy, has recently provided a new understanding of the deposition and facies distribution of the Lower Greensand.

The Lower Greensand Formation is made up of five or six transgressive–regressive sequences, separated by disconformities that pass laterally into erosion surfaces (commonly with biostratigraphical gaps) towards the basin margins The transgressive units onlap onto the surrounding platforms to variable

extents and facilitate interpretation of sea-level changes in the more complete basinal successions.

The basal Lower Greensand in the Weald and Wessex marked a major transgression of early Aptian age. This event finally drowned the brackish and fresh coastal plains of the Wealden Clays and deposited a thin coarse glauconitic sandstone called the Perna Bed across Kent, Sussex, Surrey and the Isle of Wight. The Perna Bed is a condensed transgressive lag (1–3 m thick), which locally contains abundant Jurassic clasts and fossils derived from adjacent horst blocks (Fig. 18.7). It rests with sharp disconformity upon an eroded surface of Wealden Clays, and contains a diverse fauna including robust bivalves, corals and rare ammonites.

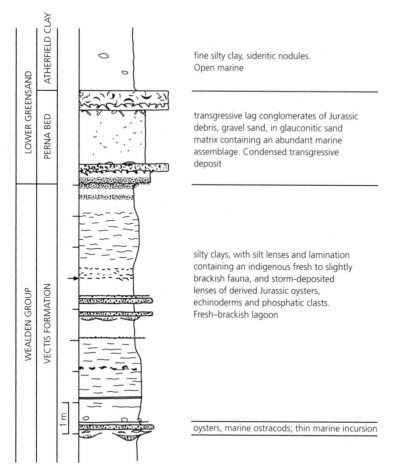

fine silty clay, sideritic nodules.
Open marine

transgressive lag conglomerates of Jurassic debris, gravel sand, in glauconitic sand matrix containing an abundant marine assemblage. Condensed transgressive deposit

silty clays, with silt lenses and lamination containing an indigenous fresh to slightly brackish fauna, and storm-deposited lenses of derived Jurassic oysters, echinoderms and phosphatic clasts. Fresh–brackish lagoon

oysters, marine ostracods; thin marine incursion

Fig. 18.7 Wealden (Vectis Formation)–Lower Greensand contact in Sandown Bay, Isle of Wight. Muds and silts, often laminated, of the Vectis Formation.

This first major marine transgression in the Early Cretaceous marked the end of the non-marine sedimentary record in southern England for the next 50 million years—until the Palaeocene. The Perna Bed transgression probably significantly reduced the topography of the horst blocks of Jurassic sediments, because it marked a cessation in supply of coarse detritus, which lasted several million years. During this time—a highstand—the silty clays of the Atherfield Formation were deposited south of the London Platform in the Weald and Wessex basins (Figs 18.7, 18.8).

The flood of sand that marks the base of the Hythe Beds (Weald Basin) and coeval Ferruginous Sands (Wessex Basin) was perhaps caused partly by regression and partly by uplift of the London–Brabant Massifs, which provided much of the detritus. The Hythe Beds and Ferruginous Sands are variably sorted, highly glauconitic, variably bioturbated and cross-bedded sands in the Isle of Wight, Sussex, Surrey and Kent (Fig. 18.8). In the eastern part of Kent, the Hythe Beds are developed in a condensed facies as alternating hard glauconitic sandy limestones and sands ('Rag and Hassock'), which contain worn phosphatic nodules at some levels. A mid-Hythe Beds hiatus marks the second regressive–transgressive cycle in the Lower Greensand.

The third major regressive–transgressive event in the Lower Greensand Formation is marked by the disconformable boundary between the Hythe and Sandgate–Bargate Beds in the Weald. The top surface of the Hythe Beds is commonly erosional, and overlain by a basal transgressive lag (coarse sands with gravel-grade clasts and pebbles) at the base of the

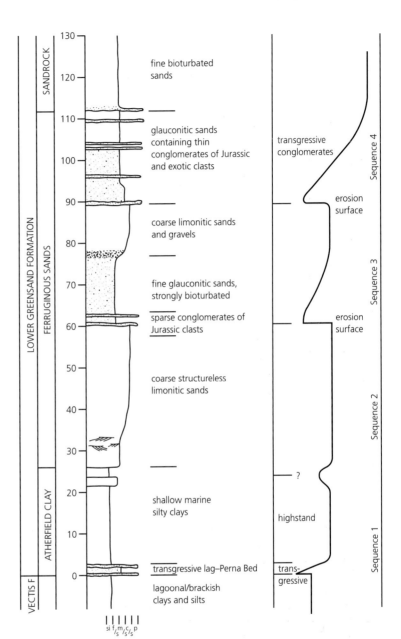

Fig. 18.8 Succession in the Ferruginous Sands, Lower Greensand, at Redcliff, Isle of Wight. Note the presence of three sedimentary cycles, separated by erosional surfaces, each with a conglomeratic base.

Sandgate Beds. One or more ammonite zones may be missing on this surface. This *'nutfieldiensis'* transgression carried sedimentation far onto the west of the London Platform, onlapping older Aptian sediments. For example, at Faringdon in Oxfordshire, a thick succession of shell and sponge gravels rests directly upon the Kimmeridge Clay. Sands such as those at Seend in Wiltshire record the feather edge of this transgression.

Close to former outcrops of Jurassic rocks—on the southern margin of the London Platform in Surrey and on the southern side of the Isle of Wight–Purbeck structure in the Isle of Wight—the lower *nutfieldiensis* zone contains abundant lithoclasts and fossils derived particularly from the Oxford and Kimmeridge Clays and ooliths from the Corallian Group (Oxfordian) limestones. The sheer abundance of this material affords evidence of uplift of Jurassic horsts at around this time.

Volcanic activity, which was probably taking place in the Netherlands, is marked in southern England (Isle of Wight, Surrey, Sussex, Bedfordshire, Berkshire) by a widespread but thin 'Fuller's Earth' in the middle part of the Sandgate Beds or their lateral equivalents (*nutfieldiensis* zone). This unit comprises from one to several metres of bentonitic clay (mostly smectite), which formed from air- and water-borne volcanic ash, including crystal and lithic tuffs. The pyroclastic fragments are locally preserved within concretions that lithified before the material altered to clay. This ash also provides one of the few good-quality Ar–Ar radiometric dates to have been obtained from the Early Cretaceous of Europe.

The highest part of the Lower Greensand comprises spectacularly cross-bedded sands, called the Folkestone Sands in the Weald and the Sandrock in the Isle of Wight. These show very strong tidal influences, which record in extraordinary detail varying tidal current strengths related to the diurnal tidal cycle and neap and spring cycles (Fig. 18.9). These are expressed as palimpset foresets, separated by mud drapes marking slack tides. In the Isle of Wight, the Sandrock comprises two units of fine shoreface and intertidal deposits, interleaved with clays, which can be interpreted as deeper water offshore deposits.

Intermittent marine connections with the sea in north-east England existed via the Bedfordshire Straits, a flat region to the west of the London–Brabant Massif (Fig. 18.10). These connections had existed since Purbeck times, and the straits were the conduit for earlier marine transgressions into the Weald and Wessex Basins such as the mid-Purbeck Cinder Bed and marine bands in the Weald Clay. During the Aptian, condensed remanié faunas of Early Cretaceous age were deposited in Bedfordshire and Cambridge, and tidally bedded sands and volcanic clays were deposited during the Late Aptian *nutfieldiensis* zone transgression at Woburn in Bedfordshire (Woburn Sands).

18.7 Before the flood: Gault and Upper Greensand

The first-order global sea-level rise, which took place through the Aptian, Albian and Cenomanian stages, is well documented in the sedimentary record of southern England. Shallow, tidally influenced sands and gravels of the Folkestone Sands, deposited in perhaps 10–20 m of water (early Albian), are overlain by the deeper water (40–60 m?) Gault Clay (Mid- and Late Albian) (Fig. 18.11). Above this, the Lower Chalk (70–100 m water depth; Cenomanian) records the initial spread of oceanic sedimentation onto the shelves—one of the 'great transgressions' of the Late Cretaceous (see Chapter 19). The progressive fining upwards of the clastic component, and decrease in clastics compared with fine carbonates, directly reflects the submergence and distancing of hinterland source areas. However, this facies succession also records changing climates probably influenced by and associated with the sea-level rise itself. For example,

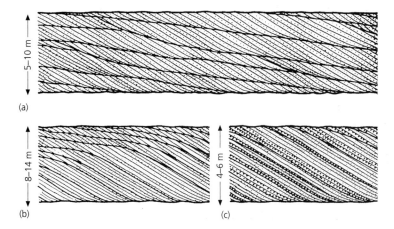

(a)

(b) (c)

Fig. 18.9 Three kinds of complex sets of cross-bedding in the Folkestone Sands of the Weald, south-east England. Three orders of erosional contact are present (top, base; surfaces of sets; reactivation surfaces within individual sets) (from Allen 1982).

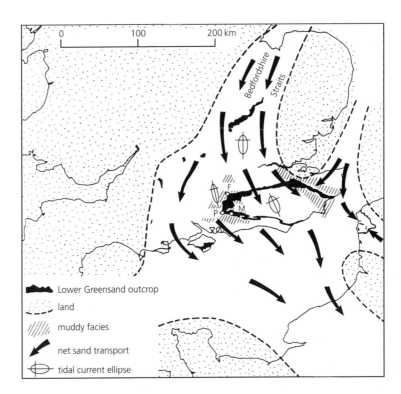

Fig. 18.10 Map showing sand dispersal patterns in the Folkestone Sands and their correlatives (after Allen 1982).

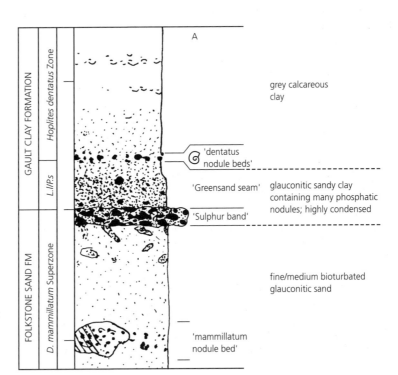

Fig. 18.11 Contact of the Lower Greensand and Gault Clay Formations, at Folkestone, Kent, to show the highly condensed concentrations of phosphatic concretions in three to four beds across the early–mid-Albian boundary.

the deposition of vast volumes of Aptian and Albian clays on a Europe-wide scale requires a very humid climatic regime.

The Gault Clay Formation records a further phase of the overall progressive marine inundation that commenced with the Perna Bed transgression in the Aptian. The formation onlaps older deposits to rest directly on Jurassic and Palaeozoic rocks of the London Platform and progressively oversteps onto the older Jurassic rocks westwards across Dorset and Devon. The Upper Gault onlaps the Lower Gault Clay northwards onto the London Platform. The basal contact of the Gault Clay with the underlying Lower Greensand is commonly very condensed, including phosphatized clasts and fossils representing several ammonite zones, set in a matrix of coarse glauconitic sand a metre or so thick (Fig. 18.11). The age of the transition from the underlying sands to Gault Clay varies considerably in basin–platform transects. In basinal settings in Sussex and northern France, the boundary is commonly within the Early Albian *D mammillatum* zone, whereas over structural highs it is usually of early mid-Albian age.

The Gault Clay Formation is of Mid- and Late Albian age and is the southern English representation of an open-marine clay-rich facies that is widely developed in basinal settings across Europe and North Africa at this time. In its typical development in south-east England, the Gault is sandwiched between the Lower and Upper Greensands and forms a distinctive narrow topographic hollow at the foot of the chalk scarp. The Gault Clay comprises silty and sandy clays and calcareous marls, which characteristically contain concretions of iron pyrites, calcium phosphate (apatite) and more rarely siderite and calcite. The numerous thin beds of reworked phosphate nodules include common internal moulds of fossils, and represent many short breaks in sedimentation. Ammonites provide a very refined stratigraphy for the Gault Clay and allow extraordinary detailed correlations to be made. However, fossils are well preserved only at particular levels in certain localities; in particular, the fine preservation of aragonite molluscs is found ubiquitously only in the lower part of the Gault Clay in the south-eastern part of the outcrop (Surrey, Sussex and Kent). Macrofossils are rare and poorly preserved in many Gault successions, particularly those in the western part of southern England (Dorey, Isle of Wight), and the biostratigra-

phy there is only sketchily known. To the west, the Gault becomes sandier and the upper part passes laterally into the Upper Greensand. The Upper Greensand Formation comprises well-bedded glauconitic and calcarenitic sands, which show evidence of a complex diagenesis in the form of carbonate and silica concretions. Storm beds are commonly present. The spread of sand from the west increased through the Albian, so the base of the Upper Greensand becomes younger in an easterly direction. The Upper Greensand was derived from Armorican and Cornubian sources, and at Haldon in Devon onlaps onto Palaeozic rocks of the Cornubian Massif.

To the north, the Gault thins over the London–Brabant Massif and in north Norfolk it passes into the Red Chalk, a thin condensed, haematite-stained sandy limestone that formed in relatively deep, sediment-starved conditions (Fig. 18.12). In the Cambridge district, the Gault is overlain by a thin unit (1 m thick) of glauconitic sand full of phosphatic nodules that immediately underlies the Chalk, called the Cambridge Greensand. There is a general increase in the amount of calcium carbonate (calcite) in the Gault towards the north and east, and towards the top of the formation, heralding the onset of chalk deposition.

The top surface of the Albian (Gault Upper Greensand) is an erosion surface everywhere in southern England, which represents a time-gap of 1–2 Myr. In Dorset and Devon, a strongly lithified hardground is present at this level, and on the Isle of Wight a conglomerate of concretions reworked from the Upper Greensand is present. A break at this level is widely found elsewhere in the world and can be related to a sea-level fall and ensuing rise (transgressive surface or sequence boundary). The large size of the gap in southern England may suggest some tectonic enhancement.

18.8 Sedimentation in the Boreal Sea: the East Midlands Shelf, Cleveland and Spilsby basins and the southern North Sea

The Market Weighton High and the London–Brabant Massif were separated by a more shallow Spilsby Basin, in which a complex succession of thin sands, clays and a limestone of Ryazanian to Aptian age accumulated. The sandy facies, called the Spilsby Sandstone, extended offshore into the southern

SW   NE

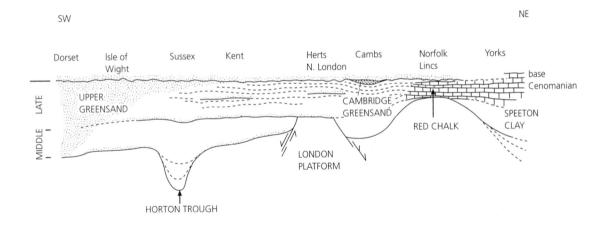

Fig. 18.12 Transect (not to scale) across the Albian deposits of north-east to south-east England.

North Sea. Over the Market Weighton High, most of the Lower Cretaceous is missing and a metre of Albian Red Chalk rests directly upon Lower Jurassic.

In the Cleveland Basin of East Yorkshire, adjacent to the North Sea Basin, marine sedimentation continued with periodic important breaks through the entire Lower Cretaceous. The faulted northern margin of the Market Weighton structure formed a southern limit to the Cleveland Basin, in which the Speeton Clay accumulated through the Ryazanian, Valanginian, Hauterivian, Barremian and Aptian stages. The Speeton Clay is a mid- to dark grey clay, with horizons containing concretions, glauconite and fossil concentrations. The basal bed of the Speeton Clay is a lag of phosphatic nodules, which rests disconformably upon the Kimmeridge Clay. An important hiatus, including the Late Valanginian Stage, is marked by the Compound Nodular Bed complex concretions representing several successive phases of sedimentation and reworking. Organic-rich laminated shales of Early Barremian age are a local representative of a widespread low-oxygen event. The Speeton Clay extends offshore into the southern North Sea Basin.

The Speeton Clay is overlain in most of eastern England by the Carstone Formation, a coarse pebbly sand, the grains of which are coated with brown and yellow oxides and hydroxides. The overlying Hunstanton Formation (Mid–Late Albian), com-monly called 'Red Chalk', is a cherry-red, haematite-stained nannofossil chalk rich in foraminifera. It passes southwards into the Gault Clay in north Norfolk. It extends across the southern North Sea and rests directly upon Jurassic strata over structures such as the mid North Sea High, contemporaneous with the final submergence of the London Platform beneath the Upper Gault Clay.

In the central North Sea, including the Central Graben, the Lower Cretaceous (Ryazanian–early Albian) is represented by the Valhall Formation (calcareous mudstones), which is a more calcareous lateral equivalent of the Speeton Clay. This is overlain by pink and red marls of the Rodby Formation (Albian), which is an expanded correlative of the Hunstanton Formation.

18.9 Summary

In southern England the base of the Cretaceous falls within a time of arid climate in which the Purbeck Limestone was deposited. At the top of the Purbeck (close to the base of the Valanginian Stage) a major climatic change brought in a humid, more seasonal climate and led to deposition of the Wealden facies. The change was perhaps initiated by uplift of massifs and the development of a drainage pattern dominated by west–east-flowing rivers. The distinctive Wealden facies comprises non-marine to brackish clays deposited on mud-plains, with many soils alternating with sands deposited by meander plains and braided outwash fans. Essentially non-marine Wealden deposition continued through the entire Valanginian,

Hauterivian and Barremian stages, into the early Aptian.

Early Aptian sea-level rise terminated the Wealden environments, and the remaining history of the Early Cretaceous (Aptian–Albian) records the progressive inundation of the massifs by rising sea across the whole of north-west Europe. On a third- and fourth-order scale, there were significant regressions and transgressions, which controlled sediment distribution and formed erosional surfaces. The first-order sea-level rise is reflected in an overall reduction in clastic grade (sand to silty clay to clay) as massifs were submerged and distanced, and finally the reduction in clay sedimentation and increase in carbonate production heralding the chalk deposition of the Late Cretaceous.

In the deeper-water North Sea Basin and the adjacent Cleveland Basin, marine deposition continued (with some breaks) through the entire Early Cretaceous. Shallow-water facies were dominated by thin condensed sandy deposition (Spilsby Sandstone) passing offshore into thicker clays (Speeton Clay). In the centre of the North Sea Basin (Central Graben) thick highly calcareous clays were deposited. Chalk deposition commenced early within the Albian in this deep-water setting with the red-coloured Hunstanton Formation.

References

Allen, J.R.L. (1982) Mud drapes in sand deposits: a physical model with applications to the Folkestone Beds (early Cretaceous, southeast England). *Philosophical Transactions of the Royal Society of London Series A* **306**, 291–370.

Allen, P. (1975) Wealden of the Weald: a new model. *Proceedings of the Geologists' Association* **86**, 389–437.

Rawson, P.F. (1992) Cretaceous. In: *The Geology of England and Wales* (eds P. McL. D. Duff & A. J. Smith), pp. 355–388. Geological Society, London.

Further reading

Allen, J.R.L. (1982) Mud drapes in sand deposits: a physical model with applications to the Folkestone Beds (early Cretaceous, southeast England). *Philosophical Transactions of the Royal Society of London Series A* **306**, 291–370. [Description of the exceptionally detailed record of tidal processes preserved in the Lower Greensand.]

Allen, P. (1975) Wealden of the Weald: a new model. *Proceedings of the Geologists' Association* **86**, 389–437. [A classic revision of the former deltaic model for Wealden sedimentation.]

Allen, P. (1981) Pursuit of Wealden models. *Journal of the Geological Society, London* **138**, 375–405. [Review of models for Wealden sedimentation.]

Casey, R. (1961) The stratigraphical palaeontology of the Lower Greensand. *Palaeontology* **3**, 487–621. [Detailed revision of the biostratigraphy of the Lower Greensand.]

Francis, J.E. (1984) The seasonal climate of the Purbeck (upper Jurassic) fossil forests. *Palaeogeography, Palaeoclimatology, Palaeoecology* **48**, 285–307. [Diagnosis of climate cycles from tree growth rings and from laminated sediments.]

Gröcke, D.R., Hesselbo, S.P. & Jenkyns, H.C. (1999) Carbon-isotope composition of Lower Cretaceous fossil wood: ocean–atmosphere chemistry and relation to sea-level change. *Geology* **27**, 155–158. [Demonstrates the correlation between carbon-isotope values in fossil wood with global rather than local environmental factors.]

Hesselbo, S.P., Coe, A.L. & Jenkyns, H.C. (1990) Recognition and documentation of depositional sequences from outcrop: an example from the Aptian and Albian of the Wessex Basin. *Journal of the Geological Society, London* **147**, 549–559. [A critical comparison of facies sequences with global sea-level curves.]

Horne, D.J. (1995) A revised ostracod biostratigraphy for the Purbeck and Wealden of England. *Cretaceous Research* **16**, 639–663. [A detailed biostratigraphic zonation for the non-marine facies of the Lower Cretaceous.]

Insole, A.N. & Hutt, S. (1994) The palaeoecology of the dinosaurs of the Wessex Formation (Wealden Group, Early Cretaceous), Isle of Wight, southern England. *Zoological Journal of the Linnean Society* **112**, 197–215. [Diagnosis of a low-productivity alluvial plain environment from the diversity of dinosaur faunas.]

Owen, H.G. (1976) The stratigraphy of the Gault and Upper Greensand of the Weald. *Proceedings of the Geologists' Association* **86**, 475–498. [Revision of the litho- and biostratigraphic correlation of this important part of the Lower Cretaceous.]

Radley, J., Gale, A.S. & Barker, M.J. (1998) Derived Jurassic fossils from the Vectis Formation (Lower Cretaceous) of the Isle of Wight, southern England. *Proceedings of the Geologists' Association* **109**, 81–91. [Evidence of syn-depositional faulting from fossil debris eroded from the uplifted footwall.]

Ruffell, A.H. & Rawson, P.F. (1994) Palaeoclimate control on sequence stratigraphic patterns in the late Jurassic to mid-Cretaceous with a case study from eastern England. *Palaeogeography, Palaeoclimatology, Palaeoecology* **110**, 43–54. [Contrasts the effects of arid and humid cycles on sequences' stratigraphic signatures.]

Ruffell, A.H. & Wach, G.D. (1991) Sequence stratigraphic

analysis of the Aptian–Albian Lower Greensand in southern England. *Marine and Petroleum Geology* 8, 342–353. [Reinterpretation of the Lower Greensand in sequence stratigraphic terms.]

Stewart, D.J., Ruffell, A., Wach, G. & Goldring, R. (1991) Lagoonal sedimentation and fluctuating salinities in the Vectis Formation (Wealden Group) of the Isle of Wight, southern England. *Sedimentary Geology* 72, 117–134. [Identification of cyclicity at the distal margin of deltas built into a Cretaceous lagoon.]

19 Late Cretaceous to Early Tertiary pelagic deposits: deposition on greenhouse Earth

A. S. GALE

19.1 Late Cretaceous palaeocontinental setting

In the Late Cretaceous, Britain was situated at the north-eastern side of a rapidly widening North Atlantic Ocean. Greenland was positioned immediately to the north of Britain and the Norwegian Sea had barely commenced to open. To the east, as far as the present Aral Sea in Central Asia, stretched a vast epicontinental sea in which chalk was deposited almost everywhere during the Coniacian to Maastrichtian. In western Europe the massifs that had begun to be flooded in the Albian (see also Fig. 18.3d, p. 344) were progressively inundated, until by the Campanian only the Baltic Shield (Norway and Sweden) and parts of the highlands of Scotland and Wales remained above sea level.

19.2 Late Cretaceous climates and sea levels

The Late Cretaceous (100–65 Ma) was a time of greenhouse climate for which no unequivocal evidence of sea-level ice in the polar regions has been demonstrated. Rather, high-latitude areas enjoyed temperate climates (e.g. mean annual temperature of about 10°C at 85°N) with high rainfall and extensive vegetation cover.

Oxygen-isotope data from the English Chalk and elsewhere provides clear evidence of the pattern of Late Cretaceous temperature change (see also Fig. 18.1, p. 340). Values rose throughout the Cenomanian to a maximum in the latest part of the stage, then began to fall gradually through the Turonian and Coniacian. The rate of the fall increased through the Campanian and particularly the Maastrichtian.

Although the isotopic values do not translate directly into absolute temperatures because of the effect of diagenetic alteration, maximum sea-surface temperatures of about 28°C are probable at the time of the late Cenomanian temperature maximum in southern England. The cause of the high global temperatures was perhaps carbon dioxide levels of four times present values, caused by ocean-floor volcanism related to rapid sea-floor spreading in the Pacific Ocean (see also Fig. 18.1, p. 340).

The Chalk succession also provides evidence that very arid conditions persisted from the UK across mid-latitude Europe to central Asia from the late Cenomanian onwards. The progressive decrease in the clay through the lower Chalk succession, and the very low levels of clay in the overlying white Chalk (mostly less than 1%), are in part a reflection of decreasing rainfall and runoff from the surrounding hinterland. This was augmented by the effect of rising sea levels, which both submerged and distanced clastic source areas and spread oceanic chalks onto the shelves. Shallow-water chalks in western France contain dolomites formed from evaporitic brines, which formed in pools along arid shorelines on the Armorican Massif.

Late Cretaceous sea levels were possibly the highest in the entire Phanerozoic. During the late Cenomanian sea-level maximum, epicontinental seas had transgressed the vast continental interiors of the USA, Europe and north Africa. Absolute values of sea levels are hard to ascertain, but maximum values of about 300 m above present levels are likely. Overall global sea levels continued to rise from the Albian (see also Fig. 18.1, p. 340), punctuated by brief and minor falls, to a maximum in the late Cenomanian–early Turonian. Sea levels fell significantly in the mid-

Turonian, but rose again through the Coniacian, Santonian and Early Campanian to a peak in the Late Campanian. According to most workers this was the highest sea–level registered in Europe, but is not so important in the Western interior of the USA, where the Rocky Mountains were rising at the time—a probable example of the effect of tectonics overriding eustatic sea-level records.

The very high sea levels in the Late Cretaceous can be attributed to high rates of sea-floor spreading, particularly in the Atlantic and south-east Pacific. Maximum highs in both sea level and temperature coincide in the Late Cenomanian; maximum rates of ocean-floor production at the time displaced vast volumes of sea water and pumped carbon dioxide into the atmosphere simultaneously. In particular, the mantle plume underlying the east Pacific hotspot has been held accountable for the exceptionally high levels of carbon dioxide.

19.3 Stratigraphical framework for the Late Cretaceous

While ammonites still afford the most refined subdivision of time in the Late Cretaceous, their usefulness is hampered by a number of factors. In the white chalk environment, aragonite underwent dissolution at shallow levels in the sediment or even on the sea floor. Consequently, ammonite shells are rarely found, except where cementation during hardground formation preceded aragonite dissolution. Thus, although ammonites were present in the seas, they are simply not preserved or else very rare in many chalk successions. A further complication was that, in the latter part of the Cretaceous, ammonites developed strong provinciality and divided into Boreal, Tethyan and Austral faunas, the precise correlation between which is not always well understood. However, in the Cenomanian Lower Chalk, a fine subdivision into seven zones can be obtained using ammonites.

Bivalves of the family Inoceramidae, which has a resistant outer shell layer made of calcite prisms, are common and well preserved in chalk successions. They evolved rapidly and many species had a nearly global distribution, independent of facies, making them excellent for biostratigraphy. Many macrofossil groups, including echinoids, crinoids, brachiopods and belemnites, have been used to develop local zonations in chalk successions. Short-lived species of the stemless crinoids *Uintacrinus* and *Marsupites* had a nearly global distribution in late Santonian chalks, providing high-resolution correlation. Microfossils are widely used in correlation of chalks, particularly the sediment-forming calcareous nannofossils and foraminiferans.

Increasingly, stable isotope ratios are used in the international correlation of Late Cretaceous chalks. In particular, $\delta^{13}C$ showed considerable change during late Cretaceous time and a number of short-lived positive and negative excursions have been identified on different continents. The strontium-isotope ratio $^{87}Sr/^{86}Sr$ underwent considerable evolution during the Coniacian to Maastrichtian, with a nearly straight-line change in isotopic ratios, of considerable use in correlations for which fossils do not provide answers. Magnetostratigraphy is of very limited use in the Late Cretaceous, because Cenomanian to Santonian time falls within the long normal polarity zone 34N.

19.4 Lithofacies of the Chalk

Chalk is a fine pelagic limestone that, at the present day, is characteristic of oceanic deep-sea environments but, in the Cretaceous, extended widely onto the shelves and into the flooded continental interiors. Chalk is the dominant sediment of the mid-latitude Cretaceous seas eastwards from the UK, across Europe as far as the Aral Sea in Central Asia. The detailed lithological succession is remarkably similar across this vast swathe, reflecting the ubiquitous effects of palaeoceanographic and climatic processes upon pelagic sedimentation.

Chalks are chemically low-Mg calcite and are composed dominantly of calcareous nannofossils (coccoliths and others) with a variable component of coarser bioclastic debris, including foraminiferans, calcispheres, bivalves (notably prisms of inoceramids), echinoderms and locally bryozoans (Fig. 19.1). Only the Cenomanian contains large quantities of clastic material, sand in shallow water and clay in deeper settings. Stratigraphically higher white chalks often contain less than 1% of insoluble residue, which is mostly clay minerals and clay-grade quartz.

Although the chalk at first appears to be a very homogeneous formation, the monotony of pelagic coccolith limestone is broken by various bedding features formed both by depositional processes and by

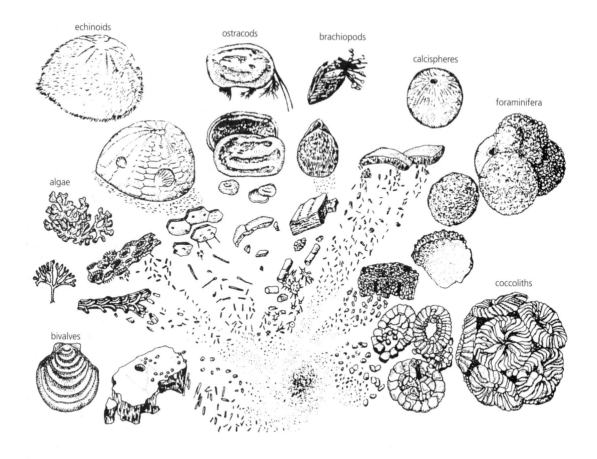

Fig. 19.1 Components of chalk; numerically dominant are tiny, 10–20 μm low-Mg-calcite coccoliths, parts of chrysophyte algae and their breakdown (submicron size) debris; coarser components are concentrated by current winnowing in calcarenitic chalks (after Bromley 1979).

early diagenesis. These features are important in the determination of environmental changes in the chalk, such as sea levels and climatic cycles, and provide important stratigraphical marker beds.

Chalk–marl alternations are conspicuous in the Lower Chalk (Cenomanian) and comprise decimetre-scale couplets of more and less marly chalks, with about 10–20% variation in the clay content (Fig. 19.2). They are products of orbitally forced climate change on a 20-kyr frequency, which affected the amount of clay brought into depositional basins by moderating rainfall and thus runoff. The couplets

provide a means of establishing an orbital time-scale in the Cretaceous and can be widely correlated (Fig. 19.3).

Thin (<20 cm) but widespread *marl beds* in the White Chalk contain up to 20% of the clay mineral smectite. Some are altered airborne ashfalls derived from mid-Atlantic volcanism. The volcanigenic beds can be identified by their rare-earth and trace-element geochemistry, and individual beds can be traced throughout England, northern France and Germany; they thus have considerable value in detailed correlation (Fig. 19.4). However, not all marl beds have this origin, some being made up entirely of detrital clay material.

Calcarenitic units (0.5 m to several metres in thickness) comprise coarse chalks made up of bioclastic detritus (echinoderm and inoceramid fragments, calcispheres). The units are mostly a product of trans-

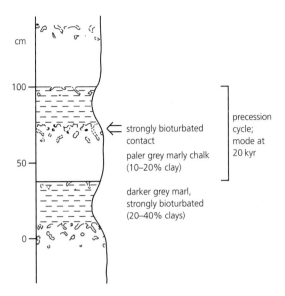

Fig. 19.2 Chalk–marl rhythmic couplets in the Lower Chalk Formation, southern England.

gressive winnowing by current action. When these units overlie major breaks in succession, these beds commonly contain glauconite and may have a basal lag of phosphatic nodules resting on an erosional surface, as in the case of the Totternhoe Stone in the Lower Chalk. In deep-water settings such as the North Sea Basin, coarser chalks form during low-stands as well as trangressive events and represent coarser material flushed from platforms.

Hardgrounds and *nodular chalks*, generally less than a metre in thickness, underwent early diagenetic cementation just beneath the sea floor. This hardening initially produced a framework of nodules within the sediment, which was locally exposed by erosion to form a rocky pavement called a hardground (Fig. 19.5). Evidence for exposure on the chalk sea floor includes the presence of borings in the surface, encrusting fauna (oysters, worm tubes) and replacement of the surface by green glauconite or brown phosphate. Once formed, hardgrounds underwent complex modification by processes of boring, physical erosion and the welding of more cemented chalk onto the surface. Hardgrounds are best developed over structural highs and often represent condensation and hiatus related to major sea-level change, at both sequence boundaries and during transgression.

Flint is a special form of chert that characteristically has a white patina and a distinctive conchoidal fracture, and occurs as diagenetic nodules composed of microcrystalline quartz. Layers of flint nodules commonly pick out bedding, even in slumped units. The sites of flint formation were thus determined shortly after deposition at the redox boundary, below which anaerobic, sulphate-reducing bacteria predominate (Fig. 19.6). A flint precursor in the form of tiny spheres of the metastable mineral cristobalite formed first. Flints formed preferentially in sites of concentrated organic matter such as *Thalassinoides* burrows, which explains their curious and complicated shapes. At about 50–100 m burial depth, the cristobalite inverted to α-quartz but relict spheres are visible in the fabric. The positions of the flint nodule beds were determined by climatic cycles because their distribution shows frequencies in the Milankovitch Band. The source of the silica was biogenic and was probably derived mostly from sponge skeletons (now commonly preserved in the chalk as pyrite- or limonite-coated composite moulds) and microfossils such as radiolarians. Individual flints may be widespread markers in chalk basins (Fig. 19.7).

Chalks underwent a very distinctive diagenetic history, in which dewatering, early cementation and flint formation occurred relatively early on at shallow depths. Once metastable minerals had been dissolved or inverted to more stable phases, the remaining low-Mg calcite was quite unreactive and retained a high porosity to considerable burial depths. At great depths, or in regions of high heat flow (e.g. Northern Ireland), recrystallization to hard white limestones took place with accompanying calcite-filled fractures. The tendency of chalks to retain high porosities has been an important factor in the development of oil reservoirs in the North Sea.

19.5 Depositional history of the Chalk

The base of the Chalk Group is represented by a major disconformity everywhere in southern England and northern France (Anglo-Paris Basin). Here a basal transgressive condensed deposit called the Glauconitic Marl rests upon either the eroded surface of the Albian Gault Clay or a hardground formed in the Upper Greensand (Fig. 19.8). This break repre-

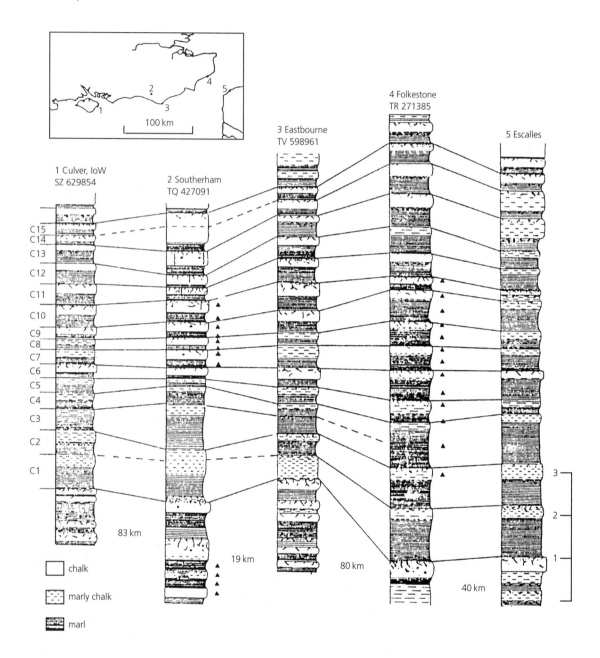

Fig. 19.3 Correlation of Cenomanian chalk–marl couplets in south-east England and northern France, using combined biostratigraphy and distinctive marker beds, characterized by trace fossil fabrics and faunas (after Gale 1989).

sents a minimum period of 2 Myr by comparison with a thick continuous basinal marl succession in southern France. The Albian–Cenomanian erosional contact represents a composite sequence boundary transgressive surface and is also a break in many other parts of the world. This break has a component of tectonic enhancement in southern England.

Above the Glauconitic Marl, the Lower Chalk

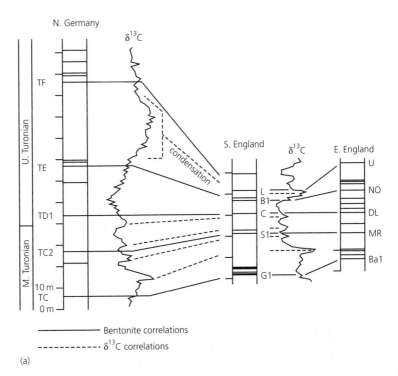

Fig. 19.4 (a) Regional correlation of marls in the Turonian chalks using carbon-isotope (δ^{13}C) values and biostratigraphy. (b) Rare-earth element traces from successive Turonian–Coniacian marl beds in northern England, showing the distinction between volcanic and detrital origins.

(a)

——————— Bentonite correlations

-------------- δ^{13}C correlations

consists of rhythmically bedded, strongly bioturbated marly chalks containing concretionary limestones full of sponges preserved as calcite. There is an overall decrease in clay content upwards, which is more sharply marked at the boundary between the Chalk Marl below and the Grey Chalk above. Five sequences related to eustatic sea-level changes can be recognized in the Lower Chalk of the Anglo-Paris Basin, within an overall pattern of rising sea level through the stage. The boundary at the base of each sequence (Fig. 19.9) is marked by an increase in clay in the marly chalks, which form a thin lowstand unit. The overlying transgressive surface is a condensed calcarenitic chalk resting on an erosional surface. Maximum flooding is represented by a decrease in clay content to purer chalks of the highstand. The transgressive surfaces come to rest directly on the sequence boundaries as the succession thins over positive structures. The sequences can be dated very precisely by ammonites and correlate with those identified in North Africa and eastern Europe.

The succession in Dorset records in detail the stepwise rise in sea level through the Cenomanian over a structural high called the Mid-Dorset Swell (Fig. 19.10). Each of the five transgressions in the Ceno-

manian onlapped the previous one, to rest directly on the Albian in Devon (Fig. 19.10(1)). A basal phosphatic lag contains ammonites that demonstrate the precise age of the base of the Chalk and track its younging to the west. The Cenomanian succession in South Devon comprises thin (one to several metres) sandy bioclastic limestones containing hardgrounds that correspond to the sequence boundaries and transgressive surfaces recognized in the basin to the east. The palaeogeography of the UK during the Cenomanian is shown in Fig. 19.11. Large areas of Ireland, Wales and Scotland (including the Shetlands Platform) remained as land, as did parts of Cornubia (Devon and Cornwall), Armorica (Brittany) and the Ardennes in Belgium. Deeper parts of the Anglo-Paris Basin and the North Sea Basin were dominated by rhythmically bedded marly chalks, whereas the shallow-water marginal successions of Cornubia and Armorica display thin developments of sands and bioclastic sands containing many breaks in deposition represented by hardgrounds.

Over the Eastern England platform, the base of the Cenomanian rests disconformably on a hardground developed in the top of the Albian Red Chalk. In the thick succession at Speeton, east Yorkshire (Sole Pit

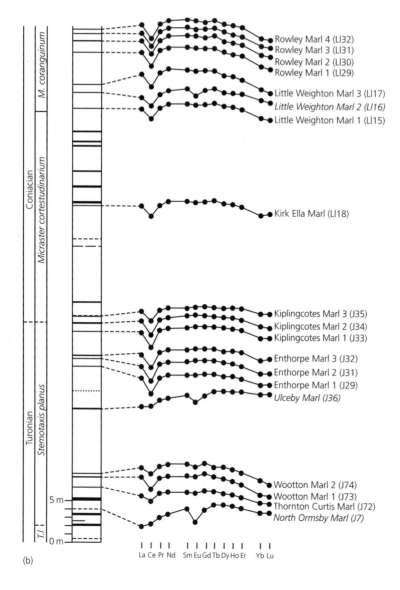

Fig. 19.4 (*Continued.*)

Basin) the contact with the underlying Albian appears to be conformable. The succession in Norfolk, Lincolnshire and Yorkshire comprises thinly bedded pure white chalks with two major erosional surfaces, overlain by condensed beds corresponding to the transgressive bases of sequences 2 and 3 in southern England. The lower of these is a calcarenitic lag called the Totternhoe Stone, which locally thickens to infill channels cut deep into the underlying chalk—the 2–3 sequence boundary. The Cenomanian in the North Sea is represented by the marly, cyclically bedded Hidra Formation, which is an expanded version of the Lower Chalk of the Anglo-Paris Basin, overlain by the thin dark organic-rich marls of the Plenus or Blodax Formation. These form a very marked gamma spike in North Sea wells.

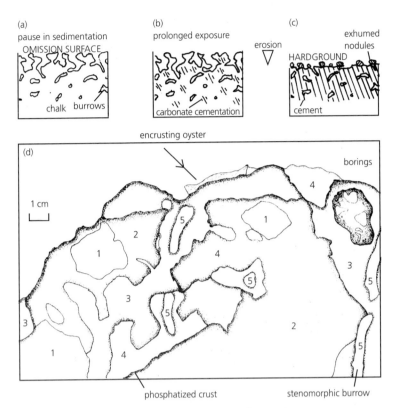

Fig. 19.5 (a)–(c) Sequence of hardground formation through time. (d) Section through a hardground surface, showing encrusters, borers, and phosphatic mineralization (stippled). Five successive sediments (1 is youngest, 5 oldest) have become cemented onto an original convolute surface. Burrowing organisms kept burrows open (5) against cementation, which narrowed the burrows.

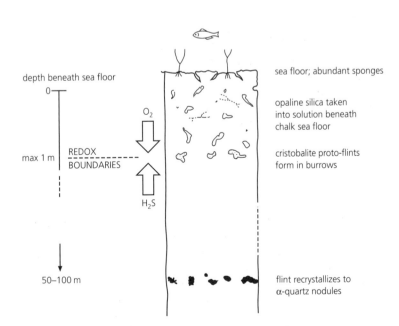

Fig. 19.6 Flint formation and early diagenetic features of chalks.

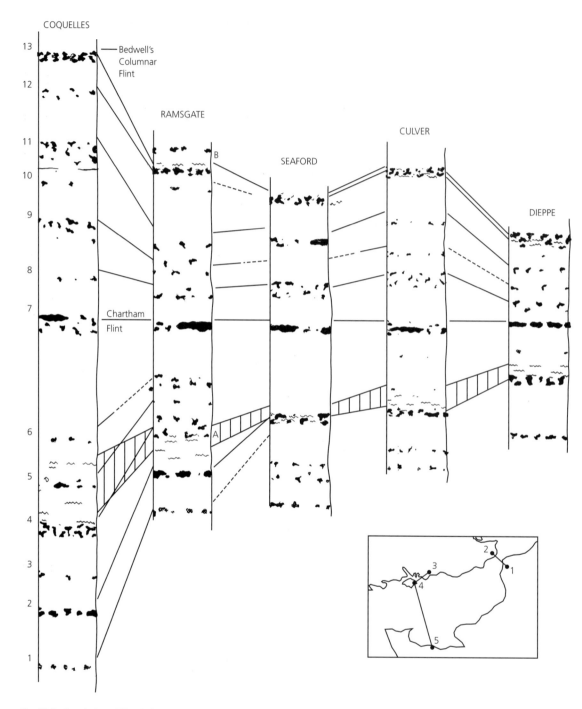

Fig. 19.7 Correlation of flints in late Coniacian and early Santonian chalks in southern England and northern France. Beds of flint nodules show remarkable lateral persistence. Two beds containing abundant inoceramid bivalves (A,B) provide biostratigraphical control. The large, lenticular Chartham Flint and the nodular Bedwell's Columnar Flint retain their distinctive morphologies over the 100 km represented in this figure.

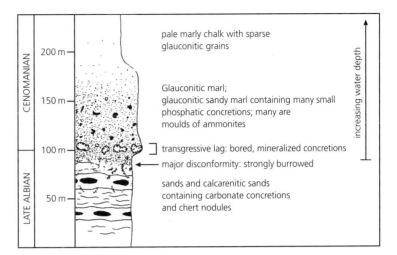

Fig. 19.8 Base of the Chalk Group, southern Isle of Wight.

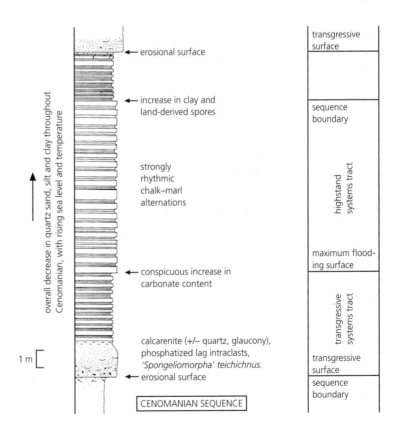

Fig. 19.9 A Cenomanian sequence in the marly rhythmically bedded Lower Chalk of the Anglo-Paris Basin.

The end of the Cenomanian (Fig. 19.12) was a plexus of environmental change on a world-wide scale, which precipitated a major faunal crisis at the Cenomanian–Turonian boundary. The changes had a root cause in the coinciding maxima of sea-level rise and temperature. These events are reflected in the sedimentology, chemostratigraphy and faunas in the UK region.

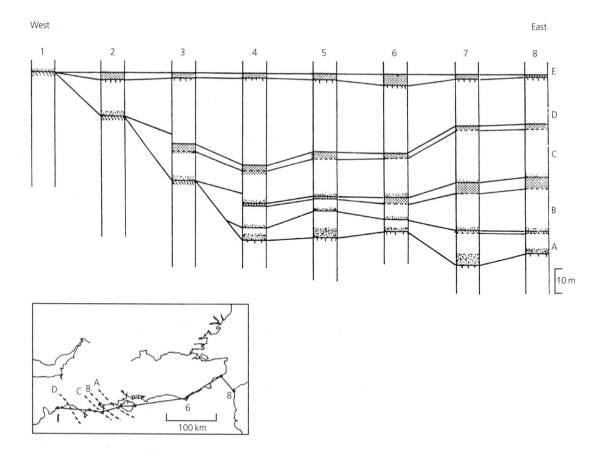

Fig. 19.10 Westward onlap of Cenomanian sequences into south-west England. Five discrete sedimentary sequences (A–E) onlap progressively onto shallower basement to the west, such that at some localities in Devon (1) chalks belonging to sequence E rest directly upon Albian Greensand. This reflects the progressive overall sea-level rise that characterized the Cenomanian Stage (after Gale 1996).

In the Anglo-Paris Basin the top unit of the Lower Chalk is a more marly chalk, rhythmically bedded, a few metres thick, called the Plenus Marl. This is overlain by a white nodular coarse chalk unit forming the base of the White Chalk Formation. The transition between the two formations is marked by the cut-off of clastic supply and a dramatic facies change, representing a major transgressive event. The change is even more emphatic in shallow-water facies where marine sands are overlain by deep-water pelagic coccolith chalks.

The base of the Plenus Marl is an erosion surface that represents a sudden sharp sea-level fall at the base of the fifth and final Cenomanian sequence. In the middle part of the Plenus Marl, sea level started to rise again. The upper part of the Plenus Marl and the base of the overlying White Chalk are coarse sediments containing abundant calcispheres and fragments of inoceramid bivalves. These units reflect winnowing of the sediment during transgression, as do the presence of numerous small intraclasts in the basal part of the White Chalk. Evidence from the shores of the Mid-European island around Dresden in Germany suggests that this transgression resulted in a sea-level rise of about 50 m.

The chalks of the Plenus Marl and the overlying beds also record a major perturbation in the carbon cycle, which is recorded world-wide (Fig. 19.12). The calcite deposited by the coccoliths and other plankton contains a higher than usual proportion of the heavy isotope of carbon (^{13}C). The resulting sharp positive excursion coincided with deposition of black

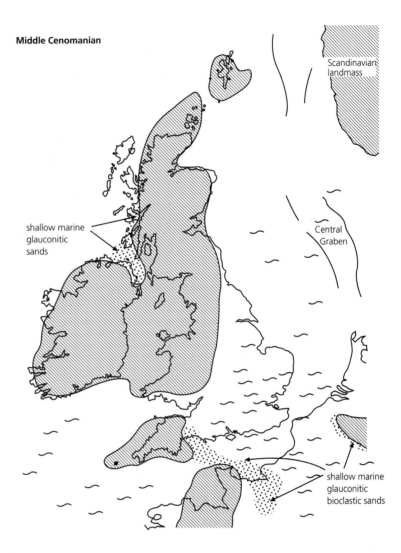

Fig. 19.11 Map of the UK and adjacent areas during the mid-Cenomanian.

organic-rich shales in deep water on the outer shelves and in the ocean basins. It is likely that the rapid rise in sea level caused a massive increase in productivity in the marine plankton at a time when, because of the high temperatures, sea water was not carrying as much oxygen as usual (oxygen solubility decreases with rising temperatures). The organic matter remained on an anoxic sea bed and was then buried, depleting the carbon reservoir of ^{12}C and creating the δ^{13}C excursion. Because of the burial of so much organic matter, the oceans were depleted in carbon and replenished this by drawing down atmospheric carbon dioxide, which started a global cooling.

In southern England the sea was too shallow and well aerated to allow any anoxic black shales to form, but in Yorkshire (Black Band) and the North Sea Basin (Blodax Fm) the end-Cenomanian event is recorded by an organic-rich (1.5% total organic carbon (TOC)) laminated shale, locally containing whole fish, which formed in very low oxygen conditions. It represents a considerable condensation.

The high sea levels of the Late Cenomanian–Early Turonian persisted through the early part of the mid-Turonian, with the deposition of fine white chalks containing thin marls and local flints. The eustatic fall in sea level that followed commonly resulted in the condensation or erosion of Turonian sediments. For

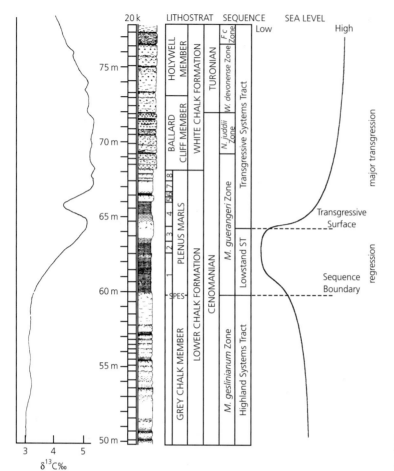

Fig. 19.12 Late Cenomanian–Early Turonian palaeoenvironmental change in the Anglo–Paris Basin.

example, in Northern Ireland the entire Turonian is represented by a single reworked ammonite preserved in younger Coniacian sediments. In southern England, the Mid- and Late Turonian is condensed over a wide area of relatively shallow basement and a group of massively lithified and strongly mineralized hardgrounds (Fig. 19.13). These hardgrounds have regionally planar surfaces and probably represent times when storm wave base reached the chalk sea floor. In basinal successions adjacent to the blocks on which hardgrounds are developed, thick wedges of redeposited lowstand chalk are found.

In the Mid-Coniacian, typical white chalk facies developed across England and far across Europe, and persisted through to the Maastrichtian (Fig. 19.14). This unit comprises soft coccolith chalks with a trace of clay and less than 10% bioclasts, in which the bedding is picked out by beds of flint nodules with an average spacing of about 1 m. The flints themselves provide evidence of climatic cyclicity by mechanisms that are not well understood, but probably represent climate changes in the Milankovitch Band (20–400 kyr). Environmental variation on a small scale is also shown by changes in oxygen-isotope values. Individual flint beds may have a distinctive morphology and be traceable over vast areas. Marls representing occasional ashfalls are found very infrequently and there is generally little evidence of the eustatic sea-level changes on the frequency of 1 Myr that can be found in coeval shallow-water deposits in Europe and elsewhere. Hardgrounds do develop over local structural highs, where sea-floor topography

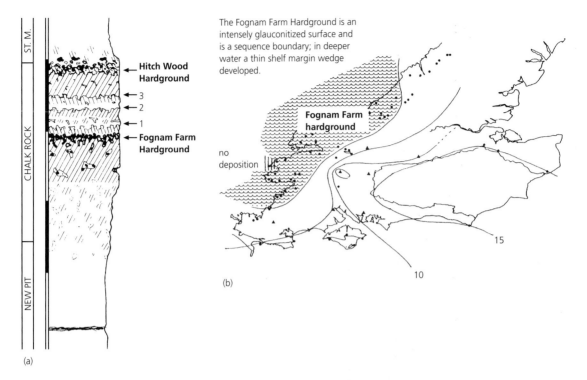

The Fognam Farm Hardground is an intensely glauconitized surface and is a sequence boundary; in deeper water a thin shelf margin wedge developed.

(a)

(b)

Fig. 19.13 Development of the Chalk Rock, a group of massive chalk hardgrounds formed over shallow basement during the mid–late Turonian sea-level lowstand (south Chilterns; a). (b) shows the distribution of the Fognam Farm Hardground (after Gale 1996).

has had the effect of accelerating bottom current velocity during sea-level changes, but these are infrequent and do not provide a cohesive picture of sea-level changes. Even more rarely, gravity slides occurred on local structures, such as the one developed at Downend in Hampshire, which underwent a complex series of movement, erosion and burial (Fig. 19.15).

In the North Sea area, chalks extend as far north as the Viking Graben (Fig. 19.16), where they pass into mudstones. Above the Plenus–Blodax Formation are the thick Hod Formation (Turonian–Campanian) and the fine white chalks of the Tor Formation (Late Campanian–Maastrichtian, Fig. 19.14). Danian (Paleocene) chalks are represented by the Ekofisk Formation, which comprises rhythmically bedded marly chalks. The chalk succession of the Central Graben (UK, Norwegian and Danish sectors)

proved to be of unexpected interest when oil reservoirs were discovered in the late 1960s and 1970s. The Central Graben was undergoing active fault-controlled subsidence through the latter part of the Late Cretaceous and 1350 m of chalk accumulated there. Although much of this chalk was deposited as pelagic rain, parts of the succession contain redeposited slides, slumps, debris flows and turbidites. The rapid deposition of these sediments caused the preservation of unusually high porosities, which enhanced their values as reservoirs. Producing wells are located on salt inversion structures and are concentrated in the Norwegian sector of the Central Graben (Ekofisk, Valhall, Hod, Elfisk, Tor).

Chalks also extend across the continental shelf to the west and south-west of the UK and are found in the Celtic Basin, the Western Approaches Basin and on the continental slope. In the North Sea and in Denmark, chalk deposition continued into the Danian (Palaeocene), with minor inversion commencing within the Late Cretaceous. In England and adjacent areas of northern Europe, structural inversion of Danian age resulted in extensive folding, uplift and erosion of chalks. The unconformity between the pelagic limestones of the Chalk Group and the conti-

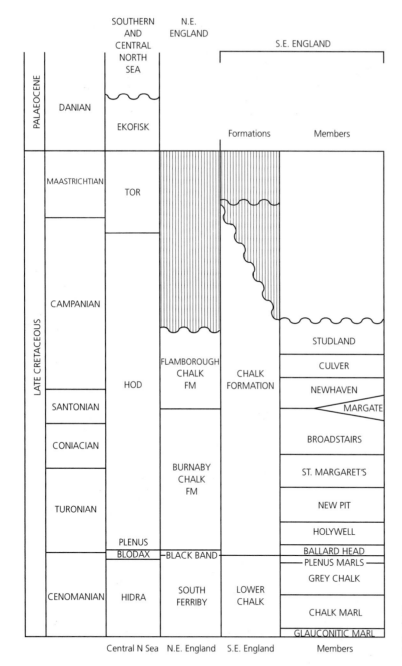

Fig. 19.14 Correlation of the Chalk formations and members of the southern–central North Sea, north-east England and southern England.

nental clays of the Reading Formation is a dramatic lithological and facies change and represents a major hiatus with a duration of about of 15 Myr. The uplift of the Chalk probably took place in the Danian because Maastrichtian chalks are present offshore, south of Dorset, and movement pre-dates deposition of the Thanet Formation in the London Basin, which is Thanetian in age. Gentle regional folding caused a slight dip (about 20°) to the south-west to develop in the Isle of Wight, as shown by westwards overstep of

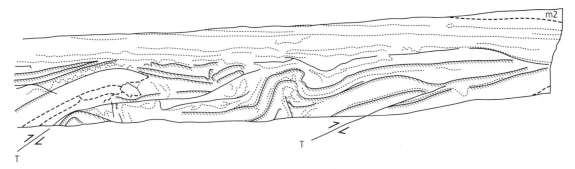

Fig. 19.15 Slump in early Campanian Chalk, Downend, Hampshire. The quarry face shows the upper part of a giant slump, 50 m thick and an estimated 500 m² in dimension. Thrusting has taken place in the compressive 'toe' of the slump, which was subsequently eroded and buried by planar bedded white chalks.

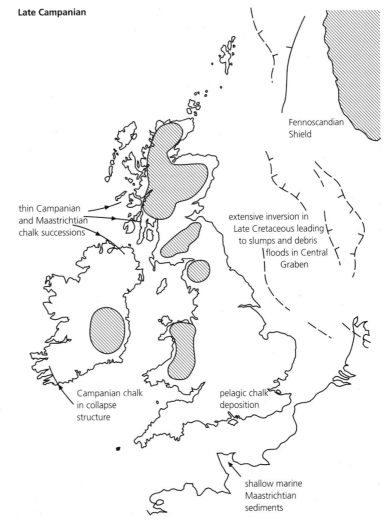

Late Campanian

Fennoscandian Shield

thin Campanian and Maastrichtian chalk successions

extensive inversion in Late Cretaceous leading to slumps and debris floods in Central Graben

Campanian chalk in collapse structure

pelagic chalk deposition

shallow marine Maastrichtian sediments

Fig. 19.16 Map of the UK and adjacent areas during the late Campanian. This probably represents the maximum sea levels of the Late Cretaceous, and only the Highlands of Scotland, and dubiously, parts of the Southern Uplands, central Wales and Ireland remained as land. Successions are entirely dominated by white chalk facies, the purity of which (95–98% $CaCO_3$) is partly a reflection of the burial of source areas under high sea-level stands. Marginal marine deposits are unknown from the entire region.

the chalk. Subsequent peneplanation removed in the order of 100 m of chalk of latest Campanian and Maastrichtian chalks across most of southern England. However, Maastrichtian chalks are preserved only in north-east Norfolk, where glacially emplaced and stacked slabs of chalk outcrop on the coast near Cromer.

19.6 Summary

The Late Cretaceous saw a progressive, blanket-like spread of oceanic nannofossil chalks across the entire north-west European continental shelf, submerging almost all land except the Baltic Shield and the highest parts of Wales and Scotland. This chalk sea extended in a mid-latitudinal belt through northern Europe into central Asia. The spread of the chalk facies was caused by a number of factors, including the very high sea levels, which drowned and distanced clastic source areas, and aridity in the land regions, which restricted runoff and hence clastic supply.

In terms of the persistence of faunal horizons and facies and the continuity of marker beds, chalk successions are remarkably similar across vast distances. This is because palaeoceanographic and climatic factors that operated very widely controlled many aspects of deposition. Tectonic factors played a rather minor role in the development of most chalk successions.

References

Bromley, R.G. (1979) University of Bergen publication.

Gale, A.S. (1989) A Milankovitch scale for Cenomanian time. *Terra Nova* **1**, 420–425.

Gale, A.S. (1996) Correlation and sequence stratigraphy of the Turonian Chalk of southern England. Sequence stratigraphy in the United Kingdom. In: *Sequence Stratigraphy in British Geology* (eds S. P. Hesselbo & D. N. Parkinson), Special Publication 103, pp. 177–195. Geological Society, London.

Further reading

Bromley, R.G. & Gale, A.S. (1982) The lithostratigraphy of the English Chalk Rock. *Cretaceous Research* **3**, 273–230. [Describes chalk hardgrounds and their correlation.]

Clayton, C.J. (1986) The chemical environment of flint formation in Upper Cretaceous chalks. In: *The Scientific Study of Chalk and Flint* (eds G. DeC. Sieveking & M. B.
Hart), pp. 43–55. Cambridge University Press, Cambridge. [Reviews the formation of flint.]

Gale, A.S. (1995) Cyclostratigraphy and correlation of the Cenomanian of Europe. In: *Orbital Forcing Timescales and Cyclostratigraphy* (eds M. R. House & A. S. Gale), Special Publication 85, pp. 177–197. Geological Society, London. [Charts the correlation and orbital significance of chalk–marl couplets.]

Gale, A.S. (1996) Correlation and sequence stratigraphy of the Turonian Chalk of southern England. Sequence stratigraphy in the United Kingdom. In: *Sequence Stratigraphy in British Geology* (eds S. P. Hesselbo & D. N. Parkinson), Special Publication 103, pp. 177–195. Geological Society, London. [Presents detailed correlation of chalk successions.]

Gale, A.S., Jenkyns, H.C., Kennedy, W.J. & Corfield, R. (1993) Chemostratigraphy versus lithostratigraphy—evidence from the Cenomanian–Turonian boundary. *Journal of the Geological Society, London* **150**, 29–32. [Documents the global extent of the carbon crisis at the end of the Cenomanian.]

Håkansson, E., Bromley, R. & Perch-Nielsen, K. (1974) Maastrichtian chalk of north-west Europe—a pelagic shelf sediment. In: *Pelagic Sediments: on land and under the sea* (eds K. J. Hsü & H. C. Jenkyns), Special Publication of the International Association of Sedimentologists 1, pp. 211–233. [Interprets the environment of chalk sedimentation.]

Hancock, J.M. (1975) The petrology of the Chalk. *Proceedings of the Geologists' Association* **86**, 499–535. [Discusses the composition of chalks.]

Hancock, J.M. (1986) Cretaceous. In: *Introduction to the Petroleum Geology of the North Sea* (ed. K. W. Glennie), pp. 255–272. Blackwell Scientific Publications, Oxford. [Reviews the stratigraphy and sedimentation of North Sea chalks.]

Jenkyns, H.C., Gale, A.S. & Corfield, R. (1994) The stable oxygen and carbon isotope stratigraphy of the English Chalk and the Italian Scaglia. *Geological Magazine* **13**, 1–34. [Describes the climate changes recorded in chalks.]

Kennedy, W.J. & Garrison, R. (1975) Morphology and genesis of nodular chalks and hardgrounds in the Upper Cretaceous of southern England. *Sedimentology* **22**, 311–386. [Describes and interprets the hardgrounds in the Chalk.]

Mortimore, R.N. (1987) Controls on Upper Cretaceous sedimentation in the South Downs, with particular reference to flint distribution. In: *The Scientific Study of Flint and Chert* (eds G. deC. Sieveking & M. B. Hart), pp. 21–42. Cambridge University Press, Cambridge. [Uses flint distribution to help map sedimentary patterns in the Chalk.]

Paul, C.R.C., Ditchfield, P., Gale, A.S., Learey, P.N., Mitchell, S. & Duane, A. (1994) Palaeoceanographic events in the middle Cenomanian of north-west Europe. *Cretaceous Research* **15**, 707–738.

[Correlates and interprets events in the Lower Chalk.]

Quine, R. & Bosence, D. (1991) Stratal geometries, facies and sea-floor erosion in Upper Cretaceous Chalk, Normandy, France. *Sedimentology* **38**, 1113–1152. [Correlates erosion surfaces sea-level lowstands on a fault footwall.]

Robaszynski, F., Gale, A.S., Juignet, P., Amédro, F. & Hardenbol, J. (1992) Sequence stratigraphy in the Upper Cretaceous of the Anglo-Paris Basin, exemplified by the Cenomanian Stage. In: *Mesozoic and Cenozoic Sequence Stratigraphy of European Mesozoic Basins* (eds J. Hardenbol, J. Thierry, M. B. Farley, Th. Jaquin, P.-C. De Graciansky & P. R. Vail), Special Publication 60, pp. 363–386. Society of Economic Paleontologists and Mineralogists, Tulsa, OK. [Sequence stratigraphic analysis of five sequences on the basis of sediment body geometry and lithology.]

Scholle, P.A. (1974) Diagenesis of Upper Cretaceous chalks from England, Northern Ireland, and the North Sea. In: *Pelagic Sediments: on Land and Under the Sea* (eds K. J. Hsü & H. C. Jenkyns), Special Publication of the International Association of Sedimentologists 1, 77–210. [Analysis of diagenetic contrasts in Chalk from east to west across Britain and Ireland.]

Scholle, P.A. (1977) Chalk diagenesis and its relation to petroleum exploration: oil from chalk, a modern miracle? *AAPG Bulletin* **61**, 982–1009. [Generic study of chalk diagenesis, with example from the North Sea.]

Scholle, P.A. & Halley, R.B. (1985) Burial diagenesis: out of sight, out of mind! In: *Carbonate Cements* (eds N. Schneidermann & P. M. Harris), Special Publication 36, pp. 309–334. Society of Economic Paleontologists and Mineralogists, Tulsa, OK. [Study of limestone burial and diagenesis, including porosity–depth data for chalks.]

20 Tertiary events: the North Atlantic plume and Alpine pulses

R. ANDERTON

20.1 The Cretaceous–Tertiary boundary

The boundary between the Cretaceous and Tertiary is one of the most significant in the geological record in that it is defined by an episode of mass extinctions. These events not only brought to the end the age of the dinosaurs but also saw the demise of numerous groups of marine organisms and some land plants. There is controversy as to whether these extinctions were the result of the catastrophic impact of a meteorite or comet or due to a brief period of rapid change in factors such as sea level, climate, volcanism and atmospheric composition. There is much evidence for a meteorite impact in the form of the buried Chicxulub Crater in the Yucatan Peninsula in Mexico, sediments consistent with catastrophic waves around the Gulf of Mexico and high iridium concentrations world wide in sediments at the Cretaceous–Tertiary boundary. However, the palaeontological record suggests that such an impact was not the sole reason for the rapid biological changes. Either way, the British geological column casts little light on this event. Where the boundary is seen at outcrop it is marked by an unconformity recording a period of uplift and erosion between the deposition of the Upper Cretaceous chalks and either the clastic sediments or lavas of the lowermost Tertiary. By contrast, sedimentation in most basinal areas was uninterrupted, and there was little significant change in the depositional environment at the boundary, similar chalks or muds being seen in both the Late Cretaceous and Early Tertiary section in many offshore wells.

20.2 Climate and dating

The stratigraphic framework of the Tertiary is better known than that of earlier parts of Earth history. Numerous ocean bottom cores together preserve a virtually continuous sedimentary record through much of the Tertiary. They have been analysed, not only for their lithological and biological contents, but also for their stable isotope and palaeomagnetic signatures. As a result it has been possible to piece together a very detailed history of the Tertiary oceans. The alternation of normal and reversed polarity intervals, preserved in sea-floor magnetic anomalies, has provided the framework for the palaeomagnetic time-scale. Radiometric age dating, mostly of volcanic horizons, together with astrochronological dating of the latest Tertiary, has allowed this framework to be tied to key dates. By assuming smoothly varying sea-floor spreading rates, it has then been possible to interpolate the dates of polarity reversals to give a very high resolution time-scale (Fig. 20.1). Biostratigraphic zonation schemes for fully marine Tertiary sediments have been developed using planktonic foraminifera (P and N zones), calcareous nannofossils (NP and NN zones) and radiolaria. Palynomorphs are useful in non-marine environments and marine areas subject to terrestrial influxes. It is not straightforward to correlate the fragmentary sedimentary record seen onshore in Europe to the detailed oceanic picture. However, with the more widespread use of stable isotope and palaeomagnetic methods together with refined biostratigraphy, it is becoming increasingly possible to tie even isolated sections into the global Tertiary picture.

Global climate changed significantly during the Tertiary. At first temperatures rose, reaching a peak in the Eocene. From then on there was progressive cooling (see also Fig. 20.8). The first ice sheet appeared in Antarctica in the Early Oligocene,

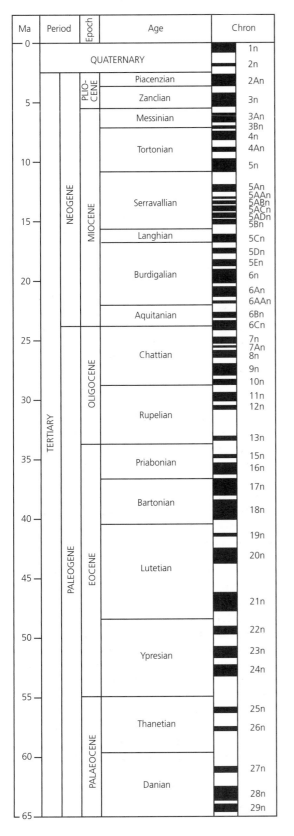

although permanent glaciation did not set in until the middle Miocene, by which time glaciers may also have become established in Greenland. In Britain, this global cooling was reinforced by a northward plate movement of about 8° during the Tertiary. Superimposed on these general trends were short-term climatic variations deduced from stable isotope studies backed up by the analysis of fossil assemblages. For example, against the background of a general warming from the Palaeocene into the Eocene, there appears to have been a very short-lived warming episode around 55.1 Ma, followed by a marked cooling centred around 54.2 Ma. It has been speculated that the former could be a greenhouse warming effect caused by the catastrophic release of gas hydrates from ocean bottom sediments, whereas the latter is likely to result from volcanic activity associated with North Atlantic rifting.

20.3 Plate evolution and Tertiary tectonics

Africa and Europe were converging during the Late Cretaceous, with sea-floor spreading taking place in the Atlantic only as far north as the Labrador Sea, which lies between Greenland and North America (Fig. 20.2). During the Paleocene, Europe–Africa convergence paused and a major hot spot developed in the Faeroe–Greenland area. This mantle plume caused thermal uplift and associated volcanism over a huge area stretching from Britain to the west coast of Greenland. Erosion of what is now the landmass of northern Britain led to the input of coarse clastics into the deep basinal areas of the adjacent North Sea and Faeroe–Shetland Basins (Fig. 20.3). At about the Paleocene–Eocene boundary, during magnetic anomaly 24r (Fig. 20.1), continental rupture took place across this thermal bulge and ocean-floor spreading started between Greenland and Europe. Thus, from Eocene times onwards north-west Europe became part of a thermally subsiding passive continental margin, which moved progressively away from the hot spot that continues today under Iceland.

Far to the south of Britain lay a complex collisional plate boundary stretching from the Pyrenees to the

Fig. 20.1 (*Left.*) Stratigraphic terminology for the Tertiary. For the palaeomagnetic or geomagnetic time-scale shown on the right (data from Cande & Kent 1995) only the normal polarity chrons shown in black are labelled, for example 24n, each of which immediately follows the corresponding reversed polarity chron, for example 24r.

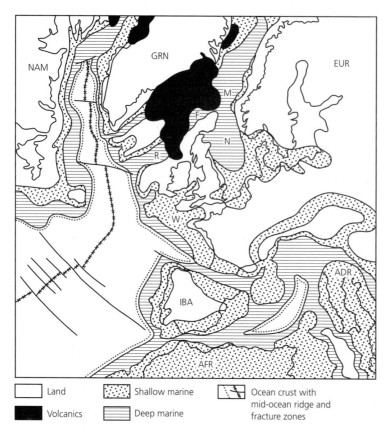

Fig. 20.2 Regional palaeogeography of the North Atlantic region during the Paleocene with present-day coastlines for reference. Abbreviations for tectonic plates: ADR, Adrian; AFR, African; EUR, European; GRN, Greenland; IBA, Iberian; NAM, North American. Abbreviations for basins around Britain: F, Faeroe–Shetland; M, Møre; N, North Sea; R, Rockall Trough; W, Western Approaches (modified from Knott *et al.* 1993).

Alps and beyond. Europe–Africa convergence started again in the Eocene and the distal waves of the Alpine orogeny were felt, in southern Britain, from then onwards through the Oligocene and Miocene. Here, the Tertiary succession is generally thinner than in the offshore areas to the north and there are long gaps in the rock record reflecting uplift and folding (Fig. 20.3).

For the purposes of description, Britain and Ireland can be divided into four broad regions during the Tertiary. The differing evolution of each of these regions reflects both their relationship to the plate boundary processes discussed above and their inherited geological structure. In the north-west, the onshore succession in north-west Scotland and Northern Ireland is dominated by Palaeocene volcanic rocks, whereas offshore a full thick Tertiary section was preserved due to the post-rift collapse of the continental margin. In south-east Britain there is a relatively thin Palaeocene to Oligocene sedimentary section. Here,

there was significant Alpine folding and the Neogene is only patchily developed due to Alpine uplift. Between these two areas lies a central swath, where the structural impact of the plate boundary processes is less dramatic, and which can be further divided into two distinct regions. In the east lies the central and northern North Sea, where a thick and complete Tertiary succession accumulated in an initially deep-water basin inherited from the end-Jurassic rifting event. This region is well known, as it contains major hydrocarbon reservoirs. By contrast, to the west, stretching from North Wales to Devon and through the offshore areas around south-west Britain, the distribution of Tertiary sediments is rather patchy and the stratigraphy is not so well understood.

20.4 North-west Britain and Ireland

By the beginning of the Tertiary, the area between Britain and Greenland had already suffered a long

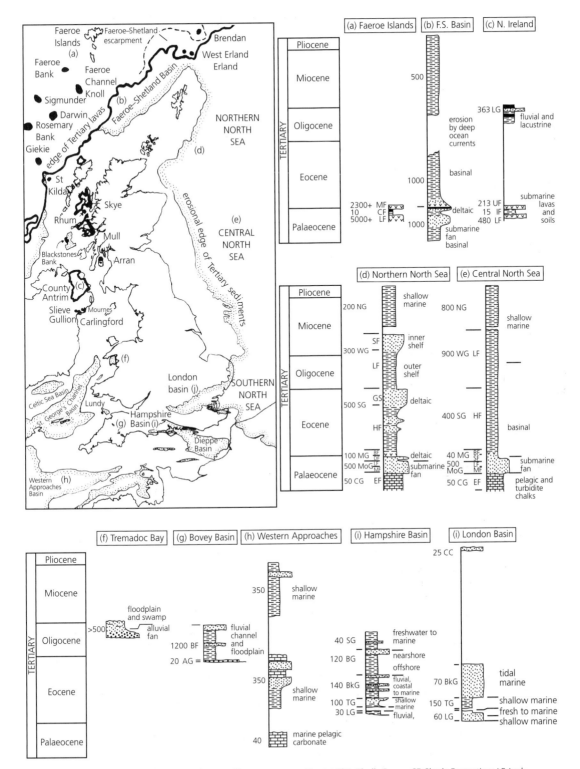

Fig. 20.3 Distribution of Tertiary sediments, lavas and igneous complexes (e.g. Darwin) around the British Isles. See map for location of representative stratigraphic columns. Abbreviations in stratigraphic columns: (a) MF, Middle and Upper Formations; CF, Coal-Bearing Formation; LF, Lower Formation; (c) LG, Lough Neagh Group; UF, Upper Formation; IF, Interbasaltic Formation; LF, Lower Formation; (d,e) NG, Nordland Group; WG, Westray Group; SG, Stronsay Group; MG, Moray Group; MoG, Montrose Group; CG, Chalk Group; SF, Skade Formation; LF, Lark Formation; GS, Grid Sandstone Member; HF, Horda Formation; BF, Balder Formation; DF, Dornoch Formation; SF, Sele Formation; LF, Lista Formation; MF, Maureen Formation; EF, Ekofisk Formation; (g) BF, Bovey Formation; AG, Aller Gravels; (i,j) SG, Solent Group; BG, Barton Group; BkG, Bracklesham Group; TG, Thames Group; LG, Lambeth Group; CC, Coralline Crag.

history of rifting stretching back through the Mesozoic and beyond (see Chapters 16–19). Along a zone running from the Møre and Faeroe–Shetland Basins to the Rockall Trough this rifting, here dominantly of Cretaceous age, had led to the extreme thinning of the continental crust, although sea-floor spreading seems only to have progressed as far north as the Central North Atlantic and Labrador Sea (Fig. 20.2). During the Palaeocene, northern Britain, western Norway and much of Greenland formed part of a huge area of thermal uplift over 2000 km across, resulting from an underlying mantle plume. This plume produced extensive igneous activity from Britain to the west coast of Greenland and up to 2 km of uplift at its centre. At about the Palaeocene–Eocene boundary this zone of uplift finally split, sea-floor spreading was initiated and the Greenland plate started to drift away from Europe to form the Norwegian–Greenland Sea. The volcanism then became concentrated at the new mid-ocean ridge, but the mantle hot spot continues to be active to the present day and is responsible for the continued existence of the volcanic island of Iceland.

Although there may have been some volcanic activity in the rift zone during the Late Cretaceous, the first major effect of the rising mantle plume was to initiate uplift (Fig. 20.4a). Areas of deep-water sedimentation in the Late Cretaceous, such as the Faeroe–Shetland Basin and the Rockall Trough, remained as basins during this uplift. But elsewhere, much of the crust was raised above sea level and subaerially eroded. Onto this surface, extensive sheets of basaltic lava (flood basalts) were erupted. These underlie Tertiary sediments in the area beyond the continental shelf edge north-west of Scotland and are found as erosional remnants both on the continental shelf and onshore (Fig. 20.3). Onshore, the lavas are best developed in the Inner Hebridean islands of Skye and Mull and in County Antrim in Northern Ireland. In the Inner Hebrides the lavas are up to 2 km thick but towards the north-west they become even thicker, reaching over 5 km on the Faeroe Islands. Individual flows are usually between 4 and 8 m thick, but exceptionally may reach tens of metres, and where exposure permits they have been traced laterally for many kilometres. They do not, however, appear to have flowed over the very long distances seen in some other flood basalt provinces. The lavas are interbedded with subaerially deposited pyroclastics

and sediments. The sediments are often plant bearing, and bauxitic to lateritic soil profiles developed on many of the lavas indicate a humid temperate climate. In some areas, where the lavas flowed into deep water such as the Faeroe–Shetland Basin, they became fragmented by reaction with sea water producing prograding clinoform packages of hyaloclastic debris. These now appear as the steep terminations to buried lavas seen on seismic profiles, such as the Faeroe–Shetland Escarpment (Fig. 20.3). These features are found today at depths of up to 3 km, indicating considerable post-Paleocene subsidence. Within the Faeroe–Shetland Basin and the Rockall Trough the igneous activity is manifested as extensive sill complexes.

Palynological data from the sediments associated with the lavas suggest that they were erupted in two main pulses at just over 58 Ma (middle Thanetian) and around 55 Ma (about the Palaeocene–Eocene boundary). For example, the Antrim lavas can be divided into the Lower and Upper Formations with an intervening Interbasaltic Formation. This consists of up to 15 m of soils and other sediments that accumulated in the 3 Myr between the main eruptive events. Locally, this Interbasaltic Formation is split by more lavas including the spectacular columnar jointed basalts of the Giant's Causeway. The 55-Ma date for the younger lavas is consistent with initiation of sea-floor spreading in the Norwegian–Greenland Sea, which started during polarity chron 24r (Fig. 20.1). Following the initiation of spreading, heat dissipation from the mantle plume would increasingly be concentrated at the oceanic ridge.

The lavas were largely erupted from fissures, which are now preserved as dykes. Locally, this magmatism produced central complexes in western Scotland and Northern Ireland, such as Skye, Mull and Slieve Gullion, which expose acid and basic plutonic masses, sills, dykes, sheets and a variety of lavas and pyroclastic sediments. Although these complexes date from the same general time interval as the lavas, they cut, and so are younger than, the adjacent lava field. The central complexes produced their own lava flows, although they are now largely eroded. Beyond the present continental shelf edge there are numerous igneous centres that are even larger than the onshore examples, although their structure is only poorly known from seismic profiles, gravity and magnetic modelling and a few boreholes. However, effects

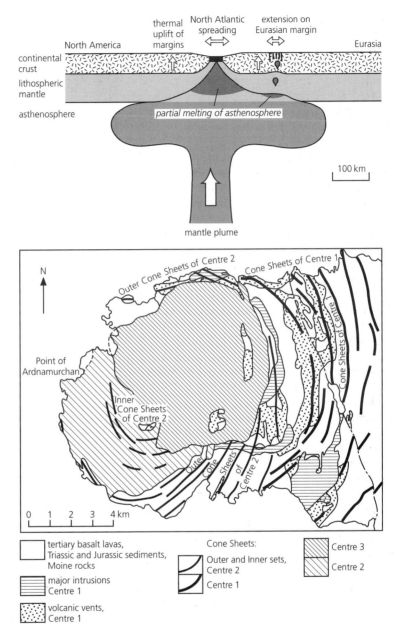

Fig. 20.4 (a) Schematic model showing how evolution of a mantle plume in the North Atlantic region in the early Tertiary resulted in partial melting of the asthenosphere, thermal uplift and extension of the North American and Eurasian continental margins, and development of volcanic centres such as those of western Scotland and County Antrim; (b) geological sketch map of the Tertiary central complex of Ardnamurchan in western Scotland.

of the thermal plume were felt a long way south, as a Tertiary granite makes up Lundy Island in the Bristol Channel (Fig. 20.3).

The central igneous complex of Ardnamurchan is an excellent example of an eroded section through a Tertiary volcanic centre (Fig. 20.4b). Volcanism commenced with eruption of basaltic lavas onto a sub-strate of Mesozoic sediments and Moine basement. This was followed by successive development of three separate volcanic centres, each dominated by mafic rocks and associated with suites of minor intrusions emplaced mainly as concentric, inward-dipping sets of cone-sheets and also as ringdykes. Centre 1 (Fig. 20.4b) was associated with highly explosive activity, involving formation of coarse agglomerates, ignimbrites and subaerial, bedded pyroclastic rocks. Extrusive activity was followed closely by intrusion of numerous microgabbro cone-sheets. These are cross-cut by the hypersthene gabbro and related intrusions which comprise Centre 2 (Fig. 20.4b). The cone sheets associated with Centre 2 comprise both basic and acid magmas which in many places have interacted to form hybrid lithologies. The youngest phase, Centre 3 (Fig. 20.4b), is a major layered mafic pluton dominated by gabbro and cut by microgabbro ring dykes.

Volumetrically, the main product of Tertiary volcanism was olivine tholeiite to alkaline basalts produced by partial melting of the upper mantle. However, in the volcanic centres, the rising basic magmas also assimilated various types of continental crust. This process, together with the complex processes of fractionation and magma mixing, resulted in a rich variety of igneous rock types such as granites and trachytes on Skye, picrites on Rhum and dacites in the Darwin and Erlend centres.

In the basinal areas not covered by flood basalts, such as the Faeroe–Shetland Basin, Upper Cretaceous marine muds pass upwards into similar Paleocene sediments. Thermal uplift is indicated by the influx of sands derived from the east. This influx started in the Danian, but reached its peak in the Mid-Thanetian, the time of the maximum thermal uplift. Some volcaniclastic sediment is included in this influx, but most consists of quartzo–feldspathic material derived both from older sandstones and basement metamorphic rocks. Submarine fan sands, which form the reservoirs of the Foinavon and Schiehallion oilfields, are found in the Mid-Thanetian. These sands are overlain by deltaic sediments with coals showing that sedimentation had filled in much of the Faeroe–Shetland Basin by the end of the Paleocene. The earliest Eocene volcanic activity is here marked only by the Balder Tuff, a unit that is also seen in the North Sea (see Section 20.5).

At the beginning of the Eocene, sea-floor spreading began between Greenland and Europe to the west of the Faeroe Islands and Hatton Bank. The adjacent continental margins then started to cool and subside. In the area of the present inner continental shelf off north-west Scotland, where the crust is of normal thickness, this subsidence eventually returned the sediment surface to near sea level. This area continued to

suffer erosion during the rest of the Tertiary, sediments being preserved only exceptionally in small shallow basins such as the Loch Neagh and Blackstones Basins, which were probably created by local Oligocene tectonic pulses resulting from plate reorganization. From the present outer shelf seawards, where extension in a whole series of rifting episodes had significantly thinned the crust, subsidence was dramatic. As the sea transgressed across the areas of subaerially erupted lavas, shallow-water sediments were briefly deposited before deep-water environments were rapidly created. By the end of the Eocene the area had evolved a complex but generally deep-water topography broadly similar to that existing today. By the Late Eocene to Oligocene the flow of cold ocean water from the Arctic into the Central Atlantic had initiated a pattern of powerful deep ocean currents that ran southwards through the Faeroe–Shetland Basin and Rockall Trough. These currents produced major erosion surfaces or channels within the deep-water sediment pile, such as the one that often separates the Eocene from younger sediments (Fig. 20.3). These currents slowed as the water deepened during the Miocene and Pliocene and deposited thick drifts of fine sediment which, in Rockall Trough, show giant climbing sediment waves. On the continental shelf the Miocene picture is complicated by tectonic uplift and sea-level changes that produced erosion surfaces. However, sedimentation had resumed here by the late Pliocene, causing the shelf edge to prograde by 50 km in places.

Onshore, and on the inner shelf, the only Tertiary sediments younger than the Paleocene lavas are upper Oligocene sediments found in small basins, usually overlying the lavas. Around Lough Neagh in Northern Ireland drilling has proved lacustrine clays, sands and lignite beds, which comprise the Lough Neagh Group. The lignites are tens of metres thick and were formed from plant debris washed into a lake that was at times much larger than the present Lough Neagh. The absence of Eocene sediments between the lavas and the Oligocene sediments shows that there must have been some Oligocene tectonism to produce the enclosed lake basin. Other smaller areas of Oligocene deposits, similar to the Lough Neagh Group, are found both onshore and offshore in the Hebridean region and in southern Britain (see Section 20.7).

20.5 Northern and Central North Sea

The structural framework of the North Sea Tertiary was essentially inherited from the Late Jurassic rifting events, which produced three deep-water basinal areas (the Viking and Central Grabens and the Moray Firth Basins) radiating from the Jurassic volcanic centre (see Chapter 17). Thermal and load-induced subsidence took place during the Cretaceous and although a considerable thickness of sediment was deposited it did not fill in the basin areas. Consequently, at the beginning of the Tertiary, the North Sea was a sediment-starved basin with deep-water troughs over the Jurassic rifts. However, partly as a result of the thermal uplift of northern Britain, by the end of the Tertiary the North Sea Basin had mostly filled with clastic sediments reaching a thickness of over 3 km at the basin centre. Within this fill are extensive Palaeocene submarine fan sands that were shed south-eastwards from the Scottish landmass into the northern and central North Sea. Less extensive Eocene fan sands show a similar dispersal pattern. These Palaeocene and Eocene sands are important clastic hydrocarbon reservoirs in the North Sea where they lie above Late Jurassic source rocks on hydrocarbon migration routes, have trapping geometries and are overlain by sealing mudstones.

The oldest Tertiary sediments are broadly similar to those of the underlying Upper Cretaceous. Danian (Early Palaeocene) pelagic chalks and marls were deposited in the central North Sea (the Ekofisk Formation), whereas further north mudstones of the Shetland Group are found. The thermal uplift of northern Britain during the Thanetian (Late Palaeocene), which led to the main influx of sands into the North Sea during the Tertiary, resulted in a river drainage pattern that funnelled sediment from a large part of the Scottish landmass into the Outer Moray Firth troughs, building a sequence of deltas. Slope failure on the delta fronts produced turbidity currents, which flowed east and south-east to form large submarine fans (Fig. 20.5). The delta slopes did not advance far into the basin, most sediment being by-passed towards the basin centre.

There are three main fan complexes in the Palaeocene, reflecting several pulses of uplift, which show the effect of the progressive filling of a complex, structurally controlled basin. The earliest fan, which deposited the Maureen Sandstone Formation, has a very odd shape controlled by the basin topography (Fig. 20.5a). It commonly incorporates chalk debris resulting from local slope failure. Later fans became more regular and symmetrical as the bathymetry was

Fig. 20.5 Palaeogeographic maps, North Sea fan systems: (a) Maureen Formation, deposited during the early Thanetian; (b) Lista Formation, mid-Thanetian; (c) Lower Sele Formation, late Thanetian; (d) Upper Sele Formation, early Ypresian. X on inset location map is line of section shown in Fig. 20.5 (based on Reynolds 1994).

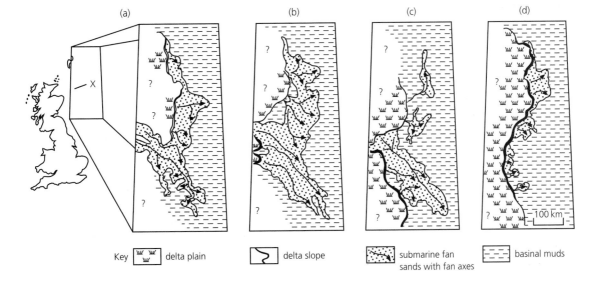

(a)   (b)   (c)   (d)

Key [delta plain]   [delta slope]   [submarine fan sands with fan axes]   [basinal muds]

progressively smoothed by deposition (Fig. 20.5b,c). There are several fan sandstone units within the Lista Formation, one of which in the middle of the formation contains tuffaceous sediment erupted during the older of the two Hebridean volcanic pulses. The lower part of the overlying Sele Formation (Fig. 20.3e) includes the most economically important of the Paleocene fan units, the Forties Sandstone (Fig. 20.5c).

The Lista–Sele boundary is marked by a significant change in mudstone facies. Lista and earlier mudstones tend to be green to grey–green and intensely bioturbated, indicating at least some oxygenation of the sea bed. The mudstones have been so homogenized by burrowing that it is difficult to deduce their origin. By contrast, Sele mudstones are grey, non-bioturbated and preserve their depositional character so well that turbiditic and hemipelagic muds can be clearly distinguished. This shows that anoxic bottom conditions set in at the beginning of Sele times (Late Thanetian) and they persisted until well into the Eocene.

In areas where the initial bathymetric relief was great enough, the fan axes were stacked vertically above each other. Eventually though, the initial depressions were filled and later fans showed patterns with fan axes developing away from the areas of thickest development of the earlier fans. In this way some fans found themselves overlying structural highs over which subsequent compaction formed good traps (e.g. the Forties, Montrose and Arbroath fields). All these Palaeocene fans are composed of beds of turbiditic sands, silts and muds. In proximal areas thick amalgamated sandy units were deposited in fan channels that can be seen on seismic sections. Inter-channel areas are more muddy and thinly bedded. Traced downstream the channels become smaller and the distal fringes of the fans are dominantly sheet like. There are many local complicating factors such as slope failure, soft-sediment deformation and the presence of salt diapirs.

By the beginning of the Eocene, fan aggradation had shallowed the basin to such an extent that it became easier for deltas to advance. The pattern of Eocene sedimentation was therefore rather different from that of the Late Palaeocene. Delta clinoforms, representing prograding delta slopes, are conspicuous throughout the Eocene and lignites are often found in the overlying delta plain deposits (Fig. 20.6). Submarine fan deposits, much less extensive than the Late Palaeocene examples, are found seawards of these deltas.

The Dornoch Formation (Fig. 20.3d) was deposited by the first of these Eocene deltas. This delta sequence coarsens-up from mudstone into sandstone overlain by lignite. It is up to 400 m thick, which, allowing for compaction, approximates the height of the delta slope. Beyond the seaward limit of this delta progradation lay a series of small submarine fans lying within the basinal muds of the upper Sele Formation (Fig. 20.5d). These fan sands pass basinwards into a turbiditic and hemipelagic mudstone succession, which contains abundant thin tuff layers. A major transgression then pushed the sea

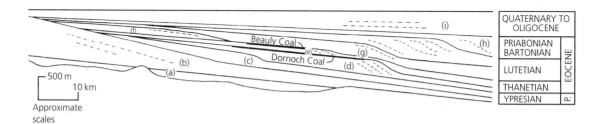

Fig. 20.6 Cross-section through the western edge of the North Sea Tertiary basin in the Outer Moray Firth; line of section shown in Fig. 20.4. Notes: (a) Earliest submarine fan unit, the Maureen Formation, filling in irregular pre-Thanetian topography. (b) Prograding slope of the Lista Formation passing east into submarine fan. (c) Lower Sele Formation delta slope. (d) Delta progrades of the Dornoch Formation capped by the Dornoch Coal Member passing seawards into the deep-water sediments of the Upper Sele Formation. (e) Beauly Member delta with coal passing seawards into the Balder Formation. (f) Early Lutetian delta deposited after maximum transgression. (g,h) Series of progrades marking progressive advance of mid- to late Lutetian deltas. (i) Oligocene to Quaternary section (based on Jones & Milton 1994).

back across this delta top until renewed delta progadation deposited another lignite-topped coarsening-upwards unit, the Beauly Member (Fig. 20.6e). This unit is much thinner than the Dornoch sequence, was deposited in shallower water, and did not prograde as far. In the Viking Graben area, coeval turbidite sands lie to the east in the laterally equivalent Balder Formation. These sands form very small fans, which are important reservoirs in, for example, the Forth and Gryphon oilfields. These fans have a totally different architecture from those seen throughout the Paleocene and up to lower Sele times. They are very small and are composed almost entirely of sand in which it is difficult to discern any bedding. They are commonly associated with enormous sandstone sill and dyke complexes and occur as a variety of odd shapes surrounded by hemipelagic mud. The Balder Formation is also important in that it records the most intense phase of volcanic activity seen in the North Sea. In the lower part of the unit there are hundreds of individual ash layers, mostly only millimetres to centimetres thick but forming a total thickness of over 8 m at the northern end of the North Sea Basin, and known informally as the Balder Tuff. This ash unit is an important marker throughout the North Sea as it produces a distinctive gamma or sonic bow on well logs. The total ash thickness declines toward the south-east, but ashes are found as far away as southern England, Germany and Denmark. The ashes are of tholeiitic–basalt composition and were probably erupted from a large volcano, somewhere along the North Atlantic rift, north-west of Britain.

This change in character of the fans during the early Eocene reflects an important change in evolution of the basin. It is no coincidence that this change took place at about the same time as the deposition of the Balder Tuff, which is thought to date the final phase of pre-rift volcanism along the North Atlantic rift and the initiation of sea-floor spreading. The pre-Balder fans were deposited during phases of repeated thermal uplift in northern Britain, which produced prograding deep-water deltas. Sediment dumped on the deltas was fed directly to the deep-water fans as a consequence of slope failure, without any significant sediment sorting in the coastal zone. From Balder times onwards Britain lay on a continental margin adjacent to ocean crust, which increasingly became the focus of magmatic activity. Pulses of uplift became more subdued, transgressive phases were more important and coastal zones with wide shallow shelves were the norm. Sediments brought into this regime suffered much reworking and sorting, possibly enhanced by tidal currents, so that the sediments supplied to submarine fans consisted of clean sands dumped in canyon heads, the fines being dispersed basinwards as muddy plumes.

Along the western margin of the North Sea Basin, above the Balder Formation, the rest of the Eocene consists of a series of prograding sandy or silty delta systems (Fig. 20.6), which make up the Stronsay Group. These deltas pass eastwards into the basinal muds of the Horda Formation, which contain isolated small sandy fans that form reservoirs in several oil and gas fields (e.g. Frigg and Alba).

By Oligocene times the distinction between shallow-water coastal and deep-water basinal environments had become blurred, as the North Sea Basin had evolved into a saucer-shaped depression without steep marginal slopes. The centre of the basin became progressively shallower throughout the Tertiary as sedimentation outpaced subsidence. Its depth must have been well over a kilometre when the Palaeocene fans were being deposited, but by the early Miocene it was probably less than 200 m. In this now rather featureless basin a monotonous Oligocene to Pliocene succession accumulated. This sequence is dominantly muddy, except towards the edges of the basin where glauconitic, shelly, shallow marine silts and sands are commonly found. The exception to this pattern was during a major eustatic sea-level fall in the late Oligocene (Chattian). Clearly, as the European continental margin subsided following North Atlantic rifting, sediment input became finer and its dispersal became controlled by wind-driven and tidal currents. North Sea sediments were derived largely from Britain until Mid-Miocene times, when input from Scandinavia first became important. Then, from late Miocene times onwards, continental Europe became a major contributor as the ancestors of the Rhine and other major European rivers started to pour sediments in from the south-east.

Although there are unconformities and onlap relationships towards the edges of the central and northern North Sea, resulting from tectonic tilting and eustatic sea-level changes, there were no major phases of uplift or tectonism in the basin centre during the Tertiary. The only widespread feature is a mid-

Miocene unconformity, which could be related to the Alpine compression evident to the south or to the North Atlantic plate reorganization recorded to the north-west.

20.6 South-east England

Although scattered occurrences are found elsewhere, the only major areas of Tertiary sediments preserved onshore in Britain are in the London and Hampshire Basins (Fig. 20.3). The Hampshire Basin extends offshore towards the French coast, where it is known as the Dieppe Basin, and the London Basin continues eastwards into the southern North Sea.

In these basins, in contrast to the northern and central North Sea, sedimentation was punctuated by several pulses of uplift. Most of these pulses can be considered as distant ripples of various phases of Alpine compression. This compression was sufficiently mild that it was largely manifested by movements along pre-existing faults. The way in which the Alpine stresses were propagated as far as southern Britain is far from clear. However, the complexities of the inherited Variscan to Early Cretaceous structural framework were probably responsible for the local variations in Tertiary structural history. The most complete Tertiary successions in southern Britain are preserved in the areas where marine deposition was most continuous: in the southern North Sea, which connected northwards with the rest of the North Sea Basin, and in the Western Approaches, which opened westwards into the North Atlantic. Transgressions advanced into the intervening area from the east and west, where the complex uplift pattern produced a more broken and discontinuous sedimentary record.

The whole of south-east England was affected by gentle uplift during an early Tertiary phase of the Alpine compressional events. Consequently, the lowest Tertiary is missing and Late Palaeocene sediments usually onlap an eroded Chalk surface. At this time, central England, Cornubia and parts of northern France formed very low relief source lands. In the intervening basins a range of shallow to marginal marine and fluvial sediments were deposited during the Paleogene, influenced by local tectonic uplift and eustatic sea-level changes. During the Neogene, uplift was even more significant and sediments are sparsely preserved. In total, the Tertiary of southern

Britain is much thinner than in the central North Sea, although it is over 500 m thick in much of the Hampshire–Dieppe Basin and reaches over 2 km in parts of the southern North Sea.

Early Tertiary uplift led to the non-deposition or erosion of Danian sediments everywhere, except in parts of the Anglo-Dutch Basin. By the late Thanetian, a shallow sea had transgressed west and south across the eroded Chalk into the London Basin. The grey–green mudstones of the Ormsby Clay Formation were deposited passing westwards into nearshore sandy facies of the Thanet Sand Formation (Fig. 20.7). These units are both fully marine, being glauconitic and intensely bioturbated. They are equivalent in part to the North Sea Lista Formation and, similarly, contain evidence of volcanic activity. They are overlain by the Upnor Formation, a higher-energy, possibly tidal, shelf sand with common oyster shells, which oversteps further west onto the Chalk. Although this unit is broadly similar to the underlying Thanet Sand Formation, heavy mineral studies imply a significant change in dispersal systems between the two as the Thanet Sand has a Scottish provenance whereas the Upnor Formation was derived from the south.

The Upnor Formation is unconformably overlain by another package consisting of fine sands, silts and clays (the Woolwich Formation), which passes generally westwards into, but has a complicated inter-digitating relationship with, more sandy facies (the Reading Formation). The Woolwich Formation contains much plant debris and a variety of facies including bioturbated marine sands, brackish-water shell beds and a local freshwater limestone. The Reading Formation is interpreted as fluvial and has suffered much pedogenesis. Clearly, a complicated area of lagoons and tidal channels lay between the rivers draining the land to the west and the open sea to the east. The Upnor and Reading–Woolwich units are equivalent to the North Sea Sele Formation.

The Harwich Formation is a varied group of sediments that rests unconformably on the Reading and Woolwich Formations. It includes offshore shelf mudstones with abundant tuff layers, equivalent to those in the North Sea Balder Formation, and more nearshore silts and sands with local flint pebble beds. Overlying all these units is the thickest and most widespread Tertiary unit in southern Britain, the London Clay Formation. It was deposited during the

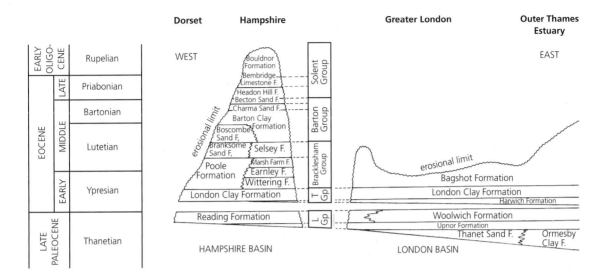

Fig. 20.7 Schematic section through the onshore stratigraphy of the Hampshire and London Basins. L Gp., Lambeth Group; T Gp. Thames Group.

Ypresian and records the most far reaching of the Tertiary transgressive phases (Fig. 20.8a). The sea extended so far landward that marginal sandy facies are poorly preserved. The Formation can be divided into five coarsening-upward divisions, showing that there were repeated transgressive pulses each of which was followed by a period when the shoreline prograded and the sea shallowed. The London Clay Formation contains a rich marine fauna and abundant drifted plant debris that gives a good insight into the character of the surrounding land. It is thought that the Tertiary climate was at its warmest at this time and that the land area of much of Britain was covered by dense tropical to temperate forest bordered by marginal mangrove swamps.

A general regression in the London Basin is indicated in early to mid-Eocene times as the London Clay passes upwards into the nearshore, tidally influenced sands of the upper Ypresian to Lutetian Bagshot Beds. This is the youngest exposed Paleogene unit, although younger Eocene marine sands and clays are found offshore. Oligocene sediments are also found offshore in the UK sector of the southern North Sea but show a patchy distribution unconformably overlying the Eocene. Further east still, the

Oligocene and Miocene are well represented in a series of basins in the onshore and offshore parts of the Netherlands. Here, Tertiary events were very different from those in the London Basin as the Netherlands basins form part of a major system of grabens that extends south-east through Germany towards the Alps. These basins subsided during the Oligocene and Miocene accumulating marine, deltaic and coastal plain sediments. In Germany, the upper Miocene sediments are significant in containing major deposits of lignites or brown coals.

The Tertiary of the Hampshire Basin has broad similarities with, but is not identical to, that of the London Basin. The oldest Paleocene sediments in the Hampshire Basin are younger than in the London Basin and are represented only by the Reading Beds (Fig. 20.7), which here consist of a thin transgressive pebbly marine sand overlain by sands and red-mottled clays. The Reading Beds were deposited by rivers that flowed from the south or south-west, as heavy minerals indicate an Armorican origin, towards the basin centre where lagoonal conditions locally prevailed. Marine Woolwich Formation facies have been found offshore to the south-east. Above the Reading Beds is the marine London Clay Formation. Here the tops of the coarsening-upward divisions are sandy with cross-bedding, suggesting deposition in nearshore tidal channels. Towards the west end of the basin, rootlet-bearing palaeosols occur near the top of the unit, in the Christchurch

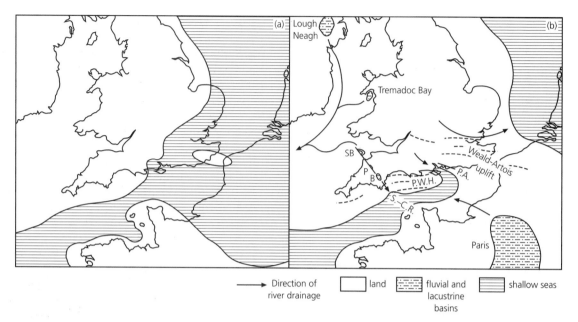

→ Direction of river drainage   ☐ land   ▦ fluvial and lacustrine basins   ▤ shallow seas

Fig. 20.8 Generalized palaeogeographic maps for southern Britain. (a) The early Eocene, a time of relatively high sea levels during deposition of the London Clay Formation. (b) The Oligocene, when the sea was less extensive and widespread fluvio-lacustrine environments are implied by the local preservation of sediments in small basins (the basins shown here were not all in existence at the same time). The Paris Basin was flooded by marine waters at times. Abbreviations for basins along the Sticklepath–Lustleigh Fault: B, Bovey; P, Petrockstow; SB, Stanley Bank. Abbreviations for uplift axes; P.A., Portsdown Anticline; P.W.H., Portland–Wight High; S.–C. R., Start–Cotentin Ridge (based on Murray 1992).

Member, recording the seawards encroachment of tidal flats. Above the London Clay a more complex succession than seen in the London Basin extends through the Eocene, and comprises the Bracklesham, Barton and Solent Groups (Fig. 20.7). These units show a complex interdigitation of marine, estuarine, lagoonal and fluvial sediments, indicating frequent oscillations of the shoreline. The open marine sediments, such as the Earnley and Selsey formations, comprise glauconitic and bioturbated sands, silts and clays with diverse marine faunas, deposited in water depths from shoreface down to a hundred metres or so. These sediments tend to alternate with clays and sands showing fine lamination, plant debris and fresh- to brackish-water faunas. Some of these units were deposited in back-barrier lagoons, others in an estuarine setting (Fig. 20.7a). To the west, coarser cross-bedded fluvial sands dominate, as in the Poole Formation, and evidence of marine influence progressively decreases. The Barton Group shows an overall upwards coarsening from marine clays to shoreface sands and is capped by a palaeosol horizon. Thus, by

Mid- to Late Eocene times, a prograding coastal plain had advanced far into the basin. The Barton Group is overlain by the Headon Hill Formation. This is a largely fresh- to brackish-water unit, deposited in back-barrier lagoons, rivers and lakes, but subjected, in the middle part, to a period of marine inundation. This is followed by the freshwater Bembridge Limestone, which contains terrestrial mollusc and mammal remains, in turn overlain by the Bouldnor Formation, which extends into the Lower Oligocene and is the youngest Tertiary unit exposed in the basin. Yet again, this unit consists of clays and sands deposited in a range of coastal environments indicated by fresh, brackish and marine faunas (Fig. 20.8b) . The heavy mineral suites from throughout the Hampshire succession show that the fully marine units tend to have a Scottish provenance, indicating longshore transport from the north during periods of relatively high sea level, whereas the coastal sediments indicate more direct input from Armorica and Cornubia during regressive phases.

The Tertiary sediments of south-east Britain

were originally much more extensive than they now are. They have suffered significant post-Oligocene uplift and folding, as shown by their locally near-vertical dips in the Portland–Wight Monocline. This uplift is usually ascribed to a late Alpine compressional phase during the Miocene. Some of the normal faults that had accomodated Mesozoic rifting were reactivated as reverse faults (Fig. 20.9). The Upper Cretaceous and Paleogene post-rift strata were draped over the tips of the reverse faults to form E–W trending folds, typically monoclinal (Figs 20.8b, 20.9b). This deformation resulted in uplift of the Portsdown Anticline, the Portland–Wight High, the Weald–Artois Axis and the Start–Cotentin Ridge (Fig. 20.8b). The correlation of transgressive–regressive cycles across the whole area, for example in the London Clay, suggests that sedimentation was strongly influenced by eustatic sea-level changes and that the London and Hampshire Basins formed a continuous seaway. However, there is good evidence from the Isle of Wight, in the form of reworked clasts and fossils, that parts of the Portland–Wight High had already risen above sea level by Selsey Formation times (Fig. 20.7) and continued to rise during the deposition of the Barton Clay. Similar uplift at these times may have affected the other positive axes. Thus, the late Alpine compression cannot be viewed as a single post-depositional event but rather as a series of repeated pulses of uplift that extended from Mid-Eocene to Miocene times.

The facies changes seen in the Hampshire Basin, where, at several levels, fluvial sediments in the west pass eastwards towards the basin centre into marine equivalents, seem to be related to the present-day outline of the basin. Although this could be taken to imply that the original basin margins were related to the present-day outcrop pattern, it must be remembered that the present outcrop is just an erosion remnant. The present basin centres may well have been the original depocentres, but the original distribution of the sediments is open to speculation. The Tertiary palaeogeography of south-east England was clearly dominated by numerous changes in the position of the coastline. Attempts have been made to relate these transgressive–regressive events to world-wide eustatic sea-level changes. Although such changes clearly must have had a significant effect on the pattern of sedimentation, the picture is complicated by the undoubted influence of local tectonic uplift. In these circumstances the reconstruction of palaeogeographies requires both very fine-scale and accurate biostratigraphy and good and extensive three-dimensional outcrop. In the absence of these requirements, very generalized palaeogeographies representing the extremes of high and and low relative sea level are shown here (Fig. 20.8).

What can be concluded from this area is that there was significant uplift and erosion some time during the Late Cretaceous to Palaeocene; shelf seas, bordered by coastal and fluvial environments, transgressed into the area during the Thanetian; local

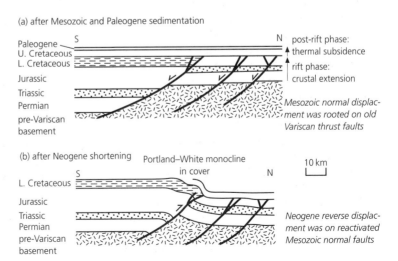

(a) after Mesozoic and Paleogene sedimentation

Paleogene
U. Cretaceous
L. Cretaceous
Jurassic
Triassic
Permian
pre-Variscan basement

post-rift phase: thermal subsidence
rift phase: crustal extension

Mesozoic normal displacment was rooted on old Variscan thrust faults

(b) after Neogene shortening

Portland–White monocline in cover

10 km

L. Cretaceous
Jurassic
Triassic
Permian
pre-Variscan basement

Neogene reverse displacment was on reactivated Mesozoic normal faults

Fig. 20.9 North-south cross-sections through the Portland-White High (Fig. 20.8b). (a) Before and (b) after Neogene deformation.

uplift was taking place by Lutetian times and major uplift and folding took place sometime during the Late Oligocene to Miocene. It is not possible to deduce much about Miocene and Pliocene sedimentation because sediments of this age are very rare. However, ferruginous sands and gravels of Miocene age, the Lenham Beds, are found in solution pipes in the Chalk in Kent and a small area of Pliocene tidal carbonate sands, the Coralline Crag, occurs in Suffolk. These remnants suggest that Miocene and Pliocene sediments were originally quite widely deposited, then removed as a result, respectively, of Late Miocene uplift and Pleistocene erosion.

20.7 South-west Britain

This region contains diverse Tertiary sedimentary basins, some very small, together with erosional remnants or outliers and a wedge of sediment that thickens westwards from the western English Channel and Celtic Sea towards the continental margin (Fig. 20.3). Sedimentation was affected by tectonic pulses that were responsible, not only for Palaeocene and Miocene uplift, but also for Eocene and Oligocene fault movements that varied in intensity and timing across the region.

The best known basins are the Tremadoc Bay Basin in North Wales (Fig. 20.3) and a series of basins lying along the Sticklepath–Lustleigh Fault in Devon and the Bristol Channel (Fig. 20.8b). There are no exposures of the Tremadoc Bay Basin, which lies largely offshore, but boreholes at the basin's eastern edge penetrate an Oligocene to Miocene succession over 500 m thick sitting unconformably on the Jurassic. Conglomerates and sandstones found in the lower part are thought to have been deposited on alluvial fans banked against an active fault scarp defining the eastern edge of the basin. These coarse clastics pass upwards into sands, silts and clays with abundant plant debris and lignite seams, deposited on fluvial floodplains with areas of waterlogged swamp. The basins along the Sticklepath–Lustleigh Fault also contain Oligocene clays, sands and lignites deposited in fluvial channel and swampy floodplain environments, but here the oldest sediments are of Eocene age. The largest of these basins is the Bovey Basin, which, although only 6 km wide, contains over 1 km of Eocene to Oligocene sediments, known as the

Bovey Formation. The upper 300 m of this is known from clay pits and shallow boreholes, the thickness of the concealed lower part being inferred from gravity data. This basin is the main British source of ball clays, kaolinite-rich clays used in the ceramics industry. These clays were produced by the deposition on river floodplains of sediment derived from the erosion of Palaeozoic sediments and granites, weathered in the warm humid Tertiary climate. Some of the lignite was formed from *in situ* plant debris, but most was also washed in from the surrounding vegetated uplands. The preservation of these basins is clearly related to the activity of the Sticklepath–Lustleigh Fault Zone. The older (Eocene), concealed part of the Bovey Basin may have accumulated in a narrow fault-bounded trough, but the younger (Oligocene) section extended well beyond the confines of active faulting and post-dates the main activity of the Sticklepath–Lustleigh Fault. This fault zone has had a complex history including:

1 an early phase of Variscan (late Carboniferous or Early Permian) dextral strike-slip movement;
2 up to 6 km of sinistral movement that produced the Tertiary pull-apart basins at left-stepping offsets in the fault zone; and
3 minor post-Oligocene dextral movements that reversed some of the faults on the basin margins.

Larger areas of Tertiary sediments are preserved offshore in south-west Britain. To the north of Cornubia, in the Celtic Sea and St George's Channel Basins, the Tertiary appears to be broadly similar to that seen in the basins along the Sticklepath–Lustleigh Fault Zone, in that it is largely of Eocene to Oligocene age and contains lignitic sands. To the south, in the western English Channel and Western Approaches, the picture is rather different. Chalk deposition was continuous from the Late Cretaceous into the Danian in the central parts of the Western Approaches Trough, showing that this area was a marine gulf that opened westwards into the open ocean. Elsewhere, the Danian was eroded during a phase of Thanetian uplift that also accounts for the poor representation of Thanetian sediments. Where present, they are thin and consist largely of marine clastics. The Eocene, however, is well represented in the centre of the Western Approaches Trough, consisting predominantly of clays at the base overlain by sandstones and limestones. The clays are partially equivalent to the London Clay, which in south-east Britain records the maximum Tertiary transgression. Therefore, it is probable that during the early Eocene (Ypresian) a seaway extended from the Western Approaches to the North Sea (Fig. 20.8a). Marine conditions continued into the later Eocene but uplift of Cornubia produced an influx of sandy sediments during the Mid-Eocene (Lutetian). More subdued relief on the Armorican side of the Channel resulted in a lack of clastic input and led to the accumulation here of bioclastic limestones.

Whereas the Oligocene is well represented to the north of Cornubia, it is largely absent from the western English Channel and Western Approaches Trough due to a pulse of early Oligocene (Late Eocene to Mid-Oligocene) uplift. Following this uplift, the sea transgressed eastwards from the continental margin. The resulting wedge of shallow marine late Oligocene to Miocene sediments is preserved only on the outer shelf and in the axis of the Western Approaches Trough. Tidal sand ridges can be seen locally on seismic sections in the upper Miocene part of this succession. It is not known how far these Oligocene to Miocene marine sediments originally extended towards the British coast, but Miocene shelly sands are found across much of north-west France. In Brittany, Late Miocene uplift was followed by the deposition of Pliocene shelly sands, similar to the Coralline Crag of Suffolk.

Attempting to synthesize the Tertiary history of this area is not easy, both because stratigraphic data are sparse compared with other regions and because the various pulses of Alpine compression have produced local rather than regional tectonic effects. So, what can be deduced? Firstly, the early Tertiary uplift, also seen in south-east Britain, can here be dated as Thanetian. This event may have produced a fairly widespread upwarping, which removed much of the Danian and Cretaceous cover. Following this uplift, subsidence and high sea levels produced a transgression in the early Eocene through the English Channel, but not to the north of Cornubia. Later in the Eocene, uplift in southern Britain shed coarser clastics both south into the English Channel and northwards into the Celtic Sea. Strike-slip movement along the Sticklepath–Lustleigh Fault Zone opened up small basins, which caught fluvial sediments from rivers draining off the Cornubian uplands. By Oligocene times the climate had become more humid and plant-rich swampy floodplain sediments were being widely deposited in valleys and lowlands, such as in Cornubia and the Tremadoc Bay area, draining into a seaway somewhere to the south-west (Fig. 20.8b). Oligocene uplift caused much erosion but only to the south of Cornubia. Miocene sediments may originally have been fairly extensive, but a fairly general late Miocene phase of uplift caused extensive erosion, reducing the Tertiary cover to a pattern of discontinuous remnants. The Palaeocene and Miocene phases of uplift equate with Alpine tectonic pulses. The Oligocene could relate to Pyrenean compression. The odd one out appears to be the Eocene, which could be related to changes in North Atlantic spreading patterns. It must not be assumed that all the tectonic processes in this region are of Mediterranean origin.

20.8 Discussion

The exposure of Tertiary rocks in Britain and Ireland is relatively poor and although there is much offshore well information, data are very biased towards those areas and horizons of economic interest. Thus, although the global stratigraphic context of the Tertiary is such that there is an enormous potential for high-resolution correlation and environmental analysis, piecing together the fragmentary regional record of Tertiary events is, paradoxically, more difficult than for some earlier periods of Earth history. The variety of the processes (see Fig. 20.10) that have clearly influenced the Tertiary rock record (e.g. continental rifting, thermal uplift, changing climate, sea-level change, compressional tectonics and oceanic circulation) and the effects of the inherited diverse geological structure mean that there is never going to be a simple explanation of Tertiary evolution in Britain and Ireland. Eustatic sea-level change may have been the most important process at certain times in certain areas. Elsewhere, fault-related tectonics or regional thermal uplift may have dominated. The effects of rapid climate change may be difficult to distinguish from tectonics. Notwithstanding the problems, the increasing precision of correlation and dating techniques is the key to disentangling the complex geological history of Tertiary time.

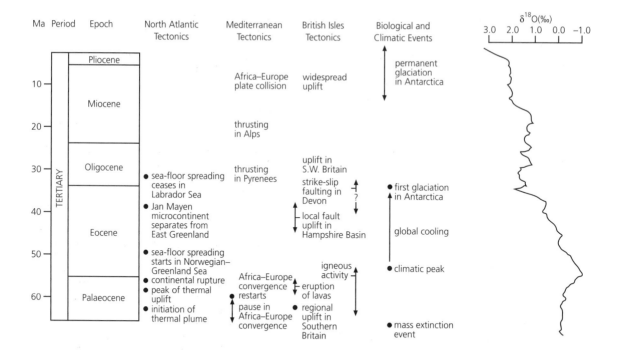

Fig. 20.10 Summary of Tertiary events. Oxygen-isotope curve (after Miller *et al.* 1987) was derived from deep-sea sediments and indicates relative global temperatures.

References

Cande, S.C. & Kent, D.V. (1995) Revised calibration of the geomagnetic polarity timescale for the Late Cretaceous and Cenozoic. *Journal of Geophysical Research* **100**, 6093–6095.

Jones, R.W. & Milton, N.J. (1994) Sequence development during uplift: Palaeogene stratigraphy and relative sea-level history of the Outer Moray Firth, UK North Sea. *Marine and Petroleum Geology* **11**, 157–165.

Knott, S.D., Burchell, M.T., Jolley, E.J. & Fraser, A.J. (1993) Mesozoic to Cenozoic plate reconstructions of the North Atlantic and hydrocarbon plays of the Atlantic margins. In: *Petroleum Geology of Northwest Europe: Proceedings of the 4th Conference* (ed. J. R. Parker), pp. 953–974. Geological Society, London.

Miller, K.G., Fairbanks, R.G. & Mountain, G.S. (1987) Tertiary isotope synthesis, sea level history, and continental margin erosion. *Paleoceanography* **2**, 1–20.

Murray, J.W. (1992) Palaeogene and Neogene. In: *Atlas of Palaeogeography and Lithofacies* (eds J. C. W. Cope, J. K. Ingham & P. F. Rawson), Memoir 13, pp. 141–147. Geological Society, London.

Reynolds, A.D. (1994) Quantitative analysis of submarine fans in the Tertiary of the North Sea Basin. *Marine and Petroleum Geology* **11**, 202–207.

Further reading

Bain, J.S. (1993) Historical overview of exploration of Tertiary plays in the UK North Sea. In: *Petroleum Geology of Northwest Europe: Proceedings of the 4th Conference* (ed. J. R. Parker), pp. 5–13. Geological Society, London. [A brief summary of the history of exploration for Tertiary reservoirs with examples of the variety of trapping mechanisms.]

Beerling, D.J. & Jolley, D.W. (1998) Fossil plants record an atmospheric $^{12}CO_2$ and temperature spike across the Paleocene–Eocene transition in NW Europe. *Journal of the Geological Society, London* **155**, 591–594. [An example of the use of a variety of data types to infer relationships between climate, atmospheric composition and tectonics.]

Bell, B.R. & Jolley, D.W. (1997) Application of palynological data to the chronology of the Palaeogene lava fields of the British Province: implications for magmatic stratigraphy. *Journal of the Geological Society, London* **154**, 701–708. [An important paper on the chronology of Tertiary volcanism.]

Edwards, R.A. (1976) Tertiary sediments and structure of the Bovey Basin, south Devon. *Proceedings of the*

Geologists' Association **87**, 1–26. [A detailed description of the stratigraphy, sedimentology and tectonic context of this basin.]

Gale, A.S., Jeffery, P.A., Huggett, J.M. & Connolly, P. (1999) Eocene inversion history of the Sandown Pericline, Isle of Wight, southern England. *Journal of the Geological Society, London* **156**, 327–339. [Discusses data on derived clasts and fossils, which provide real constraints on the palaeogeography and tectonics of the Hampshire Eocene.]

Holloway, S. & Chadwick, R.A. (1986) The Sticklepath–Lustleigh fault zone: Tertiary sinistral reactivation of a Variscan dextral strike-slip fault. *Journal of the Geological Society, London* **143**, 447–452. [Discusses the evidence for the history of this fault zone, updating earlier views on the sense of motion.]

Knox, R.W., O'B., Corfield, R.M. & Dunay, R.E. (eds) (1996) *Correlation of the Early Paleogene in Northwest Europe*. Special Publication 101. Geological Society, London. [A volume with many examples of high-resolution correlation techniques involving biostratigraphy, palaeomagnetism and sequence stratigraphy.]

Morton, A.C. (1982) Heavy minerals of Hampshire Basin Palaeogene strata. *Geological Magazine* **119**, 463–476. [Describes the use of heavy minerals in palaeogeographic reconstructions.]

Morton, A.C. & Parson, L.M. (eds) (1988) *Early Tertiary Volcanism and the Opening of the NE Atlantic*. Special Publication 39. Geological Society, London. [A volume with many useful papers; for example, White on the mantle plume and Knox & Morton on North Sea tuffs.]

Plint, A.G. (1988) Global eustacy and the Eocene sequence in the Hampshire Basin, England. *Basin Research* **1**, 11–22. [An interesting attempt to relate the alternation of transgressive and regressive episodes to the global eustatic sea-level curve.]

Ryder, G., Fastovsky, D. & Gastner, S. (eds) (1996) *The Cretaceous–Tertiary Event and Other Catastrophes in Earth History*. Geological Society of America Special Paper 307. [Includes several papers on the Cretaceous–Tertiary boundary impact hypothesis; note especially those by Sharpton *et al.* on the Chicxulub structure and Smit *et al.* on boundary sediments around the Gulf of Mexico.]

Stoker, M.S., Hitchen, K. & Graham, C.C. (1993) *United Kingdom Offshore Regional Report: the geology of the Hebrides and West Shetland shelves and adjacent deep-water areas*. HMSO for the British Geological Survey, London. [This is one of 12 reports covering the whole UK offshore area, each including a detailed description of the offshore Tertiary stratigraphy and its relationship to onshore successions.]

Wilson, H.E. & Manning, P.I. (1978) *Geology of the Causeway Coast*, 2 Volumes. Memoir of the Geological Survey of Northern Ireland, Sheet 7, HMSO for the Geological Survey of Northern Ireland, Belfast. [Includes a well-illustrated description of the lavas and intrusive igneous rocks in the Giant's Causeway area of Northern Ireland.]

21   The Quaternary: history of an ice age

N. H. WOODCOCK

21.1 The importance of the Quaternary

It might seem extravagant, after describing some gigayears of geological history, to devote a whole chapter to the last couple of million years. But the Quaternary Period is especially important in four ways.

1 Quaternary history is characterized by the effects of persistent polar ice sheets. Although earlier ice ages have left their mark indirectly, notably during Permo-Carboniferous and Late Ordovician times (see Chapters 7, 12), the Quaternary is the first time since the late Precambrian that Britain and Ireland have been inundated by large ice sheets.

2 Quaternary processes are customarily used as analogues for their counterparts in the geological past. However, the syn-glacial abnormalities of the Quaternary Earth make some such comparisons invalid. Quaternary studies help to constrain which present-day analogues are appropriate to which parts of the geological record.

3 *Homo sapiens* evolved in and lives in the Quaternary. Indeed, it may well have been the repeating glacial–interglacial cycles that stimulated rapid hominid evolution, through their effect on the balance of African rainforest and savanna environments. Without the Quaternary ice age this book might neither have been written nor read.

4 Human activity itself has had an impact on Quaternary environments, first on a local and regional scale, and increasingly on a global scale. Quaternary geology provides the baseline parameters against which to measure present and future anthropogenic global change.

This chapter begins by outlining how Quaternary time is defined and subdivided (see Section 21.2).

However, despite—or perhaps because of—the quality of the data available, Quaternary stratigraphy is fraught with problems. These are examined for the oceanic record (see Section 21.3) and for the onland record (see Section 21.4). The large-scale temporal and spatial patterns are then described (see Section 21.5), before the sequential Quaternary history of Britain and Ireland is charted (see Sections 21.6–21.9). Finally, there is a review of present-day geological processes and a look forward to the geological future of Britain and Ireland (see Section 21.10).

21.2 Definition and subdivision of the Quaternary

It was recognized in the 1830s that characteristic Tertiary sediments in north-west Europe were overlain by deposits derived from ice sheets. In 1839, Charles Lyell proposed the term Pleistocene for these ice-age deposits, which were in turn overlain by 'Recent' deposits regarded as essentially postglacial. The Recent was later renamed the Holocene, and the Pleistocene and Holocene grouped as the Quaternary Period.

The Pliocene–Pleistocene boundary, and therefore the base of the Quaternary, is formally defined in marine sedimentary rocks in southern Italy. The chosen level is where marine faunas, mostly molluscs and foraminifera, record major cooling and climatic deterioration. Palaeomagnetic data suggest that this level is just above the top of the Olduvai Normal Event on the magnetostratigraphic scale (Fig. 21.1b), radiometrically dated from the present igneous ocean crust at about 1.8 Ma. This reference level is widely accepted as the base of the Quaternary in the oceanic

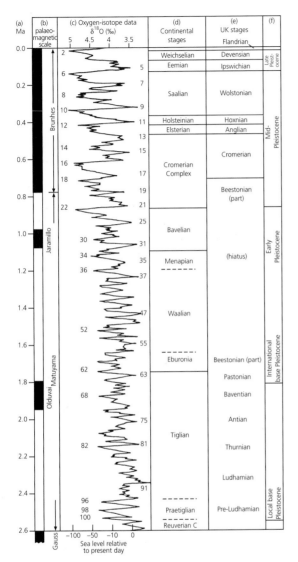

Fig. 21.1 Chart showing key stratigraphic correlation indicators through late Pliocene and Quaternary time. (a) Radiometric age. (b) Geomagnetic reversal time-scale (Valet *et al.* 1993). (c) Oxygen-isotope curve from ODP Site 677, labelled with isotope stage numbers (Shackleton *et al.* 1995) and with an inferred sea-level scale (Funnell 1995). (d) Paleotemperature curve from palaeobotanical data in Holland (Zagwijn 1985). Correlation with stratigraphic stages in the Netherlands (e) and Britain (f) (Funnell 1995). (g) Chronostratigraphy.

sediment record, where it can be correlated readily using the distinctive variation in oxygen-isotope ratios of component microfossils (Fig. 21.1c, see Section 21.3).

The Quaternary in north-west Europe has traditionally been defined, as in the Italian type sequence, by fossil evidence of the marked deterioration of Tertiary climates. Vegetation changes, deduced from preserved pollen and spores, chart an oscillating history of cold and temperate periods back through the Quaternary. However, the onland record shows rapid cooling from Pliocene climates near the top of the Gauss Normal Event of the magnetostratigraphic scale. The first major cold period would occur at about 2.4 Ma, some 600 000 years before the formally defined base to the Italian Quaternary.

Quaternary stratigraphers in north-west Europe have been reluctant to follow stratigraphic protocol and consign to the Tertiary the older (>1.8 Ma) part of their traditional sequence. This reluctance partly reflects adherence to Lyell's original concept of the Pleistocene as the 'Ice Age', a view supported by evidence from ice-rafted debris in the Atlantic Ocean that major Northern Hemisphere ice sheets were indeed initiated as early as 2.4 Ma. However, redefinition of the onland Quaternary in north-west Europe is also inhibited by the practical difficulty in correlating it with the oceanic record and with the Italian type section (Fig. 21.1, see Section 21.3) Until the present tentative correlations are clarified and supplemented, the 'early' (*c.* 2.4 Ma) base to the British and Irish Quaternary is likely to remain in informal use. It is followed in this chapter.

Most plausible correlations with the oceanic record imply substantial time gaps in the onland sequences in Britain and Ireland, and also on their surrounding continental shelves (Fig. 21.1e). It is therefore necessary to seek an uninterrupted Quaternary history in oceanic sediments, before piecing together the more fragmentary continental record.

21.3 The oceanic record: metronome of the Quaternary climate

Pelagic sediments are deposited in the oceans at rates of a few millimetres per century. Their biogenic component is sensitive to changing environmental conditions in the overlying water column. Early attempts to chart these changes used the relative proportions of foraminiferal species known to thrive in warmer or cooler water. Later it was found that the ratio of oxygen isotopes $^{18}O/^{16}O$ in the carbonate of the foraminiferal tests provides an even more sensitive environmental indicator. Part of the measured variation is due to the influence of water temperature on the $^{18}O/^{16}O$ ratio incorporated biologically in the test. However, the dominant effect is the increased tendency for the lighter ^{16}O isotope to evaporate from sea water and to be incorporated in continental ice sheets, inducing an increase in the $^{18}O/^{16}O$ ratio of the remaining ocean water during glacial episodes.

The signal of oxygen isotopes obtained from a long core through oceanic sediments is taken to record the fluctuating global volume of ice sheets (Fig. 21.1c). This hypothesis is particularly reliable if analyses come from benthic foraminifera, which are less affected than pelagic organisms by changes in local water temperature. The $^{18}O/^{16}O$ ratio is conventionally normalized against a standard sample, and shown as a parameter $\delta^{18}O$. On this scale, high values of $\delta^{18}O$ indicate large ice volumes during cold periods, and low values signify warm periods. The isotope curve is also an approximate indicator of global sea level, with high sea levels indicated by low $\delta^{18}O$ (Fig. 21.1c).

A striking feature of oxygen-isotope curves is their regular periodicity through long periods of Neogene time, with a detailed character that can be faithfully matched from one deep-sea core to another (Fig. 21.1c). The cyclic, but individually distinctive, alternations of warm and cold periods provide an ideal stratigraphic correlation tool in the oceans. To this end, the major peaks have been numbered downwards from the present day to give a set of standard oxygen-isotope stages. Even stage numbers signify cold periods and odd numbers the intervening warm periods. However, the most significant feature of the isotopic curve is that its periodicities match those predicted by the Milankovitch theory, that Earth's climate is strongly affected by the regular perturbations of its orbit around the Sun. Before 0.9 Ma, the curve is dominated by a 40 000-year cycle with a subsidiary 20 000-year cycle. These components are attributed to the variation in, respectively, the obliquity of the ecliptic and the precession of the equinoxes. Between 0.9 Ma and 0.6 Ma, the isotopic fluctuation becomes increasingly dominated by a

100 000-year cycle, attributed to changes in the eccentricity of Earth's orbit. The correspondence of the geological record to the predictions of astronomical theory points to the major controlling factor on the temporal pattern of Quaternary ice ages, if not to their ultimate origin.

The oceanic record would be of limited use in explaining the onland Quaternary record in Britain and Ireland if the two environments could not be stratigraphically correlated. This correlation cannot be achieved directly, because of the lack of suitable organisms or marker lithologies in common. The most helpful indirect method is magnetostratigraphy, particularly using the normal or reversed polarity of the remanent magnetization in Quaternary sediments (Fig. 21.1b). However, the usefulness of this technique is limited by the unsuitability of most non-marine sediments for palaeomagnetic determination and by the low stratigraphic resolution of the magnetic reversal record. Failing this, correlation must be attempted by chronometric dating, using one of the techniques listed in Table 21.1. The good radiometric calibration of the magnetostratigraphic scale means that oceanic sequences can be dated indirectly, but again data from continental successions are much more difficult to obtain. The correlations (Fig. 21.1) between oceanic and shallow marine or continental successions are therefore tentative, except at a few fixed points, particularly in the early and mid-Pleistocene.

21.4 Evidence from the onshore and shelf Quaternary

Compared with the oceanic sediment record, continental and shallow marine Quaternary successions are more laterally discontinuous and vertically more prone to unconformities representing non-deposition or erosion. They also have too much lithological variability to allow a continuous signal such as the oxygen-isotope variation to be easily extracted. The detailed fluctuations in Quaternary climate are correspondingly more difficult to discern from the continental or shallow marine record. However, the larger fluctuations have been diagnosed by combining a number of different lines of evidence (Fig. 21.2).

The first evidence that the 'Ice Age' was composite rather than a single event came from glacial sequences comprising more than one till, separated by sands and

Table 21.1 The types of evidence used in subdividing and correlating Quaternary time, with an indication of the most important techniques in deposits from different environments

Technique	Type of evidence
Absolute dating techniques	
Carbon ¹⁴C decay	c
Uranium-series decay	
Potassium–argon decay	d
Thermoluminescence	
Sediment laminations	
Tree rings	
Lichenometry	
Beryllium ¹⁰Be accumulation	
Relative dating techniques	
Oxygen-isotope variation	d
Amino-acid diagenesis	
Tephrochronology	
Palaeomagnetism	d, s, c
Flora and fauna	c
Human artefacts	
Biological evidence	
Pollen and spores	c
Plant macrofossils	
Vertebrates	
Molluscs	
Insects	c
Ostracods	
Foraminifera	d
Sedimentary evidence	
Tills	
Fluvioglacial deposits	
Periglacial deposits	
Loess	
Soils and peat	
Sediment body geometry	s
Landforms	
River terraces	
Shorelines	
Glacial trimlines	
Moraines	
Glacial erosive landforms	

Types of Quaternary evidence: key technique in deep-sea (d), shallow marine (s) and continental (c) deposits.

gravels. These apparently interglacial intercalations were particularly convincing when they contained temperate rather than arctic plants and animals. However, macrofossil remains are neither abundant nor well preserved in continental or marginal marine

Quaternary sediments, and the key fossils for constructing a more detailed history were found to be the robust and distinctive pollen remains. Vertical sampling through postulated interglacial deposits revealed a systematic variation in the proportion of different pollen types, interpreted as the sequential development and decline of the vegetation cover (Fig. 21.2). A pre-temperate phase (Substage I) dominated by pine, birch and grasses is replaced by an early-temperate phase (Substage II) including oak, elm, lime, hazel and alder. Slowly migrating forms such as hornbeam join this assemblage in the late-temperate phase (Substage III), before a rapid return to a post-temperate assemblage (Substage IV) similar to that in Substage I.

Pollen diagrams from successive temperate periods differ in detail from each other, enough to identify some interglacials without knowledge of their litho-stratigraphic context. More important, all inter-glacial pollen assemblages contrast sharply with the vegetation record during cold periods. Tree cover was then much reduced to low densities of birch, pine or willow, or replaced altogether by grassland, tundra (with mosses, algae and lichens) or arctic desert. The variation of these palaeobotanical assemblages through vertical sequences of the continental Quaternary defines alternating temperate and cold periods, analogous to those in the oxygen-isotope record in oceanic sediments (Table 21.2c,d).

Eight major alternations of cold and temperate climates have been diagnosed from the record in onshore Britain, mostly in East Anglia and the Midlands. These major stages are commonly labelled as glacials and interglacials, although only for the last three glacials — Anglian, Wolstonian and Devensian — is there unequivocal evidence of tills deposited from extensive ice sheets. Climatic variation on a yet finer scale can be recognized within some parts of the sequence. Warmer interludes, termed interstadials, within glacial periods are characterized by a short-lived return of birch, pine, spruce or herb-dominated vegetation, but without more temperate species. They separate cold periods termed stadials. Four stadial–interstadial cycles have been identified within the British Devensian (see Section 21.8).

Even with this increasing discrimination in the onshore stratigraphy, its resolution does not approach that of the oxygen-isotope record with its 50 or more warm–cold cycles. The 'missing' stages

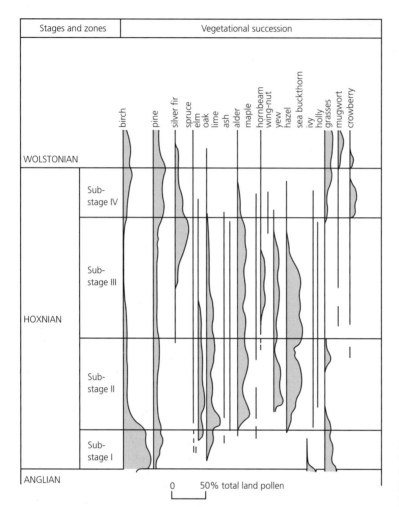

Fig. 21.2 Frequency of the main pollen types in a representative interglacial sequence, the Hoxnian lake deposit at Marks Tey, Essex (after Turner 1970, with permission from the Royal Society 1999).

partly result from the difficulties of detailed sampling in continental deposits, but mainly from the many time gaps in the sequence. Correlation of the British sequence with that in the Netherlands (Fig. 21.1, Table 21.2b,c) has revealed one major unconformity within the British Beestonian. More subtle gaps exist in both the British and Dutch successions. The more complete nature of the Dutch record has favoured its adoption for Quaternary stratigraphy in the North Sea and on the rest of the UK continental shelf. Separate stage names again are used in Ireland, and tentatively correlated with the British record (Table 21.2d).

Pollen has been the most widely used indicator of climate change in the onshore Quaternary sequences,

but multidisciplinary studies have involved a range of other groups. Beetles have proved particularly useful, because they occur in a wide variety of Quaternary habitats, are sensitive to small fluctuations in habitat due to climate change, and migrate rapidly when such changes occur. In the Late Pleistocene, it has proved possible to construct detailed curves of temperature fluctuation from beetle data (see also Fig. 21.8d and Section 21.8).

The petroleum exploration on the continental shelf of Britain and Ireland has greatly increased knowledge of the extensive offshore Quaternary deposits. The available evidence is mainly in the form of seismic reflection profiles, and is therefore different from that of both the oceanic and onshore records.

Table 21.2 The alternating succession of temperate (unshaded) and cold (shaded) stages in the British onshore Quaternary sequences, correlated with those in Ireland and with the Dutch stages used for the UK continental shelf

(a) Epoch	(b) Continental shelf stage	(c) British onshore stage	(d) Irish stage	(e) Isotope stage	(f) Age at base (Ma)
Holocene	Flandrian	Flandrian	Littletonian	1	0.01
Upper Pleistocene	Weichselian	Devensian	Midlandian	2–5d	0.11
	Eemian	Ipswichian		5e	0.15
	Saalian	Wolstonian	Munsterian	6–10	0.25
	Holsteinian	Hoxnian	Gortian	11	0.28
Middle Pleistocene	Elsterian	Anglian		12	0.35
	Cromerian	Cromerian		13	
	Complex	Beestonian (part)		14	0.78
	Bavelian	hiatus		21–31	1.00
	Menapian	hiatus		35	1.10
	Waalian	hiatus		47	1.30
	Eburonian	Beestonian (part)			1.60
	Tiglian C5–6	Pastonian		63	
Lower Pleistocene	Tiglian 4c	Pre-Pastonian/Baventian		68	
	Tiglian C1–4b	Bramertonian/Antian		75	
	Tiglian B	Thurnian		82	
	Tiglian A	Ludhamian		91	2.40
	Praetiglian	Pre-Ludhamian		96	2.50
Pliocene	Reuverian	Waltonian			

The seismic surveys reveal the cross-sectional or three-dimensional geometry of major Quaternary sediment bodies. For example, a schematic section from the southern North Sea (Fig. 21.3) shows a set of prograding bodies (Unit A), interpreted as pro-delta and delta front deposits, overlain by a basinward-thickening wedge (Unit B) interpreted as delta-top deposits. This composite deltaic division is overlain by a non-deltaic division comprising tabular sediment bodies (Units B–G and J) and the lenticular infills to incised scours (Units C and H). Fossil and palaeomagnetic dating of this North Sea section shows that the deltaic division spans the interval up to the Cromerian (middle Pleistocene), and the non-deltaic division comprises the Anglian to Flandrian (Holocene) glacial–interglacial sediments. Dating of some offshore sections is not so reliable, and correlations must be made by lateral tracing of characteristic reflectors.

21.5 Large-scale stratigraphic patterns

The stratigraphic and geographical pattern of Quaternary sedimentation in the British and Irish region (Fig. 21.4) is strongly influenced by three main factors.

1 Evidence of large Pleistocene ice sheets in the region is restricted to Anglian (*c.* 0.45 Ma) and later time, despite evidence for ice-rafted debris in North Atlantic sediments as early as 2.4 Ma. Preserved lower and lower middle Pleistocene sequences therefore record cool–temperate oscillations but lack tills (Fig. 21.4b–e).

2 The main ice sheets of mid- and late Pleistocene time advanced to different limits (Fig. 21.4a). Continental ice or grounded shallow-marine ice naturally tends to remove previously deposited Quaternary sediments, limiting the extent of the preserved record.

3 Tectonic subsidence of the North Sea and the Atlantic continental margin continued with respect to the more stable crustal areas very approximately outlined by the Quaternary, and present-day, coastlines. The subsiding areas were able to accumulate some hundreds of metres of continuous Quaternary section, compared with a few tens of metres of incomplete section in onshore or shallow marine areas.

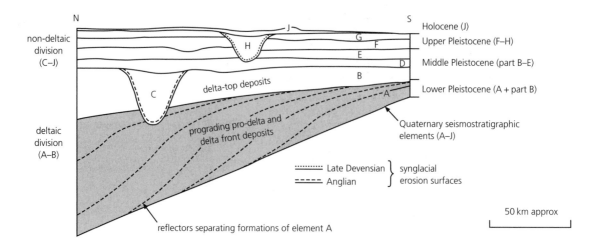

Fig. 21.3 Section across the southern North Sea, to illustrate the seismic stratigraphic elements of the Quaternary sequence (modified from Cameron *et al.* 1992 with permission from the British Geological Society, London).

Selected stratigraphic logs (Fig. 21.4b–e) show the effects of these major influences in different geographical zones of the region:

1 *Within the limit of the late Devensian ice sheet*, laterally discontinuous Devensian glacial deposits typically sit directly on an erosional land surface from which most earlier Quaternary sediments have been removed. The Irish Sea is an example of an exceptional area where thick pre-Devensian deposits are preserved in topographic basins (Fig. 21.4c), formed or deepened by the earlier Anglian ice. Pre-Anglian deposits are generally absent within this zone.

2 *Between the Devensian and Anglian ice limits*, typical sequences preserve Anglian glacial deposits overlain by periglacial or non-glacial lithologies. The Anglian tills overlie an eroded pre-Quaternary land surface, or sometimes relict patches of lower and lower middle Quaternary sediments. East Anglia contains the most complete pre-Anglian sequences (Fig. 21.4d), preserved in an area of enhanced subsidence bordering the North Sea Basin.

3 *South of the Anglian ice limit*, glacial deposits are absent. However, extensive remnants of a periglacial landscape are preserved, which began developing in Early Pleistocene or Late Neogene time.

4 *The North Sea Basin* accommodated some hundreds of metres of pre-Anglian sediment by continued tectonic subsidence (Fig. 21.4e). This is the succession illustrated on the seismic cross-section (Fig. 21.3) and described above. Part of the southern North Sea was affected by the southern edge of the Anglian (Elsterian) ice, the eastern edge of the Devensian ice, and the western edge of the Wolstonian (Saalian) ice sheet that advanced from southern Scandinavia.

5 *The North Atlantic margin* was mostly too deep to host grounded ice sheets, and glacial events are marked not by tills but by ice-rafted debris in the mudstones of the slope apron. However, the outer shelf was reached by ice, and there Anglian and Devensian tills and interglacial sediments overlie the thinned edge of the lower to middle Pleistocene slope apron sediments (Fig. 21.4b).

The Quaternary geological record is not simply that of stratigraphic sequences due to rock accumulation, as it is in most of the pre-Quaternary record. Important components of Quaternary history can be deduced from geomorphological evidence. This evidence comes from depositional, deformational or erosional landforms produced by glacial action, by fringing periglacial processes, or by the temperate processes operating during interglacial stages. This evidence will be referred to at appropriate points in the story of Quaternary history, which can now be documented.

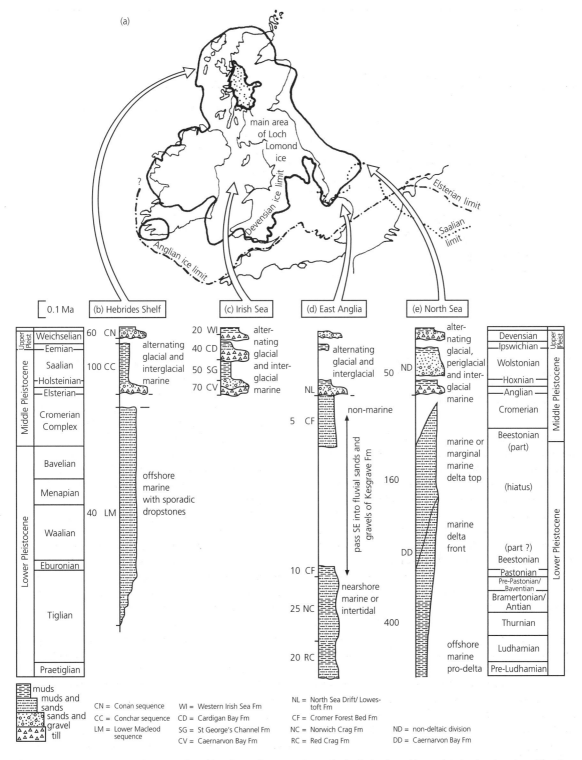

Fig. 21.4 Representative Quaternary stratigraphic columns for the British Isles and continental shelf, with ice limits of major glaciations where known. The area covered by the late Devensian ice is dominated by erosional rather than depositional features, typically with thin, patchy Devensian and Holocene deposits unconformable on pre-Quaternary rocks.

21.6 Pre-Anglian history

Only a fragmentary record of the first two million years of Quaternary time has been left onland in the British Isles, mostly in East Anglia. More substantial pre-Anglian (lower and lower middle Pleistocene) sequences are preserved on the North Atlantic margin and in the North Sea. Few of these sequences show evidence for deposition from ice sheets, which must have been restricted to upland areas where their effects were removed during the more intense Anglian to Devensian glaciations. The pre-Anglian history of the region is therefore interpreted in terms of the alternating cold and temperate conditions that followed the climatic deterioration of the earliest Pleistocene (Table 21.2).

The evidence that the onland record of marginal marine sediments in East Anglia is incomplete has already been discussed (see Section 21.4). This sequence thickens eastwards into the more continuous early to mid-Pleistocene sequence in the southern North Sea (Fig. 21.4e). Here the northward outbuilding of the major Ur-Frisia Delta is well displayed in seismic sections (Fig. 21.3) and by the progressive retreat of its marine shoreline (Fig. 21.5). The delta was fed by large sediment-laden rivers from the east and south-south-east—the Baltic River and ancestral Rhine, respectively—with smaller contributions from English rivers. Until it was diverted by the Anglian ice, the Thames flowed north-eastwards across East Anglia, depositing a sheet of fluvial sands and gravels. Early in the Mid-Pleistocene, during 'Cromerian Complex' time, sediment supply from these rivers became insufficient to match the subsidence of the delta, and its top was flooded by a shallow sea. The marine shoreline advanced southwards before its major retreat into the Anglian Glacial Stage.

Followed northwards in the North Sea, the pre-Anglian deltaic deposits pass into shelfal marine muds with sporadic sandy and shelly horizons. These muds pass north-westwards in turn into the prograding wedge of outer shelf and slope sands and muds deposited on the North Atlantic margin (Fig. 21.4b). These sequences are important in providing evidence of a pre-Anglian ice sheet in Scotland. Dropstones in the continental margin sediments represent debris transported by icebergs, although not necessarily from ice on the British or Irish landmasses. More

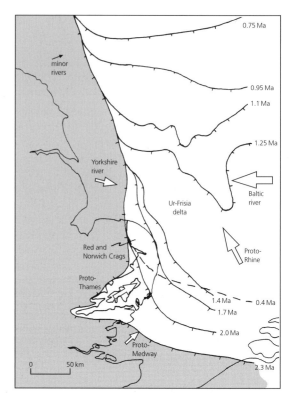

Fig. 21.5 Retreating early Pleistocene shorelines caused by northward progradation of the Ur-Frisia Delta into the southern North Sea (modified from Cameron *et al.* 1992, with permission from the British Geological Survey 2000).

definitive are subglacial tills and glaciomarine sediments in lowest middle Pleistocene successions in the Moray Firth and Forth Approaches, with clast types that indicate a local Scottish origin for the parent ice sheets.

Evidence for Early Pleistocene ice sheets in Britain and Ireland is equivocal, comprising exotic clasts within Beestonian sediments. However, the upland locations of presumed pre-Anglian ice sheets are precisely those most affected by the erosive effects of the later Anglian to Devensian ice, and preservation of evidence is therefore unlikely.

The earliest evidence for hominid colonization in southern England comprises artefacts from Boxgrove (Sussex) associated with Late Cromerian faunas. These records are consistent with evidence for early man in mainland Europe from about 500 000 years ago. However, amino acid studies have cast some doubt on the age of the Boxgrove fauna, suggesting that it may post-date the Anglian glaciation.

21.7 Anglian to Ipswichian: the older glaciations and temperate interludes

Whatever ice sheets existed on upland Britain or Ireland during the cold stages of the Early and early Mid-Pleistocene, they were small in comparison with those of the later part of mid-Pleistocene time. By the time of the Ipswichian Interglacial Stage (early late Pleistocene), ice sheets had affected the whole of Ireland, the Irish and Celtic Seas, and Britain as far south as the Severn and Thames estuaries (see also Figs 21.4, 21.7). The limits of this ice around northern Britain are less clear, because the evidence has been removed or obscured by subsequent Devensian effects. The Mid-Pleistocene ice must have reached close to the shelf edge along the Atlantic margin, and erosion surfaces in the North Sea suggest that, at times, it amalgamated there with Scandinavian ice flowing from the east. Even the Devensian ice of the last glaciation was not as extensive as that in Anglian time (see also Fig. 21.8).

The deposits of these extensive older ice sheets reveal a number of climatic alternations, the details

and correlation of which are still debated. The traditional pre-Devensian British sequence contains two cold stages, the Anglian and Wolstonian, and two interglacials, the Hoxnian and Ipswichian (Table 21.2). The reality of an Anglian glaciation is not in doubt, but the severity of the Wolstonian climate has been questioned. However, four lines of evidence point to two major glaciations.

1 Two clear horizons of pre-Devensian till or glaciomarine sediments occur in offshore sequences in the central and northern North Sea and in the Irish Sea. The tills are separated by interglacial deposits correlated with the Hoxnian.

2 Extensive erosion surfaces, including elongate valleys, were cut during the Anglian and Wolstonian stages in the same offshore areas. Although these surfaces could represent normal fluvial erosion during the times of glacially lowered sea level, the deep valleys are thought to record localized scour by ice, by meltwater below ice, or by sporadic floods of released meltwater ahead of the ice front (Fig. 21.6).

3 The Welton Till and its equivalents, in Lincolnshire and east Yorkshire, overlie Hoxnian deposits and are overlain in turn by Ipswichian sediments.

4 A widespread glaciation, including several ice advances between Hoxnian (Holsteinian) and Ipswichian (Eemian) time, is well documented in the Netherlands, North Germany and Denmark. The ice limit in the Netherlands for this Saalian glaciation can be extrapolated into eastern England south of the Welton Till (Figs 21.4, 21.7).

Fig. 21.6 Cross-section of the Quaternary sequences in the northern North Sea, showing probable syn-glacial erosion surfaces with major scours. The vertical exaggeration is about ×330 (after Johnson *et al.* 1993, with permission from the British Geological Survey 2000).

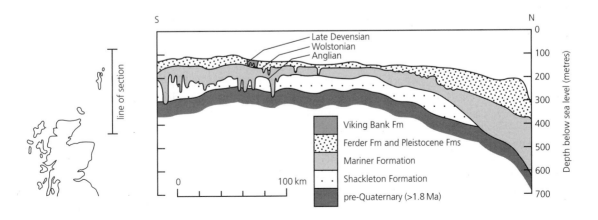

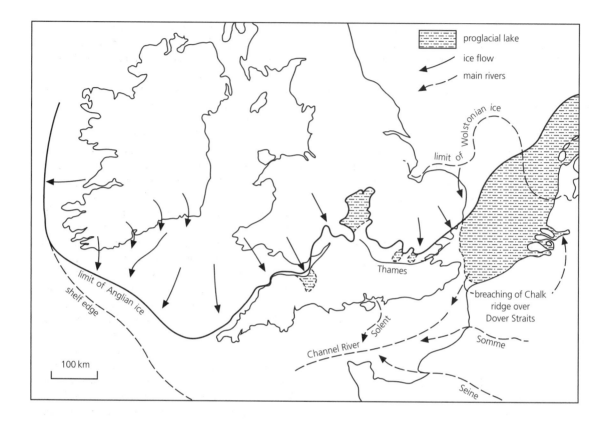

Fig. 21.7 Palaeogeographic map showing features at the southern margins of the Anglian and Wolstonian ice sheets (based on Gibbard 1988, with permission from the British Geological Survey 2000).

Even if they are accepted the arguments for a Wolstonian ice sheet over part of Britain, there remain major problems about its extent and timing.

1 The tills of the English Midlands, which provided the original type sections for the Wolstonian, may in fact correlate with the Anglian tills in East Anglia and, confusingly, may not therefore be of Wolstonian age. There is some doubt whether the extent of maximum glaciation in the Bristol Channel and western England is an Anglian or a Wolstonian ice limit (Fig. 21.7).

2 The existence of tills of Wolstonian age in Ireland is disputed. Irish tills that pre-date the last glaciation have been termed Munsterian, but some of these are now thought to correlate with the Anglian of Britain rather than with the Wolstonian.

3 As many as three stadials and two interstadials have been distinguished between the Hoxnian and the Ipswichian interglacials, both in England and on the near-continent. The relative climatic severity of these stadials is unclear, as is their correlation from Britain to the continent. Simple correlation of Wolstonian deposits in Britain with Saalian deposits on the continent is premature.

Despite the uncertainty about the older glaciations, some of their effects on the palaeogeography of southern Britain are well established (Fig. 21.7). As the Anglian ice sheet expanded southwards through eastern England, it overran the valleys of the Thames and its northern tributaries, and diverted them further south, closer to their present course. Even more significantly, ice advancing over the low-relief plain of the former Ur-Frisia Delta dammed the flow from the Thames and the Rhine and formed a large pro-glacial lake in the southern North Sea. The lowest outflow from this new lake was south-westwards, over low points of the Chalk ridge then joining south-east England with France, and on down the 'Channel River' (Fig. 21.7). Deep channels were eroded at the overflow site, initiating what are now the Straits of Dover. This feature not only determined future drainage patterns during glacial stages, but provided a direct marine connection between the Atlantic Ocean and the North Sea during interglacial times.

The Hoxnian Interglacial provides unequivocal evidence for hominid colonization of Britain in the form of artefacts and skull fragments of Neanderthals (*Homo sapiens neanderthalensis*). The rarity of artefacts during the succeeding Wolstonian cold stage is explained by its climatic inclemency. More puzzling is the apparent lack of hominid recolonization of Britain during the Ipswichian temperate stage. One explanation is that the newly flooded Straits of Dover were, by then, wide enough to bar the way to Neanderthals with little or no skill in building boats.

21.8 Devensian glaciations

Although Devensian ice sheets were not quite so extensive as those of the Anglian glaciation, they have naturally left the fullest record of erosion and deposition of any Pleistocene glaciation. Most periglacial features in Britain and Ireland are also of Devensian age. This 'last Ice Age' therefore provides a valuable analogue for earlier glacial events whose effects it has partly or wholly removed, covered or degraded. However, the Devensian record is itself tantalizingly incomplete, mainly because the most intense of a number of component stadial–interstadial cycles came in the Late Devensian, thus overprinting earlier Devensian evidence.

The climatic history of Early Devensian time is deduced by comparing the oxygen-isotope data from oceanic sediments (Fig. 21.8c) with local temperature estimates from pollen (Fig. 21.8d). The global ice volume increased rapidly from the Ipswichian Interglacial (isotope stage 5e) into the Devensian, at about 120 ka BP, with local temperatures showing a similarly rapid fall. Ice volumes fluctuated through stages 5a–5d and are mirrored by marked local temperature changes. Two interstadials—the Chelford and Brimpton—are distinguished in this interval, and tentatively correlated with stages 5c and 5a of the oceanic record. A climatic deterioration into isotope stage 4, between 80 and 70 ka BP, is indicated by both isotopic and pollen evidence. However, there are no firmly dated glacial deposits of this age. Most suggestive are the Early Devensian glaciomarine sediments in the North Sea (Fig. 21.4e) and periglacial structures in the English Midlands that pre-date deposits of the Upton Warren Interstadial.

The Mid- and Late Devensian climate is recorded by isotopic data, but in more detail by local temperature estimates from beetle faunas (Fig. 21.8d). Isotope stage 3 (about 60–30 ka BP) approximately corresponds to the Upton Warren Interstadial, before climate deteriorated again to a low point at 20 ka BP (isotope stage 2). This Dimlington Stadial marks the climax of the Devensian Glacial period, and is the probable age of its maximum ice limit in most areas (Fig. 21.9). Ice was not extensive enough to join the Scandinavian ice sheet across the North Sea Basin. Instead, lowered sea level left a wide plain affected by permafrost over much of the southern and central North Sea Basin. Further north, deeper parts of the basin retained marine water, but with a cover of sea ice. The Devensian ice reached less far south than its Anglian precursor, but it modified drainage patterns in analogous ways. The rivers Thames and Rhine again flowed out of the North Sea Basin through the Straits of Dover, whereas drainage from the English Midlands was blocked by ice to form proglacial lakes in eastern England (Fig. 21.9). Arctic desert conditions near the ice front allowed windblown sediment to be concentrated as patches of cover-sand or silt-grade loess. Structures indicating extensive permafrost developed over most of the ice-free area of southern England and Ireland, only the south-west peninsula of England being spared its severest effects.

The melting of the Dimlington Stadial ice between 17 and 13 ka BP was spasmodic, on the evidence of lines of moraines that seem to represent stasis or readvance of the ice within the overall ice limits. One of the better substantiated lines is that of the so-called Scottish Readvance in north-west England and the Drumlin Readvance in Ireland (Fig. 21.9), but many other lines are of controversial significance and age. More certain is that a Windermere Interstadial, between 13 and 11 ka BP, intervened between the main Dimlington Stadial and the Loch Lomond Stadial (Fig. 21.8). By 12.5 ka BP, the climate of England and Wales was as warm as it is today, although more continental in character.

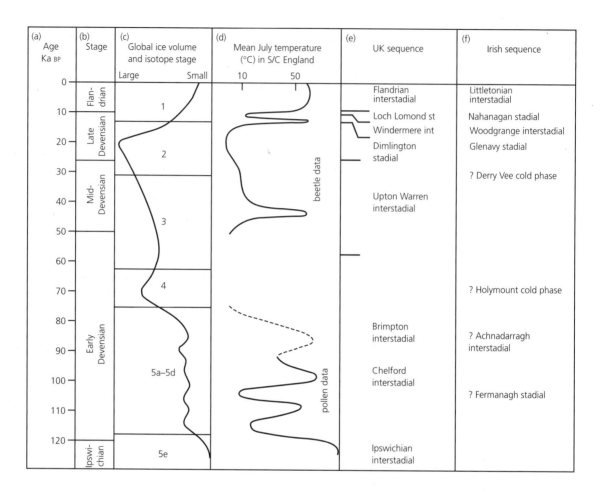

Fig. 21.8 Estimates of temperature and ice volume through the Devensian and Flandrian, correlated with stadial and interstadial period temperature data from Coope (1977) and West (1977); ice volumes from Shackleton & Opdyke (1973).

The Loch Lomond Stadial, from 11 to 10 ka BP, is clearly marked by the beetle faunas, but barely so in the isotope record (Fig. 21.8). It may therefore not reflect a global cooling, but rather the transitory effect of large volumes of meltwater being introduced into the North Atlantic from the disintegrating Laurentide and Scandinavian ice sheets. A large ice mass developed in west Scotland, and smaller cirque and valley glaciers in the uplands of southern Scotland, north-west England, Ireland and Wales (Fig. 21.9). Outside the Loch Lomond ice limits, a new phase

of periglaciation was overprinted on that from the Dimlington Stadial.

Land links between Britain and mainland Europe were re-established by the lowering of sea level during the Devensian glaciations. The pioneers were the Neanderthals, but the first skeletal remains are of modern humans (*Homo sapiens sapiens*) dated at about 30 ka BP. These colonists were seemingly undeterred by the deteriorating climate, and human remains dated close to the climax of the Dimlington ice advance have been found in caves only a few kilometres from the ice front in South Wales. Migration to Ireland was inhibited first by sea and then, through the Dimlington climax, by the Irish Sea ice. However, a brief land connection was probably established between 18 and 14 ka BP, as this ice melted and while

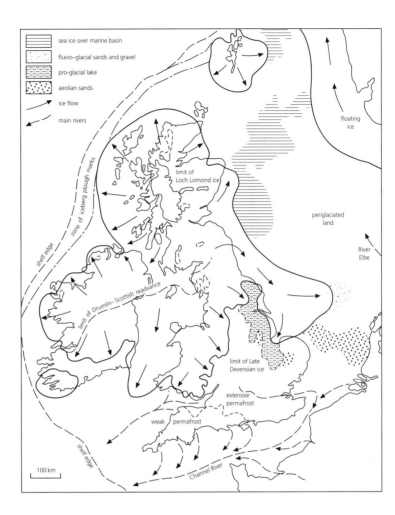

Fig. 21.9 Palaeogeography during the Devensian glaciation.

global sea level was still rising. It is therefore unclear why the first evidence for human activity in Ireland is delayed until into the Holocene.

21.9 The Holocene: the present interglacial

The beginning of the Holocene is taken at 10 ka BP, coincident with the temperature rise that marked the start of the Flandrian Interstadial (Fig. 21.8). Warm waters of the North Atlantic Drift returned rapidly to the British and Irish coasts at this time, and caused dramatic onland warming of up to 1°C per decade. Within only two centuries the climate in southern Britain had returned to that of the Windermere Interstadial. These conditions have essentially persisted to the present day, punctuated by brief cool periods from

5.5 to 5 ka BP, around 2.5 ka BP, and in the exaggeratedly named 'Little Ice Age' of about AD 1600–1830.

The most marked Holocene palaeogeographic effects have been shifts in coastlines around the British Isles. These shifts reflect a complex interplay of two consequences of deglaciation operating somewhat out of phase.

1 *Eustatic sea level* had been rising since about 20 ka BP, due to progressive melting of the large onland ice sheets remote from the British Isles. The eustatic sea-level rise had a direct and instantaneous correlation with the ice volume curve (Fig. 21.8a). It shows only a minor arrest through the time of the Loch Lomond Stadial.

2 *Isostatic rebound* was occurring in proportion to the amount of ice overburden that had depressed the

lithosphere in northern Britain. Isostatic effects are more localized than eustatic effects. Also, rather than being coincident with ice melting, they are protracted over the thousands of years required for viscous asthenosphere to flow back beneath the rising lithosphere. This mantle flow in turn depletes a peripheral bulge formed in the unglaciated areas around the ice sheet, causing lithospheric subsidence in southern Britain and Ireland. Isostatic effects are also caused by the renewed loading of sea water in marine basins, and by the residual thermal subsidence of the North Sea basin.

The differing balance of isostatic and eustatic effects has produced contrasting curves of local sea-level change in northern and southern Britain (Fig. 21.10). Most coastal sites in England and Wales record a fast sea-level rise of up to 10 mm/year between 10 ka and 6 ka BP, decaying smoothly towards about 1 mm/year over recent millennia. These southern curves are close to estimates of eustatic sea-level change. By contrast, coastal sites in Scotland record a net sea-level fall since the end of the Dimlington Stadial, because isostatic rebound in this heavily ice-loaded region has exceeded eustatic sea-

level rise. However, the ice load during the Loch Lomond Stadial temporarily reversed this trend and allowed relative sea level to rise in the period 9 ka to 6 ka BP.

The regionally variable sea levels during the 'Flandrian transgression' have produced a complex shift in shorelines around the British Isles (Fig. 21.11). These have been recorded directly from raised beaches in the north and, less readily, from submerged cliff lines in the south. Elsewhere shorelines can be estimated from theoretical crustal rebound models. It is notable that a land bridge with Ireland may have existed until 12 ka BP, that the southern North Sea remained emergent until about 10 ka BP and that Britain was isolated from France only at about 7 ka.

The improving climate during early Holocene time allowed the interglacial vegetation of birch, pine, hazel, elm, oak and alder to re-establish itself. This progressive floral invasion continued as it had in other interglacials (such as the Hoxnian, Fig. 21.2) until about 5.5 ka BP (3500 BC), when there was a sudden decline in the proportion of elm and a general decrease in the ratio of trees to herbaceous plants. These unnatural changes reflect forest clearances

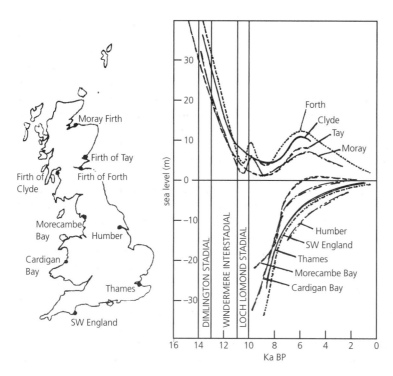

Fig. 21.10 Curves of local sea level relative to the present day at sites around the British coast (modified from Lambeck 1995, with permission from the Geological Society, London 1999).

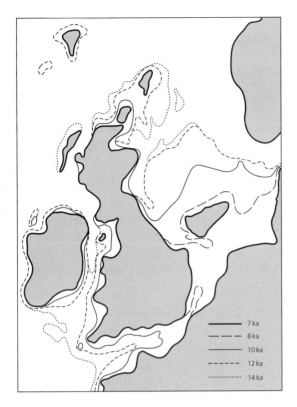

Fig. 21.11 Positions of successive coastlines around Britain and Ireland during the Flandrian (Holocene) transgression, from 14 ka to 7 ka BP (data from Lambeck 1995).

associated with early Neolithic human activity. They are followed by the appearance of cereal seeds, and other evidence of agricultural development. This human impact marks a turning point in the geological history of Britain and Ireland, at least as seen from a human perspective. In the five millennia that have followed, almost no part of the landscape and environment of Britain or Ireland has been spared human influence. Even if *Homo sapiens* is seen as just another species in a long geological record, never before has a single species had so great an impact on its contemporary geological record.

21.10 Present processes and future trends

The reconstruction of a regional geological history, such as has been attempted in this book, might once have been seen as a purely academic pursuit; something of intrinsic interest but of no practical use. This

perception has changed since the scale of human impact on regional and indeed global environments has been better appreciated. Predicting future threats such as climate change, sea-level rise or coastal erosion means extrapolating from the present-day rates and spatial organization of these variables. Estimating this baseline of present change means, in turn, comparing modern processes and geography with their counterparts in the geological past. Because of humanity's understandable preoccupation with threats on a geologically short time-scale, the most relevant geological period is the Quaternary. However, some of the human-induced perturbations to the global environment may be on a scale unrepresented in the last two million years, and we must then seek analogues in the deeper past. Yet other events of the geological future are indeed of academic interest, in that *Homo sapiens* will be long gone before these events happen.

Compiling the 'natural' modern baseline against which to measure future change is already difficult for some environmental variables, so great has been the historical impact of human activity. Some geological parameters are more robust, driven as they are by processes too forceful for human upset (Fig. 21.12):

1 *Relative sea-level change* has been estimated by fitting theoretical isostatic and eustatic models to recorded data (Fig. 21.10). An isobase map (Fig. 21.12) for 7 ka BP shows the height to which a shoreline cut at that time would now be elevated (positive values) or depressed (negative values). Averaged rates over this period reach 1.5 mm/year of sea-level fall in western Scotland, 3 mm/year of sea-level rise in East Anglia and 5 mm/year of sea-level rise in the central North Sea. Because the post-glacial sea-level changes are attenuating with time (Fig. 21.10), the present-day rates are about half to one-third of these averages.

2 *Earthquake activity* is one measure of the present tectonic deformation of the crust. A seismic hazard map (Fig. 21.12) shows the earthquake intensity that has a less than 10% chance of occurring over the next 50 years, based on the historical and instrumental earthquake record. The maximum risk is of intensity 6, at which some damage to buildings can occur. The hazard is greatest in Wales, north-west England and west Scotland, and least in Ireland, where historical epicentres are virtually unrecorded. This curious pattern, with little relationship to the gross crustal

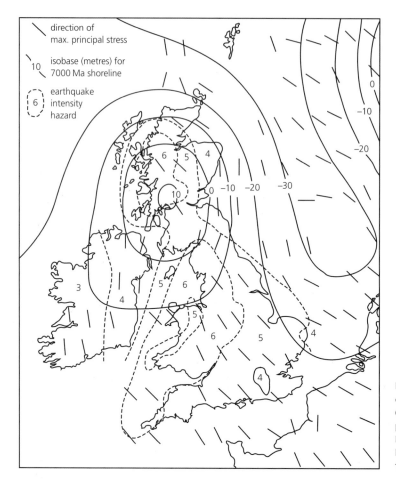

Fig. 21.12 Selected indicators of modern crustal deformation in Britain and Ireland (isobases from Lambeck 1995, with permission from the Geological Society, London (1999); stress directions from Becker & Paladini 1992; seismic hazard from Musson & Winter 1994).

zonation of Britain and Ireland, has yet to be explained.

3 *Crustal stress directions* give a clue to the forces currently causing the crust to deform, both along earthquake faults and by aseismic creep. North-westerly directions of the maximum compressive stress across Britain and Ireland (Fig. 21.12) are interpreted as the effect of the gravitationally driven push from the North Atlantic ridge crest. The stresses are weak compared with those capable of causing orogenic deformation.

Crustal stress and seismic activity are two of the parameters that demonstrate how north-west Europe is an area of only weak tectonic activity at the present time. Moreover, looking into the geological future, this continuing quiet state can be reasonably predicted for some tens of millions of years. This is the

time-scale over which major changes in plate tectonic regime take place (Fig. 21.13). Eventually, a subduction zone will need to develop at one or both sides of the North Atlantic, converting north-west Europe into a seismically and volcanically active continental margin. A potentially earlier tectonic effect would be the onset of another mantle plume in north-west Europe, similar to that which caused the thermal uplift and volcanism during Palaeocene time. The chances of another plume event are difficult to judge, given our present understanding, but being hit by two events within less than 100 million years could justifiably be regarded as bad luck.

Less far in the future, and more easily predicted, is the possible cooling and reglaciation of Northern Europe. Nothing in the oxygen-isotope curves (Fig. 21.1c) or the Milankovitch curves of orbital parame-

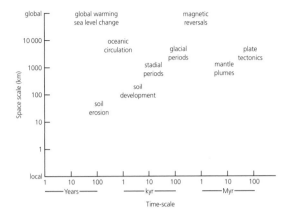

Fig. 21.13 Characteristic space and time-scales of some processes relevant to the geological future of Britain and Ireland.

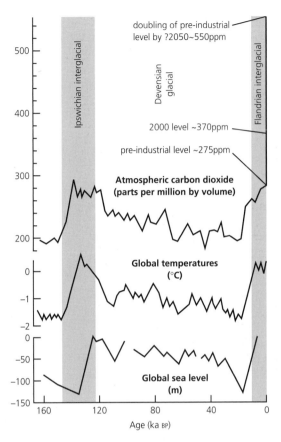

Fig. 21.14 Atmospheric carbon dioxide concentration over the past 160 ka BP measured from air bubbles in the Vostok ice core, Antarctica (data from Lorius *et al.* 1988), compared with oxygen-isotope estimates of local temperature from the same core, and with sea-level changes determined from coastal sediments in south-east Asia (data from Tooley 1993).

ters suggests that the 100-ka cyclicity of glacial and interglacial periods of the last million years has ended. Another glacial episode of Devensian severity might therefore be expected within the next 100 000 years. If the pattern of late Pleistocene glaciations is followed, this climax will be preceded by increasingly severe stadial–interstadial cycles with a period of several tens of thousands of years. The next period of permanent glaciers on the Scottish hills might be less than 10 000 years away.

Closely linked to the cycles of stadials and interstadials are the impoverishment and regrowth of land vegetation (e.g. Fig. 21.2), and the consequent loss and reaccumulation of soil. Human activities of forest clearance and agriculture have intervened in this cycle to such an extent that the survival of a thick, nutritious soil cover in Britain and Ireland is now almost entirely dependent on land management policy rather than on natural processes. Whereas it takes thousands of years to redevelop a depleted soil layer, it takes only decades of bad practice to destroy it (Fig. 21.13). Hedgerow removal, intensive arable farming and upland drainage schemes are among the techniques that threaten to exacerbate soil erosion, alter hydrological regimes and vegetation cover, and perturb local climates.

The future geological changes of most concern, however, have a time-scale of decades to centuries, and global rather than local impacts: global warming, with its attendant risks of rapid eustatic sea-level rise, disrupted ocean circulation and altered regional

climates (Fig. 21.13). Global warming is a human-induced threat, driven by the exponentially increasing emissions of carbon dioxide, methane and other greenhouse gases into the atmosphere by the world's industrialized nations. These emissions are on a scale geologically unprecedented in Quaternary and probably Tertiary history (Fig. 21.14). Carbon dioxide concentrations are now over 20% higher than at the warmest point of the Ipswichian interglacial. They are rising at about 0.4% each year, and by midway through the next century carbon dioxide concentration in the atmosphere will be about twice that of a normal interglacial period. These rates of carbon dioxide increase are predicted to cause an average global temperature rise of about 2°C, but with substantially more variable regional temperature changes—some areas of arctic Canada and Siberia are already warming at rates approaching the 1°C per decade achieved at the beginning of the Holocene. Global sea level is predicted to rise about 50 cm over the next century. This rate of 5 mm/year will double or treble the natural rate of relative sea-level rise in southern Britain, and force substantial coastline changes or require costly sea defences.

Although these extrapolated human-induced temperature and sea-level changes are impressive, they are just within the natural range for these parameters experienced during Quaternary time. Potentially more catastrophic is the unprecedented rate of increase in greenhouse gas concentrations, already rising into uncharted territory (Fig. 21.14). It is by no means certain than the ocean–atmosphere system will respond to this provocation in ways that are predictable by reference to Quaternary geological analogues. Changes in ocean circulation can occur on time-scales of hundreds of years or less. For instance, warming of water in the North Atlantic might rapidly cause the shutdown of the Gulf Stream, which maintains the temperate climate of the north-west European seaboard. Global warming might, in practice, markedly cool the climate of this region, highlighting the uncomfortable fact that our knowledge of Earth processes is inadequate to predict even the direction of potentially major regional changes. Amidst such uncertainty, the precautionary principle suggests that we would be well advised not to experiment further with the future geological history of Britain and Ireland.

References

Becker, A. & Paladini, S. (1992) Intra-plate stresses in Europe and plate-driving mechanisms. *Annales Tectonical* 6, 173–192.

Cameron, T.D.J., Crosby, A., Balson, P.S. *et al.* (1992) *The Geology of the Southern North Sea.* UK Offshore Regional Report, pp. 1–152. HMSO for British Geological Survey, London.

Coope, G.R. (1977) Fossil coleopteran assemblages as sensitive indicators of climatic changes during the Devension (last) cold stage. *Philosophical Transactions of the Royal Society* B280, 313–348.

Funnell, B.M. (1995) Global sea-level and the (pen-) insularity of late Cenozoic Britain. In: *Island Britain: a Quaternary perspective* (ed. R.C. Preece), Special Publication 96, pp. 3–13. Geological Society, London.

Gibbard, P.L. (1988) The history of the great northwest European rivers during the past three million years. *Philosophical Transactions of the Royal Society, London* B318, 559–602.

Johnson, H., Richards, P.C., Long, D. & Graham, C.C. (1993) *The Geology of the Northern North Sea.* UK Offshore Regional Report, pp. 1–110. HMSO for British Geological Survey, London.

Lambeck, K. (1995) Late Devensian and Holocene shorelines of the British Isles and North Sea from models of glacio-hydro-isostatic rebound. *Journal of the Geological Society, London* 152, 437–448.

Lorius, C., Barkov, N.I., Jouzel, J., Korotkevich, V.M. & Raynaud, D. (1988) Antarctic ice core: CO_2 and climatic change over the last climatic cycle. *Eos* 69, 681–684.

Musson, R.M.W. & Winter, P.W. (1994) Seismic hazard of the UK. *AEA Technology Report*, AEA/CS/16422000/ZJ 745/005.

Shackleton, N.J. & Opdyke, N.D. (1973) Oxygen isotope and palaeomagnetic stratigraphy of equatorial Pacific core V28-239: oxygen isotope temperatures and ice volumes on a 105 and 106 year scale. *Quaternary Research* 3, 39–55.

Shackleton, N.J., Crowhurst, S., Hagelberg, T., Pisias, N.G. & Scheider, D.A. (1995) A new Late Neogene time scale: application to Leg 138 sites. *Proceedings of the Ocean Drilling Program* 138, 73–101.

Tooley, M.J. (1993) Long term changes in eustatic sea level. In: *Climate and Sea Level Change: observations, projections and implications* (eds R. A. Warwick, E. M. Barrow & T. M. L. Wigley), pp. 81–107 Cambridge University Press, Cambridge.

Turner, C. (1970) The Middle Pleistocene deposits at Marks Tey, Essex. *Philosophical Transactions of the Royal Society, London* B257, 373–440.

Valet, J.P. & Meynadier, L. (1993) Geomagnetic field intensity and reversals during the past four million years. *Nature* 366, 234–238.

West, R.G. (1977) *Pleistocene Geology and Biology*, 2nd edn, pp. 1–440. Longman, London.

Zagwijn, W.H. (1985) An outline of the Quaternary stratigraphy of the Netherlands. *Geologie en Mijnbouw* 64, 17–24.

Further reading

Ballantyne, C. & Harris, C. (1994) *Periglaciation of Great Britain.* Cambridge University Press, Cambridge. [An excellent review of periglacial features in Britain and the processes that caused them.]

British Geological Survey (1990–5) *UK Offshore Regional Reports.* HMSO for British Geological Survey, London. [A valuable series of reports, which include well-illustrated descriptions of the Quaternary sequences on the UK continental shelf.]

Catt, J.A. (1988) *Quaternary Geology for Scientists and Engineers*. Ellis Horwood, Chichester. [A sound and succinct treatment of the principles and applications of Quaternary geology, with emphasis on the British Isles.]

Ehlers, J., Gibbard, P.L. & Rose, J. (1991) *Glacial Deposits in Great Britain and Ireland*. Balkema, Rotterdam. [A compilation of data on British and Irish glacial deposits, offering a valuable and up-to-date reference source.]

Jones, R.L. & Keen, D.H. (1993) *Pleistocene Environments in the British Isles*. Chapman & Hall, London. [A systematic description of the Pleistocene history of Britain and Ireland.]

Lowe, J.J. & Walker, M.J.C. (1984). *Reconstructing Quaternary Environments*. Longman, London. [A reliable guide to processes and methods in Quaternary geology.]

Preece, R.C. (ed.) (1995) *Island Britain—a Quaternary perspective*. Special Publication 96. Geological Society, London. [An informative collection of papers focusing on the fluctuating insularity of Britain during Quaternary sea-level changes.]

Roberts, N. (1989) *The Holocene: an environmental history*. Blackwell, Oxford. [A stimulating guide to the last 10 000 years.]

West, R.G. (1977) *Pleistocene Geology and Biology*, 2nd edn. Longman, London. [A classic book, still valuable for its clarity of style and content.]

Index